INTERNATIONAL UNION OF PURE AND APPLIED CHEMISTRY

ANALYTICAL CHEMISTRY DIVISION
COMMISSION ON SOLUBILITY DATA

SOLUBILITY DATA SERIES

Volume 30

ALKALI METAL HALATES, AMMONIUM IODATE AND IODIC ACID

SOLUBILITY DATA SERIES

SOLUBILITY DATA SERIES

Editor-in-Chief
A. S. KERTES

Volume 30

ALKALI METAL HALATES, AMMONIUM IODATE AND IODIC ACID

Volume Editors

HIROSHI MIYAMOTO
Niigata University
Niigata, Japan

MARK SALOMON
US Army ET & DL (LABCOM)
Fort Monmouth, NJ, USA

Contributors

BRUNO SCROSATI
University of Rome
Italy

GABOR JANCSO
Hungarian Academy of Sciences
Budapest, Hungary

RAUL HERRERA
Ohio State University
Columbus, OH, USA

THEODORE P. DIRKSE
Calvin College
Grand Rapids, MI, USA

ANDRZEJ MACZYNSKI
Polish Academy of Sciences
Warsaw, Poland

ALEXANDER VAN HOOK
University of Tennessee
Knoxville, TN, USA

MICHELLE C. UCHIYAMA
US Army ET & DL (LABCOM)
Fort Monmouth, NJ, USA

PERGAMON PRESS

OXFORD · NEW YORK · BEIJING · FRANKFURT
SÃO PAULO · SYDNEY · TOKYO · TORONTO

U.K.	Pergamon Press, Headington Hill Hall, Oxford OX3 0BW, England
U.S.A.	Pergamon Press, Maxwell House, Fairview Park, Elmsford, New York 10523, U.S.A.
PEOPLE'S REPUBLIC OF CHINA	Pergamon Press, Room 4037, Qianmen Hotel, Beijing, People's Republic of China
FEDERAL REPUBLIC OF GERMANY	Pergamon Press, Hammerweg 6, D-6242 Kronberg, Federal Republic of Germany
BRAZIL	Pergamon Editora, Rua Eça de Queiros, 346, CEP 04011, Paraiso, São Paulo, Brazil
AUSTRALIA	Pergamon Press Australia, P.O. Box 544, Potts Point, N.S.W. 2011, Australia
JAPAN	Pergamon Press, 8th Floor, Matsuoka Central Building, 1-7-1 Nishishinjuku, Shinjuku-ku, Tokyo 160, Japan
CANADA	Pergamon Press Canada, Suite No. 271, 253 College Street, Toronto, Ontario, Canada M5T 1R5

First edition 1987

Library of Congress has cataloged this serial title as follows:
Solubility data series. - Vol. 1 — Oxford; New York: Pergamon, c 1979–
v.; 28 cm.
Separately cataloged and classified in LC before no. 18.
ISSN 0191-5622 - Solubility data series.
1. Solubility-Tables-Collected works.
QD543.S6629   541.3'42'05–dc19   85–641351
AACR 2   MARC-S

British Library Cataloguing in Publication Data
Alkali metal halates, ammonium iodate and iodic acid.
—(Solubility data series; v. 30).
1. Alkali metal halates—Solubility
2. Iodic acid—Solubility
3. Ammonium halates—Solubility
I. Miyamoto, Hiroshi   II. Salomon, Mark
III. Scrosati, Bruno   IV. Series
546'.38   QD165

ISBN 0-08-029210-0

Printed in Great Britain by A. Wheaton & Co. Ltd., Exeter

CONTENTS

QD165
A441
1987
CHEM

SOLUBILITY DATA SERIES

Editor-in-Chief

A. S. KERTES

The Hebrew University
Jerusalem, Israel

EDITORIAL BOARD

Managing Editor
P. D. GUJRAL
IUPAC Secretariat, Oxford, UK

INTERNATIONAL UNION OF PURE AND APPLIED CHEMISTRY

IUPAC Secretariat: Bank Court Chambers, 2–3 Pound Way,
Cowley Centre, Oxford OX4 3YF, UK

FOREWORD

*If the knowledge is
undigested or simply wrong,
more is not better*

How to communicate and disseminate numerical data effectively in chemical science and technology has been a problem of serious and growing concern to IUPAC, the International Union of Pure and Applied Chemistry, for the last two decades. The steadily expanding volume of numerical information, the formulation of new interdisciplinary areas in which chemistry is a partner, and the links between these and existing traditional subdisciplines in chemistry, along with an increasing number of users, have been considered as urgent aspects of the information problem in general, and of the numerical data problem in particular.

Among the several numerical data projects initiated and operated by various IUPAC commissions, the *Solubility Data Project* is probably one of the most ambitious ones. It is concerned with preparing a comprehensive critical compilation of data on solubilities in all physical systems, of gases, liquids and solids. Both the basic and applied branches of almost all scientific disciplines require a knowledge of solubilities as a function of solvent, temperature and pressure. Solubility data are basic to the fundamental understanding of processes relevant to agronomy, biology, chemistry, geology and oceanography, medicine and pharmacology, and metallurgy and materials science. Knowledge of solubility is very frequently of great importance to such diverse practical applications as drug dosage and drug solubility in biological fluids, anesthesiology, corrosion by dissolution of metals, properties of glasses, ceramics, concretes and coatings, phase relations in the formation of minerals and alloys, the deposits of minerals and radioactive fission products from ocean waters, the composition of ground waters, and the requirements of oxygen and other gases in life support systems.

The widespread relevance of solubility data to many branches and disciplines of science, medicine, technology and engineering, and the difficulty of recovering solubility data from the literature, lead to the proliferation of published data in an ever increasing number of scientific and technical primary sources. The sheer volume of data has overcome the capacity of the classical secondary and tertiary services to respond effectively.

While the proportion of secondary services of the review article type is generally increasing due to the rapid growth of all forms of primary literature, the review articles become more limited in scope, more specialized. The disturbing phenomenon is that in some disciplines, certainly in chemistry, authors are reluctant to treat even those limited-in-scope reviews exhaustively. There is a trend to preselect the literature, sometimes under the pretext of reducing it to manageable size. The crucial problem with such preselection - as far as numerical data are concerned - is that there is no indication as to whether the material was excluded by design or by a less than thorough literature search. We are equally concerned that most current secondary sources, critical in character as they may be, give scant attention to numerical data.

On the other hand, tertiary sources - handbooks, reference books and other tabulated and graphical compilations - as they exist today are comprehensive but, as a rule, uncritical. They usually attempt to cover whole disciplines, and thus obviously are superficial in treatment. Since they command a wide market, we believe that their service to the advancement of science is at least questionable. Additionally, the change which is taking place in the generation of new and diversified numerical data, and the rate at which this is done, is not reflected in an increased third-level service. The emergence of new tertiary literature sources does not parallel the shift that has occurred in the primary literature.

With the status of current secondary and tertiary services being as briefly stated above, the innovative approach of the *Solubility Data Project* is that its compilation and critical evaluation work involve consolidation and reprocessing services when both activities are based on intellectual and scholarly reworking of information from primary sources. It comprises compact compilation, rationalization and simplification, and the fitting of isolated numerical data into a critically evaluated general framework.

The *Solubility Data Project* has developed a mechanism which involves a number of innovations in exploiting the literature fully, and which contains new elements of a more imaginative approach for transfer of reliable information from primary to secondary/tertiary sources. *The fundamental trend of the Solubility Data Project is toward integration of secondary and tertiary services with the objective of producing in-depth critical analysis and evaluation which are characteristic to secondary services, in a scope as broad as conventional tertiary services.*

Fundamental to the philosophy of the project is the recognition that the basic element of strength is the active participation of career scientists in it. Consolidating primary data, producing a truly critically-evaluated set of numerical data, and synthesizing data in a meaningful relationship are demands considered worthy of the efforts of top scientists. Career scientists, who themselves contribute to science by their involvement in active scientific research, are the backbone of the project. The scholarly work is commissioned to recognized authorities, involving a process of careful selection in the best tradition of IUPAC. This selection in turn is the key to the quality of the output. These top experts are expected to view their specific topics dispassionately, paying equal attention to their own contributions and to those of their peers. They digest literature data into a coherent story by weeding out what is wrong from what is believed to be right. To fulfill this task, the evaluator must cover *all* relevant open literature. No reference is excluded by design and every effort is made to detect every bit of relevant primary source. Poor quality or wrong data are mentioned and explicitly disqualified as such. In fact, it is only when the reliable data are presented alongside the unreliable data that proper justice can be done. The user is bound to have incomparably more confidence in a succinct evaluative commentary and a comprehensive review with a complete bibliography to both good and poor data.

It is the standard practice that the treatment of any given solute-solvent system consists of two essential parts: I. Critical Evaluation and Recommended Values, and II. Compiled Data Sheets.

The Critical Evaluation part gives the following information:

(i)   a verbal text of evaluation which discusses the numerical solubility information appearing in the primary sources located in the literature. The evaluation text concerns primarily the quality of data after consideration of the purity of the materials and their characterization, the experimental method employed and the uncertainties in control of physical parameters, the reproducibility of the data, the agreement of the worker's results on accepted test systems with standard values, and finally, the fitting of data, with suitable statistical tests, to mathematical functions;

(ii) a set of recommended numerical data. Whenever possible, the set of recommended data includes weighted average and standard deviations, and a set of smoothing equations derived from the experimental data endorsed by the evaluator;

(iii) a graphical plot of recommended data.

The Compilation part consists of data sheets of the best experimental data in the primary literature. Generally speaking, such independent data sheets are given only to the best and endorsed data covering the known range of experimental parameters. Data sheets based on primary sources where the data are of a lower precision are given only when no better data are available. Experimental data with a precision poorer than considered acceptable are reproduced in the form of data sheets when they are the only known data for a particular system. Such data are considered to be still suitable for some applications, and their presence in the compilation should alert researchers to areas that need more work.

The typical data sheet carries the following information:

(i)   components - definition of the system - their names, formulas and
        Chemical Abstracts registry numbers;
(ii)   reference to the primary source where the numerical information is
        reported. In cases when the primary source is a less common
        periodical or a report document, published though of limited
        availability, abstract references are also given;
(iii)   experimental variables;
(iv)   identification of the compiler;
(v)   experimental values as they appear in the primary source.
        Whenever available, the data may be given both in tabular and
        graphical form. If auxiliary information is available, the
        experimental data are converted also to SI units by the compiler.

Under the general heading of Auxiliary Information, the essential
experimental details are summarized:

(vi)   experimental method used for the generation of data;
(vii)   type of apparatus and procedure employed;
(viii) source and purity of materials;
(ix)   estimated error;
(x)   references relevant to the generation of experimental data as
        cited in the primary source.

This new approach to numerical data presentation, formulated at the
initiation of the project and perfected as experience has accumulated, has
been strongly influenced by the diversity of background of those whom we are
supposed to serve. We thus deemed it right to preface the
evaluation/compilation sheets in each volume with a detailed discussion of the
principles of the accurate determination of relevant solubility data and
related thermodynamic information.

Finally, the role of education is more than corollary to the efforts we
are seeking. The scientific standards advocated here are necessary to
strengthen science and technology, and should be regarded as a major effort in
the training and formation of the next generation of scientists and
engineers. Specifically, we believe that there is going to be an impact of
our project on scientific-communication practices. The quality of
consolidation adopted by this program offers down-to-earth guidelines,
concrete examples which are bound to make primary publication services more
responsive than ever before to the needs of users. The self-regulatory
message to scientists of the early 1970s to refrain from unnecessary
publication has not achieved much. A good fraction of the literature is still
cluttered with poor-quality articles. The Weinberg report (in 'Reader in
Science Information', ed. J. Sherrod and A. Hodina, Microcard Editions Books,
Indian Head, Inc., 1973, p. 292) states that 'admonition to authors to
restrain themselves from premature, unnecessary publication can have little
effect unless the climate of the entire technical and scholarly community
encourages restraint...' We think that projects of this kind translate the
climate into operational terms by exerting pressure on authors to avoid
submitting low-grade material. The type of our output, we hope, will
encourage attention to quality as authors will increasingly realize that their
work will not be suited for permanent retrievability unless it meets the
standards adopted in this project. It should help to dispel confusion in the
minds of many authors of what represents a permanently useful bit of
information of an archival value, and what does not.

If we succeed in that aim, even partially, we have then done our share in
protecting the scientific community from unwanted and irrelevant, wrong
numerical information.

                              A. S. Kertes

PREFACE

The present volume is the second of four volumes planned for inorganic metal halates. The first, on *ALKALINE EARTH METAL HALATES,* was published in 1983 (1), and two more volumes, on copper and silver halates, and on transition and rare earth metal halates are in course of preparation.

The alkali and alkaline earth metal halates have an important place in the history of both theoretical and practical analytical chemistry. In 1848, Berthelot, in France, described the use of potassium iodate as a standard titrant for the determination of iodide, and the well established method for determining phenol with excess bromate-bromine reagent in acid solution was first described by Knop in 1845, and further developed by Koppeschaar in 1875. Important practical applications of halate chemistry include their use in pyrotechnics, and in the paper pulp industry for the generation of chloric dioxide blanching agent.

In spite of the long history on the chemistry of alkali metal halates, the reader will discover that there are still considerable uncertainties in the nature of solid phases and transition temperatures for a number of systems: e.g. we can cite the binary systems $LiIO_3 - H_2O$ and $HIO_3 - H_2O$. Hopefully, this volume will serve as a guide for future studies on these systems.

The literature of the solubilities of alkali metal halates was covered through the first half of 1984, and we believe this survey to be complete. In a few instances, relevant papers were not compiled since it was not possible to obtain either reprints or other reproductions of the original publication. We were, for example, unable to obtain the paper in Ref. (2), and this publication was omitted from this volume. A number of publications were not compiled or referred to in the critical evaluations for a variety of reasons. In Ref. (3), $KClO_3$ was stated to be "appreciably soluble" in liquid SO_2, and in Ref. (4) only partial phase diagrams were given for several ternary $NaClO_3$ systems with no numerical solubility data. Some publications dealing with solubilities in non-aqueous solvents were not compiled as the authors stated various alkali metal halates were "insoluble" (5-7) without providing numerical information.

To arrive at either *recommended* or *tentative* solubility values, we generally applied a statistical treatment similar to that recommended by Cohen-Adad (8) based on the thermodynamic treatment of saturated solutions and their equilibrated solid phases (8-10) as discussed in the *INTRODUCTION TO THE SOLUBILITY OF SOLIDS IN LIQUIDS* found in this volume. These thermodynamic treatments show that for binary systems, solubilities over the complete range of ice as the solid phase to the melting of the pure solute can be expressed by

$$Y = A/(T/K) + B \, ln \, (T/K) + C + D(T/K) \qquad [1]$$

The complex Y term in eq. [1] takes different forms depending upon the concentration units employed. In the present volume, the evaluators have analyzed solubilities based on mole fraction and mass units, and in terms of mole fraction units, the complex Y term (called Y_x throughout this volume) is given by (8-10).

$$Y_x = ln \left\{ \chi^{v}(1-x)^r(v + r)^{v+r}/ \left(r^r(1 + \chi)^{v+r} \right) \right\} \qquad [2]$$

where r is the solvation number in the solid phase, v is the number of ions produced upon dissolution, and χ is the mole fraction solubility. When sufficient data were available, the evaluators used eq. [2] in a four parameter fit to eq. [1]: note that for the ice polytherm, $v = 0$ and $r = 1$.

For solubilities expressed in $mol \, kg^{-1}$ units, the evaluators used a simpler form of Y referred to as Y_m throughout this volume. Y_m is given by (see 8-11 and the INTRODUCTION to this volume):

$$Y_m = ln \, (m/m_0) - rM(m - m_0) \qquad [3]$$

where r is the solvation number of the solid, m is the molality of the saturated solution, m_0 is an arbitrarily selected reference molality (usually the molality at 298 K), and M is the molar mass of the solvent. When fitting the Y_m terms to eq. [1], the evaluators generally used a three parameter fit (i.e. the constants A, B and C were evaluated).

In fitting the solubility data for binary systems to the smoothing eq. [1], the evaluators rejected a number of data points based on the deviation from the standard error of estimate, σ: that is, when the difference between calculated and observed solubilities exceeded 2σ, the data point was rejected. For mole fraction solubilities, the standard error of estimate, σ_x is defined by:

$$\sigma_x = \left\{ \sum (x_{obsd} - x_{calcd})/(N - 4) \right\}^{1/2} \qquad [4]$$

where N is the number of data points associated with the particular polytherm being considered. A similar relation exists for the standard error of estimate for mol kg^{-1} solubilities, σ_m, but the evaluators used (N - 3) in the denominator since Y_m values were fitted to a three constant smoothing equation. In addition to reporting the standard errors σ_x and σ_m, the evaluators also reported the standard errors for the Y terms (Y_x and Y_m), denoted simply as σ_y in the evaluations.

For convenience of the users, the evaluators have prepared two computer programs written in BASIC to calculate the solubilities at any temperature. The programs called "CALC_X" and "CALC_M" are given on the pages following the references. Note that the user is requested to enter an initial estimate of the solubility to start the calculations. Since the Newton-Raphson iteration method is used, the user should be aware that a very poor initial estimate of the solubility may result in convergence at an incorrect answer. Finally, we should like to point out that both programs use double precision in the calculations (statement number 20 in both programs: DEFDBL A-H, O-Z). Using IBM-PC or compatibles with MS-BASIC, double precision is required to give at least 8-bit numerical precision.

Although an attempt has been made to locate all publications on the system under consideration through the first half of 1984, some omissions may have occurred. The editors will therefore be grateful to readers who will bring these omissions to their attention.

The editors would like to acknowledge the cooperation of the American Chemical Society and VAAP, the copyright agency of the USSR, for their permission to reproduce phase diagrams from their publications.

The editors gratefully acknowledge the advice and comments from members of IUPAC Commission V.8 (the Commission on Solubility Data), and in particular to Professors H. L. Clever, R. Cohen-Adad, J. W. Lorimer, and A. S. Kertes. We are also grateful to Dr. K. Loening of the Chemical Abstracts Service for providing Registry Numbers for numerous compounds.

One of us (H.M.) would also like to acknowledge the hospitality of Prof. H. L. Clever during his stay at the Solubility Research and Information Center at Emory University in Atlanta, GA, USA (1981-1982), and to Profs. Hideo Akaiwa (Gunma University) and Michihiro Fujii (Niigata University) for valuable comments and suggestions. We would also like to thank Ms. Karen Salomon for her help in translations. Finally the editors would like to thank Mrs. Shikako Miyamoto for her assistance with the tedious calculations of converting experimental solubility data in mass % units to S.I. units.

REFERENCES

1. Miyamoto, H.; Salomon, M. and Clever, H. L., eds. *IUPAC SOLUBILITY DATA SERIES VOLUME 14: ALKALINE EARTH METAL HALATES*. Pergamon Press, London, 1983.

2. Malyshev, A. A.; Kuz'menko, A. L.; Novikov, G. I.; Tomasheva, L. T. *Izv. Vyssh. Uchebn. Zaved., Khim. Khim. Tekhnol*. 1982, 25, 380.

3. Perkins, H.; Taft, R. *J. Phys. Chem*. 1925, 29, 1075.

4. Perel'man, F. M.; Korzhenyak, I. G. *Zh. Neorg. Khim*. 1963, 13, 277.

5. Kolthoff, I. M.; Chantooni, M. K. *J. Phys. Chem*. 1973, 77, 523.

6. Isbin, H. S.; Kobe, K. A. *J. Am. Chem. Soc*. 1945, 67, 464,

7. Miravitlles, M. L. *Ann. Fis. Quim. (Madrid)* 1945, 41, 120.

8. Cohen-Adad, R. *Pure and Appl. Chem*. 1985, 57, 255.

9. Counioux, J. -J.; Teny, R. *J. Chim. Phys*. 1981, 78, 816.

10. Tenu, R.; Counioux, J.-J. *J. Chim. Phys*. 1981, 78, 823.

REFERENCES (Continued)

11. Siekierski, S.; Mioduski, T.; Salomon, M., eds. *IUPAC SOLUBILITY DATA SERIES VOLUME 13: SCANDIUM YTTRIUM, LANTHANUM AND LANTHANIDE NITRATES.*   Pergamon Press, London, 1983.

Hiroshi Miyamoto, Niigata, Japan
Mark Salomon, Fort Monmouth, NJ, USA
November, 1986

```
10  '   CALC_M
20  DEFDBL A-H,O-Z
30  '   PROGRAM TO CALCULATE mol/kg SOLUBILITIES FOR A SPECIFIED TEMPERATURE
40  '   BASED ON THE SMOOTHING EQUATION GIVEN IN THE PREFACE
50  '
60  DIM G$[80]
70  PRINT "ENTER PROBLEM IDENTIFYING INFORMATION"
80  INPUT G$
90  PRINT
100 PRINT "ENTER CONSTANTS IN  y = A/T + B log(T) + C "
110 PRINT
120 INPUT "CONSTANT A =   ",A
130 INPUT "CONSTANT B =   ",B
140 INPUT "CONSTANT C =   ",C
150 PRINT
160 PRINT "ENTER DATA TO IDENTIFY THE POLYTHERM"
170 PRINT
180 INPUT "MOLAR MASS OF SOLVENT =   ",W
190 INPUT "SOLVATION NUMBER OF SOLID PHASE =   ",R
200 INPUT "REFERENCE MOLALITY =   ",MO
210 INPUT "CHOOSE ITERATION LIMIT FOR CALCD mol/kg SOLUBILITIES:   ",MLIM
220 PRINT
230 LPRINT
240 LPRINT G$
250 LPRINT "CONSTANT A = ";A
260 LPRINT "CONSTANT B = ";B
270 LPRINT "CONSTANT C = ";C
280 LPRINT "MOLAR MASS OF SOLVENT = ";W
290 LPRINT "SOLVATION NUMBER = ";R
300 LPRINT "REFERENCE MOLALITY = ";MO
310 LPRINT "CONVERGERNCE LIMIT SET AT "; MLIM
320 LPRINT
330 '
340 '   START CALCULATIONS
350 '
360 I = 0
370 PRINT "ENTER TEMP AND AN INITIAL GUESS FOR THE MOLALITY"
380 INPUT "T/K =   ",T
390 INPUT "GUESS FOR THE MOLALITY IS:   ",MSTART
400 I = I + 1
410 '
420 '   NEWTON-RAPHSON ITERATION
430 '
440 F0 = A/T + B*LOG(T) + C + LOG(MO/MSTART) + R*W*(MSTART - MO)/1000
450 F1 = R*W/1000 - 1/MSTART
460 MNEW = MSTART - F0/F1
470 IF ABS(MSTART - MNEW) < MLIM THEN 500
480 MSTART = MNEW
490 GOTO 440
500 PRINT
510 PRINT "FOR T/K = ";T;" , SOLUBILITY (mol/kg) = ";MNEW
520 PRINT
530 LPRINT
540 LPRINT "FOR CALCULATION No. ";I
550 LPRINT "T/K = ";T;"  or t/C = ";T-273.15
560 LPRINT "SOLUBILITY = ";MNEW;"  mol/kg"
570 LPRINT
580 PRINT "DO YOU WANT TO CALCULATE A NEW SOLUBILITY AT A NEW TEMPERATURE?"
590 INPUT "ENTER Y/N:   ",C$
600 IF C$ = "Y" OR C$ = "y" THEN 370
610 END
```

```
10   REM CALC_X
20   DEFDBL A-H, O-Z
30   REM PROGRAM TO CALCULATE MOLE FRACTION SOLUBILITIES AT A GIVEN TEMP
40   REM BASED ON THE SMOOTHING EQUATION GIVEN IN THE PREFACE
50   DIM G$[80] , C(4)
60   REM
70   PRINT"READ PROBLEM IDENTIFYING INFORMATION (80 CHARACTERS MAX)"
80   INPUT G$
90   PRINT
100  INPUT"ENTER NUMB OF CONSTANTS IN SMOOTHING EQN (3 CONSTANTS MIN): ",NC
110  IF NC = 3 THEN C(4) = 0
120  PRINT
130  FOR I = 1 TO NC
140  PRINT"ENTER VALUE OF CONSTANT NUMBER ";I
150  INPUT C(I)
160  NEXT I
170  PRINT
180  PRINT"ENTER DATA IDENTIFYING THE POLYTHERM"
190  PRINT"NOTE THAT FOR THE ICE POLYTHERM, R = 1 and V = 0"
200  PRINT"WHERE R = SOLID PHASE SOLVATION NUMBER AND V = NUMBER OF IONS"
210  PRINT
220  INPUT"ENTER SOLVATION NUMBER R:   ",R
230  INPUT"ENTER NUMBER  OF IONS  V:   ",V
240  PRINT
250  PRINT"CHOOSE THE ITERATION LIMIT FOR CALCD MOLE FRACTION SOLUBILITY"
260  INPUT"ENTER ITERATION LIMIT :   ",XLIM
270  PRINT
280  LPRINT G$
290  LPRINT
300  LPRINT"SOLVATION NUMBER R = ";R
310  LPRINT"NUMBER  OF IONS  V = ";V
320  LPRINT
330  LPRINT "CONVERGENCE LIMIT FOR MOLE FRACTION SOLUBILITY SET AT   ";XLIM
340  LPRINT
350  FOR I = 1 TO NC
360  LPRINT"CONSTANT C(";I;") = ";C(I)
370  NEXT I
380  R1 = R + V
390  REM
400  PRINT
410  INPUT"ENTER TEMPERATURE IN DEGREES C:   ",TC
420  TK = TC + 273.15
430  NITER = 0
440  INPUT"ENTER AN APPROXIMATE VALUE FOR MOLE FRACTION SOLUBILITY: ",Y
450  IF R = 0 THEN R2 = 1
460  IF R = 0 THEN 480
470  R2 = R^R
480  Y3 = C(1)/TK + C(2) * LOG(TK) + C(3) + C(4)*TK
490  REM
500  REM ITERATION BY NEWTWON-RAPHSON METHOD
510  REM
520  Y5 = R2*EXP(Y3)/R1^R1
530  F0 = Y^V*(1-Y)^R/(1+Y)^R1 - Y5
540  NITER = NITER + 1
550  P1 = (1 + Y)^R1
560  P2 = -R*Y^V*(1-Y)^(R-1) + V*Y^(V-1)*(1-Y)^R
570  P3 = -Y^V*(1-Y)^R * R1 * (1+Y)^(R1-1)
580  F1 = (P1 * P2 + P3)/(1 + Y)^2
590  Y6 = Y - F0/F1
600  IF ABS (Y6 - Y) < XLIM THEN 630
610  Y = Y6
620  GOTO 530
630  LPRINT
640  LPRINT"FOR t = ";TC;" deg C, or  T = ";TK;" K"
650  LPRINT"CALCD MOL FRACTION SOLUBILITY = ";Y
660  PRINT
670  PRINT"FOR t = ";TC;" deg C, or  T = ";TK;" K"
680  PRINT"FOR ";NITER;" ITERATIONS"
690  PRINT"CALCD MOL FRACTION SOLUBILITY = ";Y
700  PRINT
710  PRINT"DO YOU WANT TO CALCULATE ANOTHER SOLUBILITY FOR A NEW TEMPERATURE?"
720  INPUT"ENTER Y/N:   ",C$
730  IF C$ = "Y" OR C$ = "y" THEN 410
740  END
```

INTRODUCTION TO THE SOLUBILITY OF SOLIDS IN LIQUIDS

Nature of the Project

The Solubility Data Project (SDP) has as its aim a comprehensive search of the literature for solubilities of gases, liquids, and solids in liquids or solids. Data of suitable precision are compiled on data sheets in a uniform format. The data for each system are evaluated, and where data from different sources agree sufficiently, recommended values are proposed. The evaluation sheets, recommended values, and compiled data sheets are published on consecutive pages.

Definitions

A *mixture* (1, 2) describes a gaseous, liquid, or solid phase containing more than one substance, when the substances are all treated in the same way.

A *solution* (1, 2) describes a liquid or solid phase containing more than one substance, when for convenience one of the substances, which is called the *solvent*, and may itself be a mixture, is treated differently than the other substances, which are called *solutes*. If the sum of the mole fractions of the solutes is small compared to unity, the solution is called a *dilute solution*.

The *solubility* of a substance B is the relative proportion of B (or a substance related chemically to B) in a mixture which is saturated with respect to solid B at a specified temperature and pressure. *Saturated* implies the existence of equilibrium with respect to the processes of dissolution and precipitation; the equilibrium may be stable or meta-stable. The solubility of a substance in metastable equilibrium is usually greater than that of the corresponding substance in stable equliibrium. (Strictly speaking, it is the activity of the substance in metastable equilibrium that is greater.) Care must be taken to distinguish true metastability from supersaturation, where equilibrium does not exist.

Either point of view, mixture or solution, may be taken in describing solubility. The two points of view find their expression in the quantities used as measures of solubility and in the reference states used for definition of activities, activity coefficients and osmotic coefficients.

The qualifying phrase "substance related chemically to B" requires comment. The composition of the saturated mixture (or solution) can be described in terms of any suitable set of thermodynamic components. Thus, the solubility of a salt hydrate in water is usually given as the relative proportion of anhydrous salt in solution, rather than the relative proportions of hydrated salt and water.

Quantities Used as Measures of Solubility

1. *Mole fraction of substance B*, x_B:

$$x_B = n_B / \sum_{s=1}^{c} n_s \qquad [1]$$

where n_s is the amount of substance of s, and c is the number of distinct substances present (often the number of thermodynamic components in the system). *Mole per cent* of B is 100 x_B.

2. *Mass fraction of substance B*, w_B:

$$w_B = m_B' / \sum_{s=1}^{c} m_s' \qquad [2]$$

where m_s' is the mass of substance s. *Mass per cent* is 100 w_B. The equivalent terms *weight fraction* and *weight per cent* are not used.

3. *Solute mole (mass) fraction of solute B* (3, 4):

$$x_{s,B} = m_B / \sum_{s=1}^{c'} m_s = x_B / \sum_{s=1}^{c'} x_s \qquad [3]$$

$$w_{s,B} = m_B' / \sum_{s=1}^{c'} m_s' = w_B / \sum_{s=1}^{c'} w_s \qquad [3a]$$

where the summation is over the solutes only. For the solvent A, $x_{S,A} = x_A/(1 - x_A)$, $w_{S,A} = w_A/(1 - w_A)$. These quantities are called Jänecke mole (mass) fractions in many papers.

   4. Molality of solute B (1, 2) in a solvent A:

$$m_B = n_B/n_A M_A \qquad \text{SI base units: mol kg}^{-1} \qquad\qquad [4]$$

where M_A is the molar mass of the solvent.

   5. Concentration of solute B (1, 2) in a solution of volume V:

$$c_B = [B] = n_B/V \qquad \text{SI base units: mol m}^{-3} \qquad\qquad [5]$$

The symbol c_B is preferred to $[B]$, but both are used. The terms molarity and molar are not used.

   Mole and mass fractions are appropriate to either the mixture or the solution point of view. The other quantities are appropriate to the solution point of view only. Conversions among these quantities can be carried out using the equations given in Table I-1 following this Introduction. Other useful quantities will be defined in the prefaces to individual volumes or on specific data sheets.

   In addition to the quantities defined above, the following are useful in conversions between concentrations and other quantities.

   6. Density: $\rho = m/V$ \qquad SI base units: kg m^{-3} \qquad\qquad [6]

   7. Relative density: d; the ratio of the density of a mixture to the density of a reference substance under conditions which must be specified for both (1). The symbol $d_t^{t'}$ will be used for the density of a mixture at t°C, 1 bar divided by the density of water at t'°C, 1 bar. (In some cases 1 atm = 101.325 kPa is used instead of 1 bar = 100 kPa.)

   8. A note on nomenclature. The above definitions use the nomenclature of the IUPAC Green Book (1), in which a solute is called B and a solvent A In compilations and evaluations, the first-named component (component 1) is the solute, and the second (component 2 for a two-component system) is the solvent. The reader should bear these distinctions in nomenclature in mind when comparing nomenclature and theoretical equations given in this Introduction with equations and nomenclature used on the evaluation and compilation sheets.

Thermodynamics of Solubility

   The principal aims of the Solubility Data Project are the tabulation and evaluation of: (a) solubilities as defined above; (b) the nature of the saturating phase. Thermodynamic analysis of solubility phenomena has two aims: (a) to provide a rational basis for the construction of functions to represent solubility data; (b) to enable thermodynamic quantities to be extracted from solubility data. Both these are difficult to achieve in many cases because of a lack of experimental or theoretical information concerning activity coefficients. Where thermodynamic quantities can be found, they are not evaluated critically, since this task would involve critical evaluation of a large body of data that is not directly relevant to solubility. The following is an outline of the principal thermodynamic relations encountered in discussions of solubility. For more extensive discussions and references, see books on thermodynamics, e.g., (5-12).

   Activity Coefficients (1)

      (a) Mixtures. The activity coefficient f_B of a substance B is given by

$$RT \ln (f_B x_B) = \mu_B - \mu_B^* \qquad\qquad [7]$$

where μ_B^* is the chemical potential of pure B at the same temperature and pressure. For any substance B in the mixture,

$$\lim_{x_B \to 1} f_B = 1 \qquad\qquad [8]$$

      (b) Solutions.

         (i) Solute B. The molal activity coefficient γ_B is given by

$$RT \ln(\gamma_B m_B) = \mu_B - (\mu_B - RT \ln m_B)^\infty \qquad\qquad [9]$$

where the superscript ∞ indicates an infinitely dilute solution. For any solute B,

$$\gamma_B^\infty = 1 \qquad\qquad [10]$$

Activity coefficients y_B connected with concentrations c_B, and $f_{x,B}$ (called the *rational activity coefficient*) connected with mole fractions x_B are defined in analogous ways. The relations among them are (1, 9), where ρ^* is the density of the pure solvent:

$$f_B = (1 + M_A\sum_s m_s)\gamma_B = [\rho + \sum_s(M_A - M_s)c_s]y_B/\rho^* \qquad [11]$$

$$\gamma_B = (1 - \sum_s x_s)f_{x,B} = (\rho - \sum_s M_s c_s)y_B/\rho^* \qquad [12]$$

$$y_B = \rho^* f_{x,B}[1 + \sum_s(M_s/M_A - 1)x_B]/\rho = \rho^*(1 + \sum_s M_s m_s)\gamma_B/\rho \qquad [13]$$

For an electrolyte solute $B \equiv C_{\nu_+}A_{\nu_-}$, the activity on the molality scale is replaced by (9)

$$\gamma_B m_B = \gamma_\pm^\nu m_B^\nu Q^\nu \qquad [14]$$

where $\nu = \nu_+ + \nu_-$, $Q = (\nu_+^{\nu_+}\nu_-^{\nu_-})^{1/\nu}$, and $\gamma_\pm$ is the mean ionic activity coefficient on the molality scale. A similar relation holds for the concentration activity, $y_B c_B$. For the mole fractional activity,

$$f_{x,B}x_B = Q^\nu f_\pm^\nu x_\pm^\nu \qquad [15]$$

where $x_\pm = (x_+ x_-)^{1/\nu}$. The quantities x_+ and x_- are the ionic mole fractions (9), which are

$$x_+ = \nu_+ x_B/[1 + \sum_s(\nu_s - 1)x_s]; \quad x_- = \nu_- x_B[1 + \sum_s(\nu_s - 1)x_s] \qquad [16]$$

where ν_s is the sum of the stoichiometric coefficients for the ions in a salt with mole fraction x_s. Note that the mole fraction of solvent is now

$$x_A' = (1 - \sum_s \nu_s x_s)/[1 + \sum_s(\nu_s - 1)x_s] \qquad [17]$$

so that

$$x_A' + \sum_s \nu_s x_s = 1 \qquad [18]$$

The relations among the various mean ionic activity coefficients are:

$$f_\pm = (1 + M_A\sum_s \nu_s m s)\gamma_\pm = [\rho + \sum_s(\nu_s M_A - M_s)c_s]y_\pm/\rho^* \qquad [19]$$

$$\gamma_\pm = \frac{(1 - \sum_s x_s)f_\pm}{1 + \sum_s(\nu_s - 1)x_s} = (\rho - \sum_s M_s c_s)y_\pm/\rho^* \qquad [20]$$

$$y_\pm = \frac{\rho^*[1 + \sum_s(M_s/M_A - 1)xs]f_\pm}{\rho[1 + \sum_s(\nu_s - 1)x_s]} = \rho^*(1 + \sum_s M_s m_s)\gamma_\pm/\rho \qquad [21]$$

(ii) *Solvent, A:*

The *osmotic coefficient*, ϕ, of a solvent A is defined as (1):

$$\phi = (\mu_A^* - \mu_A)/RT \, M_A \sum_s m_s \qquad [22]$$

where μ_A^* is the chemical potential of the pure solvent.

The *rational osmotic coefficient*, ϕ_x, is defined as (1):

$$\phi_x = (\mu_A - \mu_A^*)/RT\ln x_A = \phi M_A \sum_s m_s/\ln(1 + M_A \sum_s m_s) \qquad [23]$$

The activity, a_A, or the activity coefficient, f_A, is sometimes used for the solvent rather than the osmotic coefficient. The activity coefficient is defined relative to pure A, just as for a mixture.

For a mixed solvent, the molar mass in the above equations is replaced by the average molar mass; i.e., for a two-component solvent with components J, K, M_A becomes

$$M_A = M_J + (M_K - M_J)x_{v,K} \qquad [24]$$

where $x_{v,K}$ is the solvent mole fraction of component K.

The osmotic coefficient is related directly to the vapor pressure, p, of a solution in equilibrium with vapor containing A only by (12, p.306):

$$\phi M_A \sum_s \nu_s m_s = -\ln(p/p_A^*) + (V_{m,A}^* - B_{AA})(p - p_A^*)/RT \qquad [25]$$

where p_A^*, $V_{m,A}^*$ are the vapor pressure and molar volume of pure solvent A, and B_{AA} is the second virial coefficient of the vapor.

The Liquid Phase

A general thermodynamic differential equation which gives solubility as a function of temperature, pressure and composition can be derived. The approach is similar to that of Kirkwood and Oppenheim (7); see also (11, 12). Consider a solid mixture containing c thermodynamic components i. The Gibbs-Duhem equation for this mixture is:

$$\sum_{i=1}^{c} x_i'(S_i'dT - V_i'dp + d\mu_i') = 0 \qquad [26]$$

A liquid mixture in equilibrium with this solid phase contains c' thermodynamic components i, where $c' \geqslant c$. The Gibbs-Duhem equation for the liquid mixture is:

$$\sum_{i=1}^{c} x_i(S_i dT - V_i dp + d\mu_i') + \sum_{i=c+1}^{c'} x_i(S_i dT - V_i dp + d\mu_i) = 0 \qquad [27]$$

Subtract [26] from [27] and use the equation

$$d\mu_i = (d\mu_i)_{T,p} - S_i dT + V_i dp \qquad [28]$$

and the Gibbs-Duhem equation at constant temperature and pressure:

$$\sum_{i=1}^{c} x_i(d\mu_i')_{T,p} + \sum_{i=c+1}^{c'} x_i(d\mu_i)_{T,p} = 0 \qquad [29]$$

The resulting equation is:

$$RT\sum_{i=1}^{c} x_i'(d\ln a_i)_{T,p} = \sum_{i=1}^{c} x_i'(H_i - H_i')dT/T - \sum_{i=1}^{c} x_i'(V_i - V_i')dp \qquad [30]$$

where

$$H_i - H_i' = T(S_i - S_i') \qquad [31]$$

is the enthalpy of transfer of component i from the solid to the liquid phase at a given temperature, pressure and composition, with H_i and S_i the partial molar enthalpy and entropy of component i.

Use of the equations

$$H_i - H_i^0 = -RT^2(\partial\ln a_i/\partial T)_{x,p} \qquad [32]$$

and

$$V_i - V_i^0 = RT(\partial\ln a_i/\partial p)_{x,T} \qquad [33]$$

where superscript o indicates an arbitrary reference state gives:

$$RT\sum_{i=1}^{c} x_i' d\ln a_i = \sum_{i=1}^{c} x_i'(H_i^0 - H_i')dT/T - \sum_{i=1}^{c} x_i'(V_i^0 - V_i')dp \qquad [34]$$

where

$$d\ln a_i = (d\ln a_i)_{T,p} + (\partial\ln a_i/\partial T)_{x,p} + (\partial\ln a_i/\partial p)_{x,T} \qquad [35]$$

The terms involving enthalpies and volumes in the solid phase can be written as:

$$\sum_{i=1}^{c} x_i'H_i' = H_s^* \qquad \sum_{i=1}^{c} x_i'V_i' = V_s^* \qquad [36]$$

With eqn [36], the final general solubility equation may then be written:

$$R\sum_{i=1}^{c} x_i' d\ln a_i = (H_s^* - \sum_{i=1}^{c} x_i'H_i^0)d(1/T) - (V_s^* - \sum_{i=1}^{c} x_i'V_i^0)dp/T \qquad [37]$$

Note that those components which are not present in both phases do not appear in the solubility equation. However, they do affect the solubility through their effect on the activities of the solutes.

Several applications of eqn [37] (all with pressure held constant) will be discussed below. Other cases will be discussed in individual evaluations.

(a) *Solubility as a function of temperature.*

Consider a binary solid compound A_nB in a single solvent A. There is no fundamental thermodynamic distinction between a binary compound of A and B which dissociates completely or partially on melting and a solid mixture of A and B; the binary compound can be regarded as a solid mixture of constant composition. Thus, with $c = 2$, $x_A' = n/(n + 1)$, $x_B' = 1/(n + 1)$, eqn [37] becomes:

$$d\ln(a_A^n a_B) = -\Delta H_{AB}^0 d(1/RT) \qquad [38]$$

where

$$\Delta H_{AB}^0 = nH_A + H_B - (n + 1)H_s^* \qquad [39]$$

is the molar enthalpy of melting and dissociation of pure solid A_nB to form A and B in their reference states. Integration between T and T_0, the melting point of the pure binary compound A_nB, gives:

$$\ln(a_A^n a_B) = \ln(a_A^n a_B)_{T=T_0} - \int_{T_0}^{T} \Delta H_{AB}^0 d(1/RT) \qquad [40]$$

(i) Non-electrolytes

In eqn [32], introduce the pure liquids as reference states. Then, using a simple first-order dependence of ΔH_{AB}^* on temperature, and assuming that the activity coefficients conform to those for a simple mixture (6):

$$RT \ln f_A = w x_B^2 \qquad\qquad RT \ln f_B = w x_A^2 \qquad [41]$$

then, if w is independent of temperature, eqn [32] and [33] give:

$$\ln\{x_B(1-x_B)^n\} + \ln\left\{\frac{n^n}{(1 + n)^{n+1}}\right\} = G(T) \qquad [42]$$

where

$$G(T) = -\left\{\frac{\Delta H_{AB}^* - T^*\Delta C_p^*}{R}\right\}\left\{\frac{1}{T} - \frac{1}{T^*}\right\}$$
$$+ \frac{\Delta C_p^*}{R} \ln(T/T^*) - \frac{w}{R}\left\{\frac{x_A^2 + n x_B^2}{T} - \frac{n}{(n + 1)T^*}\right\} \qquad [43]$$

where ΔC_p^* is the change in molar heat capacity accompanying fusion plus decomposition of the pure compound to pure liquid A and B at temperature T^*, (assumed here to be independent of temperature and composition), and ΔH_{AB}^* is the corresponding change in enthalpy at $T = T^*$. Equation [42] has the general form:

$$\ln\{x_B(1-x_B)^n\} = A_1 + A_2/(T/K) + A_3\ln(T/K) + A_4(x_A^2 + n x_B^2)/(T/K) \qquad [44]$$

If the solid contains only component B, then $n = 0$ in eqn [42] to [44].

If the infinite dilution reference state is used, then:

$$RT \ln f_{x,B} = w(x_A^2 - 1) \qquad [45]$$

and [39] becomes

$$\Delta H_{AB}^\infty = nH_A^* + H_B^\infty - (n + 1)H_s^* \qquad [46]$$

where ΔH_{AB}^∞ is the enthalpy of melting and dissociation of solid compound A_nB to the infinitely dilute reference state of solute B in solvent A; H_A^* and H_B^∞ are the partial molar enthalpies of the solute and solvent at infinite dilution. Clearly, the integral of eqn [32] will have the same form as eqn [35], with ΔH_{AB}^∞ replacing ΔH_{AB}^*, ΔC_p^∞ replacing ΔCp^*, and $x_A^2 - 1$ replacing x_A^2 in the last term.

See (5) and (11) for applications of these equations to experimental data.

(ii) Electrolytes

(a) Mole fraction scale

If the liquid phase is an aqueous electrolyte solution, and the solid is a salt hydrate, the above treatment needs slight modification. Using rational mean activity coefficients, eqn [34] becomes:

$$\ln\left\{\frac{x_B{}^{\nu}(1 - x_B)^n}{[1 + (\nu - 1)x_B]^{n+\nu}}\right\} \quad - \ln\left\{\frac{n^n}{(n + \nu)^{n+\nu}}\right\} + \ln\left\{\left[\frac{f_{B*}}{f_B}\right]^{\nu}\left[\frac{f_{A*}}{f_A}\right]^n\right\}$$

$$= -\left[\frac{\Delta H_{AB}{}^* - T^*\Delta C_p{}^*}{R}\right]\left\{\frac{1}{T} - \frac{1}{T^*}\right\} + \frac{\Delta C_p{}^*}{R}\ln(T/T^*) \tag{47}$$

where superscript * indicates the pure salt hydrate. If it is assumed that the activity coefficients follow the same temperature dependence as the right-hand side of eqn [47] (13-16), the thermochemical quantities on the right-hand side of eqn [47] are not rigorous thermodynamic enthalpies and heat capacities, but are apparent quantities only. Data on activity coefficients (9) in concentrated solutions indicate that the terms involving these quantities are not negligible, and their dependence on temperature and composition along the solubility-temperature curve is a subject of current research.

A similar equation (with $\nu = 2$ and without the heat capacity terms or activity coefficients) has been used to fit solubility data for some $MOH-H_2O$ systems, where M is an alkali metal (13); enthalpy values obtained agreed well with known values. The full equation has been deduced by another method in (14) and applied to MCl_2-H_2O systems in (14) and (15). For a summary of the use of equation [47] and similar equations, see (14).

(2) Molality scale

Substitution of the mean activities on the molality scale in eqn [40] gives:

$$\nu\ln\left[\frac{\gamma_{\pm}m_B}{\gamma_{\pm}{}^*m_B{}^*}\right] - \nu(m_B/m_B{}^* - 1) - \nu\{m_B(\phi - 1)/m_B{}^* - \phi^* + 1\}$$

$$= G(T) \tag{48}$$

where $G(T)$ is the same as in eqn [47], $m_B{}^* = 1/nM_A$ is the molality of the anhydrous salt in the pure salt hydrate and $\gamma_{\pm}$ and ϕ are the mean activity coefficient and the osmotic coefficient, respectively. Use of the osmotic coefficient for the activity of the solvent leads, therefore, to an equation that has a different appearance to [47]; the content is identical. However, while eqn [47] can be used over the whole range of composition $(0 \leqslant x_B \leqslant 1)$, the molality in eqn [48] becomes infinite at $x_B = 1$; use of eqn [48] is therefore confined to solutions sufficiently dilute that the molality is a useful measure of composition. The essentials of eqn [48] were deduced by Williamson (17); however, the form used here appears first in the *Solubility Data Series*. For typical applications (where activity and osmotic coefficients are not considered explicitly, so that the enthalpies and heat capacities are apparent values, as explained above), see (18).

The above analysis shows clearly that a rational thermodynamic basis exists for functional representation of solubility-temperature curves in two-component systems, but may be difficult to apply because of lack of experimental or theoretical knowledge of activity coefficients and partial molar enthalpies. Other phenomena which are related ultimately to the stoichiometric activity coefficients and which complicate interpretation include ion pairing, formation of complex ions, and hydrolysis. Similar considerations hold for the variation of solubility with pressure, except that the effects are relatively smaller at the pressures used in many investigations of solubility (5).

(b) Solubility as a function of composition.

At constant temperature and pressure, the chemical potential of a saturating solid phase is constant:

$$\mu_{A_nB}{}^* = \mu_{A_nB}(\text{sln}) = n\mu_A + \mu_B \tag{49}$$

$$= (n\mu_A{}^* + \nu_+\mu_+{}^{\infty} + \nu_-\mu_-{}^{\infty}) + nRT \ln f_A x_A$$

$$+ \nu RT \ln(\gamma_{\pm}m_{\pm}Q)$$

for a salt hydrate A_nB which dissociates to water (A), and a salt (B), one mole of which ionizes to give ν_+ cations and ν_- anions in a solution in which other substances (ionized or not) may be present. If the saturated solution is sufficiently dilute, $f_A = x_A = 1$, and the quantity K_s in

$$\Delta G^{\infty} = (\nu_+\mu_+{}^{\infty} + \nu_-\mu_-{}^{\infty} + n\mu_A{}^* - \mu_{AB}{}^*)$$

$$= -RT \ln K_s$$

$$= -\nu RT \ln(Q\gamma_{\pm}m_B) \qquad [50]$$

is called the *solubility product* of the salt. (It should be noted that it is not customary to extend this definition to hydrated salts, but there is no reason why they should be excluded.) Values of the solubility product are often given on mole fraction or concentration scales. In dilute solutions, the theoretical behaviour of the activity coefficients as a function of ionic strength is often sufficiently well known that reliable extrapolations to infinite dilution can be made, and values of K_S can be determined. In more concentrated solutions, the same problems with activity coefficients that were outlined in the section on variation of solubility with temperature still occur. If these complications do not arise, the solubility of a hydrate salt $C_{\nu}A_{\nu}\cdot nH_2O$ in the presence of other solutes is given by eqn [50] as

$$\nu \ln\{m_B/m_B(0)\} = -\nu\ln\{\gamma_{\pm}/\gamma_{\pm}(0)\} - n \ln\{a_A/a_A(0)\} \qquad [51]$$

where a_A is the activity of water in the saturated solution, m_B is the molality of the salt in the saturated solution, and (0) indicates absence of other solutes. Similar considerations hold for non-electrolytes.

Consideration of *complex mixed ligand equilibria* in the solution phase are also frequently of importance in the interpretation of solubility equilibria. For nomenclature connected with these equilibria (and solubility equilibria as well), see (19, 20).

The Solid Phase

The definition of solubility permits the occurrence of a single solid phase which may be a pure anhydrous compound, a salt hydrate, a non-stoichiometric compound, or a solid mixture (or solid solution, or "mixed crystals"), and may be stable or metastable. As well, any number of solid phases consistent with the requirements of the phase rule may be present. Metastable solid phases are of widespread occurrence, and may appear as polymorphic (or allotropic) forms or crystal solvates whose rate of transition to more stable forms is very slow. Surface heterogeneity may also give rise to metastability, either when one solid precipitates on the surface of another, or if the size of the solid particles is sufficiently small that surface effects become important. In either case, the solid is not in stable equilibrium with the solution. See (21) for the modern formulation of the effect of particle size on solubility. The stability of a solid may also be affected by the atmosphere in which the system is equilibrated.

Many of these phenomena require very careful, and often prolonged, equilibration for their investigation and elimination. A very general analytical method, the "wet residues" method of Schreinemakers (22), is often used to investigate the composition of solid phases in equilibrium with salt solutions. This method has been reviewed in (23), where [see also (24)] least-squares methods for evaluating the composition of the solid phase from wet residue data (or initial composition data) and solubilities are described. In principle, the same method can be used with systems of other types. Many other techniques for examination of solids, in particular X-ray, optical, and thermal analysis methods, are used in conjunction with chemical analyses (including the wet residues method).

COMPILATIONS AND EVALUATIONS

The formats for the compilations and critical evaluations have been standardized for all volumes. A brief description of the data sheets has been given in the FOREWORD; additional explanation is given below.

Guide to the Compilations

The format used for the compilations is, for the most part, self-explanatory. The details presented below are those which are not found in the FOREWORD or which are not self-evident.

Components. Each component is listed according to IUPAC name, formula, and Chemical Abstracts (CA) Registry Number. The formula is given either in terms of the IUPAC or Hill (25) system and the choice of formula is governed by what is usual for most current users: i.e., IUPAC for inorganic compounds, and Hill system for organic compounds. Components are ordered according to:
(a) saturating components;
(b) non-saturating components in alphanumerical order;
(c) solvents in alphanumerical order.

The saturating components are arranged in order according to a
18-column periodic table with two additional rows:
   Columns 1 and 2: H, alkali elements, ammonium, alkaline earth elements
         3 to 12: transition elements
      13 to 17: boron, carbon, nitrogen groups; chalcogenides, halogens
          18: noble gases
       Row 1: Ce to Lu
       Row 2: Th to the end of the known elements, in order of
           atomic number.

Salt hydrates are generally not considered to be saturating components
since most solubilities are expressed in terms of the anhydrous salt. The
existence of hydrates or solvates is carefully noted in the text, and CA
Registry Numbers are given where available, usually in the critical
evaluation. Mineralogical names are also quoted, along with their CA
Registry Numbers, again usually in the critical evaluation.

Original Measurements. References are abbreviated in the forms given
by *Chemical Abstracts Service Source Index* (CASSI). Names originally in
other than Roman alphabets are given as transliterated by *Chemical
Abstracts.*

Experimental Values. Data are reported in the units used in the
original publication, with the exception that modern *names* for units
and quantities are used; e.g., mass per cent for weight per cent;
mol dm^{-3} for molar; etc. Both mass and molar values are given. Usually,
only one type of value (e.g., mass per cent) is found in the original
paper, and the compiler has added the other type of value (e.g., mole
per cent) from computer calculations based on 1983 atomic weights (26).

Errors in calculations and fitting equations in original papers have
been noted and corrected, by computer calculations where necessary.

Method. Source and Purity of Materials. Abbreviations used in
Chemical Abstracts are often used here to save space.

Estimated Error. If these data were omitted by the original authors,
and if relevant information is available, the compilers have attempted
to estimate errors from the internal consistency of data and type of
apparatus used. Methods used by the compilers for estimating and
and reporting errors are based on the papers by Ku and Eisenhart (27).

Comments and/or Additional Data. Many compilations include this
section which provides short comments relevant to the general nature of
the work or additional experimental and thermodynamic data which are
judged by the compiler to be of value to the reader.

References. See the above description for Original Measurements.

Guide to the Evaluations

The evaluator's task is to check whether the compiled data are correct,
to assess the reliability and quality of the data, to estimate errors
where necessary, and to recommend "best" values. The evaluation takes
the form of a summary in which all the data supplied by the compiler
have been critically reviewed. A brief description of the evaluation
sheets is given below.

Components. See the description for the Compilations.

Evaluator. Name and date up to which the literature was checked.

Critical Evaluation
  (a) Critical text. The evaluator produces text evaluating all the
published data for each given system. Thus, in this section the
evaluator reviews the merits or shortcomings of the various data. Only
published data are considered; even published data can be considered only
if the experimental data permit an assessment of reliability.
  (b) Fitting equations. If the use of a smoothing equation is
justifiable the evaluator may provide an equation representing the
solubility as a function of the variables reported on all the
compilation sheets.
  (c) Graphical summary. In addition to (b) above, graphical summaries
are often given.
  (d) Recommended values. Data are *recommended* if the results of at
least two independent groups are available and they are in good
agreement, and if the evaluator has no doubt as to the adequacy and
reliability of the applied experimental and computational procedures.
Data are considered as *tentative* if only one set of measurements is

available, or if the evaluator considers some aspect of the computational or experimental method as mildly undesirable but estimates that it should cause only minor errors. Data are considered as *doubtful* if the evaluator considers some aspect of the computational or experimental method as undesirable but still considers the data to have some value in those instances where the order of magnitude of the solubility is needed. Data determined by an inadequate method or under ill-defined conditions are *rejected*. However references to these data are included in the evaluation together with a comment by the evaluator as to the reason for their rejection.

(e) References. All pertinent references are given here. References to those data which, by virtue of their poor precision, have been rejected and not compiled are also listed in this section.

(f) Units. While the original data may be reported in the units used by the investigators, the final recommended values are reported in S.I. units (1, 28) when the data can be accurately converted.

References

1. Whiffen, D.H., ed., *Manual of Symbols and Terminology for Physico-chemical Quantities and Units*. *Pure Applied Chem.* <u>1979</u>, 51, No. 1.
2. McGlashan, M.L. *Physicochemical Quantities and Units*. 2nd ed. Royal Institute of Chemistry. London. <u>1971</u>.
3. Jänecke, E. *Z. Anorg. Chem.* <u>1906</u>, 51, 132.
4. Friedman, H.L. *J. Chem. Phys.* <u>1960</u>, 32, 1351.
5. Prigogine, I.; Defay, R. *Chemical Thermodynamics*. D.H. Everett, transl. Longmans, Green. London, New York, Toronto. <u>1954</u>.
6. Guggenheim, E.A. *Thermodynamics*. North-Holland. Amsterdam. <u>1959</u>. 4th ed.
7. Kirkwood, J.G.; Oppenheim, I. *Chemical Thermodynamics*. McGraw-Hill. New York, Toronto, London. <u>1961</u>.
8. Lewis, G.N.; Randall, M. (rev. Pitzer, K.S.; Brewer, L.). *Thermodynamics*. McGraw Hill. New York, Toronto, London. <u>1961</u>. 2nd. ed.
9. Robinson, R.A.; Stokes, R.H. *Electrolyte Solutions*. Butterworths. London. <u>1959</u>. 2nd ed.
10. Harned, H.S.; Owen, B.B. *The Physical Chemistry of Electrolytic Solutions*. Reinhold. New York. <u>1958</u>. 3rd ed.
11. Haase, R.; Schönert, H. *Solid-Liquid Equilibrium*. E.S. Halberstadt, trans. Pergamon Press, London, <u>1969</u>.
12. McGlashan, M.L. *Chemical Thermodynamics*. Academic Press. London. <u>1979</u>.
13. Cohen-Adad, R.; Saugier, M.T.; Said, J. *Rev. Chim. Miner.* <u>1973</u>, 10, 631.
14. Counioux, J.-J.; Tenu, R. *J. Chim. Phys.* <u>1981</u>, 78, 815.
15. Tenu, R.; Counioux, J.-J. *J. Chim. Phys.* <u>1981</u>, 78, 823.
16. Cohen-Adad, R. *Pure Appl. Chem.* <u>1985</u>, 57, 255.
17. Williamson, A.T. *Faraday Soc. Trans.* <u>1944</u>, 40, 421.
18. Siekierski, S.; Mioduski, T.; Salomon, M. *Solubility Data Series*. Vol. 13. *Scandium, Yttrium, Lanthanum and Lanthanide Nitrates*. Pergamon Press. <u>1983</u>.
19. Marcus, Y., ed. *Pure Appl. Chem.* <u>1969</u>, 18, 459.
20. IUPAC Analytical Division. *Proposed Symbols for Metal Complex Mixed Ligand Equilibria (Provisional)*. *IUPAC Inf. Bull.* <u>1978</u>, No. 3, 229.
21. Enüstün, B.V.; Turkevich, J. *J. Am. Chem. Soc.* <u>1960</u>, 82, 4502.
22. Schreinemakers. F.A.H. *Z. Phys. Chem., Stoechiom. Verwandschaftsl.* <u>1893</u>, 11, 75.
23. Lorimer, J.W. *Can. J. Chem.* <u>1981</u>, 59, 3076.
24. Lorimer, J.W. *Can. J. Chem.* <u>1982</u>, 60, 1978.
25. Hill, E.A. *J. Am. Chem. Soc.* <u>1900</u>, 22, 478.
26. IUPAC Commission on Atomic Weights. *Pure Appl. Chem.* <u>1984</u>, 56, 653.
27. Ku, H.H., p. 73; Eisenhart, C., p. 69; in Ku, H.H., ed. *Precision Measurement and Calibration*. NBS Special Publication 300. Vol. 1. Washington. <u>1969</u>.
28. *The International System of Units*. Engl. transl. approved by the BIPM of *Le Système International d'Unités*. H.M.S.O. London. <u>1970</u>.

September, 1986
      R. Cohen-Adad,
      Villeurbanne, France

      J. W. Lorimer,
      London, Ontario, Canada

      M. Salomon,
      Fair Haven, New Jersey, U.S.A.

Table I-1

Quantities Used as Measures of Solubility of Solute B
Conversion Table for Multicomponent Systems
Containing Solvent A and Solutes s

	mole fraction $x_B =$	mass fraction $w_B =$	molality $m_B =$	concentration $c_B =$
x_B	x_B	$\dfrac{M_B x_B}{M_A + \sum\limits_s (M_s - M_A) x_s}$	$\dfrac{x_B}{M_A(1 - \sum\limits_s x_s)}$	$\dfrac{\rho x_B}{M_A + \sum\limits_s (M_s - M_A) x_s}$
w_B	$\dfrac{w_B/M_B}{1/M_A + \sum\limits_s (1/M_s - 1/M_A) w_s}$	w_B	$\dfrac{w_B}{M_B(1 - \sum\limits_s w_s)}$	$\rho w_B / M_B$
m_B	$\dfrac{M_A m_B}{1 + M_A \sum\limits_s m_s}$	$\dfrac{M_B m_B}{1 + \sum\limits_s m_s M_s}$	m_B	$\dfrac{\rho m_B}{1 + \sum\limits_s M_s m_s}$
c_B	$\dfrac{M_A c_B}{\rho + \sum\limits_s (M_A - M_s) c_s}$	$M_B c_B / \rho$	$\dfrac{c_B}{\rho - \sum\limits_s M_s c_s}$	c_B

ρ = density of solution
M_A, M_B, M_s = molar masses of solvent, solute B, other solutes s
Formulas are given in forms suitable for rapid computation; all
calculations should be made using SI base units.

COMPONENTS:	EVALUATOR:
(1) Lithium chlorate; $LiClO_3$; [13543-71-9] (2) Water; H_2O; [7732-18-5]	H. Miyamoto Department of Chemistry Niigata University, Niigata, Japan and M. Salomon US Army ET & DL Fort Monmouth, NJ, USA August, 1984

CRITICAL EVALUATION:

THE BINARY SYSTEM

Data for the solubility of $LiClO_3$ in water have been reported in five publications (1-5). The data of Mylius and Funk (1) and Treadwell and Ammann (4) can immediately be rejected because of their inconsistencies (low solubilities indicating failure to approach equilibrium), and the fact that many experimental details are absent. Although Mylius and Funk (1) suggest that the anhydrous salt is the solid phase, the value for the solubility is much too low for the experimental temperature of 291 K. Treadwell and Ammann (4) do not report the nature of the solid phases.

The remaining studies (2,3,5) all report complete phase diagrams which qualitatively appear to be in good agreement, but detailed examination of the solubility data show significant differences. All authors agree that the binary system has a tendency to form metastable solutions which probably contributes to the causes in some important differences in solubility data.

Based on the analyses of the three major works (2,3,5), the evaluators agree with Campbell and Griffiths (5) that the various solid phases present in the binary system are:

ice	$LiClO_3.1/4H_2O$
$LiClO_3.3H_2O$ [66295-75-8]	$\beta-LiClO_3$
$LiClO_3.H_2O$	$\alpha-LiClO_3$

All three major studies agree, in general, on the temperature ranges over which six solid phases exist, but disagreement exists on the composition of several of these phases. Over the temperature range of 262-317 K, Berg (3) claims the solid phase to be $LiClO_3.1/3H_2O$, Kraus and Burgess (2) claim it to be an anhydrous $\gamma-LiClO_3$, and Campbell and Griffiths (5) have shown it to be $LiClO_3.1/4H_2O$.

In their attempts to determine transition temperatures over the temperature range 262-317 K, Campbell and Griffiths experienced difficulty with metastability. Very careful dilatometric measurements over the range of 308-323 K (5) revealed no transition at 314.7 K which is the temperature claimed by Kraus and Burgess (2) for the $\gamma-LiClO_3$ $\beta-LiClO_3$ transition. X-ray diffraction patterns of anhydrous $LiClO_3$ prepared at room temperature and after heating to 353 K showed no change in the X-ray patterns (5). Campbell and Griffiths also found that Berg's photomicrographs of the so-called 1/3-hydrate were identical to the solid phase determined to be (see below) $LiClO_3.1/4H_2O$ (5). The stoichiometry of the 1/4-hydrate was unambiguously proved in (5) by studying the ternary $LiClO_3$ - $LiCl$ - H_2O system. By application of the wet residue method of Schreinemakers, Campbell and Griffiths found that the tie lines (none of which ever crossed) were very reproducible and converged to a single composition of $(LiClO_3)_4H_2O$. The evaluators therefore conclude that the stable solid phase over the temperature range of 293-315 K is the 1/4-hydrate, and that Berg's 1/3-hydrate and Kraus and Burgess' anhydrous $\gamma-LiClO_3$ are in fact the 1/4-hydrate.

To evaluate the solubility data from (2,3,5), we separately examined each polytherm in the phase diagram, and fitted the data to the smoothing equation (see eq. [1] in the PREFACE)

$$Y_x = A/(T/K) + B\ln(T/K) + C + D(T/K) \qquad [1]$$

where

$$Y_x = \ln\left\{\chi^{v(1-\chi)r}(v+r)^{v+r}/[r^r(1+\chi)^{v+r}]\right\} \qquad [2]$$

COMPONENTS:	EVALUATOR:
(1) Lithium chloride; $LiClO_3$; [13543-71-9] (2) Water; H_2O; [7732-18-5]	H. Miyamoto Department of Chemistry Niigata University Niigata, Japan and M. Salomon US Army ET & DL Fort Monmouth, NJ, USA August, 1984

CRITICAL EVALUATION:

In eq. [2], r is the hydration number in the solid phase, v is the number of ions produced upon disolution (v = 2 for $LiClO_3$), and χ is the mole fraction solubility (note that for the ice polytherm, v = 0 and r = 1). In general, when applying the solubility data from (2,3,5) to eqs. [1] and [2], a number of solubilities were rejected when the difference in calculated and observed solubilities ($\chi_{calcd} - \chi_{obsd}$) exceeded $2\sigma_x$. σ_x is the standard error of estimate defined in the usual manner by:

$$\sigma_x = [\Sigma(\chi_{obsd} - \chi_{calcd})^2/(N - NC)]^{\frac{1}{2}} \qquad\qquad [3]$$

where N is the number of data points, and NC is the number of constants adjusted in eq. [1]. For all the polytherms in the binary $LiClO_3$ - H_2O system, a four constant fit was used: i.e. NC = 4. The results of fitting the data to eq. [1] are given in Table 1, and additional details are given below. The results of these analyses were used to construct detailed graphs of portions of the polytherms in the regions of phase transitions. From these graphs the evaluators were able to determine the nine observable transition temperatures by graphical interpolation, and the results are given in Table 2. Details on the evaluation of the solubility data for each polytherm follow.

Polytherm For Ice As The Solid Phase

All 16 data points reported in (2) and (5) were used to adjust the constants in the smoothing equation (see Table 1). Mole fraction solubilities at rounded temperatures based on this analysis are designated as *recommended* values, and are listed in Table 3. Based on our graphical interpolation, we find the recommended temperature and solubility at the ice ⟶ $LiClO_3.3H_2O$ transition to be 230.55 K and 0.107, respectively. Both Kraus and Burgess (2) and Berg (3) reported this temperature as 233.2 K, and Campbell and Griffiths (5) reported it as 229.85 K.

Polytherm For $LiClO_3.3H_2O$ As The Solid Phase

Since the polytherm reaches a maximum at the congruent melting point (cmp), we analyzed this system in two parts for $\chi < 0.25$ and for $\chi > 0.25$.

    (a) $\chi < 0.25$. Twenty data points from (2) and (5) were fitted to eq. [1]. Only one data point at 281.3 K (8.1°C) from (5) was omitted. The least squares fit to eq. [1] predicts a congruent melting point of 281.16 K.

    (b) $\chi > 0.25$. All data points from (2) and (5) were considered, and all four points below 264.2 K were rejected. The resulting fit of 19 data points is given in Table 1, and the congruent melting point as calculated from eq. [1] is 281.12 K.

The solubility at the average cmp of 281.14 K is thus 0.25 mole fraction or 18.503 mol kg^{-1}. Berg (3) reported a cmp of 281.3 K, Kraus and Burgess (2) 281.2 K, and Campbell and Griffiths (5) reported 281.3 K (melting point) and 281.6 K (graphical).

Two metastable transitions involving the trihydrate were found in all three major studies. For the $LiClO_3.3H_2O$ ⟶ $LiClO_3.1/4$ transition (eutectic), the evaluators graphically determined the temperature to be 261.15 K compared to 264 K in (3) and 262.7 K in (5).

For the metastable $LiClO_3.3H_2O$ ⟶ β-$LiClO_3$ eutectic transition, a temperature of 248.5 K was found by the evaluators which is the same value reported in all three major studies (2,3,5).

All calculated solubilities based on the smoothing equation are designated as *recommended*, and values at rounded temperatures are given in Table 3.

COMPONENTS:	EVALUATOR:
(1) Lithium chlorate; $LiClO_3$; [13543-71-9] (2) Water; H_2O; [7732-18-5]	H. Miyamoto Department of Chemistry Niigata University Niigata, Japan and M. Salomon US Army ET & DL Fort Monmouth, NJ, USA August, 1984

CRITICAL EVALUATION:

Polytherm For $LiClO_3 \cdot H_2O$ As The Solid Phase.

While there is qualitative agreement between the results in (2,3,5), quantitative agreement does not exist and we are forced to select a preferred set of data. Since Berg's data, particularly at the lower temperatures, are radically different from those in (2,5) we can safely reject these results. We cannot offer convincing evidence favoring either set of data from (2) or (5), and at this time we prefer to use Campbell and Griffiths' results. The smoothed data based on (5) are given in Table 4, and are designated as *tentative* solubilities.

The evaluators have calculated a congruent melting point of 310.5 K at $\chi = 0.500$ (m = 55.508 mol kg^{-1}), but it should be noted that a melting point for $LiClO_3 \cdot H_2O$ has not been measured and that it may not melt congruently.

By graphical interpolation the evaluators find the $LiClO_3 \cdot H_2O \longrightarrow LiClO_3 \cdot 1/4H_2O$ transition temperature to be 292.85 K. This transition temperature was reported as 294.2 K, 295.2 K, and 293.65 K in (2,3,5), respectively.

Polytherm For $LiClO_3 \cdot 1/4H_2O$ As The Solid Phase

The confusion of the composition of this hydrate was discussed above. Summarizing, the evaluators agree with Campbell and Griffiths who accurately and unambiguously determined the composition as the 1/4-hydrate, and that both the γ-$LiClO_3$ phase of Kraus and Burgess and the 1/3-hydrate of Berg are in fact the same phase deduced by Campbell and Griffiths as the 1/4-hydrate.

The 1/4-hydrate easily forms metastable solutions as indicated above and from the fact that its solubility can be experimentally determined down to 261 K. In fitting the solubility data to eq. [1], again Berg's results had to be rejected, and two data points at 305.2 K and 300.8 K (32°C and 27.6°C) from Kraus and Burgess also had to be rejected. The results of fitting the remaining 24 data points to eq. [1] are given in Table 1, and smoothed solubilities designated as *tentative* values are given in Table 4. The smoothed values could not be designated as recommended because the agreement in experimental data from (2) and (5) is not particularly satisfactory as indicated by the large standard errors of estimate (σ values) in Table 1.

By graphical interpolation the evaluators found the 1/4-hydrate $\longrightarrow \beta$-$LiClO_3$ transition at 314.85 K. For this transition Berg reported a temperature of 317 K, Kraus and Burgess reported this (eutectic) temperature as 314.7 K, and according to Campbell and Griffiths, this is a peritectic transition occurring at 315.15 K.

Polytherms For Anhydrous $LiClO_3$

The three major studies (2,3,5) all agree that two anhydrous phases (β & α) exist, the latter constituting the higher temperature phase.

  β-$LiClO_3$. Forty-three data points reported in (2,3,5) were fitted to eq. [1] and the results are given in Table 1. The β-phase is stable over the range of 314.9-368.5 K (evaluators, see Table 2), but metastable solutions easily form at temperatures as low as 248 K. Because of the large standard errors of estimate obtained for this fit (see Table 1), the solubilities at rounded temperatures given in Table 4 are designated as *tentative* values.

  α-$LiClO_3$. Of the 25 data points reported in (2,3), two points from (2) at 102.5°C and 126.7°C had to be rejected. The only data point above 372.1 K reported in (5) was the melting point of 400.6-400.7 K for α-$LiClO_3$, and the value of 400.7 K was used in the least squares fit to eq. [1]. Based on this analysis (Table 1), solubilities at rounded temperatures were calculated and tabulated in Table 4: these solubilities are designated as *recommended* values.

COMPONENTS:	EVALUATOR:
(1) Lithium chlorate; $LiClO_3$; [13543-71-9] (2) Water; H_2O; [7732-18-5]	Hiroshi Miyamoto Department of Chemistry Niigata University Niigata, Japan
	Mark Salomon US Army ET & DL Fort Monmouth, NJ, USA August, 1984

CRITICAL EVALUATION:

The temperature of the $\beta \longrightarrow \alpha$ transition has proved somewhat difficult to determine. Direct thermal analyses showed marked hysteresis with transition temperatures ranging from 383 K (rising temp) to 353 K (falling temp), and a best value of 373.0 K was indicated from the falling temperature side (5). Both Berg (3) and Kraus and Burgess (2) report a value of 382 K for this transition. By graphical interpolation the evaluators determined this transition temperature as 368.45 K at a solubility of χ = 0.778. The least squares fit of the data to eq. [1] predicts a melting point of 400.72 K.

The complete phase diagram for the binary system given by Campbell and Griffiths is reproduced below following the data tables.

TERNARY SYSTEMS

The only system reported is that by Campbell and Griffiths (5) for the system $LiClO_3$ – LiCl – H_2O, and for temperatures of 276.0 K, 279.0 K, 281.7 K, and 298.2 K. Some discussion was presented above with emphasis on the confirmation of the 1/4-hydrate solid phase by Schreinemakers' method of wet residues. The four isotherms reported in (5) are reproduced below following the data tables for the binary $LiClO3$ – $H2O$ system.

REFERENCES

1. Mylius, F.; Funk, R. *Ber. Dtsch. Chem. Ges.* <u>1897</u>, *30*, 1716.

2. Kraus, C. A.; Burgess, W. M. *J. Am. Chem. Soc.* <u>1927</u>, *49*, 1226.

3. Berg, L. Z. *Anorg. Allg. Chem.* <u>1929</u>, *181*, 131.

4. Treadwell, W. D.; Ammann, A. *Helv. Chim. Acta* <u>1938</u>, *21*, 1249.

5. Campbell, A. N.; Griffiths, J. E. *Can. J. Chem.* <u>1956</u>, *34*, 1647.

COMPONENTS:	EVALUATOR:
(1) Lithium chlorate; $LiClO_3$; [13543-71-9] (2) Water; H_2O; [7732-18-5]	H. Miyamoto Department of Chemistry Niigata University Niigata, Japan and M. Salomon US Army ET & DL Fort Monmouth, NJ, USA August, 1984

CRITICAL EVALUATION:

Table 1. Constants for the smoothing equation[a]

Solid Phase	N	A	B	C	D	σ_y	σ_x
1. ICE	16	-20714.80 (1.32)	-174.9468 (0.00096)	955.5416 (0.0053)	0.372531 (2.1×10^{-5})	0.0051	0.0025
2. $LiClO_3 \cdot 3H_2O$ ($X < 0.25$)	20	29321.07 (2.96)	229.4229 (0.0021)	-1275.480 (0.012)	-0.435711 (4.6×10^{-5})	0.011	0.0030
3. $LiClO_3 \cdot 3H_2O$ ($X > 0.25$)	19	-33115.58 (1.4)	-235.5517 (0.00094)	1324.4436 (0.0053)	0.432482 (1.9×10^{-5})	0.0051	0.0028
4. $LiClO_3 \cdot H_2O$	11	208261.78 (0.69)	1454.3290 (0.00043)	-8229.9610 (0.0024)	-2.531326 (8.5×10^{-6})	0.0022	0.0026
5. $LiClO_3 \cdot 1/4H_2O$	24	63591.78 (7.17)	426.1955 (0.0095)	-2431.857 (0.026)	-0.704569 (9.1×10^{-5})	0.025	0.010
6. β-$LiClO_3$	43	-15101.43 (7.94)	-105.9926 (0.0042)	597.709 (0.024)	0.188232 (7.5×10^{-5})	0.024	0.012
7. α-$LiClO_3$	23	-24948.55 (3.28)	-170.9093 (0.0014)	972.7758 (0.0085)	0.283458 (2.2×10^{-5})	0.0082	0.0068

[a] N is the number of data points fitted to the smoothing equation
σ_y is the standard error of estimate in the function Y (see eq [2])
σ_x is the standard error of estimate for the mole fraction solubilities (see eq [3])

COMPONENTS:	EVALUATOR:
(1) Lithium chlorate; $LiClO_3$; [13543-71-9]	H. Miyamoto Department of Chemistry Niigata University Niigata, Japan
(2) Water; H_2O; [7732-18-5]	and Mark Salomon US Army ET & DL Fort Monmouth, NJ, USA August, 1984

CRITICAL EVALUATION:

Table 2.   Monovariant Equilibria in the Binary System   a
Recommended Transition Temperatures and Solubilities

Transition	T/K	Mole Fraction
ice $\longrightarrow LiClO_3.3H_2O$	230.55	0.107
$LiClO_3.3H_2O$ melt pt[b]	281.14	0.250
$LiClO_3.3H_2O \longrightarrow LiClO_3.H_2O$	271.05	0.350
$LiClO_3.H_2O \longrightarrow LiClO_3.1/4H_2O$	292.85	0.457
$LiClO_3.3H_2O \longrightarrow LiClO_3.1/4H_2O^m$	261.15	0.406
$LiClO_3.3H_2O \longrightarrow \beta-LiClO_3{}^m$	248.15	0.473
$LiClO_3.1/4H_2O \longrightarrow \beta-LiClO_3$	314.85	0.558
$\beta-LiClO_3 \longrightarrow \alpha-LiClO_3$	368.45	0.778
$\alpha-LiClO_3 \longrightarrow$ melt	400.72[c]	1.000

[a]Results of graphical interpolation by the evaluators except as noted.

[b]Congruent melting point average value obtained from smoothing equation
(see discussion in text).

[c]From least squares fit to eq. [1].

[m]metastable points.

COMPONENTS:	EVALUATOR:
(1) Lithium chlorate; $LiClO_3$; [13543-71-9] (2) Water; H_2O; [7732-18-5]	H. Miyamoto Department of Chemistry Niigata University Niigata, Japan and Mark Salomon US Army ET & DL Fort Monmouth, NJ, USA August, 1984

CRITICAL EVALUATION:

Table 3. Recommended Solubilities Calculated From
the Smoothing Equation (See Table 1)

T/K	ice polytherm mole fraction soly	$LiClO_3 \cdot 3H_2O$ $\chi \leq 0.250$	$LiClO_3 \cdot 3H_2O$ $\chi \geq 0.250$
228.15	0.1103	0.1036[m]	
233.15	0.1019	0.1071	
238.15	0.0940	0.1119	0.5043[m]
243.15	0.0859	0.1178	0.4797[m]
248.15	0.0769	0.1250	0.4562[m]
253.15	0.0662	0.1334	0.4335[m]
258.15	0.0533	0.1433	0.4111[m]
263.15	0.0377	0.1550	0.3884[m]
268.15	0.0190	0.1688	0.3644[m]
270.65	0.00826	0.1769	0.3513[m]
272.15	0.00140	0.1822	0.3429
273.15		0.1861	0.3371
278.15		0.2104	0.3014
280.15		0.2269	0.2787
281.14[a]		0.2500	0.2500

[a]Average temperatures: range ± 0.02 K (see text for discussion).

[m]Metastable equilibria

COMPONENTS:	EVALUATOR:
(1) Lithium chlorate; $LiClO_3$; [13543-71-9] (2) Water; H_2O; [7732-18-5]	H. Miyomoto Department of Chemistry Niigata University Niigata, Japan and Mark Salomon US Army ET & DL Fort Monmouth, NJ, USA August, 1984

CRITICAL EVALUATION:

Table 4.   Solubilities at Rounded Temperatures Calculated
From Eq. [1] (See Table 1)

T/K	$LiClO_3 \cdot H_2O$[a]	$LiClO_3 \cdot 1/4 H_2O$[a]	β-$LiClO_3$[a]	α-$LiClO_3$[b]
248.15			0.473^m	
258.15			0.492^m	
263.15		0.403^m	0.499^m	
273.15	0.352	0.403^m	0.511^m	
278.15	0.362	0.412^m	0.515^m	
283.15	0.382	0.424^m	0.520^m	
288.15	0.413	0.439^m	0.524^m	
293.15	0.462^m	0.457	0.529^m	
298.15		0.478	0.534^m	
303.15		0.501	0.540^m	
308.15		0.525	0.546^m	
313.15		0.549	0.554^m	
323.15			0.573	
328.15			0.585	
338.15			0.614	
348.15			0.653	
358.15			0.706	
368.15			0.776	0.777^m
373.15				0.793
378.15				0.814
383.15				0.841
388.15				0.874
393.15				0.917
398.15				0.969
400.72				1.000

[a]Tentative

[b]Recommended

COMPONENTS:	EVALUATOR:
(1) Lithium chlorate; LiClO$_3$; [13543-71-9]	Hiroshi Miyamoto
	Department of Chemistry
(2) Water; H$_2$O; [7732-18-5]	Niigata University
	Niigata, Japan
	August 1984

CRITICAL EVALUATION:

Phase Diagram for the LiClO$_3$ - H$_2$O System (5)

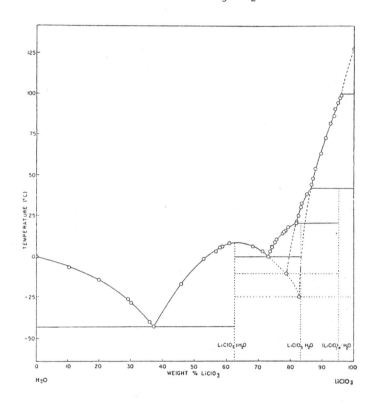

Phase Diagrams for the LiClO$_3$ - LiCl - H$_2$O System (5)

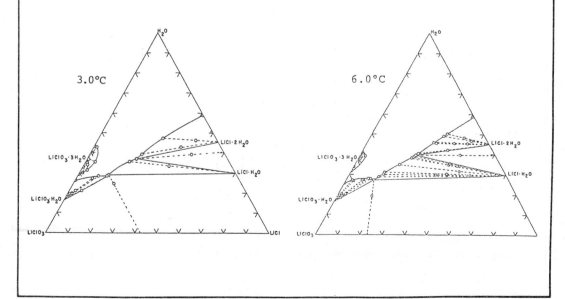

COMPONENTS:	EVALUATOR:
(1) Lithium chlorate; LiClO$_3$; [13453-71-9] (2) Wtaer, H$_2$O; [7732-18-5]	Hiroshi Miyamoto Department of Chemistry Niigata University Niigata, Japan August 1984

CRITICAL EVALUATION:

Phase Diagrams for the LiClO$_3$ - LiCl - H$_2$O System, continued (5)

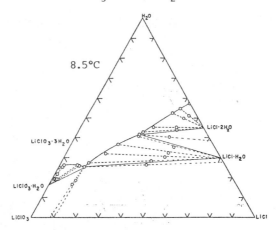

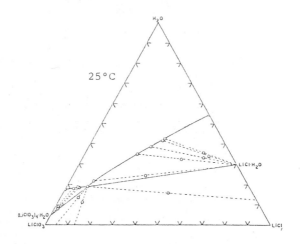

Detail of 25°C isotherm

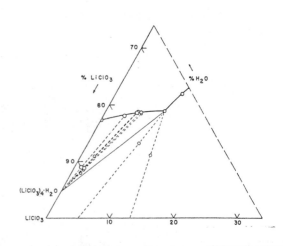

COMPONENTS:	ORIGINAL MEASUREMENTS:
(1) Lithium chlorate: LiClO$_3$; [13453-71-9]	Mylius, F.; Funk, R.
(2) Water; H$_2$O; [7732-18-5]	*Ber. Dtsch. Chem. Ges.* <u>1897</u>, *30*, 1716-25.

VARIABLES:	PREPARED BY:
T/K = 291	Hiroshi Miyamoto

EXPERIMENTAL VALUES:

The solubility of LiClO$_3$ in water at 18°C is given:

75.8 mass %	(authors)
313.5 g/100 g H$_2$O	(authors)
34.7 mol kg^{-1}	(compiler)

The density of the saturated solution was given as

1.814 g cm^{-3}

AUXILIARY INFORMATION

METHOD/APPARATUS/PROCEDURE:	SOURCE AND PURITY OF MATERIALS:
The salt and water were placed in a bottle and agitated in a constant temperature bath for a long time (time not specified). After the saturated solution settled, aliquots for analyses were withdrawn with a pipet. LiClO$_3$ was determined by evaporation to dryness.	The salt was stated to be of a "pure grade", and trace impurities stated to be absent.
	ESTIMATED ERROR: Soly: precision within 1 %. Temp: nothing specified.
	REFERENCES:

COMPONENTS:	ORIGINAL MEASUREMENTS:
(1) Lithium chlorate; $LiClO_3$; [13453-71-9] (2) Water; H_2O; [7732-18-5]	Kraus, C.A.; Burgess, W.M. J. Am. Chem. Soc. 1927, 49, 1226-35.
VARIABLES: T/K = 234.1 to 400.8	PREPARED BY: Hiroshi Miyamoto and Mark Salomon

EXPERIMENTAL VALUES:

t/°C	Water mass %	Lithium mass % (compiler)	Chlorate mol % (compiler)	mol kg^{-1} (compiler)	Nature of the
− 8.7	84.24	15.76	3.595	2.070	Ice
−13.2	79.27	20.73	4.954	2.893	"
−15.2	77.37	22.63	5.508	3.236	"
−17.4	75.56	24.44	6.056	3.578	"
−19.9	73.30	26.70	6.768	4.030	"
−23.3	71.32	28.68	7.420	4.449	"
−26.2	69.67	30.33	7.984	4.816	"
−30.5	67.29	32.71	8.832	5.378	"
−36.6	64.62	35.38	9.838	6.057	"
−39.1	63.27	36.73	10.37	6.422	"
−39.0	62.58	37.42	10.65	6.615	$LiClO_3.3H_2O$
−37.1	61.9	38.1	10.9	6.81	"
−33.9	60.95	39.05	11.32	7.088	"
−15.7	54.65	45.35	14.19	9.180	"
− 8.8	52.06	47.94	15.51	10.19	"
− 7.3	51.04	48.96	16.05	10.61	"
− 4.8	49.51	50.49	16.89	11.28	"
− 1.8	48.05	51.95	17.73	11.96	"
+ 0.5	46.73	53.27	18.51	12.61	"
2.2	45.43	54.57	19.32	13.29	"
4.8	43.34	56.66	20.67	14.46	"
6.1	41.84	58.16	21.69	15.38	"
7.3	39.85	60.15	23.13	16.70	"

continued......

AUXILIARY INFORMATION

METHOD/APPARATUS/PROCEDURE:
Solubilities were determined by thermal analysis (cooling curves). A weighed quantity of salt (12-28 g) was placed in a tube which was sealed with a rubber stopper. The stopper was fitted with a mercury sealed stirrer, a nitrogen inlet, an inlet to permit additions of water from a pycnometer with a long delivery tube, and a copper-constantan thermocouple. The solubility tube was fitted with an air jacket, and the apparatus placed in an oil or water thermostat. Nitrogen was passed through the upper part of the tube during additions of water, and experiments were carried with a slight excess pressure of N_2. After the determination of the melting point of the anhydrous salt, successive known quantities of water were added, and the solution cooled to obtain precipitation. Four independent series of experiments were carried out.

SOURCE AND PURITY OF MATERIALS:
Lithium chlorate solutions were prepd by slowly adding a 5.5 N solution of barium chlorate to a 4.5 N solution of lithium sulfate, both solutions being near the boiling point. The pptd $BaSO_4$ was removed by filtration, and the filtrate treated with barium chlorate and sulfate to insure equivalence of $LiClO_3$. $LiClO_3$ was pptd from the filtrate in several steps by slowly evaporating the solvent in vacuum over P_2O_5. Care was taken to prevent the salt in the highly concentrated solution (around 90 % of salt) from decomposing by keeping the temp below 50°C. The resulting salt was pulverized and finally dried in a desiccator over P_2O_5 under high vacuum.

Lithium sulfate prepd from the recrystd carbonate with sulfuric acid. The ppt was washed and dried at 130°C. Barium chlorate of the highest obtainable purity was recrystd from water several times and was dried at 130°C.

ESTIMATED ERROR:
Nothing specified.

COMPONENTS:	ORIGINAL MEASUREMENTS:
(1) Lithium chlorate; LiClO$_3$; [13453-71-9] (2) Water; H$_2$O; [7732-18-5]	Kraus, C.A.; Burgess, W.M. *J. Am. Chem. Soc.* 1927, 49, 1226-35.

EXPERIMENTAL VALUES: (Continued)

t/°C	Water mass %	Lithium mass % (compiler)	Chlorate mol % (compiler)	mol kg^{-1} (compiler)	Nature of the solid phase
7.85	38.49	61.51	24.16	17.68	LiClO$_3$.3H$_2$O
7.9	36.56	63.44	25.70	19.20	"
7.4	35.12	64.88	26.91	20.44	"
6.8	33.89	66.11	27.99	21.58	"
6.0	32.82	67.18	28.97	22.64	"
4.5	31.61	68.39	30.13	23.94	"
4.0	30.56	69.44	31.17	25.14	"
3.4	30.26	69.74	31.48	25.50	"
0.9	28.82	71.18	32.99	27.32	"
0.0	28.11	71.89	33.76	28.29	"
- 1.6	27.20	72.80	34.79	29.61	"
- 1.8	27.61	72.39	34.32	29.07	"
- 3.6	26.26	73.74	35.88	31.07	"
- 5.8	25.37	74.63	36.96	32.54	"
- 7.3	25.03	74.97	37.38	33.14	"
- 7.8	24.55	75.45	37.99	34.00	"
-13.6	22.75	77.25	40.36	37.57	"
+ 3.8	28.11	71.89	33.76	28.29	LiClO$_3$.H$_2$O
5.1	27.61	72.39	34.32	29.01	"
6.8	27.20	72.80	34.79	29.61	"
9.1	26.26	73.74	35.88	31.07	"
11.2	25.37	74.63	36.96	32.54	"
12.0	25.03	74.97	37.38	33.14	"
13.4	24.55	75.45	37.99	34.00	"
14.1	23.41	76.59	39.47	36.19	"
14.6	23.79	76.21	38.97	35.44	"
16.7	22.75	77.25	40.36	37.57	"
18.9	21.46	78.54	42.18	40.49	"
20.5	19.73	80.27	44.78	45.01	"
- 3.2	23.41	76.59	39.47	36.19	γ-LiClO$_3$
+ 2.9	22.41	77.59	40.83	38.30	"
8.4	21.67	78.33	41.87	39.99	"
12.8	20.74	79.26	43.24	42.28	"
16.4	20.32	79.68	43.87	43.38	"
22.1	18.32	81.68	47.05	49.32	"
27.2	17.33	82.67	48.74	52.77	"
27.6	18.33	81.67	47.03	49.29	"
32.0	17.20	82.80	48.96	53.26	"
32.0	16.09	83.91	50.97	57.69	"
32.8	16.42	83.58	50.36	56.31	"
36.2	15.48	84.52	52.11	60.40	"
36.8	14.64	85.36	53.75	64.50	"
39.6	13.73	86.27	55.60	69.51	"
43.9	12.83	87.17	57.52	75.16	"
36.9	13.73	86.27	55.60	69.51	β-LiClO$_3$
48.1	13.67	86.33	55.73	69.87	"
48.3	12.83	87.17	57.52	75.16	"
49.8	12.51	87.49	58.23	77.37	"
61.6	11.81	88.19	59.81	82.61	"
67.2	10.57	89.43	62.77	93.60	"
71.3	9.72	90.28	64.93	102.8	"
72.6	9.98	90.02	64.26	99.79	"
78.9	8.61	91.39	67.90	117.4	"
85.8	7.46	92.54	71.20	137.2	"

COMPONENTS:	ORIGINAL MEASUREMENTS:
(1) Lithium chlorate; LiClO$_3$; [13453-71-9]	Kraus, C.A.; Burgess, W.M.
(2) Water; H$_2$O; [7732-18-5]	J. Am. Chem. Soc. 1927, 49, 1226-35.

EXPERIMENTAL VALUES: (Continued)

t/°C	Water mass %	Lithium mass % (compiler)	Chlorate mol % (compiler)	mol kg^{-1} (compiler)	Nature of the solid phase
86.4	7.85	92.15	70.06	129.9	β-LiClO$_3$
86.6	7.40	92.60	71.38	138.4	"
90.0	6.55	93.45	73.98	157.8	"
90.7	7.06	92.94	72.40	145.6	"
92.3	6.32	93.68	74.71	164.0	"
95.7	6.23	93.77	75.00	166.5	"
95.7	5.65	94.35	76.90	184.7	"
100.2	4.91	95.09	79.42	214.3	α-LiClO$_3$
102.5	5.40	94.60	77.74	193.8	"
103.4	4.44	95.56	81.09	238.1	"
107.3	4.30	95.70	81.60	246.2	"
107.7	3.68	96.32	83.91	289.6	"
108.0	3.98	96.02	82.78	266.9	"
114.1	3.14	96.86	86.01	341.3	"
115.1	2.67	97.33	87.90	403.3	"
115.3	2.46	97.54	88.77	438.6	"
120.3	1.44	98.56	93.17	757.2	"
121.3	1.53	98.47	92.77	712.0	"
126.7	0.83	99.17	95.97	1322	"
127.0	0	100	100	---	"
127.6	0	100	100	---	"

COMPONENTS:	ORIGINAL MEASUREMENTS:
(1) Lithium chlorate; LiClO$_3$; [13453-71-9]	Berg, L.
(2) Water; H$_2$O; [7732-18-5]	Z. Anorg. Allg. Chem. 1929, 181, 131-6.

VARIABLES:	PREPARED BY:
T/K = 273.2 to 400.7	Hiroshi Miyamoto

EXPERIMENTAL VALUES:

t/°C	Water mass %	Lithium Chlorate mass % (compiler)	mol % (compiler)	mol kg^{-1} (compiler)	Nature of the solid phase
127.5[a]	0.0	100	100	–	α-LiClO$_3$
126[a]	0.5	99.5	97.5	2200	"
124[a]	1.0	99.0	95.2	1100	"
118.5[a]	2.4	97.6	89.0	450	"
113.5[a]	3.0	97.0	86.6	358	"
106.6	4.1	95.9	82.3	259	"
105[a]	4.5	95.5	80.9	235	"
104	4.6	95.4	80.5	229	"
100	4.9	95.1	79.5	215	"
100	5.0	95.0	79.1	210	"
95	5.5	94.5	77.4	190	β-LiClO$_3$
90	7.0	93.0	72.6	147	"
(90)	6.2	93.8	75.1	167	"
89[a]	7.0	93.0	72.6	147	"
85	7.9	92.1	69.9	129	"
84[a]	8.1	91.9	69.3	126	"
81[a]	9.0	91.0	66.8	112	"
80.9	9.0	91.0	66.8	112	"
70.0	10.8	89.2	62.2	91.4	"
68	11.2	88.8	61.2	87.7	"
60.0	11.9	88.1	59.6	81.9	"
55[a]	12.6	87.4	58.0	76.7	"
55.0	12.9	87.1	57.4	74.7	"
50.0	13.9	86.1	55.2	68.5	"
45.0	14.3	85.7	54.4	66.3	"

continued .....

AUXILIARY INFORMATION

METHOD/APPARATUS/PROCEDURE:	SOURCE AND PURITY OF MATERIALS:
Two different methods were used to determine the solubility of lithium chlorate in water. (1) Synthetic method used with visual observation of temperature of crystalization. The weighed salt and water were placed into a test-tube equipped with a stirrer and a thermocouple. The test-tube was placed in a larger test-tube which was then placed in a paraffin oil bath. The bath was slowly and evenly warmed. When the salt in the tube disappeared, the temperature of the sample solution was measured by the thermocouple. Next the bath was gently cooled, and when the salt appeared the temperature was measured again. (2) The isothermal method used to obtain an accurate liquidus curve. The salt and water were placed into an apparatus with stirrer fitted with a mercury seal. The apparatus was placed in an oil thermostat. The lithium content was determined gravimetrically (in duplicate) as lithium sulfate.	No information was given.
	ESTIMATED ERROR:
	Soly: precision within 0.2 %.
	Temp: precision ± 0.1 K (author).

COMPONENTS:	ORIGINAL MEASUREMENTS:
(1) Lithium chlorate; LiClO₃; [13453-71-9] (2) Water; H₂O; [7732-18-5]	Berg, L. Z. Anorg. Allg. Chem. 1929, 181, 131-6.

CRITICAL EVALUATION: (Continued)

t/°C	Water mass %	Lithium Chlorate mass % (compiler)	mol % (compiler)	mol kg⁻¹ (compiler)	Nature of the solid phase
42[a]	15.1	84.9	52.8	62.2	LiClO₃.1/3H₂O
40.0	15.3	84.7	52.5	61.2	"
35.0	15.8	84.2	51.5	59.0	"
30.0	16.3	83.7	50.6	56.8	"
25.0	17.1	82.9	49.1	53.6	"
20.0	17.5	82.5	48.4	52.2	"
30.0	17.7	82.3	48.1	51.4	"
25.0	18.9	81.1	46.1	47.5	"
20.0	19.6	80.4	45.0	45.4	LiClO₃.H₂O
20.0	20.5	79.5	43.6	42.9	"
18.0	22.3	77.7	41.0	38.5	"
16.2	23.2	76.8	39.8	36.6	"
12.7	24.8	75.2	37.7	33.5	"
9.2	25.9	74.1	36.3	31.7	"
8.0	26.3	73.7	35.8	31.0	"
7.4	34.4	65.6	27.5	21.1	"
6.0	42.6	57.4	21.2	14.9	"
3.0	45.9	54.1	19.0	13.0	"
0.0	29.0	71.0	32.8	27.1	LiClO₃.3H₂O(?)
0.0	46.9	53.1	18.4	12.5	LiClO₃.H₂O(?)

[a]These data obtained by synthetic method experiments. All other data from isothermal solubility determinations.

COMPONENTS:	ORIGINAL MEASUREMENTS:
(1) Lithium chlorate; $LiClO_3$; [13453-71-9] (2) Water; H_2O; [7732-18-5]	Treadwell, W.D.; Ammann, A. *Helv. Chim. Acta.* <u>1938</u>, *21*, 1249-65.
VARIABLES: One temperature: 293 K	PREPARED BY: Hiroshi Miyamoto

EXPERIMENTAL VALUES:

The solubility of lithium chlorate in water at 20°C is given as:

$$18.32 \text{ mol kg}^{-1}$$

The concentration solubility product was also given simply as the square of the solubility:

$$3.36 \times 10^2 \text{ mol}^2 \text{ kg}^{-2}$$

AUXILIARY INFORMATION

METHOD/APPARATUS/PROCEDURE:	SOURCE AND PURITY OF MATERIALS:
No information was given.	No information was given.
	ESTIMATED ERROR: Nothing specified.
	REFERENCES:

COMPONENTS:	ORIGINAL MEASUREMENTS:
(1) Lithium chlorate; $LiClO_3$; [13453-71-9] (2) Water; H_2O; [7732-18-5]	Campbell, A.N.; Griffiths, J.E. *Can. J. Chem.* <u>1956</u>, *34*, 1647-61.

VARIABLES:	PREPARED BY:
T/K = 229.9 to 400.7	Hiroshi Miyamoto

EXPERIMENTAL VALUES:

	solubility			
t/°C	mass %	mol kg^{-1} (compiler)	Method	Nature of the solid phase[a]
0.0	–	–	Thermal analysis	A
– 6.5	10.4	1.28	"	"
–14.2	20.0	2.77	"	"
–26.2	29.2	4.56	"	"
–28.6	30.2	4.79	"	"
–40.4	36.0	6.22	"	"
–43.3	37.3	6.58	"	A+B
–16.9	45.8	9.35	Solubility	B
– 1.5	52.8	12.4	"	"
+ 3.0	56.7	14.5	"	"
5.2	57.9	15.2	"	"
6.0	58.7	15.7	"	"
8.1	60.8	17.2	"	"
6.0	68.1	23.6	"	"
3.0	70.8	26.8	"	"
– 0.1	73.1	30.1	Thermal analysis	B+C
3.0	73.6	30.8	Solubility	C
5.6	74.2	31.8	"	"
6.0	74.1	31.7	"	"
8.5	75.1	33.4	"	"
10.5	75.7	34.5	"	"

continued.....

AUXILIARY INFORMATION

For solutions in equilibrium with ice, the solubilities were determined by the thermal method. The compositions of saturated solutions were determined by chemical analysis. The method of thermal analysis was also used for binary eutectics and $\alpha \rightarrow \beta$ transition, and all other solubilities were determined "in the usual way" (i.e. the isothermal method, compiler). Temperatures for thermal analyses were measured with an iron-constantan thermocouple and a potentiometer. The composition of the solid solutions was determined by chlorate analysis. Aliquots were transferred to a 250 ml iodine flask and 40 ml concentrated orthophosphoric acid added, followed by the addition of about 0.1 g sodium carbonate to exclude air from the flask. Approximately 25 ml of iodate-free potassium iodide solution (0.2 gm/ml) was added, and the stoppered flask allowed to stand at room temperature for 60 to 70 min. The free iodine was titrated with sodium thiosulphate. Standardization of the method with pure sodium chlorate indicated an accuracy of ± 0.4 %.

COMPONENTS:	ORIGINAL MEASUREMENTS:
(1) Lithium chlorate; $LiClO_3$; [13453-71-9] (2) Water; H_2O; [7732-18-5]	Campbell, A.N.; Griffiths, J.E. *Can. J. Chem.* <u>1956</u>, *34*, 1647-61.

EXPERIMENTAL VALUES: (Continued)

$t/°C$	Solubility mass %	mol kg^{-1} (compiler)	Method	Nature of the solid phase[a]
14.0	77.5	38.1	Solubility	C
15.0	78.0	39.2	"	"
16.0	78.5	40.4	"	"
18.0	79.3	42.4	"	"
20.5	81.9	50.1	"	C+D
21.5	82.0	50.4	Solubility	D
25.0	82.6	52.5	"	"
30.2	83.4	55.6	"	"
32.6	83.6	56.4	"	"
38.5	85.2	63.7	"	"
-10.5	78.7	40.9	Thermal Analysis	B+D
-25.0	82.7	52.9	"	B+E
44.2	86.7	72.1	Solubility	E
47.9	87.2	75.4	"	"
54.0	87.8	79.6	"	"
63.2	89.5	94.3	"	"
72.8	91.0	112	"	"
81.7	92.4	135	"	"
86.2	93.7	165	"	"
90.7	93.8	167	"	"
94.2	94.8	202	"	"
97.5	95.5	235	"	"
98.9	95.9	259	"	"
127.5	100.0	∞	Thermal Analysis	F

[a] A = Ice;   B = $LiClO_3 \cdot 3H_2O$;   C = $LiClO_3 \cdot H_2O$;   D = $(LiClO_3)_4 \cdot H_2O$;   E = β-$LiClO_3$; F = α-$LiClO_3$.

AUXILIARY INFORMATION

SOURCE AND PURITY OF MATERIALS:
A 1 mol dm^{-3} barium chlorate solution was heated to about 85°C, and a 1 mol dm^{-3} lithium sulfate solution was added slowly from a dropping funnel until equivalence was reached. The precipitated barium sulfate was removed by repeated filtration. The solution was evaporated slowly up to an approximated concentration of 50 % lithium chlorate. The filtered solution was transferred to a 250 ml Claissen flask and dehydration was carried out under a reduced pressure of less than 5 mm Hg, the distillate being absorbed in concentrated sulfuric acid. In this process, the temperature was kept below 85°C. Upon cooling, the solution to room temperature, the salt crystallized; it was placed under vacuum over phosphorus pentoxide. To remove the last trace of water, the salt was placed in a vacuum oven over phosphorus pentoxide , and maintained at 80°C.

ESTIMATED ERROR:

Isothermal method:
Soly: accuracy of ± 0.4 % (authors)
Temp: precision ± 0.05 K (authors).

Thermal analysis: nothing specified.

COMPONENTS:	ORIGINAL MEASUREMENTS:
(1) Lithium chloride; LiCl; [7447-41-8] (2) Lithium chlorate; LiClO$_3$; [13453-71-9] (3) Water; H$_2$O; [7732-18-5]	Campbell, A.N.; Griffiths, J.E. *Can. J. Chem.* 1956, 34, 1647-61.
VARIABLES: Composition T/K = 276.15 to 298.15.	**PREPARED BY:** Hiroshi Miyamoto

EXPERIMENTAL VALUES:

Composition of saturated solutions

t/°C	Lithium Chlorate		Lithium Chloride		Nature of the
	mass %	mol % (compiler)	mass %	mol % (compiler)	solid phase[a]
3.00	56.7[b]	20.7	–	–	A
	55.8	20.4	1.3	1.0	"
	62.2	26.0	4.3	3.8	"
	66.0	28.9	2.8	2.6	"
	68.7	31.0	1.5	1.4	"
	70.8	32.6	–	–	"
	73.6	35.7	–	–	B
	65.6	30.1	7.1	7.0	"
	62.4	27.9	9.4	9.0	"
	58.5	25.6	13.1	12.2	"
	57.5	24.9	13.7	12.6	B+E
	44.7	16.9	19.9	16.0	E
	40.9	14.9	22.1	17.2	D+E
	37.4	13.2	23.5	17.7	D
	23.9	7.43	29.1	19.3	"
	–	–	41.0	22.8	"

continued.....

AUXILIARY INFORMATION

METHOD/APPARATUS/PROCEDURE:

The isothermal method was used. Equilibrium was approached from two directions. A given solution was divided into equal parts, and one part was heated to complete disolution of the solid, and the second part was completely frozen. Both portions were then thermostated until the compositions of the liquid phases were identical. For solutions up to 10 % LiCl, equilibrium was attained after 2-3 days of stirring. For higher LiCl concentrations, 5-6 days of stirring were required to attain equilibrium.

To determine LiClO$_3$ compositions, aliquots were transferred to 250 ml iodine flasks to which 40 ml concentrated orthophosphoric acid were added followed by the addition of about 0.1 g   sodium carbonate to exclude air from the flask. Approximately 25 ml of iodate-free potassium iodide solution (0.2 gm/ml) was added, and the stoppered flask allowed to stand at room temperature for 60 to 70 min. The free iodine was titrated with sodium thiosulphate. Standardization of the method with pure sodium chlorate indicated an accuracy of ± 0.4 %.

The chloride content of solutions containing appreciable amounts of chloride was determined volumetrically with silver nitrate solution using an alcoholic solution of sodium dichlorofluoresceinate for the indicator. For solutions containing less than five percent chloride, the standard gravimetric method was used.

Compositions of the solid phases were determined by Schreinemakers' method.

COMPONENTS:	ORIGINAL MEASUREMENTS:
(1) Lithium chloride; LiCl; [7747-41-8]	Campbell, A.N.; Griffiths, J.E.
(2) Lithium chlorate; LiClO$_3$; [13453-71-9]	*Can. J. Chem.* 1956, *34*, 1647-61.
(3) Water; H$_2$O; [7732-18-5]	

EXPERIMENTAL VALUES: (Continued)

Composition of saturated solutions

t/°C	Lithium Chlorate mass %	mol % (compiler)	Lithium Chloride mass %	mol % (compiler)	Nature of the solid phase[a]
6.00	58.7[b]	22.1	–	–	A
	58.3	22.4	2.5	2.0	"
	62.7	25.6	1.8	1.6	"
	68.1	29.8	–	–	A
	74.1	36.3	–	–	B
	70.0	32.9	2.7	2.7	"
	69.7	32.9	3.4	3.4	"
	65.8	30.3	7.1	7.0	"
	64.9	29.6	7.5	7.3	"
	62.6	28.1	9.6	9.2	"
	61.4	27.4	10.7	10.2	"
	60.4	27.0	12.3	11.7	B+E
	57.5	24.9	13.7	12.6	E
	54.8	23.1	15.2	13.6	"
	53.9	22.3	15.0	13.2	"
	45.2	17.2	20.0	16.2	"
	38.2	13.6	23.7	18.1	"
	37.2	13.1	24.0	18.1	"
	36.2	12.7	24.6	18.4	D+E
	35.7	12.5	24.9	18.5	D
	34.1	11.7	25.4	18.6	"
	24.4	7.67	29.7	19.9	"
	23.5	7.33	30.0	19.9	"
	18.3	5.47	32.5	20.7	"
	16.3	4.80	33.5	21.0	"
	–	–	41.1	22.9	

AUXILIARY INFORMATION

SOURCE AND PURITY OF MATERIALS:	ESTIMATED ERROR:
	Soly: accuracy of ± 0.4 % (authors)
	Temp: precision ± 0.05 K (authors)

A 1 mol dm^{-3} barium chlorate solution was heated to about 85°C, and a 1 mol dm^{-3} lithium sulfate solution was added slowly with a dropping funnel until equivalence was reached. The precipitated barium sulfate was removed by repeated filtration. The solution was evaporated slowly up to an approximate concentration of 50% lithium chloride. The filtered solution was transferred to a 250 ml Claissen flask and dehydration was carried out under a reduced pressure of less than 5 mm Hg, the distillate being absorbed in concentrated sulfuric acid. In this process, the temperature was kept below 85°C.

Upon cooling the solution to room temperature, the salt crystallized, and was placed under vacuum over phosphorus pentoxide. To remove the last trace of water, the salt was placed in a vacuum furnace over phosphrous pentoxide, and maintained at 80°C.

The source of lithium chloride was not given.

COMPONENTS:	ORIGINAL MEASUREMENTS:
(1) Lithium chloride; LiCl; [7747-41-8]	Campbell, A.N.; Griffiths, J.E.
(2) Lithium chlorate; LiClO$_3$; [13453-71-9]	*Can. J. Chem.* **1956**, *34*, 1647-61.
(3) Water: H$_2$O; [7732-18-5]	

EXPERIMENTAL VALUES (continued)

Composition of saturated solutions

t/°C	Lithium Chlorate mass %	mol % (compiler)	Lithium Chloride mass %	mol % (compiler)	Nature of the solid phase[a]
8.50	75.0[b]	37.4	0	0	B
	73.4	36.2	1.5	1.6	"
	72.0	34.8	1.9	2.0	"
	66.1	30.9	7.7	7.7	"
	65.6	30.8	8.8	8.8	"
	64.2	30.4	11.2	11.3	B+E
	63.5	29.6	11.2	11.1	"
	62.7	29.0	11.6	11.4	"
	60.5	27.1	12.3	11.7	E
	50.8	20.5	17.4	15.0	"
	31.3	10.6	27.7	19.9	"
	30.6	10.2	27.8	19.9	D+E
	28.2	9.24	28.9	20.2	D
	22.5	7.07	32.2	21.6	"
	10.4	2.95	37.3	22.6	"
	5.6	1.54	39.7	23.2	"
	-	-	42.7	24.1	"
25.00	82.6[b]	48.6	-	-	C
	78.7	44.7	3.3	4.0	"
	76.4	42.4	5.0	5.9	"
	76.1	42.0	5.1	6.0	"
	75.7	41.7	5.6	6.6	"
	75.8	41.9	5.7	6.7	"
	75.9	42.1	5.8	6.9	"
	72.3	39.1	9.2	10.6	C+E
	71.8	38.4	9.2	10.5	"
	71.6	38.2	9.2	10.5	"
	67.7	34.0	10.7	11.5	E
	41.7	15.6	23.3	18.6	"
	33.8	11.8	27.7	20.6	"
	27.0	8.91	31.3	22.0	"
	25.5	8.28	31.7	22.0	"
	-	-	45.5	26.2	"

[a] A = LiClO$_3$.3H$_2$O;   B = LiClO$_3$.H$_2$O;   C = (LiClO$_3$)$_4$.H$_2$O;   D = LiCl.2H$_2$O;

E = LiCl.H$_2$O

[b] For the binary system the compiler computes the following:

soly of LiClO$_3$ = 14.5 mol kg^{-1} at 3.00°C

= 15.7 mol kg^{-1} at 6.00°C

= 33.2 mol kg^{-1} at 8.50°C

= 52.5 mol kg^{-1} at 25.00°C

COMPONENTS:	ORIGINAL MEASUREMENTS:
(1) Lithium chlorate; LiClO$_3$; [13453-71-9] (2) 2-Propanone (acetone); C$_3$H$_6$O; [76-64-1]	Miravitlles, Mille L. Ann. Fís. Quím. (Madrid) 1945, 41, 120-37.
VARIABLES: T/K = 288, 293 and 298	PREPARED BY: R. Herrera

EXPERIMENTAL VALUES:

Solubility[a]

t/°C	mass %	mol kg^{-1}
15	0.1563	0.01732
20	0.1502	0.01664
25	0.1424	0.01578

[a]Molalities calculated by H. Miyamoto

AUXILIARY INFORMATION

METHOD/APPARATUS/PROCEDURE:	SOURCE AND PURITY OF MATERIALS:
Saturated solutions were prepared in an Erlenmeyer flask by mixing the dried acetone with an excess of halate for two hours. The solution was constantly stirred by bubbling dry air (air was dried by passing it through CaCl$_2$ while pumping it into the solution). Air going out from the flask after bubbling through the solution carried some acetone vapor during this operation. The solution temperature was kept constant by immersing the flask in a constant temperature water bath. After two hours, the air exit was closed. The resulting pressure forced the saturated solution from the Erlenmeyer through a tube filled with cotton (which acted as a filter), and was collected in a small flask. This flask was stoppered and weighed. The halate contained in the sample was weighed after complete evaporation of acetone. In all cases, weights were reported to the fourth decimal figure.	Commercial redistilled acetone. This acetone was then dehydrated three times by leaving it in contact with calcium chloride for forty eight hours each time. Fresh CaCl$_2$ was used in each operation. Finally the dehydrated acetone was distilled at 56.3°C. Source and purity of LiClO$_3$ not specified.
	ESTIMATED ERROR: Nothing specified.
	REFERENCES:

COMPONENTS:	EVALUATOR:
(1) Sodium chlorate; $NaClO_3$; [7757-82-6] (2) Water; H_2O; [7732-18-5]	Hiroshi Miyamoto Department of Chemistry Niigata University Niigata, Japan June, 1984

CRITICAL EVALUATION:

THE BINARY SYSTEM

Data for the solubility of sodium chlorate in water have been reported in 22 publications (1-21,30). Many of these studies deal with ternary systems, and the solubility in the binary $NaClO-H_2O$ system is given as one point on a phase diagram. Many investigators (4, 6-8, 10-20) reported that the stable solid in equilibrium with the saturated solutions over the temperature range between 273 K and 373 K was anhydrous sodium chlorate. Nabiev, Tukhtaev, Musaev, Kuchrov and Shimmasov (21) measured the solubility for the binary $NaClO_3-H_2O$ system at 254.7 K, and the stable phases were $NaClO_3$ and ice.

Blanc and Schmandt (1), Bell (3) and Treadwell and Ammann (9) reported solubilities in the binary $NaClO_3-H_2O$ system only. Bittler (2), Bell (3) and Nies and Hulbert (18) reported solubilities over a wide temperature range.

A summary of solubility studies for the binary $NaClO_3-H_2O$ system is given in Table 1.

1.   Evaluation for the solubility based on mol kg^{-1} units.

Solubility at 273.2 K. The solubility has been reported in 3 publications (3, 7, 18). The value by Babaeva (7) is very distinctly larger than that of two other investigators. The arithmetic mean of two results (7, 18) is 7.465 mol kg^{-1}. The mean is designated as a tentative value.

Solubility at 283.2 K. The solubility has been reported in 2 publications (3, 18). The value recommended for the solubility at this temperature is taken as 8.220 mol kg^{-1}, which is the arithmetic mean of the two results.

Solubility at 293.2K. The solubility has been reported in 6 publications (3, 5-7, 9, 18) at 293.2 K, and in one article (1) at 293.0 K. The value of 8.994 mol kg^{-1} by Blanc and Schmandt (1) is larger than that of Nies and Hulbert (18). The reported solubilities at this temperature are widely distributed from 8.976 to 9.282 mol kg^{-1}. The tentative value was obtained as the arithmetic mean of 7 results (1, 3, 5-7, 9, 18). The value is 9.14 mol kg^{-1}, and the standard deviation is 0.14 mol kg^{-1}.

Solubility at 298.2 K. The solubility at this temperature has been reported in 10 publications (3, 8, 10-13, 17,19,20,30). The value of Vlasov and Shishkina (2) is distinctly lower than that of the other investigators, and the study of Arkhipo, Kashina and Kuzina (17) reported the highest value. Therefore, these two values are rejected. The arithmetic mean of the remaining 7 results (3,8,10-13,19,30) is 9.43 mol kg^{-1}, and the standard deviation is 0.02 mol kg^{-1}. The mean is designated as a recommended value.

Solubility at 303.2 K. The value has been reported in 3 publications (1, 3, 18). The arithmetic mean of 3 results is 9.86 mol kg^{-1}, and the standard deviation is 0.04 mol kg^{-1}. The mean is designated as a recommended value.

Solubility at 308.2 K. The solubility has been reported in one publication (3) at 308.2 K and in one publication (1) at 308.3 K. Both values are identical. The estimated error in temperature measurement is $\pm$ 0.02 K by Bell (1) and nothing specified by Blanc and Schmandt (1). The tentative value at 308.2 K is taken as 10.33 mol kg^{-1}.

Solubility at 313.2 K. The solubility has been reported in 4 publications (3, 4, 7, 18). The value of Babaeba (18) was markedly higher than those of the other researchers (3, 4, 7) and is therefore rejected. The arithmetic mean of the remaining 3 results (3, 4, 7) is 10.86 mol kg^{-1}, and the standard deviation is 0.07 mol kg^{-1}. The mean is designated as a recommended value.

Solubility at 318.2 K. The recommended value for the solubility at this temperature is taken as 11.27 mol kg^{-1} which is the arithmetic mean of two results (8, 18).

COMPONENTS:	EVALUATOR:
(1) Sodium chlorate; $NaClO_3$; [7757-82-6] (2) Water; H_2O; [7732-18-5]	Hiroshi Miyamoto Department of Chemistry Niigata University Niigata, Japan June, 1984

CRITICAL EVALUATION:

Table 1. Summary of Solubility Data for the Binary $NaClO_3-H_2O$ System

T/K	m_1/mol kg^{-1}	Reference
268.2	7.096	(18) Nies; Hulbert
273.2	7.451	(18) Nies; Hulbert
273.2	7.478	(3) Bell
273.2	7.690	(7) Babaeva
277.9	7.834	(1) Blanc; Schmandt
283.2	8.208	(18) Nies; Hulbert
283.2	8.232	(3) Bell
288.2	8.641	(8) Ricci; Yenick
291.2	8.976	(14) Windmaisser; Stockl
292.2	8.770	(11) Ricci; Weltman
293.0	8.994	(1) Blanc; Schmandt
293.2	8.976	(18) Nies; Hulbert
293.2	9.008	(3) Bell
293.2	9.20	(9) Treadwell; Ammann
293.2	9.231	(5) Di Capua; Scaletti
293.2	9.28	(6) Di Capua; Bertoni
293.2	9.282	(7) Babaeva
297.4	9.26	(4) Il'inskli
298.2	9.410	(30) Ricci; Offenbach
298.2	9.352	(20) Vlasov; Shishkina
298.2	9.402	(13) Ricci; Linke
298.2	9.417	(11) Ricci; Weltman
298.2	9.421	(10) Swenson; Ricci
298.2	9.433	(12) Ricci
298.2	9.444	(3) Bell
298.2	9.448	(8) Ricci; Yanick
298.2	9.470	(19) Arkhipov; Kashina
298.2	9.504	(17) Arkhipov; Kashina; Kuzina
303.2	9.818	(18) Nies; Hulbert
303.2	9.865	(1) Blanc; Schmandt
303.2	9.896	(3) Bell
308.2	10.33	(3) Bell
308.3	10.33	(1) Blanc; Schmandt
313.2	10.81	(18) Nies; Hulbert
313.2	10.83	(3) Bell
313.2	10.94	(4) Il'inskii
313.2	12.13	(7) Babaeva
317.9	11.25	(1) Blanc; Schmandt
318.2	11.25	(18) Nies; Hulbert
318.2	11.29	(8) Ricci; Yenick
323.2	11.71	(11) Ricci; Weltman
323.2	11.74	(10) Swenson; Ricci
323.2	11.76	(18) Nies; Hulbert
333.2	12.88	(18) Nies; Hulbert
348.2	14.79	(18) Nies; Hulbert
348.2	14.94	(8) Ricci; Yenick
371.2	18.47	(18) Nies; Hulbert
373.2	19.16	(3) Bell

COMPONENTS:	EVALUATOR:
(1) Sodium chlorate; NaClO₃; [7757-82-6] (2) Water; H₂0; [7732-18-5]	Hiroshi Miyamoto Department of Chemistry Niigata University Niigata, Japan June, 1984

CRITICAL EVALUATION:

Solubility at 323.2K. The solubility has been reported in 3 publications (10, 11, 18). The arithmetic mean of 3 results is 11.74 mol kg^{-1}, and the standard deviation is 0.03 mol kg^{-1}. The mean is designated as a recommended value.

Solubility at 348.2 K. The recommended value for the solubility at this temperature is taken as 14.87 mol kg^{-1} which is the arithmetic mean of two results (8, 18).

Solubility at other temperatures. The solubilities at 277.9 K (1), 288.2 K (8), 292.2 K (11), 297.4 K (4), 317.9 K (1), 333.2 K (18), 371.2 K (18) and 373.2 K (3) are singular values, and are designated as tentative solubilities. A summary of all solubility data in the binary for which anhydrous NaClO₃ is the sole solid phase is given in Table 1.

The recommended and tentative solubilities for the binary NaClO₃-H₂0 system for which the solid phase is the anhydrous salt is given in Table 2. The experimental mol kg^{-1} solubilities were fitted to the following smoothing equation:

$$\ln (m_1/\text{mol kg}^{-1}) = 34.97670 - 48.488690/(T/100 \text{ K}) - 31.26105 \ln (T/100 \text{ K})$$
$$+ 5.929873 (T/100 \text{ K}) \qquad \sigma = 0.08$$

The mole fraction solubilities calculated by the evaluator was fitted to the general solubility equation (see the PREFACE and eqs. [1] and [2] in the critical evaluation for the binary LiClO₃-H₂0 system):

$$Y = -4838.039/(T/K) - 27.7668 \ln(T/K) + 156.124 + 0.0521925(T/K)$$
$$\sigma_y = 0.0108 \qquad \sigma_x = 0.00099$$

The tentative solubilities in the binary system based on mol dm^{-3} based on the results of Billiter (2) were fitted by the evaluator to the following equation:

$$\ln (c_1/\text{mol dm}^{-3}) = 2.456277 - 3.660476/(T/100 \text{ K}) + 0.6567005 \ln (T/100 \text{ K})$$

$$\sigma = 0.03$$

COMPONENTS:	EVALUATOR:
(1) Sodium chlorate; NaClO$_3$; [7757-82-6] (2) Water; H$_2$O; [7732-18-5]	Hiroshi Miyamoto Department of Chemistry Niigata University Niigata, Japan June, 1984

CRITICAL EVALUATION:

Table 2. Recommended and Tentative Solubilities in the Binary
System Based on mol kg^{-1} and mole fraction units.[a]

T/K	m/mol kg^{-1}	χ/mole fraction	R or T[b]
268.2	7.096	0.1133	T
273.2	7.465	0.1185	T
277.9	7.834	0.1237	T
283.2	8.220	0.1290	R
288.2	8.641	0.1347	T
292.2	8.770	0.1364	T
293.2	9.14	0.1414	T
297.4	9.26	0.1430	T
298.2	9.43	0.1452	R
303.2	9.86	0.1508	R
308.2	10.33	0.1569	T
313.2	10.86	0.1636	R
317.9	11.25	0.1685	T
318.2	11.27	0.1688	R
323.2	11.74	0.1746	R
333.2	12.88	0.1833	T
348.2	14.87	0.2133	T
371.2	18.47	0.2497	T
373.2	19.16	0.2566	T

[a]Mole fractions calculated by the evaluator.

[b]R (recommended) and T (tentative) solubility values.

COMPONENTS:	EVALUATOR:
(1) Sodium chlorate; $NaClO_3$; [7757-82-6] (2) Water; H_2O; [7732-18-5]	Hiroshi Miyamoto Department of Chemistry Niigata University Niigata, Japan <div align="right">June, 1984</div>

CRITICAL EVALUATION:

TERNARY SYSTEMS

Many studies for the solubility of the aqueous ternary system with two saturating components have been reported. A summary of the studies is given in Tables 3-6.

Systems with alkali halides. Solubility studies of the ternary systems containing sodium chlorate and alkali halides have been reported in 11 publications (2, 5, 12, 16, 17, 20, 22, 24, 28). A summary of these studies with that of the ternary $NaClO_3-BaCl_2-H_2O$ system is given in Table 3.

Table 3. Summary of Solubility Studies in Ternary Systems
with Alkali Metal and Barium Halides.

Ternary System	T/K	Solid phase	Reference
$NaClO_3$ - NaCl - H_2O	293, 303, 333, 353, 373	Not given	2
$NaClO_3$ - NaCl - H_2O	293	Not given	5
$NaClO_3$ - NaCl - H_2O	298, 308, 318	$NaClO_3$; NaCl	16
$NaClO_3$ - NaCl - H_2O	298	$NaClO_3$; NaCl	17
$NaClO_3$ - NaCl - H_2O	293	Not given	22
$NaClO_3$ - NaCl - H_2O	247 - 373	$NaClO_3$; $NaCl.2H_2O$ NaCl; Ice	24
$NaClO_3$ - NaCl - H_2O	298	Not given	28
$NaClO_3$ - NaBr - H_2O	298	$NaClO_3$; $NaBr.2H_2O$	12
$NaClO_3$ - NaI - H_2O	298	$NaClO_3$; $NaI.2H_2O$	12
$NaClO_3$ - NaF - H_2O	298	$NaClO_3$; NaF	20
$NaClO_3$ - KCl - H_2O	298	Not given	5
$NaClO_3$ - $BaCl_2$ - H_2O	293	Not given	6

Solubilities in the ternary $NaClO_3-NaCl-H_2O$ system have been reported in 7 publications (5, 7, 16, 17, 22, 24, 28). Di Capua and Scaletti (5) and Arkhipov, Kashina and Kuzina (17) measured solubilities (mass % units) of the two saturating components ($NaClO_3$ and NaCl) over a wide concentration range at 293 and 298 K, respectively. No double salts are formed in this system.

Nallet and Paris (24) reported only one value (mass % units) at each temperature between 246 and 373 K. The details of solid phases are described on the compilation sheets. No double salts formed.

Solubilities in the aqueous ternary system based on g cm^{-3} units have been reported by Winteler (22), and by Billiter (2). Because of insufficient experimental information, it was not possible to compare these two studies.

The paper by Perel'man and Korzenyak (28) contained only a phase diagram, and was therefore not compiled.

The solubilities in ternary systems $NaClO_3-NaBr-H_2O$ and $NaClO_3-NaI-H_2O$ at 298 K were studied by Ricci (2) and the system $NaClO_3-NaF-H_2O$ at 298 was studied by Vlasov and Shishkina (20). These salt pairs formed neither double salts nor solid solutions at this temperature. The ternary $NaClO_3-NaF-H_2O$ system was of simple eutonic type.

COMPONENTS:	EVALUATOR:
(1) Sodium chlorate; $NaClO_3$; [7757-82-6] (2) Water; H_2O; [7732-18-5]	Hiroshi Miyamoto Department of Chemistry Niigata University Niigata, Japan June, 1984

CRITICAL EVALUATION:

Only one result in the $NaClO_3$-KCl-H_2O system was reported by Di Capua and Scaletti (5), and solubilities in the $NaClO_3$-$BaCl_2$-H_2O were reported by Di Capua and Bertoni (6). The solid phases in both papers were not reported.

Systems with other halates. Solubility studies of ternary systems containing sodium chlorate and other halates have been reported in 8 publications (4-6, 10, 17, 19, 23, 24). A summary of these studies with that of the ternary system $NaClO_3$-$NaClO_2$-H_2O is given in Table 4.

Table 4. Summary of Solubility Studies in Ternary
Systems with other Halates

Ternary System	T/K	Solid Phase	Reference
$NaClO_3$ - $KClO_3$ - H_2O	297, 313	$NaClO_3$; $KClO_3$	4
$NaClO_3$ - $KClO_3$ - H_2O	293	Not given	5
$NaClO_3$ - $KClO_3$ - H_2O	273, 313	Not given	23
$NaClO_3$ - $KClO_3$ - H_2O	291	$NaClO_3$; $KClO_3$; Ice	24
$NaClO_3$ - $RbClO_3$ - H_2O	298	$NaClO_3$; $RbClO_3$	17
$NaClO_3$ - $CsClO_3$ - H_2O	298	$NaClO_3$; $CsClO_3$	19
$NaClO_3$ - $Ba(ClO_3)_2$ - H_2O	298	$NaClO_3$; $Ba(ClO_3)_2$	6
$NaClO_3$ - $NaBrO_3$ - H_2O	298, 373	$NaClO_3$; $NaBrO_3$ Solid Solution	10
$NaClO_3$ - $NaClO_2$ - H_2O	288, 298, 308 318	$NaClO_3$; $NaClO_3 \cdot 3H_2O$ $NaClO_2$	15
$NaClO_3$ - $NaClO_2$ - H_2O	298	Not given	28

Solubilities in the $NaClO_3$-$KClO_3$-H_2O system were studied by Il'inskii (4) at 297.4 and 313 K, and by Di Capua and Scaletti (5) at 293 K. No double salts were formed. The results for the composition at the isothermally invariant point were reported by Munter and Brown (23).

Nallet and Paris (24) reported only one solubility at each temperature between 255.3 K and 373 K. The details of solid phases are described in the compilation. No double salts were formed.

Solubilities in the ternary $NaClO_3$-$RbClO_3$-H_2O and $NaClO_3$-$CsClO_3$-H_2O systems have been reported by Arkhipov, Kashina and Kuzina (17), and Arkhipov and Kashina (19), respectively. Solubilities in the ternary $NaClO_3$-$Ba(ClO_3)_2$-H_2O system have been reported by Di Capua and Bertoni (6). All systems were simple eutonic, and no double salts were formed.

Solubilities in the ternary $NaClO_3$-$NaBrO_3$-H_2O system were measured by Swenson and Ricci (10). The salts studied at 298 K formed a series of solid solutions which is apparently discontinuous. The solubility curve at 298 K appears to be divided into three portions corresponding to the following solid phases: (1) pure sodium bromate, (2) a sodium bromate solid solution containing up to about 5-10 mass % sodium chlorate, (3) a sodium chlorate solid solution containing from 0 to 60-65 mass % sodium bromate.

Cunningham and Oey (15) have reported solubilities in the ternary system $NaClO_3$-$NaClO_2$-H_2O. The system is simple and the solid phases were sodium chlorate and sodium chlorite trihydrate. No hydrate of sodium chlorate was formed.

COMPONENTS:	EVALUATOR:
(1) Sodium chlorate; $NaClO_3$; [7757-82-6] (2) Water; H_2O; [7732-18-5]	Hiroshi Miyamoto Department of Chemistry Niigata University Niigata, Japan June, 1984

CRITICAL EVALUATION:

The system with sodium sulfate. Solubilities in the ternary $NaClO_3$-Na_2SO_4-H_2O system
have been reported in 3 publications (7, 8, 14). A summary of the studies is given in
Table 5.

Table 5.   A Summary of Solubility Studies in the
Ternary System $NaClO_3$-Na_2SO_4-H_2O

T/K	Solid Phase	Reference
273	$NaClO_3$; $Na_2SO_4 \cdot 10H_2O$	7
288	$NaClO_3$; $Na_2SO_4 \cdot 10H_2O$	8
291	$NaClO_3$; $Na_2SO_4 \cdot 10H_2O$; Na_2SO_4	14
293	$NaClO_3$; $Na_2SO_4 \cdot 10H_2O$	7
298	$NaClO_3$; $Na_2SO_4 \cdot 10H_2O$; Na_2SO_4 $NaClO_3 \cdot 3Na_2SO_4$ (double salt)	8
313	$NaClO_3$; Na_2SO_4	7
318	$NaClO_3$; Na_2SO_4; $NaClO_3 \cdot 3Na_2SO_4$	8
348	$NaClO_3$; Na_2SO_4; $NaClO_3 \cdot 3Na_2SO_4$	8

Solubilities in this system were determined over a wide temperature range. Ricci and
Yanick (8) reported the existence of stable double salt with the formula $NaClO_3 \cdot 3Na_2SO_4$.
The double salt was formed at 298 K and above, having always a very short range of stable
existence, and persisting in metastable equilibrium over a very considerable range of
concentrations. Although Babaeva (7) measured solubilities in this system at 313 K,
the existence of the double salt was not found. Below 298 K no double salts were formed.
The solid phases in the binary systems $NaClO_3$-H_2O and Na_2SO_4-H_2O were $NaClO_3$ and
$Na_2SO_4 \cdot 10H_2O$ at 298 K and below, respectively, and $NaClO_3$ and Na_2SO_4 at 313 K and above,
respectively. The composition of the solid phase between 298 and 313 K was not studied.

Systems with the other compounds. A summary of solubility studies in the system containing the other compounds except the salts discussed above is given in Table 6.

Table 6.   Summary of Solubility Studies in Miscellaneous
Ternary Systems

Ternary System	T/K	Solid Phase	Reference
$NaClO_3$ - Na_2CO_3 - H_2O	297, 313	$NaClO_3$; $Na_2CO_3 \cdot 10H_2O$; $Na_2CO_3 \cdot 7H_2O$; Na_2CO_3	4
$NaClO_3$ - Na_2CrO_4 - H_2O	292	$NaClO_3$; $Na_2CrO_4 \cdot 6H_2O$; $Na_2CrO_4 \cdot 10H_2O$	11
$NaClO_3$ - $NaCrO_4$ - H_2O	298	$NaClO_3$; $Na_2CrO_4 \cdot 4H_2O$; $Na_2CrO_4 \cdot 6H_2O$	11
$NaClO_3$ - $NaCrO_4$ - H_2O	323	$NaClO_3$; $Na_2CrO_4 \cdot 4H_2O$	11
$NaClO_4$ - $NaNO_3$ - H_2O	298	$NaClO_3$; $NaNO_3$	12
$NaClO_3$ - Na_2MoO_4 - H_2O	298	$NaClO_3$; $Na_2MoO_4 \cdot 2H_2O$	13
$NaClO_3$ - $NaOH$ - H_2O	291	$NaClO_3$; $NaOH \cdot H_2O$	14
$NaClO_3$ - $BaBO_2$ - H_2O	254	$NaClO_3$; $NaBO_2 \cdot 4H_2O$; ice	18
$NaClO_3$ - $NaBO_2$ - H_2O	268, 273, 293, 303	$NaClO_3$; $NaBO_2 \cdot 4H_2O$	18
$NaClO_3$ - $NaBO_2$ - H_2O	318, 323	$NaClO_3$; $NaBO_2 \cdot 4H_2O$; $NaBO_4 \cdot 2H_2O$	18
$NaClO_3$ - $NaBO_2$ - H_2O	333, 348	$NaClO_3$; $NaBO_2 \cdot 2H_2O$	18

COMPONENTS:	EVALUATOR:
(1) Sodium chlorate; $NaClO_3$; [7757-82-6] (2) Water; H_2O; [7732-18-5]	Hiroshi Miyamoto Department of Chemistry Niigata University Niigata, Japan June, 1984

CRITICAL EVALUATION:

Solubilities in the $NaClO_3$-Na_2CO_3-H_2O system by Il'inskii (4), in the $NaClO_3$-Na_2CrO_4-H_2O system by Ricci and Weltmann (11), in the $NaClO_3$-$NaNO_3$-H_2O system by Ricci (12) and in the $NaClO_3$-Na_2MoO_4-H_2O system by Ricci and Linke (13) have been reported. Neither compound formation nor solid solution between any two salts were reported. Solubilities in the ternary $NaClO_3$-$NaOH$-H_2O system were reported by Windmaisser and Stockl (14), and no double salts were formed. Solubility isotherms at temperature between 353.9 and 348 K were determined for the $NaClO_2$-$NaBO_3$-H_2O system by Nies and Hulbert (18). The solid phases were $NaBO_2$.$4H_2O$, $NaBO_2$.$2H_2O$ and $NaClO_3$. No double salts formed.

OTHER MULTICOMPONENT SYSTEMS

A summary of solubility studies in other multicomponent systems is given in Table 7.

Table 7. Summary of Solubility Studies in Multicomponent Systems

Multicomponent System	T/K	Reference
$NaClO_3$ - $KClO_3$ - $NaCl$ - H_2O	293	5
$NaClO_3$ - $NaClO_2$ - $NaCl$ - H_2O	283, 293, 303	26
$NaClO_3$ - $NaClO_2$ - $NaCl$ - H_2O	298, 303, 318	27
$NaClO_3$ - $NaClO_2$ - $NaCl$ - H_2O	298	28
$NaClO_3$ - $NaCl$ - $RbClO_3$ - $RbCl$ - H_2O	298	17
$NaClO_3$ - $NaCl$ - $CsClO_3$ - $CsCl$ - H_2O	298	19
$NaClO_3$ - $NaCl$ - $KClO_3$ - KCl - H_2O	273, 313	23
$NaClO_3$ - $NaCl$ - $KClO_3$ - KCl - H_2O	247 - 373	25
$NaClO_3$ - $NaClO_2$ - $NaCl$ - Na_2CO_3 - H_2O	298	29

Only one solubility value in the $NaClO_3$-$KClO_3$-$NaCl$-KCl-H_2O system at 293 K was reported by Di Capua and Scoletti (5). No other information was given.

Solubilities in the $NaClO_3$-$NaClO_2$-$NaCl$-H_2O system were measured by Nakamori, Nagino, Hideshima and Hirai (26) at 283, 293 and 303 K, and by Oey and Cunningham (27) at 298, 303 and 318 K. No double salts formed within the temperature interval 283-318 K.

Perel'man and Korgenyak (28) reported only a phase diagram, and the paper was therefore not compiled.

The $NaClO_3$-$NaCl$-$RbClO_3$-$RbCl$-H_2O system was studied by Arkhipov, Kashina and Kuzina (17). Solubilities in the quaternary systems $NaClO_3$-$NaCl$-$RbClO_3$-H_2O and $RbClO_3$-$NaCl$-$RbCl$-H_2O, have been reported, but the solubility data in five component systems were not reported. There were four salt crystallization regions in the system: $NaCl$, $RbCl$, $NaClO_3$ and $RbClO_3$. Two ternary points were obtained corresponding to solutions saturated with: (i) $NaCl$+$NaClO_3$+$RbClO_3$; (ii) $NaCl$+$RbCl$+$RbClO_3$. The main part of the diagram is occupied by the crystallization field of rubidium chlorate (95%), followed in area by the sodium chloride field (3.5 %). The four salts did not form either double salts or solid solutions.

The $NaClO_3$-$NaCl$-$CsClO_3$-$CsCl$-H_2O system was studied by Arkhipov and Kashina (19). Solubilities in the quaternary systems $NaClO_3$-$CsClO_3$-$NaCl$-H_2O and $CsClO_3$-$CsCl$-$NaCl$-H_2O have been reported, but the solubility data for five component systems were not reported. The crystallization field of cesium chlorate occupies the greatest area on the diagram, and this is followed by the field of sodium chloride. The crystallization fields of sodium chlorate and of solid solutions of cesium and sodium chlorides are small.

COMPONENTS:	EVALUATOR:
(1) Sodium chlorate; $NaClO_3$; [7757-82-6] (2) Water; H_2O; [7732-18-5]	Hiroshi Miyamoto Department of Chemistry Niigata University Niigata, Japan June, 1984

CRITICAL EVALUATION:

Munter and Brown (23) measured solubilities in the $NaClO_3-NaCl-KClO_3-KCl-H_2O$ system at the isothermal invariant point. The solid phase at this point is simple: no double salts were formed.

Nallet and Paris (25) also measured solubilities in the $NaClO_3-NaCl-KClO_3-KCl-H_2O$ system over a wide temperature range, but only one value at each temperature was reported. No double salts or solid solutions were reported.

The $NaClO_3-NaClO_2-Na_2CO_3-NaCl-H_2O$ system was studied by Perel'man and Korzhenyak (29). The compositions of the eutectic points of the ternary and quaternary systems were determined.

REFERENCES

1. Blanc, M. Le.; Schmandt, W. *Z. Physik. Chem.* <u>1911</u>, *77*, 614.

2. Billiter, J. *Monatsh. Chem.* <u>1920</u>, *41*, 287.

3. Bell, H. C. *J. Chem. Soc.* <u>1923</u>, *123*, 2712.

4. Il'inskii, Vl. P. *J. Russ. Phys.-Chem. Soc.* <u>1923</u>, *54*, 29.

5. Di Capua, C.; Scaletti, U. *Gazz. Chim. Ital.* <u>1927</u>, *27*, 391.

6. Di Capua, C.; Bertoni, A. *Gazz. Chim. Ital.* <u>1928</u>, *58*, 249.

7. Babaeva, A. V. *Zh. Obshch. Khim.* <u>1936</u>, *6*, 1144.

8. Ricci, J. E.; Yanick, N. S. *J. Am. Chem. Soc.* <u>1937</u>, *59*, 491.

9. Treadwell, W. D.; Ammann, A. *Helv. Chim. Acta* <u>1938</u>, *21*, 1249.

10. Swenson, T.; Ricci, J. E. *J. Am. Chem. Soc.* <u>1961</u>, *61*, 1974.

11. Ricci, J. E.; Weltman, C. *J. Am. Chem. Soc.* <u>1942</u>, *64*, 2746.

12. Ricci, J. E. *J. Am. Chem. Soc.* <u>1944</u>, *66*, 1015.

13. Ricci, J. E.; Linke, W. F. *J. Am. Chem. Soc.* <u>1947</u>, *69*, 1080.

14. Windmaisser, F.; Stockl, F. *Monatsh. Chem.* <u>1951</u>, *82*, 287.

15. Cunningham, G. I.; Oey, T. S. *J. Am. Chem. Soc.* <u>1955</u>, *77*, 4498.

16. Oey, T. S.; Koopman, D. E. *J. Phys. Chem.* <u>1958</u>, *62*, 755.

17. Arkhipov, S. M.; Kashina, N. I.; Kuzina, V. A. *Zh. Neorg. Khim.* <u>1968</u>, *13*, 2872; *Russ. J. Inorg. Chem. (Engl. Transl.)* <u>1968</u>, *13*, 1476.

18. Nies, N. D.; Hulbert, R. W. *J. Chem. Eng. Data* <u>1969</u>, *14*, 14.

19. Akhipov, S. M.; Kashina, N. I. *Zh. Neorg. Khim.* <u>1970</u>, *15*, 760; *Russ. J. Inorg. Chem. (Engl. Transl.)* <u>1970</u>, *15*, 391.

20. Vlasov, G. A.; Shishkina, L. A. *Zh. Neorg. Khim.* <u>1977</u>, *22*, 2309; *Russ, J. Inorg. Chem. (Engl. Transl.)* <u>1977</u>, *22*, 1250.

COMPONENTS:	EVALUATOR:
(1) Sodium chlorate; $NaClO_3$; [7757-82-6] (2) Water; H_2O; [7732-18-5]	Hiroshi Miyamoto Department of Chemistry Niigata University Niigata, Japan June, 1984

CRITICAL EVALUATION:

REFERENCES, continued

21.   Nabiev, M. N.; Tukhtaev, S.; Mussaev, N. Yu.; Kushrov, Kh.; Shammasov, R. E. *Zh. Neorg. Khim.* 1982, 27, 2704; *Russ. J. Inorg. Chem. (Engl. Transl.)* 1982, 27

22.   Winteler, F. *Z. Electrochim.* 1900, 2, 360.

23.   Munter, P. A.; Brown, R. L. *J. Am. Chem. Soc.* 1943, 65, 2456.

24.   Nallet, A.; Paris, R. A. *Bull. Soc. Chim. Fr.* 1956, 488.

25.   Nallet, A.; Paris, R. A. *Bull. Soc. Chim. Fr.* 1956, 494.

26.   Nakamori, I.; Nagino, Y.; Hideshima, K.; Hirai, T. *Kogyo Kagaku Zasshi* 1958, 61, 147.

27.   Oey, T. S.; Cunningham, G. I.; Koopman, D. E. *J. Chem. Eng. Data* 1960, 5, 248.

28.   Perel'man, F. M.; Korzhenyak, I. G. *Zh. Neorg. Khim.* 1968, 13, 277; *Russ. J. Inorg. Chem. (Engl. Transl.)* 1968, 13, 143.

29.   Perel'man, F. M.; Korzhenyak, N. G. *Zh. Neorg. Khim.* 1968, 13, 2861; *Russ. J. Inorg. Chem. (Engl. Transl.)* 1968, 13, 1471.

30.   Ricci, J.E.; Offenbach, J.A. *J. Am. Chem. Soc.* 1951, 73, 1597.

COMPONENTS:	ORIGINAL MEASUREMENTS:
(1) Sodium chlorate; $NaClO_3$; [7775-09-9] (2) Water; H_2O; [7732-18-5]	Blanc, M.L.; Schmandt, W. Z. Pysik. Chem. <u>1911</u>, 77, 614-638.

VARIABLES:	PREPARED BY:
T/K = 277.93, 293.00, 303.18, 308.25 and 317.87	Hiroshi Miyamoto

EXPERIMENTAL VALUES:

Solubility of $NaClO_3$[a]

$t/°C$	mass %	mol kg^{-1}
4.78	45.47	7.834
19.85	48.91	8.994
30.05	51.22	9.865
35.10	52.36	10.33
44.72	54.50	11.25

[a] Molalities calculated by the compiler.

AUXILIARY INFORMATION

METHOD/APPARATUS/PROCEDURE:	SOURCE AND PURITY OF MATERIALS:
Nothing specified.	Nothing specified.
	ESTIMATED ERROR: Nothing specified.
	REFERENCES:

COMPONENTS:	ORIGINAL MEASUREMENTS:
(1) Sodium chlorate; $NaClO_3$; [7775-09-9] (2) Water; H_2O; [7732-18-5]	Bell, H.C. J. Chem. Soc. 1923, 123, 2712-3.
VARIABLES: T/K = 273 - 373	PREPARED BY: Hiroshi Miyamoto

EXPERIMENTAL VALUES:

Solubility of $NaClO_3$ [a]

t/°C	mass %	mol % (compiler)	mol kg^{-1} (compiler)
0	44.32	11.87	7.478
10	46.70	12.91	8.232
20	48.95	13.96	9.008
25	50.13	14.54	9.444
30	51.30	15.13	9.896
35	52.38	15.69	10.33
40	53.54	16.32	10.83
100	67.10	25.66	19.16

[a] Nature of the solid phase not specified.

AUXILIARY INFORMATION

METHOD/APPARATUS/PROCEDURE:	SOURCE AND PURITY OF MATERIALS:
A solution of sodium chlorate, saturated at the boiling point, was stirred vigorously in an electrically controlled thermostat at the required temperature for about three hours. A weighed sample was evaporated in a conical flask by immersing this in a bath at 100°C and gently distilling benzene on to the surface of the liquid. In about fifteen minutes the salt was obtained in a thin crust over the bottom of the flask. A shallow layer of benzene was then poured into the flask, which was heated in an air oven at 120°C until its weight was constant. The necessary buoyancy corrections were made to the observed weighings.	Nothing specified.
	ESTIMATED ERROR: Soly: nothing specified. Temp: precision ± 0.02 K.
	REFERENCES:

COMPONENTS:	ORIGINAL MEASUREMENTS:
(1) Sodium chlorate; $NaClO_3$; [7775-09-9] (2) Water; H_2O; [7732-18-5]	Treadwell, W.D.; Ammann, A. *Helv. Chim. Acta.* <u>1938</u>, *21*, 1249-56.
VARIABLES: One temperature: 293 K	PREPARED BY: Hiroshi Miyamoto

EXPERIMENTAL VALUES:

The solubility of sodium chlorate in water at 20°C was given as:

$$9.20 \text{ mol kg}^{-1}$$

The concentration solubility product was also given simply as the square of the solubility:

$$8.46 \times 10^1 \text{ mol}^2 \text{ kg}^{-2}$$

AUXILIARY INFORMATION

METHOD/APPARATUS/PROCEDURE:	SOURCE AND PURITY OF MATERIALS:
No information was given.	No information was given.
	ESTIMATED ERROR: Nothing specified.
	REFERENCES:

COMPONENTS:	ORIGINAL MEASUREMENTS:
(1) Sodium metaborate; $NaBO_2$; [7775-19-1]	Nies, N.P.; Hulbert, R.W.
(2) Sodium chlorate; $NaClO_3$; [7775-09-9]	J. Chem. Eng. Data 1969, 14, 14-6.
(3) Water; H_2O; [7732-18-5]	

VARIABLES:	PREPARED BY:
Composition	Hiroshi Miyamoto
T/K = 254 to 371 K	

EXPERIMENTAL VALUES: Composition of saturated solutions

t/°C	NaBO_2 mass %	NaBO_2 mol % (compiler)	NaClO_3 mass %	NaClO_3 mol % (compiler)	Nature of the solid phase[c]
-19.3	5.01	2.03	34.73	8.707	Ice+A+C
- 5	13.2[a]	3.997	0.00	0.00	A
	9.99	3.31	12.08	2.471	"
	7.57	2.80	24.61	5.624	"
	5.72	2.42	37.36	9.757	A+C
	0.00	0.00	43.03[d]	11.33	C
0	14.5	4.44	0.00	0.00	A
	10.92	3.632	11.74	2.414	"
	8.37	3.087	23.70	5.404	"
	6.12	2.618	37.98	10.04	A+C
	1.12	0.484	43.12	11.52	C
	0.00	0.000	44.23[d]	11.83	"
10	0.00	0.000	46.63[d]	12.88	C
20	20.0	6.41	0.00	0.00	A
	16.46	5.604	9.18	1.93	"
	13.02	4.847	20.47	4.711	"
	9.06	4.112	39.83	11.17	A+C
	0.00	0.000	48.86[d]	13.92	C

continued....

AUXILIARY INFORMATION

METHOD/APPARATUS/PROCEDURE:
Solutions of about 200g containing $NaBO_2$ and $NaClO_3$ were prepd in polypropylene bottles, brought to the operating temp, usually seeded with about 50 g of the solid phases desired, and agitated for several hours to several days in a water or brine bath.
At least three samples from each mixture were analyzed, and the averages are shown in the table and the figure. In some experiments the solid phases were detd by X-ray diffraction.
Na_2O and B_2O_3 were detd by titrn with 0.5 mol dm^{-3} HCl using methyl red followed by addition of mannitol and titration to pheno-lphthalain with 0.5 mol dm^{-3} NaOH which had been standardized against recrystd dry boric acid.
$NaBO_3$ mass % calcd from the percent of B_2O_3.
Chlorate was detd either by boiling with SO_2 followed by analysis of the resulting chloride by the Volhard method, or by addition of excess $FeSO_4$ with H_2SO_4, boiling, and back-titrating with $Na_2Cr_2O_7$ using barium diphenylamine sulfonate indicator. The $FeSO_4$ solution was standardized with $K_2Cr_2O_7$ in the presence of H_3PO_4.

SOURCE AND PURITY OF MATERIALS:
Photographic grade sodium metaborate dihydrate and tetrahydrate (United States Borax & Chem. Corp.) were used. The results of typical analysis were given in the following: 0.007 and 0.002 % SO_4, 0.05 and 0.04 % Cl, 0.003 and 0.002 % Ca, 1 and 1.5 ppm Fe, respectively, and 10 ppm Al. Reagent grade $NaClO_3$ (J.T. Baker Chem Co) was used, assay 100.0 %, analysis 0.01 % BrO_3 and 0.003 % or less Ca, Mg and NH_4OH precipitate, Cl, N, SO_4 and Fe.   Distilled water was used.

ESTIMATED ERROR:
Soly: nothing specified.
Temp: precision $\pm$ 0.1 K.

REFERENCES:

COMPONENTS:	ORIGINAL MEASUREMENTS:
(1) Sodium metaborate; NaBO$_2$; [7775-19-1]	Nies, N.P.; Hulbert, R.W.
(2) Sodium chlorate; NaClO$_3$; [7775-09-9]	J. Chem. Eng. Data 1969, 14, 14-6.
(3) Water; H$_2$O; [7732-18-5]	

EXPERIMENTAL VALUES: (Continued)

Composition of saturated solutions

t/°C	NaBO$_2$		NaClO$_3$		Nature of
	mass %	mol % (compiler)	mass %	mol % (compiler)	the solid phase[c]
30	23.6	7.80	0.00	0.00	A
	18.77	6.746	12.26	2.724	"
	14.78	5.931	25.33	6.284	"
	12.02	5.567	38.70	11.08	A+C
	5.76	2.707	45.22	13.14	C
	0.00	0.000	51.10[d]	15.03	"
40	27.9	9.58	0.00	0.00	–
	22.97	8.610	12.37	2.866	A
	19.09	8.004	25.08	6.500	"
	16.90	8.042	36.34	10.69	A+C[b]
	7.64	3.71	45.79	13.74	C
	0.00	0.00	53.5[a,d]	16.30	"
41.6	18.43	8.747	34.82	10.22	A+B+C
45	30.8[a]	10.86	0.00	0.00	A
	26.62	10.10	10.24	2.402	"
	24.12	9.750	17.76	4.438	"
	21.81	9.791	27.89	7.739	A+B[b]
	21.04	9.576	29.57	8.320	B
	18.44	8.900	35.97	10.73	"
	8.56	4.216	45.97	14.00	C
	0.00	0.00	54.5[d]	16.86	"

continued.....

COMPONENTS:	ORIGINAL MEASUREMENTS:
(1) Sodium metaborate; $NaBO_2$; [7775-19-1]	Nies, N.P.; Hulbert, R.W.
(2) Sodium chlorate; $NaClO_3$; [7775-09-9]	J. Chem. Eng. Data 1969, 14, 14-6.
(3) Water; H_2O; [7732-18-5]	

EXPERIMENTAL VALUES: (Continued)

Composition of saturated solutions

t/°C	NaBO₂		NaClO₃		Nature of
	mass %	mol % (compiler)	mass %	mol % (compiler)	the solid phase[c]
50	34.1	12.41	0.00	0.00	A
	30.04	11.92	11.06	2.713	"[b]
	29.65	11.99	12.97	3.243	A+B[b]
	29.18	11.86	13.79	3.465	B[b]
	25.22	10.86	21.76	5.790	"
	23.99	10.55	24.45	6.646	B
	18.67	9.172	36.97	11.23	B+C
	9.78	4.89	45.87	14.17	C
	0.00	0.00	55.6[a,d]	17.49	"
60	38.3	14.53	0.00	0.00	−
	29.58	12.52	16.65	4.356	B
	19.74	10.03	38.28	12.03	B+C
	11.07	5.704	46.74	14.89	C
	0.00	0.00	57.82[d]	18.83	"
75	42.2	16.7	0.00	0.00	B
	33.90	14.74	14.93	4.012	"
	26.56	13.01	29.86	9.040	"
	22.99	12.30	38.66	12.78	B+C
	10.03	5.510	51.61	17.53	C
	0.00	0.00	61.15[d]	21.04	"
98	0.00	0.00	66.28[d]	24.96	C

[a] Interpolated;   [b] Identified by X-ray diffraction

[c] A = $Na_2O.B_2O_3.8H_2O$ or $NaBO_2.4H_2O$;   B = $Na_2O.B_2O_3.4H_2O$;   C = $NaClO_3$

C = $NaClO_3$

[d] For the binary system the compiler computes the following:

t/°C	soly NaClO₃/mol kg⁻¹	t/°C	soly NaClO₃/mol kg⁻¹
−5	7.096	45	11.25
0	7.451	50	11.76
10	8.208	60	12.88
20	8.976	75	14.79
30	9.818	98	18.47
40	10.81		

continued.....

COMPONENTS:	ORIGINAL MEASUREMENTS:
(1) Sodium metaborate; $NaBO_2$; [7775-19-1]	Nies, N.P.; Hulbert, R.W.
(2) Sodium chlorate; $NaClO_3$; [7775-09-9]	J. Chem. Eng. Data 1969, 14, 14-6.
(3) Water; H_2O; [7732-18-5]	

COMMENTS AND/OR ADDITIONAL DATA: (Continued)

Solubility isotherms in the $NaBO_2$ - $NaClO_3$ - H_2O systems at -5° to 75°C are given below:

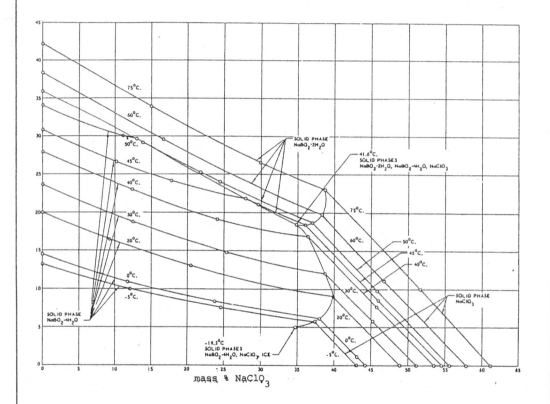

COMPONENTS:	ORIGINAL MEASUREMENTS:
(1) Sodium nitrate; $NaNO_3$; [7631-99-4]	Ricci, J.E.
(2) Sodium chlorate; $NaClO_3$; [7775-09-9]	*J. Am. Chem. Soc.* <u>1944</u>, *66*, 1015-6.
(3) Water; H_2O; [7732-18-5]	

VARIABLES:	PREPARED BY:
Composition at 298.15 K	Hiroshi Miyamoto

EXPERIMENTAL VALUES: Composition of saturated solutions

	$NaClO_3$			$NaNO_3$		Density	Nature of the
	mass %	mol % (compiler)	mass %		mol % (compiler)	g cm^{-3}	solid phase[a]
	50.10[b]	14.52	0		0	1.432	A
	43.98	13.25	9.26		3.49	1.481	"
	38.82	12.17	17.47		6.859	1.517	"
	35.72	11.52	22.65		9.149	1.528	"
	34.28[c]	11.36	25.96		10.78	1.549	A+B
	34.29	11.37	25.95		10.77	1.557	"
	34.28	11.35	25.90		10.74	1.552	"
	34.28	11.35	25.91		10.75	–	"
	34.28	11.36	25.96		10.78	1.554	"
(Av)	34.28	11.36	25.94		10.76	1.553	"
	32.15	10.47	27.08		11.05	1.548	B
	27.34	8.590	29.72		11.69	1.505	"
	20.96	6.353	33.94		12.88	1.468	"
	13.85	4.040	38.66		14.12	1.440	"
	6.93	1.95	43.27		15.25	–	"
	0	0	47.87		16.29	1.389	"

[a] $A = NaClO_3$; $B = NaNO_3$

[b] For the binary system the compiler computes the following:

soly of $NaClO_3$ = 9.433 mol kg^{-1}.

[c] Isothermally invariant solution saturated with two salts.

AUXILIARY INFORMATION

METHOD/APPARATUS/PROCEDURE:	SOURCE AND PURITY OF MATERIALS:
Complexes were stirred for at least two days at 25°C. Equilibrium was established in several instances by constancy of composition upon repeated analysis. The analysis of the saturated aqueous sln involved argentometric titration of the chloride with eosin as absorption indicator, determination of water in a separate sample by evaporation, and calculation of the sodium chlorate by difference. A few of the chloride determinations for the isothermally invariant points were verified by the Volhard method.	C.p. grade $NaClO_3$ and $NaNO_3$ were used without further purification.
The solubilities of the individual salts were determined both volumetrically and by evaporation, with very close agreement between the two methods.	ESTIMATED ERROR: Soly: nothing specified. Temp: precision ± 0.05 K.
	REFERENCES:

AMH—C

COMPONENTS:	ORIGINAL MEASUREMENTS:
(1) Sodium sulfate; Na_2SO_4; [7757-82-6] (2) Sodium chlorate; $NaClO_3$; [7775-09-9] (3) Water; H_2O; [7732-18-5]	Babaeva, A.V. *Zh. Obshch. Khim.* 1936, 6, 1144-6.
VARIABLES: Composition at 273, 293 and 313 K	PREPARED BY: Hiroshi Miyamoto

EXPERIMENTAL VALUES: Composition of saturated solutions

t/°C	Sodium Sulfate mass %	mol % (compiler)	Sodium Chlorate mass %	mol % (compiler)	Nature of the solid phase[a]
0	4.82	0.638	–	–	A
	1.65	0.243	14.77	2.897	"
	1.28	0.196	19.23	3.926	"
	0.97	0.16	29.45	6.674	"
	0.36	0.072	43.96	11.78	A+C
	–	–	45.01[b]	12.17	C
20	16.25	2.402	–	–	A
	13.05	1.980	6.01	1.22	"
	9.40	1.53	16.45	3.564	"
	8.29	1.50	27.34	6.606	"
	6.30	1.20	33.81	8.617	"
	5.75	1.15	38.10	10.18	A+C
	4.72	0.988	42.46	11.86	C
	2.41	0.518	46.86	13.45	"
	–	–	49.70[b]	14.33	"
40	32.50	5.755	–	–	B
	24.30	4.273	8.01	1.88	"
	27.71	4.897	4.86	1.15	"
	19.09	3.446	15.73	3.789	"
	15.25	2.937	25.06	6.440	"
	12.84	2.698	34.23	9.599	"

continued....

AUXILIARY INFORMATION

METHOD/APPARATUS/PROCEDURE:	SOURCE AND PURITY OF MATERIALS:
The compiler assumes that the isothermal method was used. Equilibrium was reached in 2 to 3.5 hours. The sodium chlorate content was determined volumetrically by addition of iron (II) sulfate solution to the sample solution, and back-titrating the excess Fe(II) with potassium permanganate solution. The solution containing sodium chlorate and sodium sulfate was heated with sulfuric acid and then succesively heated to dryness. The sodium sulfate content was calculated by difference. The composition of the solid phase was identified by microscopy and direct analysis.	"Chemically pure" grade sodium chlorate and sulfate were recrystallized.
	ESTIMATED ERROR: Nothing specified.
	REFERENCES:

COMPONENTS:	ORIGINAL MEASUREMENTS:
(1) Sodium sulfate; Na_2SO_4; [7757-82-6]	Babaeva, A.V.
(2) Sodium chlorate; $NaClO_3$; [7775-09-9]	*Zh. Obshch. Khim.* 1936, 6, 1144-6.
(3) Water: H_2O; [7732-18-5]	

EXPERIMENTAL VALUES: (Continued)

Composition of saturated solutions

t/°C	Sodium Sulfate		Sodium Chlorate		Nature of the solid phase[a]
	mass %	mol % (compiler)	mass %	mol % (compiler)	
40	11.42	2.632	42.13	12.96	B+C
	7.43	1.71	46.34	14.26	C
	1.70	0.392	52.39	16.12	"
	–	–	56.35[b]	17.93	"

[a] A = $Na_2SO_4 \cdot 10H_2O$;   B = Na_2SO_4;   C = $NaClO_3$

[b] For the binary system the compiler computes the following:

soly of $NaClO_3$ = 7.690 mol kg^{-1} at 0°C

= 9.282 mol kg^{-1} at 20°C

= 12.13 mol kg^{-1} at 40°C

COMMENTS AND/OR ADDITIONAL DATA:

The phase diagram is given below (based on mass % units).

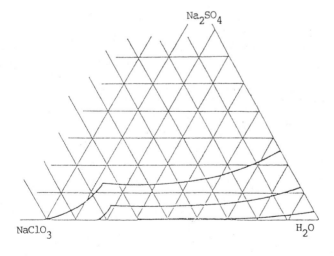

COMPONENTS:	ORIGINAL MEASUREMENTS:
(1) Sodium sulfate; Na_2SO_4; [7757-82-6]	Ricci, J.E.; Yanick, N.S.
(2) Sodium chlorate; $NaClO_3$; [7775-09-9]	J. Am. Chem. Soc. 1937, 59, 491-6.
(3) Water; H_2O; [7732-18-5]	

VARIABLES:	PREPARED BY:
Composition at 288.2, 298.2, 313.2 and 343.2 K	Hiroshi Miyamoto

EXPERIMENTAL VALUES: Composition of saturated solutions

t/°C	NaClO₃ mass %	NaClO₃ mol % (compiler)	Na₂SO₄ mass %	Na₂SO₄ mol % (compiler)	Density gm⁻³	Nature of the solid phase[a]
15	0.00	0.00	11.60	1.637	1.106	A
	19.86	4.272	5.52	0.890	1.200	"
	34.75	8.702	4.06	0.762	1.323	"
	36.89	9.499	4.15	0.801	1.348	"
	39.37	10.45	4.02	0.799	1.372	"
	41.16	11.17	3.92	0.797	-	"
	43.07	11.99	3.89	0.811	-	"
	43.67	12.25	3.90	0.820	-	"
	41.14	11.18	4.03	0.820	1.422	A+C
	44.10	12.48	4.09	0.868	1.422	"
	44.12	12.49	4.06	0.861	1.422	"
	35.93	9.750	8.91	1.81	1.393	B(m)
	38.37	10.62	7.98	1.66	1.408	"
	41.77	11.86	6.52	1.39	-	B(m)+C(m)
	41.92	11.90	6.36	1.35	1.424	"
	41.85	11.88	6.44	1.37	1.424	"
	42.66	12.10	5.59	1.19	-	C(m)
	44.34	12.55	3.83	0.812	1.422	"
	45.86	12.94	2.19	0.463	1.423	"
	47.91[b]	13.47	0.00	0.00	1.406	"

continued.....

AUXILIARY INFORMATION

METHOD/APPARATUS/PROCEDURE:
Weighed complexes of known composition were equilibrated by stirrring in a large water bath. The time required for attainment of equilibrium was determined by analysis, and required several weeks. The order of mixing of the components, and the process of seeding or inoculations for required phases had to be varied in accordance with the phase sought.

Chlorate was determined by the method of Peters and Deutshlander (1): to the chlorate sample (containing about 0.11g of ClO_3^-) was added a definite volume (50 cm³) of 0.05 mol dm⁻³ arsenious oxide solution; after the addition of a trace of KBr, the solution was acidified strongly with HCl and boiled for ten minutes. The excess arsenious oxide was then titrated by means of 0.033 mol dm⁻³ KBrO₃ solution using indigo sulfonic acid as an indicator.

SOURCE AND PURITY OF MATERIALS:
Nothing specified.

ESTIMATED ERROR:
Soly: nothing specified.
Temp: precision ± 0.02 K.

REFERENCES:
1. Kolthoff, I.M.; Furman, N.H. Volumetric Analysis Vol 2, 1929, John Wiely and Sons, New York, p. 465.

COMPONENTS:	ORIGINAL MEASUREMENTS:
(1) Sodium sulfate; Na_2SO_4; [7757-82-6] (2) Sodium chlorate; $NaClO_3$; [7775-09-9] (3) Water; H_2O; [7732-18-5]	Ricci, J.E.; Yanick, N.S. *J. Am. Chem. Soc.* 1937, 59, 491-6.

EXPERIMENTAL VALUES (Continued)

Composition of saturated solutions

__	$NaClO_3$		Na_2SO_4	Density	Nature of
mass %	mol % (compiler)	mass %	mol % (compiler)	g cm^{-3}	the solid phase[a]
0.00	0.00	21.78	3.411		A
6.58	1.42	18.20	2.935		"
12.30	2.739	15.77	2.631		"
18.05	4.192	13.90	2.419		"
23.45	5.712	12.64	2.307		"
27.36	6.938	12.06	2.292		"
28.92	7.494	12.21	2.371		A+B
28.87	7.459	12.03	2.329		"
(Av) 28.90	7.478	12.12	2.350		"
29.29	7.628	12.23	2.387		A(m)
29.52	7.708	12.20	2.387		"
29.90	7.850	12.25	2.410		A(m)+E(m)
29.89	7.849	12.27	2.414		"
29.90	7.851	12.26	2.412		"
0.00	0.000	33.97	6.125		B(m)
6.03	1.46	28.62	5.186		"
17.09	4.227	19.89	3.686		"
28.02	7.209	12.53	2.416		"
32.47	8.531	9.86	1.94		B
38.07	10.38	7.21	1.47		"
42.39	11.94	5.37	1.13		"
44.76	12.88	4.60	0.992		"
46.28	13.50	4.02	0.878		B+E
46.26	13.49	4.02	0.878		"
46.40	13.55	3.99	0.873		"
(Av) 46.31	13.51	4.01	0.877		"

AUXILIARY INFORMATION   continued.....

METHOD/APPARATUS/PROCEDURE:	SOURCE AND PURITY OF MATERIALS:
The total solid was determined by evapora-tion to dryness at 100°C followed by heat-ing to 250°C, and the sulfate was then calculated by difference. For the identification of known solid phases, microscopic examination and algebraic extrapolation of tie-lines suf-ficed. The densities reported for some saturated solutions were obtained by means of volu-metric pipets calibrated for delivery.	
	ESTIMATED ERROR:
	REFERENCES:

COMPONENTS:	ORIGINAL MEASUREMENTS:
(1) Sodium sulfate; Na_2SO_4; [7757-82-6] (2) Sodium chlorate; $NaClO_3$; [7775-09-9] (3) Water; H_2O; [7732-18-5]	Ricci, J.E.; Yanick, N.S. J. Am. Chem. Soc. 1937, 59, 491-6.

EXPERIMENTAL VALUES: (Continued)

Composition of saturated solutions

t/°C	NaClO₃		Na₂SO₄		Density	Nature of
	mass %	mol % (compiler)	mass %	mol % (compiler)	g cm⁻³	the solid phase[a]
25	25.26	6.548	15.72	3.054		E(m)
	27.00	7.062	14.73	2.887		"
	30.80	8.142	11.90	2.358		"
	31.65	8.397	11.35	2.257		"
	32.71	8.715	10.65	2.126		"
	33.85	9.064	9.93	1.993		"
	34.36	9.221	9.61	1.933		"
	36.08	9.762	8.56	1.736		"
	37.31	10.17	7.91	1.62		"
	39.75	11.01	6.70	1.39		"
	41.19	11.51	5.99	1.25		"
	44.10	12.61	4.76	1.02		"
	44.55	12.81	4.75	1.02		"
	46.57	13.60	3.83	0.838		E+C
	46.63	13.64	3.89	0.853		"
	46.68	13.65	3.83	0.839		"
	46.62	13.63	3.87	0.848		"
	46.63	13.64	3.88	0.850		B(m)+C(m)
	46.64	13.64	3.85	0.843		"
	46.64	13.64	3.86	0.846		"
	47.62	13.90	2.80	0.612		C
	50.14[b]	14.54	0.00	0.000		"
45	0.00	0.00	32.08	5.652		B
	17.88	4.333	17.52	3.181		"
	31.36	8.034	9.03	1.73		"
	36.12	9.553	6.87	1.36		"
	37.97	10.18	6.09	1.22		"
	41.84	11.57	4.61	0.955		"
	45.88	13.21	3.55	0.766		"
	48.64	14.41	2.80	0.621		"
	49.76	14.92	2.53	0.568		B+E
	49.66	14.88	2.60	0.584		"
	49.71	14.90	2.57	0.577		"
	51.46	15.79	2.38	0.547		B(m)
	20.10	5.078	18.68	3.537		E(m)
	28.23	7.295	12.66	2.452		"
	33.73	8.919	9.13	1.81		"
	37.67	10.19	7.00	1.42		"
	40.14	11.04	5.85	1.21		"
	43.56	12.30	4.45	0.942		"
	46.18	13.36	3.57	0.774		"
	49.48	14.81	2.67	0.599		"
	50.22	15.13	2.40	0.542		E
	51.79	15.92	2.21	0.509		"
	52.57	16.29	1.97	0.458		"
	53.16	16.58	1.80	0.421		E+C
	53.02	16.51	1.85	0.432		"
	53.12	16.53	1.70	0.396		"
	53.10	16.54	1.77	0.413		"
	54.59[b]	16.91	0.00	0.000		C

COMPONENTS:	ORIGINAL MEASUREMENTS:
(1) Sodium sulfate; Na_2SO_4; [7757-82-6]	Ricci, J.F.; Yanick, N.S.
(2) Sodium chlorate; $NaClO_3$ [7775-09-9]	J. Am. Chem. Soc. 1937, 59, 491-6.
(3) Water; H_2O; [7732-18-5]	

EXPERIMENTAL VALUES: (Continued)

Composition of saturated solutions

t/°C	NaClO₃		Na₂SO₄		Density g cm^{-3}	Nature of the solid phase[a]
	mass %	mol % (compiler)	mass %	mol % (compiler)		
75	0.00	0.00	30.33	5.233		B
	6.26	1.447	24.70	4.278		"
	27.19	6.749	10.56	1.964		"
	35.05	9.144	6.88	1.35		"
	45.51	12.97	3.19	0.681		"
	50.00	15.01	2.39	0.538		"
	51.85	15.93	2.09	0.481		"
	53.63	16.88	1.92	0.453		"
	54.59	17.39	1.73	0.413		"
	55.78	18.05	1.57	0.381		B+E
	55.62	17.94	1.51	0.365		"
	55.74	18.01	1.49	0.361		"
	55.71	18.00	1.52	0.368		"
	57.81	19.35	1.61	0.404		B(m)
	41.42	11.47	5.11	1.06		E(m)
	42.98	12.05	4.50	0.946		"
	46.82	13.60	3.27	0.712		"
	49.91	14.99	2.51	0.565		"
	51.15	15.59	2.28	0.521		"
	52.84	16.44	1.94	0.452		"
	53.20	16.65	1.99	0.467		"
	54.90	17.60	1.84	0.442		"
	56.25	18.36	1.62	0.396		E
	57.26	18.89	1.30	0.321		"
	58.34	19.62	1.37	0.345		"
	59.79	20.48	1.05	0.270		"
	60.10	20.73	1.14	0.295		"
	60.56	21.02	1.05	0.273		E+C
	60.80	21.14	0.93	0.242		"
	60.73	21.12	1.00	0.261		"
	61.40[b]	21.21	0.00	0.000		C

[a]   A = $Na_2SO_4.10H_2O$;   B = Na_2SO_4;   C = $NaClO_3$;   E = double salt, $NaClO_3.3Na_2SO_4$

m = metastable

[b] For the binary system the compiler computes the following:

soly of $NaClO_3$ = 8.641 mol kg^{-1} at 15°C

= 9.448 mol kg^{-1} at 25°C

= 11.29 mol kg^{-1} at 45°C

= 14.94 mol kg^{-1} at 75°C

COMPONENTS:	ORIGINAL MEASUREMENTS:
(1) Sodium sulfate; Na_2SO_4; [7757-82-6]	Ricci, J.E.; Yanick, N.S.
(2) Sodium chlorate; $NaClO_3$; [7775-09-9]	J. Am. Chem. Soc. 1937, 59, 491-6.
(3) Water; H_2O; [7732-18-5]	

COMMENTS AND/OR ADDITIONAL DATA: (Continued)

The phase diagrams are given below (based on mass % units)

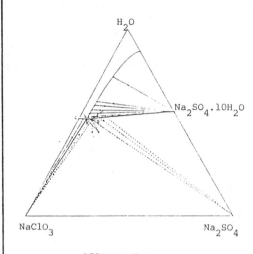

15°C Isotherm

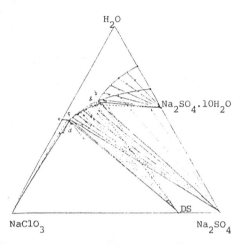

25°C Isotherm

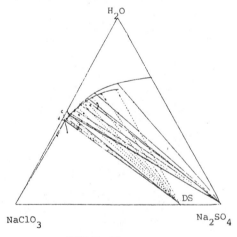

45°C Isotherm

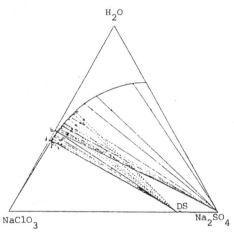

75°C Isotherm

COMPONENTS:	ORIGINAL MEASUREMENTS:
(1) Sodium sulfate; Na_2SO_4; [7757-82-6] (2) Sodium chlorate; $NaClO_3$; [7775-09-9] (3) Water; H_2O; [7732-18-5]	Windmaisser, F.; Stockl, F. *Monatsh. Chem.* 1951, *82*, 287-94.
VARIABLES: Composition at 291 K	PREPARED BY: Hiroshi Miyamoto

EXPERIMENTAL VALUES:

Composition of saturated solutions

Sodium Chlorate		Sodium Sulfate		Nature of
mass %	mol % (compiler)	mass %	mol % (compiler)	the solid phase[a]
–	–	14.04	2.030	A
9.49	1.92	9.91	1.51	"
14.77	3.097	8.04	1.26	"
20.96	4.628	6.79	1.12	"
25.31	5.819	6.14	1.06	"
29.97	7.231	5.67	1.03	"
34.80	8.885	5.50	1.05	"
40.00	10.93	5.54	1.13	"
39.90	10.89	5.57	1.14	"
42.71	12.12	5.56	1.18	"
43.14	12.33	5.68	1.22	A+B
44.60	12.94	5.28	1.15	B+C
44.09	12.78	5.71	1.24	A(m)
46.88	13.40	2.12	0.454	C
48.86[b]	13.92	–	–	"

[a] A = $Na_2SO_4 \cdot 10H_2O$;   B = Na_2SO_4;   C = $NaClO_3$;   m = metastable.

[b] For the binary system the compiler computes the following:

soly of $NaClO_3$ = 8.976 mol kg^{-1}

AUXILIARY INFORMATION

METHOD/APPARATUS/PROCEDURE:	SOURCE AND PURITY OF MATERIALS:
Complexes of salts and water placed in a Jena glass bottle. The bottles were shaken in a thermostat for 24 hours. The liquid and solid phases were separated by filtration. Barium chloride was added to the sample solution containing the sulfate to precipitate barium sulfate. The chlorate content was determined iodometrically by the method of Dietz (ref 1).	No information was given in the paper.
	ESTIMATED ERROR: Soly: nothing specified. Temp: precision ± 0.1 K.
	REFERENCES: 1. Dietz, H. *Chem. Ztg.* 1901, 727.

COMPONENTS:	ORIGINAL MEASUREMENTS:
(1) Sodium fluoride; NaF; [7681-49-4] (2) Sodium chlorate; $NaClO_3$; [7775-09-9] (3) Water; H_2O; [7732-18-5]	Vlasov, G.A.; Shishkina, L.A. *Zh. Neorg. Khim.* 1977, *22*, 2309-11; *Russ. J. Inorg. Chem.* (*Engl. Transl.*) 1977, *22*, 1250-1.

VARIABLES:	PREPARED BY:
Composition at 298 K	Hiroshi Miyamoto

EXPERIMENTAL VALUES: Composition of saturated solutions at 25°C

Sodium Fluoride			Sodium Chlorate			Nature of
mol kg^{-1}	mass %	mol % (compiler)	mol kg^{-1}	mass %	mol % (compiler)	the solid phase[a]
0.928	3.75	1.64	0	0	0	A
0.855	3.38	1.51	0.248	2.49	0.439	"
0.744	2.88	1.31	0.507	4.98	0.895	"
0.692	2.62	1.22	0.756	7.26	1.33	"
0.569	2.11	0.998	1.030	9.68	1.81	"
0.446	1.56	0.773	1.606	15.11	2.955	"
0.351	1.17	0.603	2.290	19.38	3.941	"
0.303	0.95	0.51	3.072	24.42	5.220	"
0.235	0.70	0.40	3.767	28.43	6.333	"
0.133	0.33	0.21	6.421	40.48	10.35	"
0.088	0.19	0.14	8.929	48.65	13.84	A+B
0	0	0	9.352	49.90	14.43	B

[a] A = NaF;   B = $NaClO_3$

AUXILIARY INFORMATION

METHOD/APPARATUS/PROCEDURE:	SOURCE AND PURITY OF MATERIALS:
Mixtures of sodium fluoride, sodium chlorate, and water were kept for one month at room temperature in tightly closed polyethylene flasks, and then placed in a thermostat at 25°C. The mixtures were stirred using magnetic stirring. Equilibrium was established after 6-8 hours in the thermostat. The chlorate content was determined by adding excess Fe^{2+} and back-titrating with permanganate. Fluoride was determined by the zirconium alizarin photocolorimetric method. The water content was found by difference.	"Analytically pure" grade $NaClO_3$, highly pure grade NaF, and CO_2-free distilled water were used.
	ESTIMATED ERROR: Soly: nothing specified. Temp: precision $\pm$ 0.5 K.
	REFERENCES:

COMPONENTS:	ORIGINAL MEASUREMENTS:
(1) Sodium chloride; NaCl; [7647-14-5] (2) Sodium chlorate; $NaClO_3$; [7775-09-9] (3) Water; H_2O; [7732-18-5]	Winteler, F. Z. Electrochem. 1900, 2, 360-2.

VARIABLES:	PREPARED BY:
T/K = 293 Concentration of NaCl	Hiroshi Miyamoto and Mark Salomon

EXPERIMENTAL VALUES:

Composition of saturated solutions[a]

conc NaCl		soly NaClO3		Density
g dm^{-3}	c_1/mol dm^{-3}	g dm^{-3}	c_2/mol dm^{-3}	g cm^{-3}
5	0.09	668	6.28	1.426
10	0.17	661	6.21	1.424
15	0.26	653	6.13	1.423
20	0.34	645	6.06	1.421
25	0.43	638	5.99	1.419
30	0.51	630	5.92	1.418
35	0.60	622	5.84	1.417
40	0.68	615	5.78	1.415
45	0.77	607	5.70	1.414
50	0.86	599	5.63	1.412
55	0.94	590	5.54	1.411
60	1.0	582	5.47	1.409
65	1.1	574	5.39	1.408
70	1.2	566	5.32	1.406
75	1.3	559	5.25	1.405
80	1.4	551	5.18	1.404
85	1.4_5	544	5.11	1.402
90	1.5_4	537	5.05	1.401
95	1.6	529	4.97	1.399
100	1.71	522	4.90	1.398
105	1.80	514	4.83	1.396
110	1.88	507	4.76	1.394

continued.....

AUXILIARY INFORMATION

METHOD/APPARATUS/PROCEDURE:	SOURCE AND PURITY OF MATERIALS:
Mixtures of salts and water were thermostated at 20°C for several days, and shaken frequently. Aliquots of the saturated solution were acidified with nitric acid and then titrated with silver nitrate using potassium chromate as an indicator. The compiler assumes that the total salt concentration of the solution was determined gravimetrically, and the chlorate content was determined by difference. It appears that the NaCl concentrations given in the above data table are initial concentrations (compilers).	No information was given.
	ESTIMATED ERROR: Nothing specified.
	REFERENCES:

COMPONENTS:	ORIGINAL MEASUREMENTS:
(1) Sodium chloride; NaCl; [7647-14-5]	Winteler, F.
(2) Sodium chlorate; NaClO$_3$; [7775-09-9]	Z. *Electrochem.* 1900, 2, 360-2.
(3) Water; H$_2$O; [7732-18-5]	

EXPERIMENTAL VALUES: (Continued)

Composition of saturated solutions[a]

concn NaCl		soly NaClO$_3$		Density
g dm^{-3}	c_1/mol dm^{-3} (compiler)	g dm^{-3}	c_2/mol dm^{-3} (compiler)	g cm^{-3}
115	1.97	499	4.69	1.392
120	2.05	491	4.61	1.391
125	2.14	484	4.55	1.389
130	2.22	476	4.47	1.387
135	2.31	467	4.39	1.385
140	2.40	459	4.31	1.383
145	2.48	451	4.24	1.381
150	2.57	442	4.15	1.379
155	2.65	432	4.06	1.377
160	2.74	423	3.97	1.374
165	2.82	414	3.89	1.372
170	2.91	403	3.79	1.369
175	2.99	393	3.69	1.365
180	3.08	382	3.59	1.362
185	3.17	371	3.49	1.359
190	3.25	360	3.38	1.355
195	3.34	349	3.28	1.350
200	3.42	338	3.18	1.345
205	3.51	326	3.06	1.340
210	3.59	315	2.96	1.335
215	3.68	302	2.84	1.330
220	3.76	287	2.70	1.324
225	3.85	270	2.54	1.319
230	3.94	257	3.41	1.313
235	4.02	243	2.28	1.307
240	4.11	228	2.14	1.301
245	4.19	211	1.98	1.295
250	4.28	197	1.85	1.289
255	4.36	184	1.73	1.283
260	4.45	170	1.60	1.276
265	4.53	150	1.41	1.270
270	4.62	135	1.27	1.263
275	4.71	120	1.13	1.256
280	4.79	105	0.986	1.249
285	4.88	91	0.85	1.241
290	4.96	78	0.73	1.235
295	5.05	67	0.63	1.226
300	5.13	55	0.52	1.217

[a] Composition of the solid phases not given.

COMPONENTS:	ORIGINAL MEASUREMENTS:
(1) Sodium chlorate; $NaClO_3$; [7775-09-9] (2) Sodium chloride; NaCl; [7647-14-5] (3) Water; H_2O; [7732-18-5]	Billiter, J. *Monatsh. Chem.* <u>1920</u>, *41*, 287-95.
VARIABLES: T/K = 293 to 373 Concentration of NaCl	PREPARED BY: Hiroshi Miyamoto

EXPERIMENTAL VALUES:

	concn NaCl		soly $NaClO_3$	
t/°C	g/100 cm^3	c_2/mol dm^{-3}	g/100 cm^3	c_1/mol dm^{-3}
20	0	0	72.2	6.78
	10	1.7	66	6.2
	20	3.4	57.4	5.39
	32	5.5	41.8	3.93
30	0	0	77	0.72
40	0	0	82	7.7
	10	1.7	75	7.0
	20	3.4	65	6.1
	32	5.5	42	3.9
50	0	0	86.6	8.14
60	0	0	91.3	8.58
	10	1.7	83.5	7.84
	20	3.4	70	6.58
	32	5.5	42.4	3.98
70	0	0	96	9.0
80	0	0	100.2	9.41
	10	1.7	92	8.6
	20	3.4	77	7.2
	32	5.5	43.3	4.07
90	0	0	106	9.96
100	0	0	111	10.4
	10	1.7	102	9.58
	20	3.4	87	8.2
	32	5.5	44	4.1

AUXILIARY INFORMATION

METHOD/APPARATUS/PROCEDURE:	SOURCE AND PURITY OF MATERIALS:
The apparatus is shown in Fig. 1. 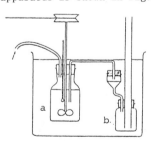 The vessel "a" equipped with a stirrer was sunk in a thermostat and the mixture of salts and water were placed in the vessel. The saturated solution was filtered in a receiver "b" through a siphon-tube. The aliquots of the saturated solution were withdrawn with a pipet. For determination of chlorate, the aliquot was added to excess acidic $FeSO_4$ solution and titrated with potassium permanganate solution.	No information was given.
	ESTIMATED ERROR: Nothing specified.
	REFERENCES:

COMPONENTS:	ORIGINAL MEASUREMENTS:
(1) Sodium chloride; NaCl; [7647-14-5] (2) Sodium chlorate; NaClO$_3$; [7775-09-9] (3) Water; H$_2$O; [7732-18-5]	Di Capua, C.; Scaletti, U. *Gazz. Chim. Ital.* 1927, 27, 391-9.

VARIABLES:	PREPARED BY:
T/K = 293	B. Scrosati and H. Miyamoto

EXPERIMENTAL VALUES: Composition of saturated solutions at 20°C[a]

mass %	mol % (compiler)	mass %	mol % (compiler)
0	0	49.56[b]	14.26
4	2	43	12
7.4	3.6	38	10
10	4.7	33.4	8.65
12.75	5.840	28.75	7.231
12.86	5.976	29.82	7.609
14.30	6.372	25.2	6.17
16.06	7.079	22.82	5.523
16.91	7.426	21.8	5.26
17.8	7.58	18.4	4.30
18.04	7.670	18.08	4.221
19.35	8.044	15	3.4
21	8.5	11	2.4
22.1	8.80	8.75	1.91
23.6	9.20	5.5	1.2
25	9.5	2.4	0.50
26.80	10.14	0	0

[a] Composition of solid phases not specified.

[b] For the binary system the compiler computes the following:

Soly of NaClO$_3$ = 9.231 mol kg^{-1}.

AUXILIARY INFORMATION

METHOD/APPARATUS/PROCEDURE:	SOURCE AND PURITY OF MATERIALS:
Mixtures of salts and water were stirred in a thermostat for 7 days. Samples of saturated solution were removed with a pipet and weighed. The chlorate ion concentration was determined by the Volhard method after reduction to chloride with zinc and acetic acid. The sodium content was determined by precipitation as the triple acetate of sodium, uranyl and magnesium, according to the method described by Kling and Lasieur (ref 1).	No information given.

ESTIMATED ERROR:
Large error may be related to the method used for the determ of sodium. The method was tested by the authors and errors ranging from +0.5 % to -32 % were found.

REFERENCES:

1. Kling and Lasieur. *Giorn. Chim. Ind. Applicata* 1925, 7.

COMPONENTS:	ORIGINAL MEASUREMENTS:
(1) Sodium chloride; NaCl; [7647-14-5] (2) Sodium chlorate; $NaClO_3$; [7775-09-9] (3) Water; H_2O; [7732-18-5]	Nallet, A.; Paris, R.A. *Bull. Soc. Chim. Fr.* 1956, 488-94.

VARIABLES:	PREPARED BY:
Composition T/K = 246.90 to 373	Hiroshi Miyamoto

EXPERIMENTAL VALUES:

Composition of saturated solutions

t/°C	Sodium Chloride			Sodium chlorate			Density g cm⁻³	Nature of the solid phase[a]
	g/100gH₂O	mass % (compiler)	mol % (compiler)	g/100gH₂O	mass % (compiler)	mol % (compiler)		
-26.25	23.9	19.3	8.89	31.3	23.8	6.03	1.3125	I+A+C
-19.2	25.9	20.1	9.48	33.1	24.9	6.43	1.320	A+C
-19.2	5.45	5.17	2.40	56.8	36.2	9.24	1.340	I+C
- 9.8	27.0	21.3	10.4	36.0	26.5	7.08	1.334	A+C
- 5.7	27.7	21.7	10.7	37.3	27.2	7.37	1.3385	A+B+C
+10	24.9	19.9	10.5	49.9	33.3	9.62	1.3664	B+C
+30	21.25	17.53	10.10	70.6	41.4	13.1	1.4089	"
+50	17.85	15.15	9.552	95.8	48.9	16.9	1.456	"
+70	14.95	13.01	8.899	123.8	55.3	20.8	1.506	"
+100	12.45	11.07	8.884	185	64.9	28.6	1.587	"

a A = NaCl.2H₂O; B = NaCl; C = NaClO₃; I = Ice

METHOD/APPARATUS/PROCEDURE:

Mixtures of salts and water were placed in bottles and agitated in a thermostat for 2 hours at 100°C, and for 2 hours or more at a lower temperature.

Equilibrium was approached from super-saturation.

The chloride ion concentration was determined by a potentiometric method using silver nitrate solution. After the determination of the chloride, the chlorate was reduced with Mohr's salt in mineral acids, and the excess Fe(II) titrated with potassium dichromate solution.

The sodium content was determined in duplicate by flame photometry.

The nature of the solid phase was determined by Schreinemakers' residues method.

The densities of the saturated solutions were also determined.

SOURCE AND PURITY OF MATERIALS:

Sodium chlorate was recrystallized twice and dried. The purity of the sodium chlorate was 99.9 %. Sodium chloride was prepared by passing HCl gas through sodium carbonate solution.

ESTIMATED ERROR:

Soly: precision 0.5 % (compiler).
Temp: nothing specified.

REFERENCES:

COMPONENTS:	ORIGINAL MEASUREMENTS:
(1) Sodium chloride; NaCl; [7647-14-5]	Oey, T.S.; Koopman, D.E.
(2) Sodium chlorate; NaClO$_3$ [7775-09-9]	J. Phys. Chem. 1958, 62, 755-6.
(3) Water; H$_2$0; [7732-18-5]	

VARIABLES:	PREPARED BY:
Composition	Hiroshi Miyamoto
T/K = 298, 308, 318	

EXPERIMENTAL VALUES: Composition of saturated solutions

t/°C	x^a	w^b	NaClO$_3$^{c} mol kg^{-1}	NaCl^c mol kg^{-1}	Sp. gr.	Nature of the solid phase^d
25	0.0000	9.01	0.000	6.161	1.200	B
	0.1593	8.18	1.081	5.705	1.240	"
	0.2142	7.95	1.496	5.487	1.255	"
	0.2696	7.64	1.959	5.307	1.271	"
	0.3867	7.01	3.062	4.856	1.309	"
	0.4394	6.66	3.662	4.672	1.327	"
	0.4722	6.57	3.990	4.459	1.340	"
	0.6175	5.55	6.176	3.826	1.402	A+B
	0.6940	5.75	6.700	2.954	1.408	A
	0.7478	5.82	7.132	2.405	1.414	"
	0.8362	5.79	8.017	1.570	1.423	"
	0.9163	5.82	8.739	0.798	1.429	"
	1.000	5.88	9.440	0.000	1.440	"
35	0.0000	8.96	0.000	6.195	1.201	B
	0.0948	8.48	0.621	5.925	1.224	"
	0.1808	8.03	1.250	5.663	1.246	"
	0.2265	7.79	1.614	5.512	1.259	"
	0.3333	7.22	2.562	5.126	1.289	"
	0.4382	6.62	3.674	4.711	1.325	"
	0.5932	5.67	5.807	3.983	1.388	"
	0.6754	5.14	7.294	3.505	1.430	A+B
	0.7060	5.18	7.565	3.150	1.433	A
	0.8133	5.26	8.583	1.970	1.444	"
	0.8659	5.29	9.086	1.407	1.451	"

AUXILIARY INFORMATION   continued.....

METHOD/APPARATUS/PROCEDURE:	SOURCE AND PURITY OF MATERIALS:
Original method described in (1). Mixtures of known composition were prepared from the solid salts and distilled water in Pyrex solubility tubes, and were equilibrated by rotation in a large thermostated water-bath at various temperatures for periods of 120 hours or longer. The liquid sample was passed through a glass wool filter without taking the solubility tube or the filter out of the thermostated water-bath. Aliquots of saturated solution were withdrawn with a calibrated pipet having small stopcocks at each end. Procedures for the analysis of chlorate, chloride and alkali were as described in ref (2). The water content was determined by difference. The nature of solid phases was determined by the Schreinemakers' wet residue method.	"Analytical reagent" grade sodium chlorate and chloride were used. The impurities in this grade were deemed much too small to affect the solubility determinations. Distilled water was used in all of the experiments.
	ESTIMATED ERROR: Soly: nothing specified. Temp: precision ± 0.1 K.
	REFERENCES: 1. Cunningham, G.L.; Oey, T.S. J. Am. Chem. Soc. 1955, 77, 799. 2. White, J.F. Am. Dyestuff Reporter 1942, 31, 484.

COMPONENTS:	ORIGINAL MEASUREMENTS:
(1) Sodium chloride; NaCl; [7647-14-5]	Oey, T.S.; Koopman, D.E.
(2) Sodium chlorate; NaClO3; [7775-09-9]	J. Phys. Chem. 1958, 62, 755-6.
(3) Water; H2O; [7732-18-5]	

EXPERIMENTAL VALUES: (Continued)

Composition of saturated solutions

t/°C	x^a	w^b	NaClO3[c] mol kg^{-1}	NaCl[c] mol kg^{-1}	Sp. Gr.	Nature of the solid phase[d]
35	0.8942	5.31	9.348	1.106	1.453	A
	1.000	5.35	10.38	0.000	1.467	"
45	0.0000	8.82	0.0000	6.294	1.201	B
	0.1042	8.29	0.6977	5.998	1.226	"
	0.1560	8.04	1.077	5.827	1.240	"
	0.2600	7.50	1.924	5.477	1.267	"
	0.3917	6.81	3.193	4.958	1.308	"
	0.4702	6.39	4.084	4.602	1.336	"
	0.6158	5.47	6.249	3.899	1.398	"
	0.7228	4.75	8.447	3.239	1.458	"
	0.7562	4.69	8.950	2.886	1.462	A+B
	0.8723	4.90	9.882	1.447	1.476	A
	0.9202	4.89	10.45	0.906	1.481	"
	1.0000	4.90	11.33	0.000	1.491	"

[a] The x function is the moles of sodium chlorate divided by the sum of the moles of
sodium chlorate and the moles of sodium chloride.

[b] The w function is the moles of water divided by the sum of the moles of sodium chlorate
and the moles of sodium chloride.

[c] Molalities calculated by the compiler.

[d] A = NaClO3;   B = NaCl

25°C

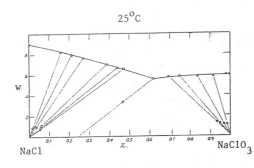

35°C

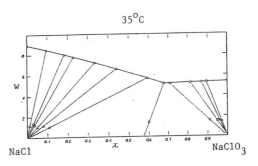

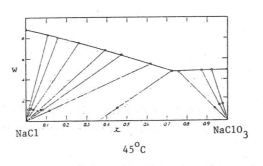

45°C

COMPONENTS:	ORIGINAL MEASUREMENTS:
(1) Sodium chloride; NaCl; [7647-14-5]	Arkhipov, S.M.; Kashina, N.I.; Kuzina, V.A.
(2) Sodium chlorate; NaClO₃; [7775-09-9]	*Zh. Neorg. Khim.* <u>1968</u>, *13*, 2872-6; *Russ. J. Inorg. Chem. (Engl. Transl.)*
(3) Water; H₂0; [7732-18-5]	<u>1968</u>, *13*, 1476-9.

VARIABLES:	PREPARED BY:
Composition at 298.2 K	Hiroshi Miyamoto

EXPERIMENTAL VALUES:

Composition of saturated solutions

Sodium Chloride		Sodium Chlorate		Nature of the
mass %	mol % (compiler)	mass %	mol % (compiler)	solid phase[a]
26.56	10.03	---	---	A
23.80	9.313	5.73	1.23	"
22.09	8.966	10.56	2.353	"
18.51	7.862	17.60	4.104	"
16.19	7.274	24.30	5.994	"
12.43	5.980	32.90	8.691	"
11.82	5.794	34.83	9.374	A+B
11.90	5.844	34.91	9.413	"
11.01	5.436	36.05	9.772	B
7.73	3.85	39.50	10.81	"
5.24	2.68	43.40	12.18	"
2.13	1.10	47.04	13.39	"
---	---	50.29[b]	14.62	"

[a] A = NaCl;   B = NaClO₃

[b] For the binary system the compiler computes the following:

soly of NaClO₃ = 9.504 mol kg^{-1}

AUXILIARY INFORMATION

METHOD/APPARATUS/PROCEDURE:	SOURCE AND PURITY OF MATERIALS:
The isothermal method was used. Equilibrium was reached in 30 hours. Samples of the solid and liquid phases were analyzed. Chlorate was found by adding an excess of iron(II) sulfate to an aliquot of saturated solution and back-titrating with potassium permanganate. Chloride was determined argentometrically. Sodium was determined by difference. The solid phases were identified by the method of residues and by X-ray diffraction.	Sodium chlorate and chloride had a purity of 99.9 % or better.

SOURCE AND PURITY OF MATERIALS:
Sodium chlorate and chloride had a purity of 99.9 % or better.

ESTIMATED ERROR:
Soly: nothing specified.
Temp: precision ± 0.1 K.

COMMENTS AND/OR ADDITIONAL DATA:
The phase diagram is given below (based on mass % units).

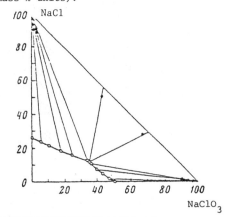

COMPONENTS:	ORIGINAL MEASUREMENTS:
(1) Sodium chlorite; $NaClO_2$; [7758-19-2] (2) Sodium chlorate; $NaClO_3$; [7775-09-9] (3) Water; H_2O; [7732-18-5]	Cunningham, G.L.; Oey, T.S. J. Am. Chem. Soc. 1955, 77, 4498-9.
VARIABLES: Composition T/K = 288.2 to 318.2	PREPARED BY: Hiroshi Miyamoto

EXPERIMENTAL VALUES: Composition of saturated solutions

Molalities[c]

t/°C	x^a	w^b	$NaClO_3$ mol kg^{-1}	$NaClO_2$ mol kg^{-1}	Sp. Gr.	Nature of the solid phase[d]
15	0.0000	8.18	0.000	6.786	1.327	A
	0.1422	7.23	1.092	6.585	1.361	"
	0.2142	6.55	1.815	6.659	1.383	"
	0.2537[e]	6.30	2.235	6.576	1.394	"
	0.4066	5.12	4.408	6.433	1.457	"
	0.4448	4.78	5.165	6.447	1.482	A+C
	0.5063	4.78	5.880	5.733	1.483	"
	0.5273	5.09	5.750	5.155	1.460	"
	0.7051	5.55	7.052	2.949	1.439	"
	0.8574	5.86	8.122	1.351	1.424	"
	1.0000	6.49	8.553	0.000	1.409	"
25	0.0000	6.50	0.000	8.540	1.375	A
	0.0426	6.22	0.3802	8.544	1.394	"
	0.0598	5.98	0.5551	8.727	1.391	"
	0.0788	5.97	0.7327	8.565	1.402	"
	0.1374	5.58	1.367	8.581	1.421	"
	0.1692	5.41	1.736	8.524	1.463	"
	0.2351	4.88	2.674	8.701	1.461	"
	0.2594	4.72	3.051	8.710	1.474	"
	0.3241	4.22	4.263	8.891	1.508	"
	0.3652	3.86	3.252	9.129	1.535	A+C

continued.....

AUXILIARY INFORMATION

METHOD/APPARATUS/PROCEDURE:

Method similar to that described in (1) where mixtures of known composition were prepared from the solid salts and distilled water in Pyrex solubility tubes. The mixtures were equilibrated by rotation in a large thermostated water-bath at various temperatures for periods of 120 hours or longer. The liquid sample was passed through a glass wool filter without taking the solubility tube or the filter out of the thermostated water-bath. Aliquots of saturated solution were withdrawn with a calibrated pipet having small stopcocks at each end. Procedures for the analysis of chlorite, chlorate and alkali were described in ref (2). The water content was determined by difference. The nature of solid phases was determined by the Schreinemakers' wet residue method.

SOURCE AND PURITY OF MATERIALS:

C.p. grade sodium chlorate was used. Technical grade sodium chlorite (Mathieson Chemical Co.) was recrystallized three times from distilled water as the trihydrate, and then stored in a cool place in amber bottles. Anal. Found: $NaClO_2$, 58.50 %, NaCl, 0.00 %, $NaClO_3$, 0.00 %; alkalinity as Na_2O, 0.06 %; water by difference, 41.44 %. Distilled water was used.

ESTIMATED ERROR:
Soly: nothing specified.
Temp: precision ± 0.1 K (authors).

REFERENCES:
1. Cunningham, G.L.; Oey, T.S. J. Am. Chem. Soc. 1955, 77, 799.
2. White, J.F. Am Dyestuff Reporter 1942, 31, 484.

COMPONENTS:	ORIGINAL MEASUREMENTS:
(1) Sodium chlorite; $NaClO_2$; [7758-19-2]	Cunningham, G.L.; Oey, T.S.
(2) Sodium chlorate; $NaClO_3$; [7775-09-9]	J. Am. Chem. Soc. 1955, 77, 4498-9.
(3) Water; H_2O; [7732-18-5]	

EXPERIMENTAL VALUES: (Continued)

Composition of saturated solutions

Molalities[c]

t/°C	x^a	w^b	$NaClO_3$ mol kg^{-1}	$NaClO_2$ mol kg^{-1}	Sp. Gr.	solid phase[d]
25	0.3941	4.22	5.184	7.970	1.520	C
	0.5004	4.56	6.091	6.082	1.498	"
	0.6278	4.87	7.156	4.242	1.472	"
	0.7269	5.07	7.958	2.990	1.461	"
	0.8053	5.24	8.531	2.063	1.456	"
	0.8486	5.30	8.888	1.586	1.450	"
	0.9046	5.50	9.130	0.963	1.446	"
	1.0000	5.88	9.440	0.000	1.444	"
35	0.0000	4.95	0.000	11.21	1.406	A
	0.0464	4.54	0.567	11.66	1.478	"
	0.1202[d]	4.06	1.643	12.03	1.515	"
	0.2276	3.56	3.549	12.05	1.563	"
	0.2918	3.42	4.736	11.49	1.571	"
	0.3177	3.15	5.598	12.02	1.595	A+C
	0.3171	3.13	5.624	12.11	1.595	"
	0.4475	3.82	6.503	8.028	1.540	C
	0.5411	4.17	7.203	6.109	1.516	"
	0.7103	4.55	8.665	3.534	1.490	"
	0.8657	4.89	9.827	1.525	1.473	"
	1.0000	5.06	10.97	0.000	1.467	"
45	0.0000	4.28	0.000	12.97	1.501	B
	0.1482	3.64	2.260	12.99	1.543	"
	0.2550	3.25	4.355	12.72	1.586	"
	0.3524	2.85	6.864	12.61	1.621	B+C
	0.4112	3.16	7.223	10.34	1.590	C
	0.5141	3.54	8.061	7.619	1.558	"
	0.6397	3.97	8.944	5.038	1.529	"
	0.7745	4.18	10.29	2.995	1.510	"
	1.0000	4.41	12.59	0.000	---	"

[a] The x function is the moles of sodium chlorate divided by the sum of the moles of sodium chlorate and the moles of sodium chlorite.

[b] The w function is the moles of water divided by the sum of the moles of sodium chlorate and the moles of sodium chlorite.

[c] Molalities calculated by the compiler.

[d] A = $NaClO_2 \cdot 3H_2O$;   B = $NaClO_2$;   C = $NaClO_3$.

[e] The solubility tube put in a water-bath for 5 and 10 days.

COMPONENTS:	ORIGINAL MEASUREMENTS:
(1) Sodium chlorate; $NaClO_3$; [7775-09-9]	Ricci, J.E.
(2) Sodium bromide; NaBr; [7647-15-6]	J. Am. Chem. Soc. 1944, 66, 1015-6.
(3) Water; H_2O; [7732-18-5]	

VARIABLES:	PREPARED BY:
Composition at 298.15 K	Hiroshi Miyamoto

EXPERIMENTAL VALUES: Composition of saturated solutions

	$NaClO_3$		NaBr		Nature of the
mass %	mol % (compiler)	mass %	mol % (compiler)		solid phase[a]
50.10^b	14.52	0	0		A
37.93	10.94	11.86	3.537		"
29.54	8.566	20.72	6.215		"
20.87	6.165	30.75	9.397		"
16.29	4.912	36.77	11.47		"
15.37	4.662	38.10	11.95		"
13.87^c	4.251	40.32	12.78		A+B
13.89	4.256	40.28	12.77		"
13.97	4.280	40.18	12.73		"
13.85	4.247	40.36	12.80		"
14.03	4.297	40.11	12.71		"
13.98	4.283	40.18	12.74		"
(Av)13.89	4.256	40.28	12.77		"
12.38	3.758	41.16	12.92		B
8.07	2.387	43.74	13.39		"
7.22	2.126	44.28	13.49		"
0	0	48.49	14.15		"

[a] A = $NaClO_3$;   B = $NaBr \cdot 2H_2O$

[b] For the binary system the compiler computes the following:

soly of $NaClO_3$ = 9.433 mol kg^{-1}

[c] Isothermally invariant solution saturated with two salts, the density of the solution = 1.583 g cm^{-3}.

AUXILIARY INFORMATION

METHOD/APPARATUS/PROCEDURE:	SOURCE AND PURITY OF MATERIALS:
Complexes were stirred for at least two days at 25°C. Equilibrium was established in several instances by constancy of composition upon repeated analysis. The analysis of the saturated aqueous solution involved argentometric titration of the chloride with eosin as adsorption indicator, determination of water in a separate sample by evaporation, and calculation of the sodium chlorate by difference. A few of the chloride determinations for the isothermally invariant points were verified by the Volhard method. The solubilities of the individual salts were determined both volumetrically and by evaporation, with very close agreement between the two methods.	C.p. grade $NaClO_3$ and NaBr were used without further purification.
	ESTIMATED ERROR: Soly: nothing specified. Temp: precision ± 0.05 K.
	REFERENCES:

COMPONENTS:	ORIGINAL MEASUREMENTS:
(1) Sodium chlorate; $NaClO_3$; [7775-09-9] (2) Sodium bromate; $NaBrO_3$; [7789-38-0] (3) Water; H_2O; [7732-18-5]	Swenson, T.; Ricci, J.E. J. Am. Chem. Soc. 1939, 61, 1974-7.

VARIABLES:	PREPARED BY:
Composition at 298 and 323 K	Hiroshi Miyamoto

EXPERIMENTAL VALUES: Composition of saturated solutions

t/°C	NaBrO₃ mass %	NaBrO₃ mol % (compiler)	NaClO₃ mass %	NaClO₃ mol % (compiler)	Nature of the solid phase[a]
25	28.29[b]	4.498	0	0	A
	16.46	2.816	18.91	4.586	"
	13.96	2.466	24.21	6.062	"
	12.20	2.208	28.03	7.191	"
	8.68	1.68	36.75	10.06	"
	7.14	1.43	40.98	11.62	"
	7.00	1.41	41.47	11.82	SSI
	6.54	1.33	42.62	12.26	"
	5.99	1.22	43.66	12.64	"
	6.05	1.24	43.55	12.60	SSI+SSII
	5.33	1.09	44.64	12.98	SSII
	5.07	1.04	44.98	13.09	"
	4.49	0.922	45.66	13.30	"
	3.79	0.779	46.46	13.54	"
	3.69	0.759	46.56	13.57	"
	2.84	0.584	47.42	13.81	"
	1.89	0.388	48.36	14.07	"
	0.96	0.20	49.16	14.27	"
	0.79	0.16	49.36	14.33	"
	0	0	50.07[b]	14.51	B

continued.....

AUXILIARY INFORMATION

METHOD/APPARATUS/PROCEDURE:	SOURCE AND PURITY OF MATERIALS:
The solubilities detd by the usual procedures of stirring, sampling, filtering and temperature control. Starting with complexes of known composition, and analyzing the saturated solutions at equilibrium, the solid phases were determined by the methods of graphical or algebraic extrapolation and occasional analyses of wet and centrifuged residues. The analytical method for the saturated solutions depended on the combined percentage of the $NaClO_3$ and $NaBrO_3$. For large $NaBrO_3$ compositions, solutions were analyzed by evaporation, and iodometric titration of the bromate with thiosulfate solution, thus allowing the calculation of the percentage of the chlorate by difference. In the presence of a large amount of chlorate, small quantitites of bromate were determined as follows: to about 100 ml of solution was added sodium iodide, 5 g (20 ml of 25 % solution) giving a concentration of 0.33N after dilution to 100 ml and 1.5 ml of concentrated HCl (0.18 to 0.2N after dilution). After waiting 1.5 min, the sln	Sodium bromate was purified by recrystallization. Sodium chlorate contained small amounts of the corresponding bromate; this bromate content was determined by iodometric titration, and the necessary corrections were then made when the dry chlorates are weighed out for the preparation of the ternary complexes.
	ESTIMATED ERROR: Soly: precision 0.05 %. Temp: nothing specified.
	REFERENCES:

continued.....

COMPONENTS:	ORIGINAL MEASUREMENTS:
(1) Sodium chlorate; NaClO3; [7775-09-9]	Swenson, T.; Ricci, J.E.
(2) Sodium bromate; NaBrO3; [7769-38-0]	J. Am. Chem. Soc. 1939, 61, 1974-7.
(3) Water; H2O; [7732-18-5]	

EXPERIMENTAL VALUES: (Continued)

Composition of saturated solutions

t/°C	NaBrO3 mass %	NaBrO3 mol % (compiler)	NaClO3 mass %	NaClO3 mol % (compiler)	Nature of the solid phase[a]
50	35.50	6.166	0	0	A
	27.3	4.87	10.8	2.73	"
	11.2	2.31	38.9	11.4	A or SS
	7.09	1.56	47.7	14.9	SS
	6.75	1.49	48.2	15.1	"
	5.80	1.28	49.1	15.4	"
	4.53	1.01	51.1	16.1	"
	2.83	0.632	53.0	16.8	"
	2.62	0.586	53.3	16.9	"
	1.35	0.301	54.4	17.2	"
	0	0	55.54[b]	17.45	B

[a] A = NaBrO3;   B = NaClO3

   SSI = sodium bromate solid solution containing up to 5 - 10 % sodium chlorate

   SSII = sodium chlorate solid solution containing from 0 to 60-65 % sodium bromate

   SS = solid solution, the composition is not given.

[b] For binary systems the compiler computes the following:

   soly of NaClO3 = 9.421 mol kg^{-1} at 25°C

                  = 11.74 mol kg^{-1} at 50°C

   soly of NaBrO3 = 2.614 mol kg^{-1} at 25°C

                  = 3.648 mol kg^{-1} at 50°C

AUXILIARY INFORMATION

METHOD/APPARATUS/PROCEDURE:	COMMENTS AND/OR ADDITIONAL DATA:
(Continued) was titrd with 0.2 N sodium thiosulfate solution. The same procedure using a 0.02N sodium thiosulfate solution for titration could be used for the detection of quantities as small as 0.001(± 0.0005) % of bromate in chlorate.	The phase diagram is given below (based on mass % units). 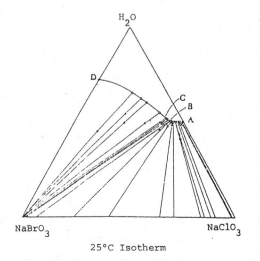

25°C Isotherm

COMPONENTS:	ORIGINAL MEASUREMENTS:
(1) Sodium chlorate; $NaClO_3$; [7775-09-9] (2) Sodium iodide; NaI; [7681-82-5] (3) Water; H_2O; [7732-18-5]	Ricci, J.E. J. Am. Chem. Soc. <u>1944</u>, 66, 1015-6.

VARIABLES:	PREPARED BY:
Composition at 298.15 K	Hiroshi Miyamoto

EXPERIMENTAL VALUES: Composition of saturated solutions

$NaClO_3$		NaI		Nature of the
mass %	mol % (compiler)	mass %	mol % (compiler)	solid phase[a]
50.10[b]	14.52	0	0	A
38.72	11.51	12.40	2.618	"
27.62	8.522	25.23	5.528	"
18.67	6.036	36.53	8.387	"
10.28	3.584	48.76	12.08	"
7.11	2.614	54.63	14.26	"
5.44	2.095	58.56	16.01	"
4.50[c]	1.808	61.52	17.55	A+B
4.28	1.720	61.74	17.62	"
4.08	1.635	61.79	17.58	"
4.20	1.684	61.73	17.58	"
4.51	1.815	61.61	17.61	".
4.32	1.735	61.68	17.59	"
2.83	1.126	62.65	17.70	B
1.43	0.566	63.67	17.88	"
1.22	0.484	64.00	18.02	"
0	0	64.80	18.12	"

[a] A = $NaClO_3$;   B = $NaI \cdot 2H_2O$

[b] For the binary system the compiler computes the following:

soly of $NaClO_3$ = 9.433 mol kg^{-1}

[c] Isothermally invariant solution saturated with two salts, the density of the solution = 1.911 g cm^{-3}

AUXILIARY INFORMATION

METHOD/APPARATUS/PROCEDURE:	SOURCE AND PURITY OF MATERIALS:
Complexes were stirred for at least two days at 25°C. Equilibrium was established in several instances by constancy of composition upon repeated analysis. The analysis of the saturated aqueous solution involved argentometric titration of the chloride with eosin as adsorption indicator, determination of water in a separate sample by evaporation, and calculation of the sodium chlorate by difference. A few of the chloride determinations for the isothermally invariant points were verified by the Volhard method. The solubilities of the individual salts were determined both volumetrically and by evaporation, with very close agreement between the two methods.	C.p. grade $NaClO_3$ and NaI were used without further purification.
	ESTIMATED ERROR: Soly: nothing specified. Temp: precision ± 0.05 K.
	REFERENCES:

COMPONENTS:	ORIGINAL MEASUREMENTS:
(1) Sodium chlorate; $NaClO_3$; [7775-09-9] (2) Sodium iodate; $NaIO_3$; [7681-55-2] (3) Water; H_2O; [7732-18-5]	Ricci, J.E. J. Am. Chem. Soc. 1938, 60, 2040-3.
VARIABLES: Composition at 298.15 K and 323.15 K	PREPARED BY: Hiroshi Miyamoto and Mark Salomon

EXPERIMENTAL VALUES: Composition of saturated solutions

t/°C	NaIO$_3$ mass %	NaIO$_3$ mol % (compiler)	NaClO$_3$ mass %	NaClO$_3$ mol % (compiler)	Density g cm^{-3}	Nature of the solid phase[a]
25	8.57[b]	0.846	0.0	0.0	1.075	A
	4.51	0.462	8.36	1.591	1.098	"
	3.14	0.343	16.50	3.347	1.146	"
	2.43	0.286	24.67	5.402	1.204	"
	1.97	0.252	32.57	7.748	1.273	"
	1.69	0.232	38.66	9.862	1.332	"
	1.52	0.220	42.99	11.57	-	"
	1.46	0.216	44.56	12.23	1.396	"
	1.39	0.210	46.37	13.03	1.404	"
	1.33	0.206	48.13	13.85	1.425	"
	1.30	0.204	49.19	14.37	1.440	"
	1.29	0.203	49.42	14.48	1.445	A+C
	1.29	0.203	49.40	14.47	-	"
	1.29	0.203	49.32	14.43	1.441	"
	1.29	0.203	49.44	14.49	1.446	"
	1.29	0.203	49.32	14.43	1.444	"
	1.29	0.203	49.40	14.47	-	"
	1.29	0.203	49.38	14.46	1.444(av)	"
	1.16	0.183	49.52	14.50	1.444	C
	0.0	0.0	50.14	14.54	-	"

continued.....

AUXILIARY INFORMATION

METHOD/APPARATUS/PROCEDURE:

Isothermal method. At 25°C complexes of known compn seeded and stirred for up to 60d, and mean error in compn of solid phases was 1 %. At 50°C equil was readily attained and mean error in solid phase compn was 0.09 %. More precise solid phase compns at 25°C obtained by first dissolving $NaIO_3.H_2O$ followed by addn of $NaClO_3$, seeding with the monohydrate, and stirring for at least 6 d. At 50°C metastability for anhyd and hydrated $NaIO_3$ easily maintained, in the first case by starting with anhyd salt and not seeding, and in the second case by starting with the hydrate and seeding.
Filtered samples of satd sln analyzed for iodate by titrn with std thiosulfate in the presence of excess KI and acetic acid: titrn error was 1 part in 3000. Total solids detd by evapn to dryness, and $NaClO_3$ detd by difference. Solid phase compn detd by algebraic extrapolation of tie-lines. The mean error of 1 % in compn at 25°C indicates existence of the anhyd salt even after 60 d of stirring. This problem was eliminated by first preparing the sln with the hydrate as described above. (continued)

SOURCE AND PURITY OF MATERIALS:

C.p. grade sodium iodate recrystallized, and dried at 100-110°C. Analysis by titrn with std thiosulfate sln showed it to be 100.0 % pure. C.p. grade sodium chlorate was powdered and dried at 150-200°C.

ESTIMATED ERROR:

Soly: precision ± 0.04 %.
Solid phase compn: see discussion at left.
Temp: precision ± 0.01 K.

METHOD/APPARATUS/PROCEDURE: (Continued)

Densities of satd slns at 25°C detd by means of pipets calibrated for delivery.

COMPONENTS:	ORIGINAL MEASUREMENTS:
(1) Sodium chlorate; $NaClO_3$; [7775-09-9] (2) Sodium iodate; $NaIO_3$; [7681-55-2] (3) Water; H_2O; [7732-18-5]	Ricci, J. E. J. Am. Chem. Soc. 1938, 60, 2040-3.

EXPERIMENTAL VALUES: (Continued)

Composition of saturated solutions

t/°C	NaIO₃ mass %	NaIO₃ mol % (compiler)	NaClO₃ mass %	NaClO₃ mol % (compiler)	Density g cm⁻³	Nature of the solid phase[a]
50	13.49	1.400	0.00	0.000		A
	7.67	0.824	10.02	2.002		"
	5.69	0.639	16.56	3.457		"
	4.91	0.570	20.61	4.448		"
	3.23	0.424	33.33	8.131		"
	2.41	0.357	43.71	12.030		"
	2.12	0.336	48.95	14.432		A(m)
	1.92	0.323	53.20	16.66		"
	1.87	0.321	54.58	17.44		A(m)+C
	1.87	0.322	54.61	17.46		"
	1.87	0.322	(av)54.59	17.45		"
	2.50	0.369	43.41	11.91		B(m)
	(2.2)[c]	0.330	(45)[c]	12.56		A+B
	2.14	0.334	47.86	13.90		B
	1.75	0.297	53.83	16.97		"
	1.71	0.294	54.69	17.46		B+C
	1.68	0.289	54.74	17.48		"
	1.69	0.290	(av)54.71	17.47		"
	1.26	0.216	54.98	17.50		C
	0.0	0.0	55.74	17.57		"

[a] $A = NaIO_3.H_2O$; $B = NaIO_3$; $C = NaClO_3$

[b] Interpolated

[m] Metastable

For the binary system the compiler computes the following

$$soly \ of \ NaIO_3 = 0.474 \ mol \ kg^{-1} \ at \ 25°C$$
$$= 0.7880 \ mol \ kg^{-1} \ at \ 50°C$$

COMMENTS AND/OR ADDITIONAL DATA:

Isotherms based on mass % units are reproduced below.

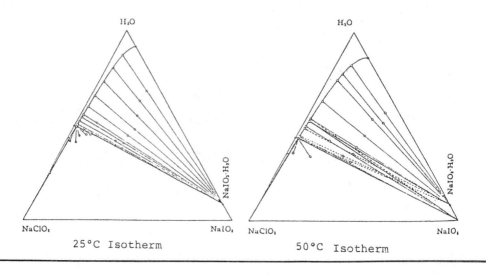

25°C Isotherm   50°C Isotherm

COMPONENTS:	ORIGINAL MEASUREMENTS:
(1) Sodium chlorate; $NaClO_3$; [7775-09-9] (2) Sodium hydroxide; NaOH; [1310-73-2] (3) Water; H_2O; [7732-18-5]	Windmaisser, F.; Stockl, F. *Monatsh. Chem.* 1951, *82*, 287-94.

VARIABLES:	PREPARED BY:
Composition at 291 K	Hiroshi Miyamoto

EXPERIMENTAL VALUES: Composition of saturated solutions at 18°C.

Sodium		Sodium Chlorate		Nature of
mass %	mol % (compiler)	mass %	mol % (compiler)	the solid phase[a]
–	–	48.86[b]	13.92	A
6.55	4.54	37.86	9.866	"
15.25	9.706	25.10	6.003	"
19.93	12.32	19.45	4.516	"
33.34	19.99	7.90	1.78	"
41.58	25.36	3.98	0.912	"
44.56	27.69	3.65	0.852	"
46.90	29.64	3.56	0.845	"
48.73	31.20	3.46	0.832	A+B
51.43	32.29	–	–	B

[a] $A = NaClO_3$; $B = NaOH \cdot H_2O$

[b] For the binary system the compiler computes the following:

soly of $NaClO_3$ = 8.976 mol kg^{-1}

AUXILIARY INFORMATION

METHOD/APPARATUS/PROCEDURE:	SOURCE AND PURITY OF MATERIALS:
The details of the solubility determinations were not given in the original paper, but see the compilation for the $NaClO_3$-Na_2SO_4-H_2O system by these authors.	No information given.
	ESTIMATED ERROR: Nothing specified.
	REFERENCES:

COMPONENTS:	ORIGINAL MEASUREMENTS:
(1) Sodium chlorate; $NaClO_3$; [7775-09-9] (2) Potassium chlorate; $KClO_3$; [3811-04-9] (3) Water; H_2O; [7732-18-5]	Di Capua, C.; Scaletti, U. *Gazz. Chim. Ital.* 1927, 27, 391-9.
VARIABLES: T/K = 293	PREPARED BY: B. Scrosati and H. Miyamoto

EXPERIMENTAL VALUES:

Composition of saturated solutions at 20°C (solid phases not specified)

	$NaClO_3$		$KClO_3$	
mass %	mol % (compiler)	mass %	mol % (compiler)	
49.56[a]	14.26	0	0
3.01	0.543	4.25	0.666
6.01	1.11	3.65	0.584
8.42	1.59	4.13	0.678
14.93	2.988	3.56	0.619
22.34	4.827	3.65	0.685
26.33	5.924	3.40	0.664
32.87	7.896	2.62	0.547
34.93	8.587	2.50	0.534
40.05	10.49	2.50	0.569
40.35	10.63	2.60	0.595
42.57	11.54	2.57	0.605
0	0	6.75[a]	1.05
47.43	13.31	0.34	0.083
47.82	13.49	0.31	0.076
48.50	13.80	0.24	0.059
48.40	13.73	0.14	0.034
48.60	13.83	0.20	0.049
48.84	13.95	0.20	0.050

[a] For the binary systems the compiler computes the following:

soly of $NaClO_3$ = 9.231 mol kg^{-1}

soly of $KClO_3$ = 0.591 mol kg^{-1}

AUXILIARY INFORMATION

METHOD/APPARATUS/PROCEDURE:	SOURCE AND PURITY OF MATERIALS:
Mixtures of salts and water were stirred in a thermostat for 7 days. Samples of saturated solution were withdrawn with a pipet and weighed. The chlorate ion concentration was determined by the Volhard method after reduction to chloride with zinc and acetic acid. The sodium content was determined by precipitation as the triple acetate of sodium, uranyl and magnesium, according to the method described by Kling and Lasieur (ref 1).	No information is given.

ESTIMATED ERROR:
Large error may be related to the method used for the determ of sodium. The method was tested by the authors and errors ranging from +0.5 % to -32 % were found.

REFERENCES:

1. Kling and Lasieur. *Giorn. Chom. Ind. Applicata* 1925, 7.

COMPONENTS:	ORIGINAL MEASUREMENTS:
(1) Sodium chlorate; NaClO$_3$ [7775-09-9] (2) Potassium chlorate; KClO$_3$; [3811-04-9] (3) Water; H$_2$O; [7732-18-5]	Munter, P.A.; Brown, R.L. J. Am. Chem. Soc. 1943, 65, 2456-7.

VARIABLES:	PREPARED BY:
Composition at 273 K and 313 K	Hiroshi Miyamoto and Mark Salomon

EXPERIMENTAL VALUES:

Composition at the isothermally invariant points

t/°C	Sodium Chlorate mass %	mol % (compiler)	Potassium Chlorate mass %	mol % (compiler)	Water mass %	mol % (compiler)
0	44.21	11.90	0.44	0.10	55.35	88.00
40	51.75	16.19	3.41	0.927	44.85	82.88

AUXILIARY INFORMATION

METHOD/APPARATUS/PROCEDURE:	SOURCE AND PURITY OF MATERIALS:
At 0°C mixts were sealed in Pyrex bottles which were fastened to a rotor suspended in a glycerol/water bath. At 40°C mixts were placed in 250 ml 3-neck flasks and thermo-stated in a water bath. The slns were stir-red with glass stirrers provided with mercury seals. Preliminary experiments identified mixtures which result in satd solutions, several of which were used to prepare the solutions reported in the data table above. Only the compositions of two solutions were reported. Equilibrated slns were sampled by with-drawing aliquots with pipets fitted with cotton plugs. After determining densities the solutions were diluted for analyses. Chlorate detd by the method of Dietz as described in (1). Sodium was detd by pptn with zinc uranyl nitrate, and potassium detd by calculation based on stoichiometry. Water was found by difference.	C.p. grade salts were used without further purification. The chlorates were found to be 99.9 % pure.

	ESTIMATED ERROR: Soly: nothing specified but probably poor due to method of analysis of Na. Temp: at 0°C precision = ± 0.1 K. at 40°C precision = ± 0.05 K.
	REFERENCES: 1. Kolthoff, I.M.; Furman, N.H. Volumetric Analysis, Vol II. 1929, 388.

COMPONENTS:	ORIGINAL MEASUREMENTS:
(1) Sodium chlorate; $NaClO_3$; [7775-09-9]	Nallet, A.; Paris, R.A.
(2) Potassium chlorate; $KClO_3$; [3811-04-9]	*Bull. Soc. Chim. Fr.* 1956, 488-94.
(3) Water; H_2O; [7732-18-5]	

VARIABLES:	PREPARED BY:
Composition T/K = 255.30 to 373	Hiroshi Miyamoto

EXPERIMENTAL VALUES:

Composition of saturated solutions

t/°C	Sodium chlorate			Potassium chlorate			Density g cm^{-3}	Nature of the solid phase[a]
	g/100gH$_2$O	mass % (compiler)	mol % (compiler)	g/100gH$_2$O	mass % (compiler)	mol % (compiler)		
-17.85	65.9	39.7	10.1	0.665	0.661	0.146	1.356	A+B+I
- 9.8	71.7	41.8	11.0	1.00	0.990	0.226	1.3735	A+B
- 9.8	35.3	26.1	5.69	0.822	0.815	0.154	1.217	B+I
- 4	14.62	12.76	2.441	1.16	1.15	0.191	1.095	"
+10	88.1	31.6	7.47	2.38	2.32	0.476	1.4173	A+B
+30	106.1	51.48	16.25	5.14	4.89	1.35	1.4625	"
+50	127.3	56.01	20.72	9.97	9.07	2.91	1.510	"
+70	153.9	60.61	28.00	18	15	6.1	1.566	"
+100	212	67.9	60.8	40.6	28.9	22.4	1.661	"

[a] A = NaClO$_3$; B = KClO$_3$; I = Ice.

METHOD/APPARATUS/PROCEDURE:
Mixtures of salts and water were placed in bottles and agitated in a thermostat for 2 hours at 100°C, and for 2 hours or more at a lower temperature.
Equilibrium was approached from supersaturation.
The chlorate was reduced with Mohr's salt in mineral acids, and the excess Fe(II) titrated with potassium dichromate solution. The analyses of cations were performed in duplicate.
The potassium and sodium contents were determined by flame photometry, and also the potassium was determined gravimetrically with sodium tetraphenylborate.
The nature of the solid phase was determined by Schreinemakers' residues method.
The densities of the saturated solutions were also determined.

SOURCE AND PURITY OF MATERIALS:
Sodium and potassium chlorate were recrystallized twice. The purity of the chlorates was 99.9 %.

ESTIMATED ERROR:
Soly: precision 0.5 % (compiler).
Temp: nothing specified.

REFERENCES:

COMPONENTS:	ORIGINAL MEASUREMENTS:
(1) Sodium chlorate; NaClO$_3$; [7775-09-9] (2) Rubidium chlorate; RbClO$_3$; [13446-71-4] (3) Water; H$_2$O; [7732-18-5]	Arkhipov, S.M.; Kashina, N.I.; Kuzina, V.A. *Zh. Neorg. Khim.* 1968, *13*, 2872-6; *Russ. J. Inorg. Chem. (Engl. Transl.)* 1968, *13*, 1476-9.
VARIABLES: Composition at 298.2 K	PREPARED BY: Hiroshi Miyamoto

EXPERIMENTAL VALUES:

Composition of saturated solutions

Rubidium Chlorate		Sodium Chlorate		Nature of
mass %	mol % (compiler)	mass %	mol % (compiler)	the solid phase[a]
6.42[b]	0.726	---	---	A
4.61	0.532	4.15	0.760	"
2.69	0.329	12.55	2.437	"
2.20	0.294	21.83	4.624	"
1.83	0.273	32.22	7.616	"
1.80	0.303	42.27	11.31	"
1.67	0.300	47.00	13.38	"
1.70	0.316	49.44	14.58	A+B
1.68	0.312	49.41	14.56	"
1.13	0.209	49.81	14.63	B
---	---	50.29[b]	14.62	"

[a] A = RbClO$_3$;   B = NaClO$_3$

[b] For binary systems the compiler computes the following:

soly of RbClO$_3$ = 0.406 mol kg^{-1}

soly of NaClO$_3$ = 9.504 mol kg^{-1}

AUXILIARY INFORMATION

METHOD/APPARATUS/PROCEDURE:

The isothermal method was used. Equilibrium reached in 30 hours. Samples of solid and liquid phases were analyzed. Rubidium was determined as the tetraphenylborate or when at low concentration, by flame photometry. Chlorate was found by adding an excess of iron(II) sulfate to an aliquot of saturated solution and back-titrating with potassium permanganate. Sodium was determined by difference.
The solid phases were identified by the method of residues, and by X-ray diffraction.

SOURCE AND PURITY OF MATERIALS:
Sodium chlorate and rubidium chlorate had a purity of 99.9 % or more.

ESTIMATED ERROR:
Soly: nothing specified.
Temp: precision ± 0.1 K.

COMMENTS AND/OR ADDITIONAL DATA:
The phase diagram is given below (based on mass % units).

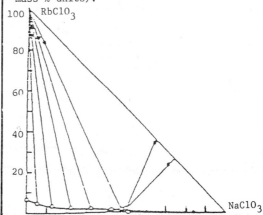

COMPONENTS:	ORIGINAL MEASUREMENTS:
(1) Sodium chlorate; $NaClO_3$; [7775-09-9]	Arkhipov, S.M.; Kashina, N.I.
(2) Cesium chlorate; $CsClO_3$; [13763-67-2]	*Zh. Neorg. Khim.* 1970, *15*, 760-4. *Russ. J. Inorg. Chem. (Engl. Transl.)* 1970, *15*, 391-2.
(3) Water; H_2O; [7732-18-5]	

VARIABLES:	PREPARED BY:
Composition at 298.2 K	Hiroshi Miyamoto

EXPERIMENTAL VALUES:
Composition of saturated solution at 25°C

Cesium Chlorate		Sodium Chlorate		Nature of the
mass %	mol %	mass %	mol %	solid phase[a]
	(compiler)		(compiler)	
7.24[b]	0.646	---	---	A
3.74	0.346	7.86	1.48	"
2.78	0.278	17.07	3.470	"
2.48	0.273	26.68	5.976	"
2.17	0.277	39.58	10.29	"
2.18	0.294	43.54	11.92	"
2.17	0.299	45.25	12.68	"
2.15	0.315	49.64	14.79	A+B
2.13	0.312	49.58	14.76	"
---	---	50.20[b]	14.57	B

[a] A = $CsClO_3$;   B = $NaClO_3$

[b] For binary systems the compiler computes the following:

soly of $NaClO_3$ = 9.470 mol kg^{-1}

soly of $CsClO_3$ = 0.361 mol kg^{-1}

AUXILIARY INFORMATION

METHOD/APPARATUS/PROCEDURE:
Solubilities were determined by the isothermal method by mixing the solid and
liquid phases in glass test-tubes and thermostating in a water bath. Samples of
liquid and solid phases were analyzed for
the anions and cesium.

Chlorate was found by adding excess iron(II)
sulfate to an aliquot of saturated solution
and back-titrating with potassium permanganate solution. Cesium was determined
gravimetrically as cesium tetraphenylborate.
Sodium was found by difference. The solid
phases were identified by the method of
residues, and X-ray diffraction.

SOURCE AND PURITY OF MATERIALS:
C.p. grade $NaClO_3$ and $CsClO_3$ with a purity
of 99.5 % or better were used.

ESTIMATED ERROR:
Soly: nothing specified.
Temp: precision ± 0.1 K.

COMMENTS AND/OR ADDITIONAL DATA:
The phase diagram is given below (based on
mass % units).

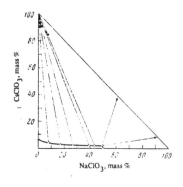

COMPONENTS:	ORIGINAL MEASUREMENTS:
(1) Sodium chlorate; $NaClO_3$; [7775-09-9]	Di Capua, C.; Bertoni, A.
(2) Barium chlorate; $Ba(ClO_3)_2$; [13477-00-4]	*Gazz. Chim. Ital.* 1928, *58*, 249-53.
(3) Water; H_2O; [7732-18-5]	

VARIABLES:	PREPARED BY:
T/K = 293 Composition	B. Scrosati, H. Miyamoto and M. Salomon

EXPERIMENTAL VALUES:

Solubilities in the $NaClO_3$-$(BaClO_3)_2$-H_2O ternary system at 20°C.[a]

$NaClO_3$		$Ba(ClO_3)_2$	
mass %	mol kg^{-1}	mass %	mol kg^{-1}
4.97	9.283[b]	0	0
45.	7.84	1.05	0.0640
43.2	7.506	2.73	0.166
36.5	5.696	3.30	0.180
29.52	4.218	4.73	0.236
25.32	3.47	6.13	0.294
15.52	1.908	8.05	0.346
8.5	0.983	10.29	0.416
4.52	0.540	16.91	0.707
0	0	23.75	1.024[c]

[a] Molalities calculated by the compilers.

[b] Author gives 9.228 mol kg^{-1}.

[c] Author gives 1.068 mol kg^{-1}.

AUXILIARY INFORMATION

METHOD/APPARATUS/PROCEDURE:	SOURCE AND PURITY OF MATERIALS:
The method and the procedure for preparing the saturated solutions were not reported in the original publication. Chloride was determined by the Mohr method, and chlorate was determined by the Volhard method after reduction with zinc and acetic acid. The barium content was determined gravimetrically as the sulfate, and the sodium content was determined by difference after the mass of water was determined. Nature of solid phases not specified.	Nothing specified.
	ESTIMATED ERROR: No estimates possible due to insufficient experimental details.
	REFERENCES:

COMPONENTS:	ORIGINAL MEASUREMENTS:
(1) Sodium chlorate; $NaClO_3$; [7775-09-9]	Ricci, J.E.; Weltman, C.
(2) Sodium chromate; Na_2CrO_4; [7775-11-3]	J. Am. Chem. Soc. 1942, 64, 2746-8.
(3) Water; H_2O; [7732-18-5]	

VARIABLES:	PREPARED BY:
Composition T/K = 293, 298 and 323	Hiroshi Miyamoto

EXPERIMENTAL VALUES: Composition of saturated solutions

	Sodium chromate		Sodium chlorate		Nature of
t/°C	mass %	mol % (compiler)	mass %	mol % (compiler)	the solid phase[a]
19	0.00	0.00	48.28[b]	13.64	A
	6.43	1.20	41.91	11.93	"
	14.56	2.738	33.59	9.611	"
	27.00	5.170	21.57	6.285	"
	35.05	6.915	15.01	4.506	A+C
	35.03	6.911	15.03	4.512	"
	35.04	6.913	15.02	4.509	"
	37.26	7.146	10.70	3.123	C
	40.60	7.571	5.14	1.459	"
	42.26	7.766	2.31	0.646	"
	43.63	7.926	0.00	0.000	B
25	0.00	0.00	(50.06)[b]	14.50	A
	5.95	1.14	43.88	12.75	"
	12.45	2.381	37.06	10.79	"
	20.42	3.949	29.30	8.623	"
	28.51	5.583	21.50	6.407	"
	35.18	7.021	15.65	4.753	"
	36.43	7.283	14.43	4.390	A+D
	36.44	7.287	14.44	4.394	"
	36.43	7.283	14.43	4.390	"
	36.43	7.283	14.43	4.390	"
	39.47	7.734	9.82	2.93	D

continued...

AUXILIARY INFORMATION

METHOD/APPARATUS/PROCEDURE:	SOURCE AND PURITY OF MATERIALS:
Mixtures prepd by weight and rotated in a thermostat at the specified temperature. About three days were required to reach equilibrium. Samples for analysis withdrawn with pipets fitted with filter paper. Sodium chromate in the presence of sodium chlorate was detd volumetrically as follows: the chromate was pptd by addn of barium chloride. The precipitate was filtered, dissolved in HNO_3, and the chromate titrd with thiosulfate solution. Sodium chlorate detd by difference from the percentage of total solid obtained by evaporation of the satd solution at 110°C. To supplement the indirect detn of chlorate, direct gravimetric analysis carried out by reduction of chlorate with SO_2 followed by pptn of chloride as AgCl. The solubility result given in parenthesis in the above table was determined by evaporation.	C.p. grade sodium chlorate was used and found to be 100.0 % pure by reduction and precipitation. Sodium chromate tetrahydrate (Mackay Co.) was used; the percentage of Na_2CrO_4 found by titration was 69.15 % and by dehydration 69.25 % as compared with the theoretical value of 69.21 %.
	ESTIMATED ERROR: Soly: accuracy within ± 0.05 % (authors). Temp: precision ± 0.02 K.
	REFERENCES:

COMPONENTS:	ORIGINAL MEASUREMENTS:
(1) Sodium chlorate; NaClO$_3$; [7775-09-9]	Ricci, J.E.; Weltman, C.
(2) Sodium chromate; Na$_2$CrO$_4$; [7775-11-3]	J. Am. Chem. Soc. 1942, 64, 2746-8.
(3) Water; H$_2$O; [7732-18-5]	

EXPERIMENTAL VALUES: (Continued)

Composition of saturated solutions

t/°C	Sodium chromate		Sodium chlorate		Nature of the solid phase[a]
	mass %	mol % (compiler)	mass %	mol % (compiler)	
25	41.04	7.949	7.34	2.16	D
	45.59	8.525	0.00	0.00	C
50	0.00	0.00	55.49[b]	17.42	A
	6.36	1.31	48.49	15.18	"
	18.37	3.842	36.71	11.68	"
	31.45	6.665	23.55	7.594	"
	40.80	8.968	15.81	5.238	"
	43.13	9.566	13.87	4.681	A+D
	43.15	9.571	13.85	4.675	"
	43.14	9.569	13.86	4.678	"
	44.21	9.619	11.54	3.821	D
	47.32	9.969	6.20	1.988	"
	50.66	10.25	0.00	0.000	"

[a] A = NaClO$_3$;   B = Na$_2$CrO$_4$.10H$_2$O;   C = Na$_2$CrO$_4$.6H$_2$O;   D = Na$_2$CrO$_4$.4H$_2$O

[b] For the binary system the compiler computes the following:

soly of   NaClO$_3$ = 8.770 mol kg^{-1} at 19°C
= 9.417 mol kg^{-1} at 25°C
= 11.71 mol kg^{-1} at 50°C

COMMENTS AND/OR ADDITIONAL DATA:

The phase diagram is given below (based on mass % units).

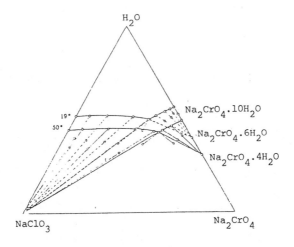

COMPONENTS:	ORIGINAL MEASUREMENTS:
(1) Sodium chlorate; $NaClO_3$; [7775-09-9]	Ricci, J.E.; Linke, W.F.
(2) Disodium (I-4)-tetraoxomolybdate (2-) (sodium molybdate); Na_2MoO_4; [7631-95-0]	J. Am. Chem. Soc. 1947, 69, 1080-3.
(3) Water; H_2O; [7732-18-5]	

VARIABLES:	PREPARED BY:
Composition at 298.15 K	Hiroshi Miyamoto

EXPERIMENTAL VALUES: Composition of saturated solutions at 25.0°C

| Na_2MoO_4 | | $NaClO_3$ | | Density | Nature of the |
mass %	mol % (compiler)	mass %	mol % (compiler)	g cm^3	solid phase[a]
39.38	5.378	0.00	0.00	1.432	A
36.11	4.972	4.23	1.13	1.441	"
32.42	4.509	9.04	2.43	1.441	"
28.53	4.011	14.12	3.840	1.440	"
22.83	3.278	21.94	6.093	1.442	"
17.95	2.643	29.14	8.301	1.453	"
14.59	2.196	34.39	10.02	1.466	"
13.04	1.990	37.05	10.94	1.472	"
11.77	1.817	39.21	11.71	1.478	A+B
11.75	1.814	39.25	11.72	1.479	"
11.81	1.823	39.17	11.70	1.481	"
11.77	1.817	39.21	11.71	1.479	"
11.74	1.813	39.29	11.74	1.476	B
8.87	1.358	41.85	12.40	1.465	"
5.72	0.868	44.70	13.12	1.456	"
2.60	0.392	47.60	13.87	1.438	"
0.00	0.000	50.02[b]	14.49	1.433	"

[a] A = $Na_2MoO_4 \cdot 2H_2O$;   B = $NaClO_3$

[b] For the binary system the compiler computes the following:

$$\text{soly of } NaClO_3 = 9.402 \text{ mol kg}^{-1}$$

continued.....

AUXILIARY INFORMATION

METHOD/APPARATUS/PROCEDURE:	SOURCE AND PURITY OF MATERIALS:
The solubilities were determined by stirring complexes of known compositions in Pyrex tubes and sampling the equilibrated solutions by means of calibrated pipets fitted with filtering tips. One sample of saturated solution was analyzed by evaporating and drying to constant weight at 125°C to obtain the combined percentage of the two salts. A second sample was used for the determination of molybdate by precipitation of silver molybdate followed by a Volhard titration of the excess silver in the filtrate.	C.p. grade sodium molybdate dihydrate was used. The salt was completely dehydrated by heating at 180°C, and stored at 150°C. The purity of this anhydrous salt was found to be 100 %. C.p. grade sodium chlorate was found to be pure within 1/1000 by reduction to chloride and the determination of the chloride by the Volhard method.
	ESTIMATED ERROR: Soly: nothing specified. Temp: precision ± 0.04 K.
	REFERENCES:

COMPONENTS:	ORIGINAL MEASUREMENTS:
(1) Sodium chlorate; $NaClO_3$; [7775-09-9]	Ricci, J.E.; Linke, W.F.
(2) Disodium (I-4)-tetraoxomolybdate (2-) (sodium molybdate); Na_2MoO_4; [7631-95-0]	J. Am. Chem. Soc. 1947, 69, 1080-3.
(3) Water; H_2O; [7735-18-5]	

COMMENTS AND/OR ADDITIONAL DATA:

The phase diagram is given below (based on mass % units).

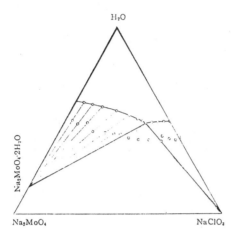

AUXILIARY INFORMATION

METHOD/APPARATUS/PROCEDURE:	SOURCE AND PURITY OF MATERIALS:
	ESTIMATED ERROR:
	REFERENCES:

COMPONENTS:	ORIGINAL MEASUREMENTS:
(1) Sodium chlorate; $NaClO_3$; [7757-82-6] (2) Calcium nitrate; $Ca(NO_3)_2$; [10124-37-5] (3) Water ; H_2O ; [7732-18-5]	Musaev, N.Yu.; Tukhtaev, S.; Shammasov, R.E.; Kucharov, Kh. *Zh. Neorg. Khim.* 1984, 29, 1342-4; *Russ. J. Inorgan. Chem. (Engl. Transl.)* 1984, 29, 770-1.
VARIABLES: T/K = 228 - 323 Composition	PREPARED BY: Mark Salomon

EXPERIMENTAL VALUES:

	$NaClO_3$		$Ca(NO_3)_2 \cdot 4H_2O$	$Ca(NO_3)_2$ [a]		
t/°C	mass %	mole %	mass %	mass %	mole %	solid phase composition [b]
-18.5	41.9	10.08	---	---	---	ice + A
-19.5	36.2	9.622	10.1	7.018	1.210	"
-19.9	34.5	9.22$_5$	13.0	9.033	1.567	"
-21.0	30.0	8.178	21.0	14.592	2.580	"
-44.4	18.5	5.857	50.4	35.020	7.193	"
-28.7	---	---	62.1	43.15	7.692	ice + B
-34.4	8.9	2.607	56.3	39.12	7.433	"
-44.6	18.6	5.88$_5$	50.2	34.881	7.159	ice + A + B
-6.0	18.6	6.163	54.1	37.591	8.079	"
11.8	18.0	6.860	65.7	45.651	11.286	"
25.2	17.2	7.346	74.5	51.766	14.342	"

[a]Calculated by the compiler.

[b]Solid phases: A = $NaClO_3$;   B = $Ca(NO_3)_2 \cdot 4H_2O$

For the binary $NaClO_3-H_2O$ systems, the compiler computes the following:

soly $NaClO_3$ at -18.5°C = 6.775 mol kg^{-1}

soly $NaClO_3$ at -28.7°C = 4.626 mol kg^{-1}

AUXILIARY INFORMATION

METHOD/APPARATUS/PROCEDURE:	SOURCE AND PURITY OF MATERIALS:
"Visual-polythermal" method used: i.e. probably the synthetic method (compiler). The original publication contains a phase diagram. In the temperature range studied, neither solid solutions nor new compounds are formed: i.e. the systems are of the simple eutonic type.	"C.p." grade $NaClO_3$ and $Ca(NO_3)_2$ were recrystallized two times. No other information was given.
	ESTIMATED ERROR: Nothing specified.
	REFERENCES:

COMPONENTS:	ORIGINAL MEASUREMENTS:
(1) Sodium chlorate; NaClO$_3$; [7775-09-9] (2) Silver chlorate; AgClO$_3$; [7783-92-8] (3) Water; H$_2$O; [7732-18-5]	Ricci, J. E.; Offenbach, J. A. J. Am. Chem. Soc. 1951, 73, 1597-9.
VARIABLES: T/K = 298 Composition	PREPARED BY: H. Miyamoto

EXPERIMENTAL VALUES:

The equilibrium results for the ternary system AgClO$_3$-NaClO$_3$-H$_2$O are given.

Composition of Saturated Solutions

mass % NaClO$_3$	mol % NaClO$_3$ (compiler)	mass % AgClO$_3$	mol % AgClO$_3$ (compiler)	Nature of solid phase*
0.00	0	14.46	1.567	AgClO$_3$
8.11	1.630	10.02	1.121	SSI
17.49	3.762	7.48	0.895	SSI
27.53	6.463	5.56	0.726	SSI
34.39	8.610	4.23	0.589	SSI
41.78	11.276	2.85	0.428	SSI
46.57	13.275	2.14	0.339	SSI + SSII
46.54	13.263	2.15	0.341	SSI + SSII
46.55	13.268	2.15	0.341	SSI + SSII
47.52	13.628	1.66	0.265	SSII
49.23	14.220	0.56	0.090	SSII
50.04	14.495	0.00	0	NaClO$_3$

*SSI = ∿37% NaClO$_3$ in solid phase
 SSII = ∿26% AgClO$_3$ in solid phase

The compiler calculates the solubility of AgClO$_3$ in water as 0.755$_8$ mol kg^{-1}, and the solubility of NaClO$_3$ as 9.410 mol kg^{-1}.

continued.......

AUXILIARY INFORMATION

METHOD/APPARATUS/PROCEDURE:

Ternary mixtures, AgClO$_3$-NaClO$_3$-H$_2$O, of known composition were allowed to come to equili-brium at 25°C after two weeks of stirring. The results were unchanged after 1 to 3 weeks of further stirring. The saturated liquid solution was filtered and sampled for analy-sis. One sample was titrated for silver with standard KSCN solution and one was evaporated to dryness at 110-125°C, for total salt con-tent whereupon NaClO$_3$ was calculated by dif-ference.

SOURCE AND PURITY OF MATERIALS:

AgClO$_3$ was made from C.P. AgNO$_3$ and C.P. NaClO$_3$. After three recystallizations, the product was 99.72 % pure (on the basis of gravimetric determination of silver as AgCl after reduction with NaNO$_3$ in the presence of some NaCl).

ESTIMATED ERROR:
Nothing specified in original article.
Solubility: ± 0.03 mass % (compiler).

Temp: precision probably better than ± 0.1 K
(compiler).

REFERENCES:

COMPONENTS:	ORIGINAL MEASUREMENTS:
(1) Sodium chlorate; $NaClO_3$; [7775-09-9]	Ricci, J. E.; Offenbach, J. A.
(2) Silver chlorate; $AgClO_3$; [7783-92-8]	J. Am. Chem. Soc. 1951, 73, 1597-9.
(3) Water; H_2O; [7732-18-5]	

EXPERIMENTAL VALUES: (Continued)

The phase diagram is presented below.

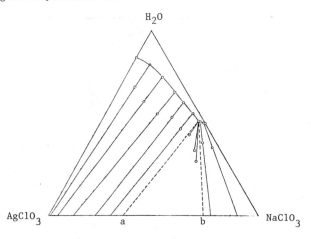

The limiting compositions of SSI and SSII are estimated as ∿37 mass % $NaClO_3$ in SSI and
∿26 mass % $AgClO_3$ in SSII. The composition of the isothermally invariant liquid
saturated with these two limiting solid solutions is 2.15 mass % $AgClO_3$ and 46.55 mass %
$NaClO_3$.

COMPONENTS:	ORIGINAL MEASUREMENTS:
(1) Sodium chlorate; NaClO₃; [7775-09-9] (2) Urea; CH₄N₂O; [57-13-6] (3) Water; H₂O; [7732-18-5]	Nabiev, M.N.; Tukhtaev, S.; Musaev, N.Yu.; Kuchrov, Kh.; Shammasov, R.E. *Zh. Neorg. Khim.* 1982, *27*, 2704-6; *Russ. J. Inorg. Chem. (Engl. Transl.)* 1982, *27*, 1533-4.
VARIABLES: T/K = 248.4 to 354.8 Composition	PREPARED BY: Hiroshi Miyamoto and Mark Salomon

EXPERIMENTAL VALUES: Composition of saturated solutions

	Urea		Sodium Chlorate		Nature of the
t/°C	mass %	mol % (compiler) .	mass %	mol % (compiler)	solid phase[a]
-11.2	32.0	12.4	-	-	I+A
-15.8	28.9	12.8	14.2	3.54	"
-24.8	21.5	11.0	31.5	9.07	"
-18.5	-	-	41.9[b]	10.9	I+B
-22.0	12.9	6.29	35.4	9.73	"
-25.2	21.0	10.7	32.0	9.22	I+A+B
-13.5	27.1	14.9	32.0	9.95	A+B
2.2	34.0	20.6	32.0	10.9	"
29.8	46.3	33.9	32.1	13.3	"
48.0	54.2	46.0	32.2	15.4	"
81.6	65.7	77.2	34.3	22.8	"

[a] I = Ice;   A = CO(NH₂)₂;   B = NaClO₃.

[b] For the binary system at - 18.5°C the compiler computes the following:

   soly of NaClO₃ = 6.78 mol kg^{-1}

AUXILIARY INFORMATION

METHOD/APPARATUS/PROCEDURE: The method of isothermal sections was used. Eight internal sections were employed. No other information given.	
SOURCE AND PURITY OF MATERIALS: "Chemically pure" grade potassium chlorate and urea were twice recrystallized from water. No other information given.	
ESTIMATED ERROR: Nothing specified.	

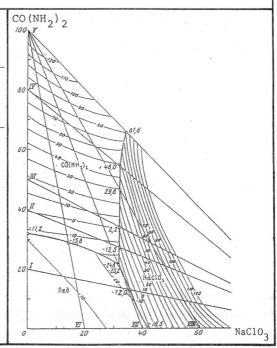

AMH—D*

COMPONENTS:	ORIGINAL MEASUREMENTS:
(1) Sodium carbonate; Na_2CO_3; [497-19-8] (2) Sodium chloride; NaCl; [7647-14-5] (3) Sodium chlorate; $NaClO_3$; [7775-09-9] (4) Water; H_2O; [7732-18-5]	Perel'man, F.M.; Korzhenyak, N.G. *Zh. Neorg. Khim.* 1968, *13*, 2861-4; *Russ. J. Inorg. Chem. (Engl. Transl.)* 1968, *13*, 1471-2.
VARIABLES: T/K = 298	PREPARED BY: Hiroshi Miyamoto

EXPERIMENTAL VALUES:

Composition of saturated solutions at the eutonic points[a]

Sodium Chlorate		Sodium Chloride		Sodium Carbonate	
mass %	mol %	mass %	mol %	mass %	mol %
25.0	6.52	10.5	4.99	8.5	2.23
34.9	9.41	11.9	5.84	–	–
42.0	12.6	–	–	10.4	3.13
–	–	15.4	6.32	17.2	3.89

[a] Mol % data calculated by the compiler.

AUXILIARY INFORMATION

METHOD/APPARATUS/PROCEDURE:	SOURCE AND PURITY OF MATERIALS:
No information was given.	No information was given.
	ESTIMATED ERROR: Nothing specified.
	REFERENCES:

COMPONENTS:	ORIGINAL MEASUREMENTS:
(1) Sodium chloride; NaCl; [7647-14-5] (2) Sodium chlorite; NaClO$_2$; [7758-19-2] (3) Sodium chlorate; NaClO$_3$; [7775-09-9] (4) Water; H$_2$O; [7732-18-5]	Nakamori, I.; Nagino, Y.; Hideshima, K.; Hirai, T. *Kogyo Kagaku Zasshi* 1958, 61, 147-9.
VARIABLES: Composition at 283, 293 and 303 K	PREPARED BY: Hiroshi Miyamoto

EXPERIMENTAL VALUES: Composition of saturated solutions

t/°C	NaCl mole fraction[a]	NaClO$_2$ mole fraction[a]	NaClO$_3$ mole fraction[a]	Moles H$_2$O[b]	Nature of the solid phase[c]
10	0.445	0.555	0.000	6.87	A+B
	0.392	0.520	0.888	6.35	"
	0.339	0.490	0.171	5.92	"
	0.475	0.000	0.525	6.19	A+C
	0.434	0.074	0.492	6.00	"
	0.318	0.284	0.398	5.52	"
	0.007	0.515	0.478	5.54	B+C
	0.115	0.475	0.410	5.30	"
	0.150	0.457	0.393	5.11	"
	0.239	0.425	0.336	5.08	A+B+C
20	0.248	0.598	0.154	5.19	A+B
	0.191	0.552	0.257	4.60	"
	0.328	0.672	0.000	6.12	"
	0.359	0.106	0.535	5.53	A+C
	0.412	0.000	0.588	5.80	"
	0.246	0.340	0.414	5.04	B+C
	0.068	0.558	0.374	4.42	"
	0.112	0.540	0.348	4.30	"
	0.157	0.527	0.316	4.35	A+B+C

continued.....

AUXILIARY INFORMATION

METHOD/APPARATUS/PROCEDURE:	SOURCE AND PURITY OF MATERIALS:
Mixtures were placed in glass bottles and agitated in a thermostat at the desired temperature for 1.5 to 5.5 h. After equilibrium was established the slns were allowed to settle in the thermostat for one h or more. Aliquots were analyzed for Cl$^-$, ClO$_2^-$, and ClO$_3^-$. The solution was weighed, and chloride detd by pptn using silver nitrate sln. The chlorite concn in slns containing chlorite and chlorate was detd by iodometric titration after addn of dilute acetic acid. To another sample of solution, sulfuric acid and Fe(II) sulfate solution were added and the excess Fe(II) titrd with potassium permanganate sln, and the chlorate content calculated by difference. The weight of NaCl, NaClO$_2$ and NaClO$_3$ was calculated from the solubility data, and the water content was determined by difference.	"Chemically pure" grade sodium chloride and chlorite were used without further purification. Sodium chlorate of purity 85 % or better was recrystallized, and the product NaClO$_3$.3H$_2$O obtained.
	ESTIMATED ERROR: Nothing specified.
	REFERENCES:

COMPONENTS:	ORIGINAL MEASUREMENTS:
(1) Sodium chloride; NaCl; [7647-14-5]	Nakamori, I.; Nagino, Y.;
(2) Sodium chlorite; NaClO$_2$; [7758-19-2]	Hideshima, K.; Hirai, T.
(3) Sodium chlorate; NaClO$_3$; [7775-09-9]	*Kogyo Kagaku Zasshi* 1958, *61*, 147-9.
(4) Water; H$_2$O; [7732-18-5]	

EXPERIMENTAL VALUES: (Continued)

Composition of saturated solutions

t/°C	NaCl mole fraction[a]	NaClO$_2$ mole fraction[a]	NaClO$_3$ mole fraction[a]	Moles H$_2$O[b]	Nature of the solid phase[c]
30	0.141	0.733	0.126	4.27	A+B
	0.112	0.688	0.200	3.85	"
	0.198	0.802	0.000	4.98	"
	0.350	0.000	0.650	5.38	A+C
	0.213	0.342	0.445	4.53	"
	0.142	0.497	0.361	4.12	"
	0.020	0.679	0.301	3.55	B+C
	0.085	0.649	0.266	3.57	A+B+C

[a] Mole fraction based on total moles of solutes.

[b] Mole of water/1 mol of the solute

[c] A = NaCl;   B = NaClO$_2$;   C = NaClO$_3$

COMPONENTS:	ORIGINAL MEASUREMENTS:
(1) Sodium carbonate; Na_2CO_3; [497-19-8] (2) Sodium chlorite; $NaClO_2$; [7758-19-2] (3) Sodium chlorate; $NaClO_3$; [7775-09-9] (4) Water; H_2O; [7732-18-5]	Perel'man, F.M.; Korzhenyak, N.G. Zh. Neorg. Khim. 1968, 13, 2861-4; Russ. J. Inorg. Chem. (Engl. Transl.) 1968, 13, 1471-2.
VARIABLES: T/K = 298	PREPARED BY: Hiroshi Miyamoto

EXPERIMENTAL VALUES:

Composition of saturated solutions at the eutonic points[a]

Sodium Chlorite		Sodium Chlorate		Sodium Carbonate	
mass %	mol %	mass %	mol %	mass %	mol %
32.0	11.7	22.0	6.86	2.2	0.69
34.8	13.22	23.7	7.65	–	–
35.4	10.7	–	–	7.0	1.81
–	–	42.0	12.59	10.4	3.13

[a]Mole % data calculated by the compiler.

AUXILIARY INFORMATION

METHOD/APPARATUS/PROCEDURE:	SOURCE AND PURITY OF MATERIALS:
No information was given	No information was given.
	ESTIMATED ERROR: Nothing specified.
	REFERENCES:

COMPONENTS:	ORIGINAL MEASUREMENTS:
(1) Sodium chloride; NaCl; [7647-14-5] (2) Sodium chlorite; NaClO$_2$; [7758-19-2] (3) Sodium chlorate; NaClO$_3$; [7775-09-9] (4) Water; H$_2$O; [7732-18-5]	Perel'man, F.M.; Korzhenyak, I.G. *Zh. Neorg. Khim.* 1968, *13*, 277-80; *Russ. J. Inorg. Chem. (Engl. Transl.)* 1968, *13*, 143-5.
VARIABLES: T/K = 298	PREPARED BY: Hiroshi Miyamoto

EXPERIMENTAL VALUES:

The details of solubility data were not described in the original article. The experimental and calculated solubilities were shown in figures only.

The phase diagrams of the eutectic point of the quatanary NaClO$_3$-NaClO$_2$-NaCl-H$_2$O system are given as below (based on mass %).

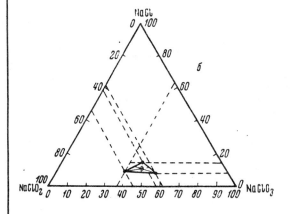 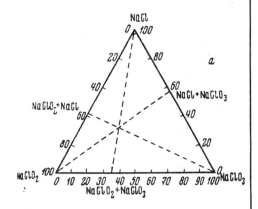

AUXILIARY INFORMATION

METHOD/APPARATUS/PROCEDURE:	SOURCE AND PURITY OF MATERIALS:
Probably, the isothermal method was used. The ions ClO$_3^-$, ClO$_2^-$ and Cl$^-$ were determined in the presence of one another as follows: the chlorite content was determined iodometrically, the sum of the ClO$_2^-$ and ClO$_3^-$ ion concentrations by permanganate in a strongly acidic medium. The chlorate content was determined by difference. The chloride ion concentration was determined in a neutral medium by Mohr's method.	No information was given in the original paper.
	ESTIMATED ERROR: Nothing specified.
	REFERENCES:

COMPONENTS:	ORIGINAL MEASUREMENTS:
(1) Sodium chloride; NaCl; [7647-14-5] (2) Sodium chlorite; $NaClO_2$; [7758-19-2] (3) Sodium chlorate; $NaClO_3$; [7775-09-9] (4) Water; H_2O; [7732-18-5]	Oey, T.S.; Cunningham, G.I.; Koopman, D.E. J. Chem. Eng. Data 1960, 5, 248-50.
VARIABLES: Composition at 298, 303 and 318 K	PREPARED BY: Hiroshi Miyamoto

EXPERIMENTAL VALUES: Composition of saturated solutions

t/°C	$NaClO_2$ moles	mol % (compiler)	$NaClO_3$ moles	mol % (compiler)	NaCl moles	mol % (compiler)	H_2O w^a	sp gr	Nature of the solid phase[b]
25	0.1763	2.020	0.1361	1.559	0.6885	7.889	7.72	1.273	D
	0.2488	3.001	0.1418	1.710	0.6095	7.351	7.29	1.290	"
	0.3899	4.824	0.1170	1.448	0.4984	6.166	7.04	1.321	"
	0.5340	6.407	0.1351	1.621	0.4091	4.909	6.73	1.356	"
	0.2461	3.141	0.2124	2.711	0.5408	6.903	6.84	1.314	"
	0.3096	4.076	0.2124	2.796	0.4788	6.303	6.59	1.332	"
	0.4210	5.937	0.1928	2.719	0.3863	5.448	6.09	1.361	"
	0.5546	8.631	0.1806	2.811	0.2641	4.110	5.43	1.412	"
	0.1170	1.386	0.1095	1.297	0.7737	9.165	7.44	1.249	"
	0.1252	1.650	0.3680	4.848	0.5068	6.677	6.59	1.338	"
	0.1160	1.278	0.0997	1.098	0.7851	8.649	8.07	1.249	"
	0.1718	2.191	0.2784	3.551	0.5498	7.013	6.84	1.318	"
	0.1752	2.440	0.3798	5.289	0.4451	6.199	6.18	1.358	"
	0.1771	2.020	0.1096	1.250	0.7132	8.133	7.77	1.265	"
	0.1166	1.780	0.6861	10.48	0.1972	3.011	5.55	1.429	C
	0.2304	3.652	0.5743	9.103	0.1951	3.093	5.31	1.443	"
	0.3272	5.490	0.4863	8.159	0.1865	3.129	4.96	1.459	"
	0.4209	7.455	0.4078	7.223	0.1706	3.022	4.65	1.481	"
	0.1270	1.971	0.5899	9.154	0.2837	4.403	5.44	1.424	"
	0.1849	2.930	0.5483	8.689	0.2669	4.229	5.31	1.432	"

continued.....

AUXILIARY INFORMATION

METHOD/APPARATUS/PROCEDURE:	SOURCE AND PURITY OF MATERIALS:
Method similar to that described in (1) where mixtures of known composition were prepared from the solid salts and distilled H_2O. Pyrex solubility tubes used. The mixtures of solid and liquid were equilibrated by rotation in a large thermostated water-bath at various temperatures for periods of 120 hours or longer. The liquid sample was passed through a glass wool filter without taking the solubility tube or the filter out of the thermostated water-bath. Aliquots of saturated solution were removed by means of a calibrated pipet having small stopcocks at each end. Procedures for the analysis of chloride, chlorite, chlorate and alkali were described in ref (2). The Schreinemakers' wet residue method was used to detn solid phase compositions.	"Analytical reagent" grade sodium chlorate and chloride were used. Technical grade sodium chlorite (Mathieson Chemical Co.) was recrystallized three times from distilled water as the trihydrate, and then stored in a cool place in amber bottles. Distilled water was used.
	ESTIMATED ERROR: Soly: nothing specified. Temp: precision ± 0.1 K (authors).

	REFERENCES:
	
	1. Cunningham, G.L.; Oey, T.S.
	   J. Am. Chem. Soc. 1955, 77, 799.
	2. White, J.F. Am. Dyestaff Reporter
	   1942, 31, 484.

COMPONENTS:	ORIGINAL MEASUREMENTS:
(1) Sodium chloride; NaCl; [7647-14-5]	Oey, T.S.; Cunningham, G.I.; Koopman, D.E.
(2) Sodium chlorite; NaClO$_2$; [7758-19-2]	
(3) Sodium chlorate; NaClO$_3$; [7775-09-9]	J. Chem. Eng. Data 1960, 5, 248-50.
(4) Water; H$_2$O; [7732-18-5]	

EXPERIMENTAL VALUES: (Continued)

Composition of saturated solutions

t/°C	NaClO$_2$ moles	NaClO$_2$ mol % (compiler)	NaClO$_3$ moles	NaClO$_3$ mol % (compiler)	NaCl moles	NaCl mol % (compiler)	H$_2$O w^a	sp gr	Nature of the solid phase^b
25	0.1980	3.128	0.6149	9.713	0.1872	2.957	5.33	1.439	C
	0.2979	5.050	0.4566	7.740	0.2454	4.160	4.90	1.452	"
	0.1106	1.658	0.7947	11.91	0.0947	1.420	5.67	1.438	"
	0.6970	12.06	0.1921	3.324	0.1109	1.919	4.78	1.451	A
	0.8023	12.07	0.1225	1.842	0.0751	1.129	5.65	1.420	"
	0.7599	12.01	0.1832	2.894	0.0568	0.897	5.33	1.447	"
	0.4781	8.930	0.3395	6.341	0.1831	3.420	4.35	1.481	C+D
	0.3144	5.374	0.4338	7.415	0.2519	4.306	4.85	1.446	"
	0.4209	7.615	0.3728	6.745	0.2058	3.723	4.53	1.468	"
	0.5276	10.05	0.3069	5.846	0.1655	3.152	4.25	1.493	"
	0.0564	0.869	0.5875	9.053	0.3560	5.486	5.49	1.410	"
	0.1645	2.611	0.5306	8.421	0.3050	4.841	5.30	1.427	"
	0.4122	7.436	0.3950	7.126	0.1934	3.489	4.54	1.478	"
	0.4626	8.516	0.3775	6.949	0.1696	3.122	4.38	1.493	"
	0.5952	12.05	0.3296	6.671	0.0754	1.526	3.94	1.532	A+C
	0.6124	12.13	0.3479	6.890	0.0396	0.784	4.05	1.531	"
	0.6469	11.35	0.1782	3.126	0.1749	3.068	4.70	1.466	A+D
	0.6798	11.37	0.1130	1.890	0.2071	3.464	4.98	1.434	"
	0.6336	12.21	0.3664	7.060	0.0000	0.000	4.19	1.533	A+C
	0.5709	11.89	0.3139	6.539	0.1153	2.402	3.80	1.534	A+D+C
30	0.7512	11.63	0.0000	0.000	0.2488	3.851	5.46	1.4150	D
	0.5520	7.132	0.0000	0.000	0.4480	5.788	6.74	1.3245	"
	0.3160	3.575	0.0000	0.000	0.6840	7.738	7.84	1.2642	"
	0.0850	0.876	0.0000	0.000	0.9150	9.433	8.70	1.2152	"
	0.0000	0.000	0.0000	0.000	1.0000	9.990	9.01	1.1955	"
	0.5380	10.15	0.4620	8.717	0.0000	0.000	4.30	1.5142	C
	0.1804	2.953	0.8196	13.41	0.0000	0.000	5.11	1.4621	"
	0.0000	0.000	1.0000	15.41	0.0000	0.000	5.49	1.4560	"
	1.0000	7.364	1.0000	7.364	0.0000	0.000	5.79	1.4120	A
	0.8743	13.64	0.0000	0.000	0.1257	1.961	5.41	1.4310	"
	0.9190	14.68	0.0910	1.453	0.0000	0.000	5.20	1.4444	"
	0.7890	14.75	0.2110	3.944	0.0000	0.000	4.35	1.5075	"
	0.0000	0.000	0.6459	10.08	0.3541	5.524	5.41	1.4091	C+D
	0.0611	0.970	0.6050	9.603	0.3339	5.300	5.30	1.4235	"
	0.1972	3.281	0.5201	8.654	0.2827	4.704	5.01	1.4415	"
	0.3407	6.052	0.4352	7.730	0.2241	3.980	4.63	1.4695	"
	0.5111	10.24	0.3470	6.954	0.1419	2.844	3.99	1.5111	"
	0.8025	13.49	0.0000	0.000	0.1975	3.319	4.95	1.4395	A+D
	0.6790	15.12	0.3210	7.149	0.0000	0.000	3.49	1.5645	A+C
	0.5994	13.29	0.3140	6.962	0.0866	1.920	3.51	1.5360	A+C+D
45	0.1367	1.571	0.1379	1.585	0.7254	8.338	7.70	1.269	D
	0.2425	3.251	0.2658	3.563	0.4917	6.591	6.46	1.338	"
	0.3124	4.889	0.3665	5.736	0.3211	5.025	5.39	1.409	"
	0.7818	12.01	0.0000	0.000	0.2182	3.352	5.51	1.444	"
	0.6712	9.616	0.0000	0.000	0.3288	4.711	5.98	1.378	"
	0.5338	6.888	0.0000	0.000	0.4662	6.015	6.75	1.329	"
	0.2769	3.115	0.0000	0.000	0.7231	8.134	7.89	1.262	"
	0.1476	1.574	0.0000	0.000	0.8524	9.087	8.38	1.234	"
	0.0000	0.000	0.0000	0.000	1.0000	9.990	9.01	1.200	"

continued.....

COMPONENTS:	ORIGINAL MEASUREMENTS:
(1) Sodium chloride; NaCl; [7647-14-5]	Oey, T.S.; Cunningham, G.I.; Koopman, D.E.
(2) Sodium chlorite; NaClO$_2$; [7758-19-2]	
(3) Sodium chlorate; NaClO$_3$; [7775-09-9]	*J. Chem. Eng. Data* <u>1960</u>, 5, 248-50.
(4) Water; H$_2$O; [7732-18-5]	

EXPERIMENTAL VALUES: (Continued)

Composition of saturated solutions

| t/°C | NaClO$_2$ | | NaClO$_3$ | | NaCl | | H$_2$O w^a | sp gr | Nature of the solid phase[b] |
	moles	mol % (compiler)	moles	mol % (compiler)	moles	mol % (compiler)			
45	0.1605	2.821	0.6682	11.74	0.1713	3.011	4.69	1.487	C
	0.3187	6.117	0.6055	11.62	0.0758	1.455	4.21	1.526	"
	0.7340	16.03	0.1842	4.022	0.0818	1.786	3.58	1.561	B
	0.8012	16.59	0.1303	2.698	0.0685	1.418	3.83	1.543	"
	1.0000	18.32	0.0000	0.000	0.0000	0.000	4.46	1.508	"
	0.9244	17.31	0.0000	0.000	0.0756	1.416	4.34	1.504	"
	0.8710	16.28	0.0000	0.000	0.1290	2.411	4.35	1.501	B+D
	0.1382	2.550	0.6375	11.76	0.2243	4.138	4.42	1.487	C+D
	0.2845	5.589	0.5418	10.64	0.1737	3.413	4.09	1.515	"
	0.4932	10.96	0.4013	8.918	0.1055	2.344	3.50	1.574	"
	0.0000	0.000	0.7228	12.57	0.2772	4.821	4.75	1.458	"
	0.6979	15.51	0.2069	4.598	0.0952	2.12	3.50	1.569	B+D
	0.7567	16.03	0.1359	2.879	0.1074	2.275	3.72	1.543	"
	0.8710	16.16	0.0000	0.000	0.1290	2.393	4.39	1.561	"
	0.6107	15.00	0.3412	8.382	0.0483	1.19	3.07	1.621	B+C
	0.6312	15.51	0.3688	9.061	0.0000	0.000	3.07	1.646	"
	0.5953	14.70	0.3305	8.162	0.0740	1.83	3.05	1.620	B+C+D

[a] The w function is the moles of water divided by the sum of the moles of sodium chlorate, sodium chlorite and sodium chloride.

[b] A = NaClO$_2$.3H$_2$O;   B = NaClO$_2$;   C = NaClO$_3$;   D = NaCl

[c] For the binary system the compiler computes the following:

      soly of NaClO$_3$ = 1.711 mol kg^{-1} at 30°C

COMMENTS AND/OR ADDITIONAL DATA:

The phase diagrams are given below (based on mass % units).

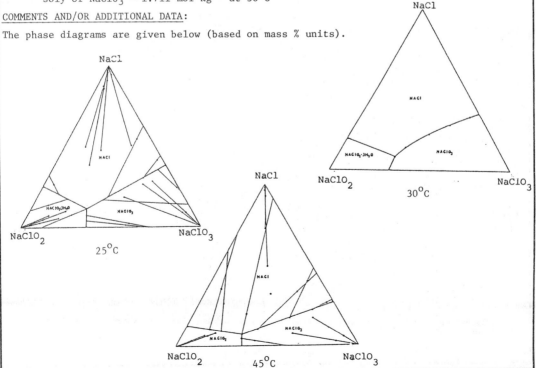

COMPONENTS:	ORIGINAL MEASUREMENTS:
(1) Sodium chloride; NaCl; [7647-14-5]	Munter, P.A.; Brown, R.L.
(2) Sodium chlorate; NaClO$_3$; [7775-09-9]	*J. Am. Chem. Soc.* 1943, 65, 2456-7.
(3) Potassium chloride; KCl; [7447-40-7]	
(4) Potassium chlorate; KClO$_3$; [3811-04-9]	
(5) Water; H$_2$O; [7732-18-5]	

VARIABLES:	PREPARED BY:
Composition at 273 K and 313 K	Hiroshi Miyamoto and Mark Salomon

EXPERIMENTAL VALUES:

Mass % compositions of saturated solutions at isothermally invariant points

t/°C	chloride	chlorate	sodium	potassium	water	density/g cm^{-3}	solid phase[a]
0	16.54	0.91	8.63	3.99	69.93	1.235	A+B+C
	9.52	19.64	11.39	0.34	59.11	1.349	A+B+D
40	17.17	2.84	7.60	7.34	65.05	1.257	A+B+C
	5.60	33.41	12.02	1.39	47.58	1.450	A+B+D

[a] Solid phases: A = KClO$_3$;   B = NaCl;   C = KCl;   D = NaClO$_3$

AUXILIARY INFORMATION

METHOD/APPARATUS/PROCEDURE:	SOURCE AND PURITY OF MATERIALS:
At 0°C mixts were sealed in Pyrex bottles which were fastened to a rotor suspended in a glycerol/water bath. At 40°C mixts were placed in 250 ml 3-neck flasks and thermostated in a water bath. The slns were stirred with glass stirrers provided with mercury seals.	C.p. grade salts were used without further purification. The chlorates were found to be 99.9 % pure.

Preliminary experiments identified mixtures which result in satd solutions, several of which were used to prepare the solutions reported in the data table above. The compositions of four solutions were reported.

Equilibrated slns were sampled by withdrawing aliquots with pipets fitted with cotton plugs. After determining densities the solutions were diluted for analyses.

Chloride was detd by a modified Volhard method (1), and chlorate detd by the method of Dietz as described in (2). Sodium was detd by pptn with zinc uranyl nitrate, and potassium was detd by calculation based on stoichiometry. Water was found by difference.

ESTIMATED ERROR:
Soly: nothing specified but probably poor due to method of analysis of Na.
Temp: at 0°C, precision is ± 0.1 K.
      at 40°C, precision is ± 0.05 K.

REFERENCES:
1. Caldwell, J.R.; Moyer, H.V. *Ind. Eng. Chem. Anal. Ed.* 1935, 7, 38.
2. Kolthoff, I.M.; Furman, N.H. *Volumetric Analysis*, Vol. II 1929, 388.

COMPONENTS:	ORIGINAL MEASUREMENTS:
(1) Sodium chlorate; $NaClO_3$; [7775-09-9]	Nallet, A.; Paris, R.A.
(2) Sodium chloride; NaCl; [7647-14-5]	*Bull. Soc. Chim. Fr.* <u>1956</u>, 494-7.
(3) Potassium chlorate; $KClO_3$; [3811-04-9]	
(4) Potassium chloride; KCl; [7447-40-7]	
(5) Water; H_2O; [7732-18-5]	

VARIABLES:	PREPARED BY:
T/K = 246.9 to 393	Hiroshi Miyamoto
Composition	

EXPERIMENTAL VALUES:

t/°C	Mole fraction of ion in mixture of anhydrous salt				gH_2O/ mol of mixture of anhydrous salt	Density g cm^{-3}	Nature of the solid phase[a]
	Anion		Cation				
	Cl^-	ClO_c^-	Na^+	K^+			
-26.3	0.583	0.417	0.995	0.0047	142.2	1.313	A+C+D+I
-23.15	0.992	0.0084	0.814	0.1865	173.4	1.208	A+E+C+E
-19.2	0.991	0.0092	0.708	0.292	198	1.192	E+D+I
-19.2	0.990	0.0099	0.807	0.1935	167	1.219	A+E+D
-19.2	0.146	0.854	0.992	0.0078	159.5	1.342	D+C+I
-19.2	0.579	0.421	0.994	0.0064	134	1.323	A+D+C
- 9.8	0.574	0.426	0.991	0.0093	124	1.337	A+C+D
- 9.8	0.986	0.0145	0.804	0.196	154	1.230	A+E+D
- 5.85	0.572	0.428	0.989	0.0108	119.8	1.342	A+B+C+D
- 2.55	0.981	0.0191	0.80	0.20	144.8	1.238	A+B+E+D
+10	0.469	0.531	0.981	0.0187	109.5	1.3731	B+C+D
+10	0.972	0.0283	0.752	0.248	139.3	1.2414	B+D+E
30	0.340	0.660	0.965	0.353	93	1.4231	B+D+C
30	0.949	0.051	0.676	0.324	130	1.2496	B+D+E
50	0.235	0.765	0.942	0.0585	78.7	1.481	B+D+C
50	0.916	0.0842	0.602	0.398	118.2	1.263	B+D+E
70	0.1583	0.842	0.912	0.0885	61.5	1.546	B+D+C
70	0.867	0.1328	0.534	0.466	105.9	1.281	B+D+E
100	0.763	0.237	0.453	0.547	87	1.329	B+D+E
100	0.0904	0.910	0.859	0.141	42.4	1.656	B+E+C

a A = $NaCl.2H_2O$;   B = NaCl;   C = $NaClO_3$;   D = $KClO_3$;   E = KCl;   I = Ice

COMPONENTS:

(1)   Sodium chlorate; $NaClO_3$; [7775-09-9]

(2)   Sodium chloride; NaCl; [7647-14-5]

(3)   Potassium chlorate; $KClO_3$; [3811-04-9]

(4)   Potassium chloride; KCl; [7447-40-7]

(5)   Water; H_2O; [7732-18-5]

ORIGINAL MEASUREMENTS:

Nallet, A.; Paris, R.A.

Bull. Soc. Chim. Fr. 1956, 494-7.

EXPERIMENTAL VALUES:

AUXILIARY INFORMATION

METHOD/APPARATUS/PROCEDURE:

The procedure of equilibration and the method for analysis of the saturated solutions were not described in the original paper, but the compiler assumes that the procedure and the method were similar to that given in ref (1).
See the compilations for the
$NaClO_3$ - $KClO_3$ - H_2O and
$NaClO_3$ - NaCl - H_2O systems for complete descriptions of the experimental method.

SOURCE AND PURITY OF MATERIALS:

The source and purity of materials were not given in the original paper, but probably similar to that described in ref (1).

ESTIMATED ERROR:

Nothing specified.

REFERENCES:

1. Nallet, A.; Paris, R.A.
   Bull. Chem. Soc. Fr. 1956, 488.

COMPONENTS:	ORIGINAL MEASUREMENTS:
(1) Sodium chloride; NaCl; [7647-14-5]	Arkhipov, S.M.; Kashina, N.I.; Kuzina, V.A.
(2) Sodium chlorate; NaClO$_3$; [7775-09-9]	Zh. Neorg. Khim. 1968, 13, 2872-6;
(3) Rubidium chloride; RbCl; [7791-11-9]	Russ. J. Inorg. Chem. (Engl. Transl.)
(4) Rubidium chlorate; RbClO$_3$; [13446-71-4]	1968, 13, 1476-9.
(5) Water; H$_2$O; [7732-18-5]	

VARIABLES:	PREPARED BY:
Composition at 298.2 K	Hiroshi Miyamoto

EXPERIMENTAL VALUES: Composition of saturated solutions[a]

t/°C	NaCl mass %	NaCl mol %	NaClO$_3$ mass %	NaClO$_3$ mol %	RbCl mass %	RbCl mol %	RbClO$_3$ mass %	RbClO$_3$ mol %	Nature of the solid phase
25	2	1.08	-	-	48.20	12.63	0.82	0.15	A+B
	2.25	1.18	-	-	45.80	11.62	0.82	0.15	"
	4.45	2.31	-	-	43.18	10.83	0.84	0.15	"
	7.07	3.62	-	-	40.07	9.908	0.85	0.15	"
	9.74	4.95	-	-	37.33	9.165	0.91	0.16	"
	13.32	6.634	-	-	33.95	8.172	-	-	C+A
	13.59	6.807	-	-	33.12	8.018	0.98	0.17	A+B+C
	13.64	6.862	-	-	33.38	8.116	0.99	0.17	"
	13.52	6.771	-	-	33.19	8.033	0.95	0.16	"
	15.98	7.495	-	-	26.08	5.912	1.15	0.187	C+B
	18.91	8.252	-	-	17.69	3.731	1.37	0.207	"
	21.31	8.886	-	-	10.77	2.171	2.43	0.351	"
	23.69	9.637	-	-	5.19	1.02	3.83	0.539	"
	25.47	10.24	-	-	-	-	6.40	0.890	"
	24.06	9.727	3.49	0.774	-	-	4.58	0.640	"
	22.71	9.308	7.80	1.76	-	-	2.91	0.413	"
	21.00	8.777	11.79	2.706	-	-	2.16	0.312	"
	18.57	8.092	17.80	4.259	-	-	1.82	0.274	"
	15.79	7.195	23.95	5.992	-	-	1.71	0.270	"

continued.....

AUXILIARY INFORMATION

METHOD/APPARATUS/PROCEDURE:

The isothermal method was used. Equilibrium was reached in 30 hours.
Samples of solid and liquid phases were analyzed, rubidium being determined as the tetraphenylborate or when at lower concentration, by flame photometry, and chloride determined argentometrically. Chlorate was found by adding an excess of iron(II) sulfate to an aliquot of saturated solution and back-titrating with potassium permanganate. Sodium was determined by difference. The solid phases were identified by the method of residues and by X-ray diffraction.

SOURCE AND PURITY OF MATERIALS:
The NaCl, NaClO$_3$, RbCl and RbclO$_3$ had a purity of <99.9 %.

ESTIMATED ERROR:
Soly: nothing specified.
Temp: precision ± 0.1 K.

REFERENCES:

COMPONENTS:	ORIGINAL MEASUREMENTS:
(1) Sodium chloride; NaCl; [7647-14-5]	Arkhipov, S.M.; Kashina, N.I.; Kuzina, V.A.
(2) Sodium chlorate; NaClO₃; [7775-09-9]	
(3) Rubidium chloride; RbCl; [7791-11-9]	*Zh. Neorg. Khim.* 1968, *13*, 2872-6;
(4) Rubidium chlorate; RbClO₃; [13446-71-4]	*Russ. J. Inorg. Chem. (Engl. Transl.)*
(5) Water; H₂0; [7732-18-5]	1968, *13*, 1476-9.

EXPERIMENTAL VALUES: (Continued)

Composition of saturated solutions[a]

t/°C	NaCl mass %	NaCl mol %	NaClO₃ mass %	NaClO₃ mol %	RbCl mass %	RbCl mol %	RbCl₃ mass %	RbCl₃ mol %	Nature of the solid phase[b]
	11.82	5.793	34.82	9.370	–	–	–	–	C+D
	11.79	5.887	34.53	9.466	–	–	1.59	0.275	C+D+B
	11.66	5.825	34.61	9.493	–	–	1.65	0.285	"
	11.76	5.899	34.85	9.599	–	–	1.64	0.285	"
	8.71	4.47	39.07	11.02	–	–	1.68	0.299	D+B
	5.66	3.04	44.81	13.22	–	–	1.67	0.310	"
	3.15	1.65	45.27	13.04	–	–	1.65	0.300	"
	---	---	49.44	14.58	–	–	1.70	0.316	"

[a] Mole fractions calculated by the compiler.

[b] A = RbCl;   B = RbClO₃;   C = NaCl;   D = NaClO₃

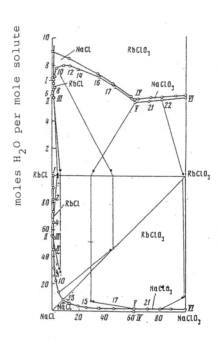

The Na[+], Rb[+] ‖Cl[-],ClO[-] - H₂O system at 25°C.
Circles: composition of liquid phase;
black points: composition of residues;
crosses: composition of soild phases.

COMPONENTS:	ORIGINAL MEASUREMENTS:
(1) Sodium chloride; NaCl; [7647-14-5] (2) Sodium chlorate; NaClO$_3$; [7775-09-9] (3) Cesium chloride; CaCl; [7647-17-8] (4) Cesium chlorate; CsClO$_3$; [13763-67-2] (5) Water; H$_2$O; [7732-18-5]	Arkhipov, S.M.; Kashina, N.I. *Zh. Neorg. Khim.* <u>1970</u>, *15*, 760-4; *Russ. J. Inorg. Chem. (Engl. Transl.)* <u>1970</u>, *15*, 391-2.

VARIABLES:	PREPARED BY:
Composition and temperature T/K = 298.2, 323.2 and 348.2 K	Hiroshi Miyamoto

EXPERIMENTAL VALUES:

The phase diagram is given below, and numerical data follow on the next two pages.

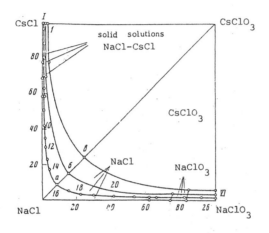

The Cs$^+$, Na$^+$ || Cl$^-$, ClO$_3^-$ – H$_2$O systems at 25°C (a), 50°C (b) and 75°C (c)

continued.....

AUXILIARY INFORMATION

METHOD/APPARATUS/PROCEDURE:	SOURCE AND PURITY OF MATERIALS:
Solubility was investigated by the isothermal method by mixing the solid and liquid phases in glass test-tubes in a water thermostat. Samples of liquid and solid phases were analyzed for the anions and cesium. Chloride was titrated with silver nitrate solution by using potassium chromate as indicator. Chlorate was found by adding excess iron(II) sulfate to an aliquot of saturated solution and back-titrating with potassium permanganate solution. Cesium was determined gravimetrically as cesium tetraphenylborate. Sodium was found by difference. The solid phases were identified by the method of residues, and X-ray diffraction.	"Chemically pure" grade NaCl, NaClO$_3$, CsCl and CsClO$_3$ with a purity of 99.5 % or better were used.
	ESTIMATED ERROR: Soly: nothing specified. Temp: precision ± 0.1 K.
	REFERENCES:

COMPONENTS:	ORIGINAL MEASUREMENTS:
(1) Sodium chloride; NaCl; [7647-14-5]	Arkhipov, S.M.; Kashina, N.I.
(2) Sodium chlorate; $NaClO_3$; [7775-09-9]	*Zh. Neorg. Khim.* <u>1970</u>, *15*, 760-4;
(3) Cesium chloride; CsCl; [7647-17-8]	*Russ. J. Inorg. Chem.* (*Engl. Transl.*)
(4) Cesium chlorate; $CsClO_3$; [13763-67-2]	<u>1970</u>, *15*, 391-2.
(5) Water; H_2O; [7732-18-5]	

EXPERIMENTAL VALUES: (Continued)

t/°C	Sodium Chloride mass %	Sodium Chloride mol %	Sodium Chlorate mass %	Sodium Chlorate mol %	Cesium Chloride mass %	Cesium Chloride mol %	Cesium Chlorate mass %	Cesium Chlorate mol %	Nature of the solid phase [a]
25	--	--	--	--	65.83	17.30	0.54	0.11	A+B
	0.65	0.50	--	--	65.70	17.41	0.54	0.11	B+E
	2.66	1.89	--	--	60.64	14.95	0.65	0.12	"
	3.18	2.26	--	--	60.25	14.85	0.65	0.12	"
	4.75	3.30	--	--	57.88	13.94	0.67	0.13	"
	8.75	5.49	--	--	49.41	10.77	0.80	0.14	"
	9.73	5.97	--	--	47.34	10.09	0.85	0.14	"
	11.10	6.531	--	--	44.72	9.134	--	--	C+E
	11.22	6.695	--	--	44.45	9.208	0.97	0.16	C+B+E
	11.35	6.825	--	--	44.72	9.335	1.04	0.169	"
	13.61	7.708	--	--	39.23	7.713	1.23	0.188	C+B
	15.63	8.098	--	--	31.72	5.705	1.49	0.209	"
	17.32	8.434	--	--	25.81	4.363	1.82	0.239	"
	18.59	8.677	--	--	21.68	3.513	1.90	0.240	"
	20.45	9.097	--	--	15.59	2.408	2.88	0.346	"
	22.03	9.411	--	--	9.76	1.45	4.24	0.489	"
	24.89	10.27	--	--	1.01	0.145	7.81	0.870	"
	24.86	10.24	--	--	--	--	8.71	0.969	"
	23.22	9.545	5.35	1.21	--	--	4.91	0.545	"
	21.39	8.880	9.57	2.18	--	--	3.27	0.367	"
	19.57	8.311	13.38	3.120	--	--	3.01	0.345	"
	17.21	7.771	21.50	5.331	--	--	2.15	0.262	"
	14.15	6.717	28.29	7.373	--	--	1.93	0.247	"
	11.86	5.819	34.87	9.394	--	--	--	--	C+D
	11.75	5.889	34.28	9.433	--	--	2.06	0.279	C+D+B
	11.80	5.914	34.32	9.445	--	--	1.99	0.269	"
	10.54	5.330	35.92	9.973	--	--	2.08	0.284	D+B
	5.16	2.69	42.62	12.18	--	--	1.95	0.274	"
	2.53	1.36	46.84	13.79	--	--	2.02	0.293	"
	--	--	49.60	14.77	--	--	2.13	0.312	"

continued.....

COMPONENTS:	ORIGINAL MEASUREMENTS:
(1) Sodium chloride; NaCl; [7647-14-5]	Arkhipov, S.M.; Kashina, N.I.
(2) Sodium chlorate; NaClO$_3$; [7775-09-9]	Zh. Neorg. Khim. 1970, 15, 760-4;
(3) Cesium chloride; CsCl; [7647-17-8]	Russ. J. Inorg. Chem. (Engl. Transl.)
(4) Cesium chlorate; CsClO$_3$; [13763-67-2]	1970, 15, 391-2.
(5) Water; H$_2$O; [7732-18-5]	

EXPERIMENTAL VALUES: (Continued)

Composition of saturated solutions

t/°C	Sodium Chloride mass %	Sodium Chloride mol %	Sodium Chlorate mass %	Sodium Chlorate mol %	Cesium Chloride mass %	Cesium Chloride mol %	Cesium Chlorate mass %	Cesium Chlorate mol %	Nature of the solid phase[a]
50	--	--	--	--	68.98	19.90	1.43	0.321	A+B
	8.47	5.69	--	--	54.06	12.61	--	--	C+E
	8.23	5.58	--	--	52.73	12.41	1.93	0.354	B+C+E
	8.54	4.65	45.08	13.47	--	--	--	--	C+D
	8.14	4.61	43.37	13.48	--	--	4.24	0.648	C+D+B
	--	--	53.90	17.78	--	--	4.26	0.691	D+B
75	--	--	--	--	68.94	20.57	2.81	0.652	A+B
	5.93	4.56	--	--	62.50	16.68	--	--	C+E
	6.33	4.92	--	--	58.92	15.89	3.63	0.762	C+B+E
	5.97	3.63	54.36	18.14	--	--	--	--	C+D
	5.45	3.59	50.53	18.27	--	--	8.11	1.44	C+D+B
	--	--	57.99	21.97	--	--	7.79	1.45	D+B

a A = CsCl;   B = CsClO$_3$;   C = NaCl;   D = NaClO$_3$;   E = (Na,Cs)Cl solid solution

COMPONENTS:	ORIGINAL MEASUREMENTS:
(1) Sodium chlorate; NaClO₃; [7775-09-9] (2) 1,2-Ethandiol (ethylene glycol); $C_2H_6O_2$; [107-21-1]	Isbin, H.S.; Kobe, K.A. J. Am. Chem. Soc. 1945, 67, 464-5.

VARIABLES:	PREPARED BY:
T/K = 298	Hiroshi Miyamoto

EXPERIMENTAL VALUES:

The solubility of $NaClO_3$ in ethylene glycol at 25°C is

$$16.0g/100g \text{ solvent} \quad \text{(authors)}$$

$$1.50 \text{ mol kg}^{-1} \quad \text{(compiler)}$$

AUXILIARY INFORMATION

METHOD/APPARATUS/PROCEDURE:	SOURCE AND PURITY OF MATERIALS:
The solvent and excess solid were sealed in a soft glass test-tube and rotated for at least one week in a water thermostat at 25°C. All analyses were made on a weight basis by use of weighing pipets. Both the standard gravimetric determination of chloride and the volumetric method, using dichlorofluoresein as an indicator, were used. The chlorate was reduced to the chloride by boiling with excess sulfurous acid.	Technical grade ethylene glycol (Carbide and Carbon Chem. Co) was used, and purified by fractionation. Analytical grade NaClO₃ was used.
	ESTIMATED ERROR: Soly: precision within 0.5 %. Temp: precision ± 0.08 K.
	REFERENCES:

COMPONENTS:	ORIGINAL MEASUREMENTS:
(1) Sodium chlorate; $NaClO_3$; [7775-09-9] (2) 2-Aminoethanol(monoethanolamine) C_2H_7NO; [141-43-5]	Isbin, H.S.; Kobe, K.A. J. Am. Chem. Soc. 1945, 67, 464-5.

VARIABLES:	PREPARED BY:
T/K = 298	Hiroshi Miyamoto

EXPERIMENTAL VALUES:

The solubility of $NaClO_3$ in monoethanolamine at 25°C is

$$19.7g/100g \ solvent \qquad (authors)$$

$$1.85 \ mol \ kg^{-1} \qquad (compiler)$$

AUXILIARY INFORMATION

METHOD/APPARATUS/PROCEDURE:	SOURCE AND PURITY OF MATERIALS:
The solvent and excess solid were sealed in soft glass test-tubes and rotated for at least one week in a water thermostat. All analyses were made on a weight basis by use of weighing pipets. Both the standard gravimetric determination of chloride and the volumetric method, using dichloro-fluoresein as an indicator, were used. The chlorate was reduced to the chloride by boiling with excess sulfurous acid.	Technical grade monoethanolamine (Carbide and Carbon Chem Co) was used, and purified by careful fractionation. Analytical grade $NaClO_3$ was used.
	ESTIMATED ERROR: Soly: precision within 0.5 %. Temp: precision ± 0.08 K.
	REFERENCES:

COMPONENTS:	ORIGINAL MEASUREMENTS:
(1) Sodium chlorate; $NaClO_3$; [7775-09-9] (2) 2-Propanone (acetone); C_3H_6O; [76-64-1]	Miravitlles, Mille L. Ann. Fis. Quim. (Madrid) 1945 41, 120-37.

VARIABLES:	PREPARED BY:
T/K = 288, 293 and 298	H. Herrera

EXPERIMENTAL VALUES:

Solubility[a]

t/°C	mass %	mol kg^{-1}
15	0.1038	0.009762
20	0.0961	0.009037
25	0.0943	0.008868

[a]Molalities calculated by H. Miyamoto

AUXILIARY INFORMATION

METHOD/APPARATUS/PROCEDURE:

Saturated solutions were prepared in an Erlenmeyer flask by mixing the dried acetone with an excess of halate for two hours. The solution was constantly stirred by bubbling dry air (air was dried by passing it through $CaCl_2$ while pumping it into the solution). Air going out from the flask after bubbling in the solution carried some acetone vapor during this operation. The solution temperature was kept constant by immersing the flask in a constant temperature water bath. After two hours, the air exit was closed. The resulting pressure forced the saturated solution from the Erlenmeyer through a tube filled with cotton (which acted as a filter) and was collected in a small flask. This flask was stoppered and weighed. The halate contained in the sample was weighed after complete evaporation of acetone. In all cases, weights were reported to the fourth decimal figure.

SOURCE AND PURITY OF MATERIALS:

Commercial redistilled acetone. This acetone was then dehydrated three times by leaving it in contact with calcium chloride for forty eight hours each time. Fresh $CaCl_2$ was used in each operation. Finally the dehydrated acetone was distilled at 56.3°C.

Source and purity of $NaClO_3$ not specified.

ESTIMATED ERROR:

Nothing specified.

REFERENCES:

COMPONENTS:	ORIGINAL MEASUREMENTS:
(1) Sodium chlorate; $NaClO_3$; [7775-09-9] (2) 1,2-Ethanediamine(ethylene- diamine); $C_2H_8N_2$; [107-15-3]	Isbin, H.S.; Kobe, K.S. J. Am. Chem. Soc. 1945, 67, 464-5.
VARIABLES: T/K = 298	PREPARED BY: Hiroshi Miyamoto

EXPERIMENTAL VALUES:

The solubility of $NaClO_3$ in ethylenediamine at 25°C is

       52.8g/100g solvent   (authors)

       4.96 mol kg^{-1} (compiler)

AUXILIARY INFORMATION

METHOD/APPARATUS/PROCEDURE:	SOURCE AND PURITY OF MATERIALS:
The solvent and excess solid were sealed in soft glass test-tubes and rotated for at least one week in a water thermostat at 25°C. All analyses were made on a weight basis by use of weighing pipets. Both the standard gravimetric determination of chloride and the volumetric method, using dichlorofluoresein as an indicator, were used. The chlorate was reduced to the chloride by boiling with excess sulfurous acid.	Ethylenediamine was dehydrated and purified by the method given in ref. 1. Analytical grade $NaClO_3$ was used.
	ESTIMATED ERROR: Soly: precision within 0.5 %. Temp: precision ± 0.08 K.
	REFERENCES: 1. Putnam, G.L.; Kobe, K.A. Trans. Electrochem. Soc. 1938, 74, 609.

COMPONENTS:	ORIGINAL MEASUREMENTS:
(1) Sodium chlorate; $NaClO_3$; [7775-09-9] (2) Tetrahydrothiophene 1,1-dioxide (sulfolane); $C_4H_8O_2S$; [126-33-0]	Starkovich, J.A.; Janghorbani, M. *J. Inorg. Nucl. Chem.* <u>1972,</u> 34, 789-91.

VARIABLES:	PREPARED BY:
T/K = 313.2	Hiroshi Miyamoto

EXPERIMENTAL VALUES:

The authors reported results for two solubility determinations at 40°C:

$$40 \pm 2 \text{ mmol dm}^{-3}$$

$$33 \pm 2 \text{ mmol dm}^{-3}$$

The mean of the two values is

$$36 \pm 5 \text{ mmol dm}^{-3}$$

AUXILIARY INFORMATION

METHOD/APPARATUS/PROCEDURE:
Excess salt and solvent were sealed in 5 ml glass ampules and equilibrated at 40°C for 250-300 hours with periodic shaking. 1 ml aliquots were transferred in the laboratory atmosphere to 1/2 dram polyethylene vials and sealed. These 1/2 dram vials were then sealed in 2 dram vials, and the chloride content determined by neutron activation (^{38}Cl activity). Each sample was irradiated twice for 30 minutes at neutron fluxes of 2.8×10^{10} and 5.6×10^9 neutrons cm^{-2} sec^{-1}. A calibration plot of ^{38}Cl activity vs chloride concentration was used for the analyses.

After each activation the 1/2 dram vials were placed in new 2 dram vials, and γ-radiation counted in a NaI(Tl) well detector coupled to a 400 channel analyzer. Both the 1.64 and 2.16 MeV peaks were used for the analyses, and were corrected for Comptom scattering and decay. Where interferences were noted, only one γ-ray was used.

SOURCE AND PURITY OF MATERIALS:
Sulfolane (Shell Chemical Co.) was distilled twice under vacuum at temperatures less than 100°C. The purified solvent was found to contain less than 0.02 mass % water by Karl Fischer titration.

Reagent grade $NaClO_3$ was used.

ESTIMATED ERROR:
Soly: precision about $\pm$ 15 % (compiler).
Temp: precision $\pm$ 0.5 K.

REFERENCES:

COMPONENTS:	ORIGINAL MEASUREMENTS:
(1) Sodium chlorate; $NaClO_3$; [7775-09-9] (2) Dimethylformamide; C_3H_7NO; [68-12-2]	Paul, R.C.; Sreenathan, B.R. *Indian J. Chem.* <u>1966</u>, *4*, 382-6.

VARIABLES:	PREPARED BY:
One temperature: 298.2 K	Mark Salomon and Hiroshi Miyamoto

EXPERIMENTAL VALUES:

   The solubility of $NaClO_3$ in $HCON(CH_3)_2$ was reported as

   23.4 g/100 g solvent (2.198 mol kg^{-1}, compiler)

   The solid phase is the anhydrous salt.

AUXILIARY INFORMATION

METHOD/APPARATUS/PROCEDURE:	SOURCE AND PURITY OF MATERIALS:
Isothermal method used. Excess salt and 10 ml of solvent were placed in a Pyrex test tube, sealed, and rotated in a constant temperature water-bath for 24-30 hours. The seal was broken in a dry box and the slurry quickly filtered. The authors state that the metal was estimated from a known quantity of the saturated (filtered) solution, but no details were given. The saturated solution was colorless, and the heat of solution estimated to be less than 10 kcal mol^{-1} (42 kJ mol^{-1}): method used to estimate the heat of solution was not described.	Dimethylformamide (Baker "analyzed" grade) was further purified as described in (1). A.R. grade $NaClO_3$ was warmed and placed under vacuum for 6-8 hours.
	ESTIMATED ERROR: Soly: nothing specified. Temp: precision $\pm$ 0.1 K.
	REFERENCES: 1. Paul, R.C.; Guraya, P.S.; Sreenathan, B.R. *Indian J. Chem.* <u>1963</u>, *1*, 335.

COMPONENTS:	ORIGINAL MEASUREMENTS:
(1) Sodium chlorate; $NaClO_3$; [7775-09-9]	Welsh, T.W.B.; Broderson, H.J.
(2) Hydradzine; N_2H_4; [302-01-2]	J. Am. Chem. Soc. <u>1915</u>, 37, 816-24.

VARIABLES:	PREPARED BY:
Room temperature (Compiler's assumption)	Mark Salomon and Hiroshi Miyamoto

EXPERIMENTAL VALUES:

The solubility of $NaClO_3$ in hydrazine at room temperature was given as:

$$0.66 \text{ g/1 cm}^3 \text{ } N_2H_4$$

The authors stated that the chief object of the research was to obtain qualitative and approximate quantitative data, and the temperature was not kept constant.

AUXILIARY INFORMATION

METHOD/APPARATUS/PROCEDURE:	SOURCE AND PURITY OF MATERIALS:
The solubility vessel was a glass tube to which a U-shaped capillary tube was attached to the bottom. A stopcock at the end of the capillary permitted the adjustment of the rate of flow of dry nitrogen. About 1 cc of anhydrous hydrazine was placed in the tube, and small amounts of $NaClO_3$ added from weighing bottle. After each addition of $NaClO_3$, a loosely fitting cork was placed in the top of the solubility tube. Nitrogen was bubbled through solution until the salt dissolved. The process was repeated until no more salt would dissolve. Temperature was not kept constant. The accuracy in this method is very poor. In addition the authors stated that it was difficult to prevent the oxidation of hydrazine.	Anhydrous hydrazine was prepd by first partially dehydrating commercial hydrazine with sodium hydroxide according to the method of Raschig (1). Further removal of water by distillation from barium oxide after the method of de Bruyn (2). The distillation apparatus employed and the procedure followed in the respective distillation were those described by Welsh (3). The product was found on analysis to contain 99.7 % hydrazine. The hydrazine was stored in 50 cm^3 sealed tubes. Sodium chlorate was the ordinary pure chemical of standard manufacture.

ESTIMATED ERROR:
Soly: accuracy $\pm$ 50 % at best (compilers).

REFERENCES:
1. Raschig, F. Ber. Dtsch. Chem. Ges.<u>1927</u>, 43, 1927.: Hale, C.F.; Shetterly, F.F. J. Am. Chem. Soc. <u>1911</u>, 33, 1071.
2. de Bruyn, L. Rec. Trav. Chim. Pays-Bas. <u>1895</u>, 14, 458.
3. Welsh, T.W. J. Am. Chem. Soc. <u>1915</u>, 37, 497.

COMPONENTS:	EVALUATOR:
(1) Potassium chlorate; $KClO_3$; [3811-04-9] (2) Water; H_2O; [7732-18-5]	H. Miyamoto Department of Chemistry Niigata University Niigata, Japan and M. Salomon US Army ET & DL Fort Monmouth, NJ, USA July, 1984

CRITICAL EVALUATION: THE BINARY SYSTEM

Data for the solubility of potassium chlorate in water have been reported in 23 studies
(1-23).   Five studies (2, 5, 9, 15, 16) deal solely with the binary system, three
studies (3, 7, 11) report solubilities in water-organic solvent mixtures, and the remaining studies deal with multicomponent systems. Most studies are based on isothermal
measurements with chemical analyses either gravimetrically (3, 5, 7, 11) or by titration
of chloride after reduction of the chlorate (6, 11-13, 20-23). One high temperature
study by Benrath et al. (15) used the synthetic method.

Mellor (24) has cited a number of studies which are unavailable to the compilers and
evaluators, and hence have not been included in this volume. The studies cited by
Mellor are: Gay Lussac (25), Mulder (26), Geradin (27), Nordenshjold (28), Schlosing (29),
Blarez (30), Arrhenius (31), Etarde (32) and Calvert (33).

In all studies including those for multicomponent systems, no hydrates of $KClO_3$ were
reported. It is therefore concluded that the anhydrous salt is the solid phase over the
temperature range of 273 - 578 K. A summary of all the binary solubility data is given
in Table 1. In this table we list the solubilities in mole fraction units (calculated
by the evaluators), and the solubilities in mol kg^{-1} units can be found in the compilations. Also included in Table 1 are the weighting factors used in fitting the data to
the smoothing equations. A weight of (1) or (0) was given depending whether the specified result was included or omitted, respectively. We considered giving higher weight
factors to data of higher precision (13, 14, 17-19, 22), but doing so would have resulted
in the rejection of important data of less precision, mainly those of Pawlewsky (1).
Data were rejected from the smoothing equation fit when the difference between the calculated and observed mole fraction solubilities exceeded twice the standard error of
estimate: i.e. when

$$abs \ [\chi_{obsd} - \chi_{calcd}] \ > \ 2\sigma_x$$

Two smoothing equations were used. For mole fractions we used

$$Y_x = A/(T/K) + Bln(T/K) + C + D(T/K) \qquad\qquad [1]$$

and for molalities we used

$$Y_m = A/(T/K) + Bln(T/K) + C \qquad\qquad [2]$$

The complex Y terms in eqs. [1] and [2] are defined in the PREFACE to this volume and
in the $LiClO_3$-H_2O and $RbClO_3$-H_2O critical evaluations. The resulting smoothing equations based on fitting only those solubilities between 273-373 K are given in eqs. [3]
and [4] below. The smoothed solubilities at rounded temperatures are given in Table 2
and are designated as *recommended* values.

$$Y_x = -60986.171/(T/K) - 328.4915ln(T/K) + 1925.070 + 0.4819091(T/K) \qquad [3]$$

$$\sigma_y = 0.023 \qquad\qquad \sigma_x = 0.00023$$

$$Y_m = -6472.065/(T/K) - 11.3302ln(T/K) + 86.2555 \qquad\qquad [4]$$

$$\sigma_y = 0.020 \qquad\qquad \sigma_m = 0.049$$

COMPONENTS:	EVALUATOR:
(1) Potassium chlorate; KClO₃; [3811-04-9] (2) Water; H₂O; [7732-18-5]	H. Miyamoto Department of Chemistry Niigata University Niigata, Japan and M. Salomon US Army ET & DL Fort Monmouth, NJ, USA July, 1984

CRITICAL EVALUATION:

Table 1.   Summary of Experimental Solubilities in the Binary System[a]

T/K	χ	(weight)	ref.	T/K	χ	(weight)	ref.
273.2	0.004594	(1)	1	313.2	0.01902	(0)	12
273.2	0.004603	(1)	12	313.2	0.01919	(1)	1
278.2	0.005584	(1)	1	313.2	0.02007	(1)	3
278.2	0.005810	(1)	19	318.2	0.02153	(0)	1
281.2	0.006543	(1)	5	318.2	0.02318	(1)	13
283.2	0.006499	(0)	1	323.2	0.02482	(0)	12
288.2	0.007803	(0)	1	323.2	0.02571	(1)	1
288.2	0.008769	(1)	13	323.2	0.02692	(0)	20
288.2	0.008871	(1)	9	326.2	0.02998	(0)	2
290.8	0.00953	(1)	11	328.2	0.02894	(1)	1
293.0	0.01040	(1)	5	333.2	0.03328	(1)	1
293.2	0.01034	(0)	16	338.2	0.03613	(1)	1
293.2	0.01053	(1)	1	341.2	0.04263	(0)	2
293.2	0.01048	(1)	7	343.2	0.04111	(1)	1
293.2	0.01053	(1)	8	348.2	0.04625	(1)	1
293.2	0.01055	(1)	10[b]	353.2	0.05149	(1)	1
293.2	0.01058	(1)	12	354.2	0.04328	(0)	2
293.2	0.01060	(1)	9	358.2	0.05730	(1)	1
298.2	0.01187	(0)	1	359.2	0.06050	(0)	2
298.2	0.01219	(1)	6	363.2	0.06348	(1)	1
298.2	0.01240	(1)	22	368.2	0.07024	(1)	1
298.2	0.01244	(1)	17	372.2	0.07769	(0)	5
298.2	0.01245	(1)	13	373.2	0.07656	(0)	7
298.2	0.01246	(1)	14	373.2	0.07548	(0)	1
298.2	0.01247	(1)	21	450.2	0.2152	---	15
298.2	0.01249	(1)	21	468.2	0.2618	---	15
298.2	0.01250	(1)	23	476.2	0.2854	---	15
298.2	0.01261	(1)	20	485.2	0.3153	---	15
298.2	0.01262	(1)	9	495.2	0.3533		15
303.2	0.01343	(0)	1	515.2	0.4266	---	15
303.2	0.01473	(1)	3	550.2	0.5891	---	15
303.2	0.01475	(1)	12	557.2	0.6347	---	15
303.2	0.01487	(1)	5	578.2	0.7659	---	15
308.2	0.01658	(0)	1				

[a]Original units were mainly mass %, and conversions to mole fractions were calculated by the evaluators.

[b]See Volume 14 for the compilations of Mazzetti's paper.

COMPONENTS:	EVALUATOR:
(1) Potassium chlorate; $KClO_3$; [3811-04-9] (2) Water; H_2O; [7732-18-5]	H. Miyamoto Department of Chemistry Niigata University Niigata, Japan and M. Salomon US Army ET & DL Fort Monmouth, NJ, USA July, 1984

CRITICAL EVALUATION:

Table 2.   Recommended Solubilities at Rounded Temperatures Calculated from the Smoothing
            Equations [3] and [4].

T/K	m/mol kg^{-1}	χ
273.2	0.259	0.00457
278.2	0.323	0.00576
283.2	0.397	0.00715
288.2	0.485	0.00873
293.2	0.585	0.01051
298.2	0.699	0.01250
303.2	0.828	0.01470
308.2	0.973	0.01712
313.2	1.134	0.01976
318.2	1.311	0.02265
323.2	1.505	0.02578
328.2	1.716	0.02918
333.2	1.944	0.03287
338.2	2.188	0.03688
343.2	2.449	0.04126
348.2	2.725	0.04604
353.2	3.016	0.05128
358.2	3.322	0.05704
363.2	3.641	0.06341
368.2	3.972	0.07049
373.2	4.314	0.07839

The data of Benrath et al. (15) were also fitted to the smoothing eq. [1]:

$$Y_x = -6275.39/(T/K) - 6.4129\ln(T/K) + 49.9386 + 0.0024454(T/K) \qquad [5]$$

$$\sigma_y = 0.009 \qquad\qquad \sigma_x = 0.003$$

We tried to extrapolate eq. [5] to the melting point of $KClO_3$ (i.e. calculate the temperature for $\chi = 1.00$), but found a value of 608.4 K. The literature value of the melting
point of $KClO_3$ is 641.6 K (45). The solubility at 373.2 K calculated from eq. [5] is
$\chi = 0.0749$ which is much too low (see Table 2), and again suggests caution in using eq.
[5] to compute solubilities outside the range of temperatures used in the least squares
fit to this equation. As indicated in Table 1, the temperature range used to derive eq.
[5] is 450-578 K.

COMPONENTS:	EVALUATOR:
(1) Potassium chlorate; $KClO_3$; [3811-04-9]	Hiroshi Miyamoto Department of Chemistry Niigata University
(2) Water; H_2O; [7732-18-5]	Niigata, Japan
	July, 1984

CRITICAL EVALUATION:

TERNARY SYSTEMS

Many studies for solubilities of potassium chlorate in aqueous ternary systems with saturating inorganic compounds have been reported. Some studies for solubilities in aqueous-
organic solvent mixtures and in the presence of non-saturating components have also been
reported. A summary of studies on aqueous ternary systems is given in Table 3. In
general, the solubility of potassium chlorate is depressed by the addition of other potassium salts or by the addition of other chlorates.

1. One Saturating Component

Bronsted (5) measured solubilities in aqueous potassium hydroxide solutions over the KOH
concentration range from 4.71 to 15.02 mol dm^{-3}. The solubility of potassium chlorate in
aqueous KOH solution decreases with increasing concentration of KOH.

The solubility of potassium chlorate in ethanol-water mixtures has been measured by
Taylor (10) at 303 and 313 K, and by Wright (7) at 293 and 373 K. The solubility in
acetone-water mixtures have been studied by Taylor (3) at 303 and 313 K, and by Hartley
(11) at 298 K. The solubility in glycerol containing low concentrations of water at 292 K
was reported by Holms (33). The solubility of potassium chlorate in these mixtures decreases with increasing concentration of organic solvent.

Schnellbach and Rosin (37) measured the solubility of potassium chlorate in aqueous
glycine solution. The solubility of potassium chlorate in aqueous solution containing
glycine is higher than that of potassium chlorate in water.

2. Two Saturating Components

Systems with halides. The data for the aqueous ternary systems have been reported in 12
publications (4, 8, 10, 12, 14, 20, 23, 39-41, 42, 45) (see Table 3).

Solubilities in the NaCl - $KClO_3$-H_2O system were reported by Di Capua and Scaletti (8)
and Nallet and Paris (42). Only one point at each temperature investigated was reported
in both papers. Above 283 K the composition of the solid phase was pure $KClO_3$ and NaCl.

Solubilities in the ternary $KClO_3$-KCl-H_2O system have been reported in 9 publications
(4, 8, 12, 20, 23, 39-41). Neither double salts nor solid solutions are formed.

Donald (39), Munter and Brown (41), and Nallet and Paris (42) studied the compositions of
the solutions only at the ternary isothermally invariant point. These results and those
of other investigators (8, 12, 20, 23) are summarized in Table 4. Mutter and Brown (41)
stated that the result of Di Capua and Scaletti (8) have mistaken the sharp curvature
of the potassium chlorate solubility isotherm in the range of low potassium chloride concentration as an indication of the location of the invariant point. The evaluators agree
that there is a serious error in (8), and the data for this system have been rejected and
the paper was not compiled.

Benrath and Braun (40) measured solubility in the ternary $KClO_3$-KCl-H_2O system at 473,
and 523 K. The solid phases at these temperatures were similar to those at temperatures
of 373 K or below.

Ricci (14) measured solubilities in the ternary systems $KClO_3$-KBr-H_2O and $KClO_3$-KI-H_2O
at 298 K. The only solid phases in these systems are the pure anhydrous salts.

Systems with other halates. Mazzetti (10) measured solubilities for the ternary
$KClO_3$-Ca(ClO_3)$_2$-H_2O system at 293 K, and Kirgintsev, Kashina, Vulikh and Korotkevich (21)
for the ternary systems $KClO_3$-$RbClO_3$-H_2O and $KClO_3$-$CsClO_3$-H_2O at 298 K. Neither double
salts nor solid solutions form in these systems.

COMPONENTS:	EVALUATOR:
(1) Potassium chlorate; $KClO_3$; [3811-04-9] (2) Water; H_2O; [7732-18-5]	Hiroshi Miyamoto Department of Chemistry Niigata University Niigata, Japan July, 1984

CRITICAL EVALUATION:

Table 3.   A Summary of Solubility Studies in the Ternary Systems with Halides

Ternary system	T/K	Solid phase	Reference
$KClO_3$ + NaCl + H_2O	293	Not given	8
"	251.2, 254.0 263.4	$KClO_3$; $NaCl.2H_2O$; Ice	42
"	269	$KClO_3$; Ice	42
"	271.8	$KClO_3$; $NaCl.2H_2O$; NaCl	42
"	283, 303, 323, 343, 373	$KClO_3$; NaCl	42
$KClO_3$ + KCl + H_2O	293	Not given	4
"	293	Not given	8
"	293	$KClO_3$; KCl	10
"	273, 293, 303 313, 323	$KClO_3$; KCl	12
"	298, 323	$KClO_3$; KCl	20
"	298	$KClO_3$; KCl	23
"	273, 323, 348	Not given	39
"	423, 448, 473	$KClO_3$; KCl	40
"	273	Not given	39
"	262.3, 263.3	$KClO_3$; KCl; Ice	42
"	269	$KClO_3$; Ice	42
"	283, 303, 323, 343, 373	$KClO_3$; KCl	42
"	323	$KClO_3$; KCl	45
$KClO_3$ + KBr + H_2O	298	$KClO_3$; KBr	14
$KClO_3$ + KI + H_2O	298	$KClO_3$; KI	14

Swenson and Ricci (17) studied solubilities in the ternary system $KClO_3$-$KBrO_3$-H_2O at 298 K.   In the system two salts dissolve to a limited extent in each other, forming two solid solutions containing up to 3 % $KClO_3$ in $KBrO_3$ and up to 5% $KBrO_3$ in $KClO_3$.

Karnakhov, Lepeshkov and Fursova (44) measured solubilities in the ternary $KClO_3$-$KClO_4$-H_2O system at 298 K.   Potassium perchlorate has a great tendency to form solid solutions with potassium chlorate.

Table 4.   Composition at Isothermally Invariant Points in the System $KClO_3$-KCl-H_2O

T/K	Composition at isothermally invariant point (mol %)		Reference
	KCl	$KClO_3$	
273	6.135	0.13	12
273	6.204	0.13	41
283	7.004	0.197	42
293	7.430	0.285	12
293	7.58	0.288	39
298	7.901	0.444	20
298	8.025	0.361	23

COMPONENTS:	EVALUATOR:
(1) Potassium chlorate; $KClO_3$; [3811-04-9] (2) Water; H_2O; [7732-18-5]	Hiroshi Miyamoto Department of Chemistry Niigata University Niigata, Japan July, 1984

CRITICAL EVALUATION:

Table 4.   Composition at Isothermally Invariant Point
in the System $KClO_3-KCl-H_2O$ (Continued)

T/K	Composition at isothermall invariant point (mol %)		Reference
	KCl	$KClO_3$	
303	7.959	0.429	26
303	8.14	0.431	42
313	8.281	0.600	12
323	8.800	0.870	12
323	9.204	0.983	20
323	9.14	0.848	39
323	9.21	0.840	42
343	9.94	1.51	42
348	9.99	1.68	42
373	10.8	3.38	42

MULTICOMPONENT SYSTEMS

Although Di Capua and Scaletti (8) studied solubility in the quaternary system $KClO_3-KCl-NaCl-H_2O$ at 298 K, they reported only one value.

Mazzetti (10) studied the $KClO_3-KCl-Ca(ClO_3)_2-H_2O$ system at 293 K, but solubility data were only reported for ternary and quaternary systems. In these systems, the solid phases found were the pure components $KClO_3$, KCl, $Ca(ClO_3)_2 \cdot 2H_2O$, $CaCl_2 \cdot 6H_2O$ and $CaCl_2 \cdot 4H_2O$.

Arkhipov, Kashina and Kuzina (23) studied the $KClO_3-KCl-RbClO_3-RbCl-H_2O$ system at 298 K but only reported solubility data for the quaternary systems $KClO_3-KCl-RbClO_3-H_2O$, $KClO_3-KCl-RbCl-H_2O$ and $KCl-RbCl-RbClO_3-H_2O$, and for the ternary systems $KClO_3-KCl-H_2O$, $KClO_3-RbClO_3-H_2O$ and $RbCl-RbClO_3-H_2O$. In the 5 component system, there were three fields on the solubility diagram for the crystallization of the salts $KClO_3$, $RbClO_3$, and for the solid solution of potassium and rubidium chloride.

Karnaukhov, Lepeshkov and Fursova (44) studied solubilities in the quaternary $KClO_3-KCl-KClO_4-H_2O$ system, and also for the ternary systems $KCl-KClO_4-H_2O$, $KClO_3-KCl-H_2O$ and $KClO_3-KClO_4-H_2O$. Solid phases found are the pure components KCl, $KClO_3$ and $KClO_4$, and the solid solutions $nKCl \cdot mKClO_4$ and $nKClO_4 \cdot mKClO_3$.

COMPONENTS:	EVALUATOR:
(1) Potassium chlorate; $KClO_3$; [3811-04-9]	Hiroshi Miyamoto Department of Chemistry Niigata University Niigata, Japan
(2) Water; H_2O; [7732-18-5]	
	July, 1984

CRITICAL EVALUATION:

REFERENCES:

1. Pawlewsky, B. *Ber. Dtsch, Chem. Ges.* <u>1899</u>, *32*, 1040.

2. Tschungaeff, L.; Chlopin, W. *Z. Anorg. Allg. Chem.* <u>1914</u>, *45*, 154.

3. Tayler, A. E. *J. Phys. Chem.* <u>1897</u>, *1*, 718.

4. Winteler, F. *Z. Electrochem.* <u>1900</u>, *7*, 360.

5. Calzolari, F. *Gazz. Chim. Ital.* <u>1912</u>, *42*, 85.

6. Toda, S. *Nippon Kagaku Kaishi* (*J. Chem. Soc. Jpn.*) <u>1922</u>, *43*, 320; *Coll. Sci. Kyoto Imp. Univ.* <u>1922</u>, 377.

7. Wright, R. *J. Chem. Soc.* <u>1927</u>, 1334.

8. Di Capua, C.; Scaletti, U. *Gazz. Chim. Ital.* <u>1927</u>, *27*, 391.

9. Flottman, F. *Z. Anal. Chem.* <u>1928</u>, *73*, 1.

10. Mazzetti, C. *Ann. Chim. Appl.* <u>1929</u>, *19*, 273. (see Volume 14).

11. Hartley, G. S. *Trans. Faraday Soc.* <u>1931</u>, *27*, 10.

12. Fleck, J. *Bull. Soc. Chem. Fr.* <u>1937</u>, Ser. 5, *4*, 558; *Bull. Soc. Chem. Fr.* <u>1936</u>, Ser. 5, *3*, 350.

13. Ricci, J. E.; Yanick, N. S. *J. Am. Chem. Soc.* <u>1937</u>, *59*, 491.

14. Ricci, J. E. *J. Am. Chem. Soc.* <u>1937</u>, *59*, 866.

15. Benrath, A.; Gjedebo, F.; Schiffer, B.; Wunderlich, H. *Z. Anorg. Allg. Chem.* <u>1937</u>, *231*, 285.

16. Treadweel, W. D.; Ammann, A. *Helv. Chim. Acta* <u>1938</u>, *21*, 1249.

17. Swenton, T.; Ricci, J. E. *J. Am. Chem. Soc.* <u>1939</u>, *61*, 1974.

18. Chang, T. L.; Hsieh, Y. Y. *Sci. Repts. Natl. Tsing Hua Univ.* <u>1948</u>, A5, 252.

19. Noonan, E. C. *J. Am. Chem. Soc.* <u>1948</u>, *70*, 2915.

20. Turnetskaya, A. F.; Lepeshkov, I. *Zh. Neorg. Khim.* <u>1965</u>, *10*, 2163; *Russ. J. Inorg. Chem.* (*Engl. Transl.*) <u>1965</u>, *10*, 1176.

21. Kirgintsev, A. N.; Kashina, N. I.; Vulikh, A. I.; Korotkevich, B. I. *Zh. Neorg. Khim.* <u>1965</u>, *10*, 1225; *Russ. J. Inorg. Chem.* (*Engl. Transl.*) <u>1965</u>, *10*, 662.

22. Kirgintsev, A. N.; Kozitskii, V. P. *Zh. Neorg. Khim.* <u>1968</u>, *13*, 3342; *Russ. J. Inorg. Chem.* (*Engl. Transl.* <u>1968</u>, *13*, 1723.

23. Arkhipov, S. M.; Kashina, N. I.; Kuzina, V. A. *Zh. Neorg. Khim.* <u>1969</u>, *14*, 567; *Russ. J. Inorg. Chem.* (*Engl. Transl.*) <u>1969</u>, *14*, 294.

24. Mellor, J. W. *A Comprehensive Treaties on Inorganic and Theoretical Chemistry* Vol. II Longmans, Green and Co., London, <u>1937</u>, p324-70.

25. Gay Lussac, J. L. *Ann. Chim. Phys.* <u>1819</u>, (2) *11*, 314.

26. Mulder, G. J. *Bijdragen tot de geschieden van het scherkunding gebonder water*, Rotterdam, <u>1864</u>, 143.

COMPONENTS:	EVALUATOR:
(1) Potassium chlorate; $KClO_3$; [3811-04-9] (2) Water; H_2O; [7732-18-5]	Hiroshi Miyamoto Department of Chemistry Niigata University Niigata, Japan July 1984

CRITICAL EVALUATION:

REFERENCES: (Continued)

27. Gerardin, C. A. *Ann. Chim. Phys.* 1865, (4) *5*, 148.

28. Nordenskjold, N. G. *Pogg. Ann.* 1869, *136*, 213.

29. Schlosing, T. *C. R. Hebd. Seances Acad. Sci.* 1871, *73*, 1271.

30. Blarez, C. *C. R. Hebd. Seances Acad. Sci.* 1891, *112*, 1213.

31. Arrhenius, S. *Z. Physik. Chem.* 1893, *11*, 397.

32. Etard, A. *C. R. Hebd. Seances Acad. Sci.* 1899, *108*, 176.

33. Calvert, H. T. *Z. Physik. Chem.* 1901, *38*, 513.

34. Dean, R. B.; Dixon, W. J. *Anal. Chem.* 1951, *23*, 636.

35. Bronsted, J. N. *J. Am. Chem. Soc.* 1920, *42*, 1448.

36. Holm, K. *Pharm. Weekblad* 1921, *58*, 1033; *Pharm. Weekblad* 1921, *58*, 860.

37. Schnellbach, W.; Rosin, J. *J. Am. Pharm. Assoc.* 1931, *20*, 227.

38. Shineider, H. *Z. Anal. Chem.* 1953, *135*, 191.

39. Donald, M. B. *J. Chem. Soc.* 1937, 1325.

40. Benrath, A.; Braun, A. *Z. Anorg. Allg. Chem.* 1940, *244*, 348.

41. Munter, P. A.; Brown, R. L. *J. Am. Chem. Soc.* 1945, *65*, 2456.

42. Nallet, A.; Paris, R. A. *Bull. Soc. Chim. Fr.* 1956, 488.

43. Nallet, A.; Paris, R. A. *Bull. Soc. Chim. Fr.* 1956, 494.

44. Karnakhov, A. S.; Lepeshkov, I. N.; Fursova, A. F. *Zh. Neorg. Khim.* 1969, *14*, 2211; *Russ. J. Inorg. Chem. (Engl. Transl.)* 1969, *14*, 1160.

45. J. A. Dean, Ed. *Lange's Handbook of Chemistry: Twelfth Edition.* McGraw-Hill, NY. 1979.

COMPONENTS:	ORIGINAL MEASUREMENTS:
(1) Potassium chlorate; KClO$_3$; [3811-04-9] (2) Water; H$_2$0; [7732-18-5]	Pawlewski, B. *Ber. Dtsch. Chem. Ges.* 1899, *32*, 1040-1.

VARIABLES:	PREPARED BY:
T/K = 273 to 373	Hiroshi Miyamoto

EXPERIMENTAL VALUES:

Solubility of KClO$_3$ [a]

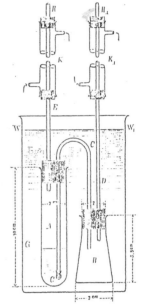

t/°C	mass %	g/100 gH$_2$0	mol kg^{-1}
0	3.06	3.14	0.256
5	3.67	3.82	0.312
10	4.27	4.45	0.363
15	5.11	5.35	0.437
20	6.76	7.22	0.589
25	7.56	8.17	0.667
30	8.46	9.26	0.756
35	10.29	11.47	0.936
40	11.75	13.31	1.086
45	13.16	14.97	1.222
50	15.18	17.95	1.465
55	16.85	20.27	1.654
60	18.97	23.42	1.911
65	20.32	25.50	2.081
70	22.55	29.16	2.379
75	24.82	32.99	2.692
80	26.97	36.93	3.013
85	29.25	41.35	3.374
90	31.36	46.11	3.763
95	33.76	51.39	4.193
100	35.83	55.54	4.532

[a] Molalities calculated by the compiler.

There are a number of inconsistencies between the
experimental g/100 gH$_2$0 solubilities and the author's
calculations of mass %. We assume the author made several mistakes in calculation.

AUXILIARY INFORMATION

METHOD/APPARATUS/PROCEDURE:	SOURCE AND PURITY OF MATERIALS:
The apparatus for the solubility measurement is shown in the Figure above. The water and potassium chlorate were placed in test tube A. The tube A was equipped with a condenser K and a siphon glass tube C, and connected with a weighing bottle B equipped with a condenser K'. The apparatus was placed into a large thermostated glass beaker. To mix the water and potassium chlorate, air was bubbled through the mixture. After equilibrium was established, the saturated solution in the tube A was filtered into the weighing tube B through the siphon tube C equipped with a cotton wool filter. The apparatus was removed from the large beaker, cooled and/or dried, and bottle B weighed. KClO$_3$ was determined gravimetrically after evaporation of the solvent.	No information was given.
	ESTIMATED ERROR:
	Nothing specified.
	REFERENCES:

COMPONENTS:	ORIGINAL MEASUREMENTS:
(1)　Potassium chlorate; $KClO_3$; [3811-04-9] (2)　Water; H_2O; [7732-18-5]	Calzolari, F. *Gazz. Chem. Ital.* 1912, *42*, 85-92.

VARIABLES:	PREPARED BY:
T/K = 281 to 372	B. Scrosati and H. Miyamoto

EXPERIMENTAL VALUES:

Solubility

t/°C	g/100gH$_2$O	mol kg^{-1} (compiler)
8	4.48	0.366
19.8	7.15	0.583
30	10.27	0.838
99	57.3	4.675

AUXILIARY INFORMATION

METHOD/APPARATUS/PROCEDURE:	SOURCE AND PURITY OF MATERIALS:
Method of equilibration not specified, but probably the isothermal method was employed. Aliquots of saturated solution for analysis were withdrawn with a pipet. The aliquots were placed in platinum dishes and the water evaporated. The residues were dried at 120°C to constant weight.	Potassium chlorate was prepared by treating potassium sulfate with barium chlorate. The product was repeatedly recrystallized until no trace of sulfate and barium was detected. The purity of the salt was checked by volumetrically determining chlorine in the anhydrous chloride dried at 150-160°C. The result was not given.

ESTIMATED ERROR:
Not possible to estimate due to insufficient data.

REFERENCES:

COMPONENTS:	ORIGINAL MEASUREMENTS:
(1) Potassium chlorate; KClO₃; [3811-04-9] (2) Water; H₂O; [7732-18-5]	Tschugaeff, L.; Chlopin, W. Z. Anorg. Chem. 1914, 86, 154-62.

VARIABLES:	PREPARED BY:
T/K = 326 to 341	Hiroshi Miyamoto

EXPERIMENTAL VALUES:

Solubility of $KClO_3$[a]

t/°C	mass %	mol kg^{-1}
53	17.37	1.715
68	23.25	2.472
81	28.53[b]	3.258
86[c]	30.46	3.574

[a] Molalities computed by the compiler.

[b] Original value of 23.53 mass % is obviously a typographical error as correct value (28.53 mass %) is given in Figure 4 of the original publication.

[c] Original value of 68°C is obviously a typographical error. Figure 4 shows the correct temperature to be 86°C.

METHOD/APPARATUS/PROCEDURE:

The apparatus used to determine solubilities at high temperatures is shown in Figs. 1 and 2. A saturation vessel A with a condenser C was connected to an aspirator to reduce the pressure. The constancy and the value of the pressure were regulated by a mercury-regulator R.

Very fine crystals of potassium chlorate and water were placed in the vessel A. After reaching a desired pressure by aspirating the system, the vessel A was dipped in an oil-bath whose temperature was kept at a temperature 5-10°C above the boiling point. After the solution boiled and reached saturation, an aliquot for analysis was removed through stopcock C by admitting air through the condenser. The concentration of the solution was determined by evaporation of the solvent or by another method. Details of the other method were not reported.

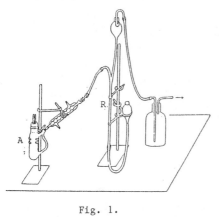

Fig. 1.

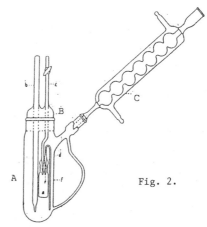

Fig. 2.

SOURCE AND PURITY OF MATERIALS:	ESTIMATED ERROR:
Potassium chlorate was repeatedly recrystallized from distilled water.	Nothing specified.

COMPONENTS:	ORIGINAL MEASUREMENTS:
(1) Potassium chlorate; $KClO_3$; [3811-04-9]	Flottman, F.
(2) Water; H_2O; [7732-18-5]	Z. Anal. Chem. 1928, 73, 1-39.

VARIABLES:	PREPARED BY:
T/K = 288, 293 and 298	Hiroshi Miyamoto

EXPERIMENTAL VALUES: Solubility of potassium chlorate[a]

t/°C	mass %	mol kg^{-1}	density/g cm^{-3}
15	5.7381		
	5.7390		
	(Av)5.739	0.497	1.0363
20	6.7927		
	6.7963		
	6.7907		
	(Av)6.793 (σ=0.003)	0.595	1.0420
25	8.0046		
	8.0055		
	8.0120		
	7.9742		
	(Av)7.999 (σ=0.017)	0.709	1.0484

[a]Molalities and standard deviations calculated by the compiler.

AUXILIARY INFORMATION

METHOD/APPARATUS/PROCEDURE:	SOURCE AND PURITY OF MATERIALS:
An excess of $KClO_3$ and double distilled water were placed into a shaking bottle. The bottle was agitated in a thermostat for about 10 hours. Equilibrium was established from both undersaturation and supersaturation. The saturated solution and solid phase were separated by filtration. Two analytical methods were used to determine the chlorate content in the saturated solution. (1) An aliquot of saturated solution was concentrated by evaporation, and the residue dried at 110°C. (2) The chlorate in an aliquot of saturated solution was reduced to chloride by addition of sulfuric acid. The solution was evaporated and the KCl heated in an open flame to constant weight.	The purest commercial $KClO_3$ (Kahlbaum, Berlin) was dissolved in distilled water, and the solution decantated three times to remove any impurity. The recrystallized $KClO_3$ was used for the solubility determinations.
	ESTIMATED ERROR: Soly: standard deviation is given in the above data table (compiler). Temp: precision $\pm$ 0.02 K (author).
	REFERENCES:

COMPONENTS:	ORIGINAL MEASUREMENTS:
(1) Potassium chlorate; $KClO_3$; [3811-04-9] (2) Water; H_2O; [7732-18-5]	Benrath, A.; Gjedebo, F.; Schiffers, B.; Wunderlich, H. Z. Anorg. Allgem. Chem. 1937, 231, 285-97.
VARIABLES: T/K = 450 to 578	PREPARED BY: Hiroshi Miyamoto

EXPERIMENTAL VALUES:

Solubility of $KClO_3$

t/°C	mass %	mol kg^{-1} (compiler)
177	65.1	15.2
195	70.7	19.7
203	73.1	22.2
212	75.8	25.6
222	78.8	30.3
242	83.5	41.3
277	90.7	79.6
284	92.2	96.5
305	95.7	181

AUXILIARY INFORMATION

METHOD/APPARATUS/PROCEDURE:	SOURCE AND PURITY OF MATERIALS:
Synthetic method used with visual observation of temperatures of crystallization and solubilization (ref 1). The weighed salt and water were placed in a small tube. The tubes were set in an oven equipped with a mica window. A thermometer was immersed in the oven.	No information was given.
	ESTIMATED ERROR: Nothing specified.
	REFERENCES: 1. Janencke, E. Z. Physik. Chem. 1936, A177, 7.

COMPONENTS:	ORIGINAL MEASUREMENTS:
(1) Potassium chlorate; $KClO_3$; [3811-04-9] (2) Water; H_2O; [7732-18-5]	Treadwell, W.D.; Ammann, A. *Helv. Chim. Acta.* 1938, *21*, 1249-56.

VARIABLES:	PREPARED BY:
One temperature; 293 K	Hiroshi Miyamoto

EXPERIMENTAL VALUES:

The solubility of potassium chlorate in water at 20°C was given as:

$$0.58 \text{ mol kg}^{-1}$$

The concentration solubility product was also given simply as the square of the solubility:

$$3.36 \times 10^{-1} \text{ mol}^2 \text{ kg}^{-2}$$

AUXILIARY INFORMATION

METHOD/APPARATUS/PROCEDURE:	SOURCE AND PURITY OF MATERIALS:
No information was given.	No information was given.
	ESTIMATED ERROR: Nothing specified.
	REFERENCES:

COMPONENTS:	ORIGINAL MEASUREMENTS:
(1) Potassium chlorate; $KClO_3$; [3811-04-9] (2) Water-d_2; D_2O; [7789-20-0] (3) Water; H_2O; [7732-18-5]	Noonan, E.C. J. Am. Chem. Soc. 1948, 70, 2915-8.

VARIABLES:	PREPARED BY:
T/K = 278.15	G. Jancso and H. Miyamoto

EXPERIMENTAL VALUES:

Water-d_2 mass %	Sodium Chlorate moles/100 moles of solvent
0	0.5845
91.43	0.5182
100	0.5120[a]

[a] Extrapolated by author.

AUXILIARY INFORMATION

METHOD/APPARATUS/PROCEDURE:	SOURCE AND PURITY OF MATERIALS:
Solubilities were determined by equilibrating solutions with excess salt, evaporating a filtered weighed portion of solution to dryness, and weighing the remaining salt to ± 0.05 mg. Equilibrium was approached from above. The ampules were rotated end over end twelve to forty-eight hours in a water bath. All solubility determinations were performed in duplicate.	C.p. grade potassium chlorate was recrystallized from two to five times. Heavy water was purified by consecutive distillation from alkaline permanganate and then from crystals of potassium dichromate or chromic anhydride. Deuterium content of the heavy water mixture was determined from density measurements.
	ESTIMATED ERROR: Soly: precision better than 0.5 %. Temp: precision ± 0.05 K (author).
	REFERENCES:

COMPONENTS:	ORIGINAL MEASUREMENTS:
(1) Potassium chlorate; $KClO_3$; [3811-04-9] (2) Water-d_2; D_2O; [7789-20-0] (3) Water; H_2O; [7732-18-5]	Chang, T.L.; Hsieh, Y.Y. *Sci. Repts. Natl. Tsing Hua Univ.* <u>1948</u>, A5, 252-9.

VARIABLES:	PREPARED BY:
T/K = 298.15	G. Jancso and H. Miyamoto

EXPERIMENTAL VALUES:

t/°C	Water-d_2 mass %	Potassium Chlorate moles/55.51 moles of H_2O-D_2O mixture
25	0	0.7085 0.707 (Av)0.708
	32.9	0.690 0.690 (Av)0.690
	68.0	0.679 0.678 (Av)0.679
	100	0.662[a]

[a] The solubility in 100 % D_2O was obtained from the solubilities in the H_2O-D_2O mixtures by linear extrapolation.

AUXILIARY INFORMATION

METHOD/APPARATUS/PROCEDURE:	SOURCE AND PURITY OF MATERIALS:
Saturated solutions of $KClO_3$ in the H_2O-D_2O mixture were prepared by the method of supersaturation. The supersaturated solutions were made by agitating the excess salt with the mixture for one hour at 60°C; the time of agitation in the 25°C bath was 2 hours. A sample of the clear solution was delivered into a weighing bottle, then the solvent evaporated and the residual pure salt was dried in vacuum at 100°C and weighed.	Baker's analyzed c.p. grade potassium chlorate was dried over calcium chloride in a desiccator for several days before use. D_2O content of the water mixture was determined by pycnometer both before and after each measurement. The mole percentage was calculated from the specific gravity at 25°C (ref 1).
	ESTIMATED ERROR: Soly: accuracy about 1 % (authors). Temp: precision ± 0.03 K (authors).
	REFERENCES: 1. Swift, E. Jr. *J. Am. Chem. Soc.* <u>1939</u>, *61*, 198.

COMPONENTS:	ORIGINAL MEASUREMENTS:
(1) Sodium chloride; NaCl; [7647-14-5] (2) Potassium chlorate; KClO$_3$; [3811-04-9] (3) Water; H$_2$O; [7732-18-5]	Nallet, A.; Paris, R.A. *Bull. Soc. Chim. Fr.* 1956, 494-7.

VARIABLES:	PREPARED BY:
Composition T/K = 251.2 to 373	Hiroshi Miyamoto

EXPERIMENTAL VALUES:

Composition of saturated solutions

t/°C	Sodium chloride g/100gH$_2$O	Sodium chloride mass % (compiler)	Sodium chloride mol % (compiler)	Potassium chlorate g/100gH$_2$O	Potassium chlorate mass % (compiler)	Potassium chlorate mol % (compiler)	Density g cm^{-3}	Nature of the solid phase[a]
- 22.0	30.3	23.3	8.77	2.63	2.56	0.461	1.205	A+C+I
- 19.2	31.3	23.7	9.00	2.91	2.83	0.512	1.2085	A+C
- 19.2	27.1	21.3	7.93	2.82	2.74	0.486	1.188	C+I
- 9.8	33.4	25.0	9.71	3.93	3.78	0.700	1.221	A+C
- 9.8	14.8	12.9	4.50	3.40	3.29	0.548	1.119	C+I
- 4	5.61	5.31	1.75	3.54	3.42	0.538	1.061	"
- 1.35	35.75	26.34	10.45	5.01	4.77	0.902	1.233	A+B+C
+ 10	36.0	26.5	10.7	6.90	6.45	1.245	1.235	B+C
+ 30	36.3	26.6	11.3	11.52	10.33	2.087	1.2437	"
+ 50	37.0	27.0	12.2	18.33	15.49	3.344	1.260	"
+ 70	37.9	27.5	13.6	28.3	22.1	5.22	1.283	"
+100	39.3	28.2	16.9	51.4	33.9	9.69	1.338	"

a A = NaCl.2H$_2$O; B = NaCl; C = KClO$_3$; I = Ice

METHOD/APPARATUS/PROCEDURE:
The procedure of equilibration and the method for analysis of the saturated solution were not described in the original paper, but the compiler assumes that the procedure and the method were similar to that given in ref (1). See the compilation of this paper for the KCl-KClO$_3$-H$_2$O system.

SOURCE AND PURITY OF MATERIALS:
The source and purity of materials were not reported in the original paper, but probably similar to that described in ref (1).

ESTIMATED ERROR:
Nothing specified.

REFERENCES:
1. Nallet, A.; Paris, R.A.   *Bull. Soc. Chim. Fr.* 1956, 488.

COMPONENTS:	ORIGINAL MEASUREMENTS:
(1) Potassium nitrate; KNO_3; [7757-79-1]	Toda, S.
(2) Potassium chlorate; $KClO_3$; [3811-04-9]	*Nippon Kagaku Kaishi* (*J. Chem. Soc. Japan*), 1922, 43, 320-28; *Mem. Coll. Sci. Kyoto Imp. Univ.* 1922, 377-82.
(3) Water; H_2O; [7732-18-5]	

VARIABLES:	PREPARED BY:
Composition T/K = 298 K	Hiroshi Miyamoto

EXPERIMENTAL VALUES: Composition of saturated solutions

Potassium Chlorate		Potassium Nitrate		Nature of the
mass %	mol % (compiler)	mass %	mol % (compiler)	solid phase[a]
7.745[b]	1.219	–	–	A
7.65	1.21	0.68	0.13	C
7.07	1.12	1.15	0.220	"
6.52	1.05	3.59	0.699	"
5.76	0.949	7.12	1.42	"
5.10	0.881	12.81	2.682	"
4.39	0.800	18.97	4.190	"
3.90	0.771	27.14	6.503	"
3.90	0.771	27.12	6.496	C+B
3.90	0.771	27.14	6.503	"
3.90	0.771	27.16	6.509	"
3.61	0.712	27.21	6.503	B
1.63	0.315	27.57	6.468	"
–	–	27.24	6.254	"

[a] A = $KClO_3$;;   B = KNO_3;   C – Solid solution $K(ClO_3,NO_3)$

[b] For the binary system the compiler computes the following:

soly of $KClO_3$ = 0.6850

AUXILIARY INFORMATION

METHOD/APPARATUS/PROCEDURE:	SOURCE AND PURITY OF MATERIALS:
Mixtures of the salts were placed in Erlenmeyer flasks of capacity about 30 cm^3 with well ground stoppers. Flasks were rotated in a thermostat for about 2 days. When equilibrium was attained, the solutions were permitted to settle in the thermostat and satd solution removed by pipet through a short glass tube with a purified cotton wool filter. Exactly 10 cm^3 of the solution was diluted with about 30 cm^3 of water, treated with 40 cm^3 of 10 % aqueous ferrous fulfate and boiled for about 15 minutes. Aqueous ammonia which was absolutely free from chlorine was added to the solution, and the solution gently boiled until the excess ammonia was expelled. The ppt was filtered and washed 5 times with hot water. The filtrate was used for the determination of chloride by a modified Volhard method (1). Potassium was detd as the sulfate by sulfuric acid as described in (2). The composition of the solid phase was also determined by the same method.	Both potassium chlorate and nitrate (Japan Pharmacopeia) were recrystallized three times. Distilled water was used.
	ESTIMATED ERROR: Nothing specified.

	REFERENCES:
	1. Rothumund, V.; Burgstaller, A.
	   Z. Anorg. Chem. 1909, 63, 330.
	2. Treadwell, F.P.; Hall, W.T.
	   Analytical Chemistry Vol II.
	   1915. p41.

COMPONENTS:	ORIGINAL MEASUREMENTS:
(1) Potassium sulfate; K_2SO_4; [7778-80-5]	Ricci, J.E.; Yanick, N.S.
(2) Potassium chlorate; $KClO_3$; [3811-04-9]	J. Am. Chem. Soc. 1937, 59, 491-6.
(3) Water; H_2O; [7732-18-5]	

VARIABLES:	PREPARED BY:
Composition and temperature T/K = 288.15, 298.15, 318.15	Hiroshi Miyamoto

EXPERIMENTAL VALUES: Composition of saturated solutions

t/°C	KClO₃ mass %	mol % (compiler)	K₂SO₄ mass %	mol % (compiler)	Density g cm⁻³	Nature of the solid phase[a]
15	0.00	0.00	9.258	1.044	1.076	A
	3.29	0.537	7.86	0.901	1.085	A+B
	3.29	0.537	7.86	0.901	1.084	"
	3.29	0.537	7.86	0.901	1.085	"
	5.676[b]	0.877	0.00	0.000	1.032	B
25	0.00	0.00	10.76	1.231	1.083	A
	1.80	0.295	9.93	1.15	1.089	"
	3.30	0.547	9.43	1.10	1.099	"
	4.95	0.827	8.66	1.02	1.102	A+B
	4.96	0.828	8.62	1.01	1.100	"
	4.96	0.828	8.62	1.01	1.099	"
	4.96	0.828	8.64	1.01	1.100	"
	5.06	0.842	8.19	0.958	1.099	B
	5.77	0.942	5.57	0.639	1.080	"
	6.72	1.08	2.73	0.307	1.063	"
	7.897[b]	1.245	0.00	0.00	1.048	"
45	0.00	0.00	13.53	1.592		A
	9.80	1.73	9.13	1.13		A+B
	9.80	1.73	9.12	1.13		"
	9.80	1.73	9.13	1.13		"
	13.90[b]	2.318	0.00	0.00		B

AUXILIARY INFORMATION

METHOD/APPARATUS/PROCEDURE:	SOURCE AND PURITY OF MATERIALS:
Weighed mixtures of known composition were brought to equilibrium by stirring at the desired temperature. The time required for attainment of equilibrium was determined by analysis, and required several days. The order of mixing of the components, and the process of seeding or inoculation for required phases had to be varied in accordance with the phase sought. In one sample of the saturated solution, chlorate was determined by the method of Peters and Deutshlander (1): to the chlorate sample (containing about 0.11g of ClO_3^-) is added a definite volume (50 cm³) of 0.05 mol dm⁻³ arsenious oxide solution. After the addition of a trace of KBr, the solution is acidified strongly with HCl and boiled for ten minutes. The excess arsenious oxide is then titrated by means of 0.033 mol dm⁻³ KBrO₃ using indigo sulfonic acid indicator. continued.....	Nothing specified.

ESTIMATED ERROR:
Soly: nothing specified.
Temp: precision ± 0.02 K.

REFERENCES:

1. Kolthoff, I.M.; Furman, N.H. Volumetric Analysis, Vol. 2 John Wiley and Sons. New York. 1929. p 465.

COMPONENTS:	ORIGINAL MEASUREMENTS:
(1) Potassium sulfate; K_2SO_4; [7778-80-5]	Ricci, J.E.; Yanick, N.S.
(2) Potassium chlorate; $KClO_3$; [3811-04-9]	J. Am. Chem. Soc. 1937, 59, 491-6.
(3) Water; H_2O; [7732-18-5]	

EXPERIMENTAL VALUES: (Continued)

[a] $A = K_2SO_4$; $B = KClO_3$

[b] For the binary system the compiler computes the following:

$$\text{soly of } KClO_3 = 0.4910 \text{ mol kg}^{-1} \text{ at } 15°C$$

$$= 0.6996 \text{ mol kg}^{-1} \text{ at } 25°C$$

$$= 1.317 \text{ mol kg-1 at } 45°C$$

METHOD/APPARATUS/PROCEDURE: (Continued)

In other samples the total solid was determined by evaporation to dryness at 100°C followed by 250°C, and the sulfate was then calculated by difference.
For the identification of known solid phases, microscopic examination and algebraic extrapolation of tie-lines sufficed.
The densities reported for some of the isotherms were obtained by means of volumetric pipets calibrated for delivery.

COMMENTS AND/OR ADDITIONAL DATA:

The phase diagram is given below (based on mass % units).

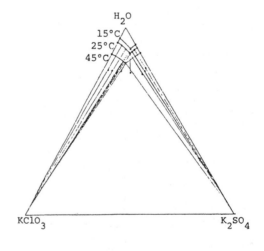

COMPONENTS:	ORIGINAL MEASUREMENTS:
(1) Potassium chlorate; $KClO_3$; [3811-04-9]	Winteler, F.
(2) Potassium chloride; KCl; [7446-40-7]	Z. Electrochem. 1900, 7, 360-2.
(3) Water; H_2O; [7732-18-5]	

VARIABLES:	PREPARED BY:
T/K = 293 Concentration of KCl (see comments below)	Hiroshi Miyamoto and Mark Salomon

EXPERIMENTAL VALUES: Composition of saturated solutions at 20°C

concn KCl		soly $KClO_3$		Density
g dm^{-3}	c_2/mol dm^{-3} (compiler)	g dm^{-3}	c_1/mol dm^{-3} (compiler)	g cm^{-3}
0	0	71.1	0.580	1.050
10	0.134	58	0.47	1.050
20	0.268	49	0.40	1.050
30	0.402	43	0.35	1.050
40	0.537	39.5	0.322	1.054
50	0.671	36.5	0.298	1.058
60	0.804	34	0.28	1.064
70	0.939	32	0.26	1.070
80	1.07	30	0.24	1.075
90	1.21	28	0.23	1.081
100	1.34	27	0.22	1.086
110	1.48	25.5	0.208	1.091
120	1.61	24.5	0.200	1.098
130	1.74	23.5	0.192	1.103
140	1.88	22.5	0.184	1.108
150	2.01	21.5	0.175	1.113
160	2.15	21.0	0.171	1.119
170	2.28	20.5	0.167	1.124
180	2.41	20.0	0.163	1.130
190	2.55	20.0	0.163	1.135
200	2.68	20	0.16	1.140
210	2.82	20	0.16	1.145
220	2.95	20	0.16	1.150
230	3.09	20	0.16	1.156
240	3.22	20	0.16	1.161
250	3.35	20	0.16	1.168

The composition of the solid phase is not given in the original paper.

METHOD/APPARATUS/PROCEDURE:	SOURCE AND PURITY OF MATERIALS:
Mixtures of salts and water were placed into a thermostat at 20°C for several days and shaken frequently. Aliquots of the saturated solution were acidified with nitric acid and then titrated with silver nitrate using potassium chromate as an indicator. The compiler assumes that the total salt concentration of the solution was determined gravimetrically, and that the chlorate content was determined by difference. It appears that the concentrations of KCl given in the above data are initial concentrations (compilers).	No information was given.
	ESTIMATED ERROR: Nothing specified.
	REFERENCES:

COMPONENTS:	ORIGINAL MEASUREMENTS:
(1) Potassium chloride; KCl; [7447-40-7] (2) Potassium chlorate; $KClO_3$; [3811-04-9] (3) Water; H_2O; [7732-18-5]	Di Capua, C.; Scaletti, U. *Gazz. Chim. Ital.* 1927, *27*, 391-9.

VARIABLES:	PREPARED BY:
T/K = 293	B. Scrosati and H. Miyamoto

EXPERIMENTAL VALUES:

Composition of saturated solutions at 20°C[a]

	KCl		$KClO_3$	
mass %	mol % (compiler)		mass %	mol % (compiler)
0	0		6.75[b]	1.05
1	0.3		6	0.9
2	0.5		5	0.8
3	0.8		4	0.6
5	1		3.2	0.50
6.5	1.7		3	0.5
9	2		2.5	0.40
12	3.3		2.2	0.36
15	4.2		2	0.3
19	5.5		1.85	0.324
22	6.5		1.5	0.27
25	7.5		1.2	0.22
26.08	7.856		0	0

[a] Nature of solid phases not reported.

[b] For the binary system the compiler computes the following:

$$\text{soly of } KClO_3 = 0.591 \text{ mol kg}^{-1}$$

AUXILIARY INFORMATION

METHOD/APPARATUS/PROCEDURE:	SOURCE AND PURITY OF MATERIALS:
The mixtures of salts and water were stirred in a thermostat for 7 days. Samples of saturated solutions were withdrawn with a pipet and weighed. The chlorate ion concentration was determined by the Volhard method after reduction to chloride ion with zinc and acetic acid. Probably, the potassium content was determined by precipitation as the triple acetate of potassium, uranyl and magnesium, according to the method described by Kling and Lasieur (ref 1).	No information given.
	ESTIMATED ERROR: Nothing specified.
	REFERENCES: 1. Kling and Lasieur. *Giorn. Chim. Ind. Applicata* 1925, *7*.

COMPONENTS:	ORIGINAL MEASUREMENTS:
(1) Potassium chloride; KCl; [7447-40-7]	Fleck, J.
(2) Potassium chlorate; KClO$_3$; [3811-04-9]	*Bull. Soc. Chem. Fr.* <u>1937</u>, *Ser. 5,* *4,* 558-60 (*see also Bull. Soc. Chem. Fr.* <u>1936</u>, *Ser. 5, 3,* 350).
(3) Water; H$_2$O; [7732-18-5]	

VARIABLES:	PREPARED BY:
Composition T/K = 273.2 to 323.2	Hiroshi Miyamoto

EXPERIMENTAL VALUES: Composition of saturated solutions

t/°C	KCl mass %	KCl mol % (compiler)	KClO$_3$ mass %	KClO$_3$ mol % (compiler)	Density g cm^{-3}	Nature of the solid phase[a]
0	0.00	0.00	3.05[b]	0.460	1.022	A
	8.47	2.21	1.09	0.173	1.068	"
	16.21	4.502	0.82	0.14	1.121	"
	21.16	6.135	0.71	0.13	1.157	A+B
	21.90	6.346	0.00	0.00	1.153	B
20	0.00	0.00	6.78[b]	1.06	1.044	A
	2.84	0.735	5.18	0.815	1.051	"
	7.44	1.97	3.74	0.603	1.070	"
	13.47	3.711	2.44	0.409	1.106	"
	20.26	5.888	1.75	0.309	1.153	"
	24.58	7.421	1.55	0.285	1.184	A+B
	24.63	7.440	1.55	0.285	1.185	"
	25.17	7.588	0.89	0.16	1.183	B
	25.70	7.714	0.00	0.00	1.176	"
30	0.00	0.00	9.24[b]	1.47	1.058	A
	4.67	1.24	6.62	1.07	-	"
	9.45	2.57	4.82	0.799	1.088	"
	11.03	3.032	4.38	0.733	1.097	"
	13.06	3.634	3.82	0.647	1.108	"
	17.86	5.151	3.12	0.547	1.140	"

continued.....

AUXILIARY INFORMATION

METHOD/APPARATUS/PROCEDURE:	SOURCE AND PURITY OF MATERIALS:
The system was studied by the isothermal method. The KClO$_3$ was added to the solution, saturated with KCl and stirred to establish equilibrium. The chloride content was determined by Volhard's method. For determination of chlorate, a weighed amount of saturated solution was added to excess FeSO$_4$ solution and titrated with permanganate solution. The densities were also determined.	Potassium chlorate and chloride were purchased from Poulence. No other information was given in the paper.

	ESTIMATED ERROR:
	Soly: nothing specified. Temp: precision ± 0.1 K.

	REFERENCES:

COMPONENTS:	ORIGINAL MEASUREMENTS:
(1) Potassium chloride; KCl; [7447-40-7]	Fleck, J.
(2) Potassium chlorate; $KClO_3$; [3811-04-9]	*Bull. Soc. Chem. Fr.* 1937, *Ser.* 5, 4, 558-60 (*see also Bull. Soc. Chem. Fr.* 1936, *Ser.* 5, 3, 350).
(3) Water; H_2O; [7732-18-5]	

EXPERIMENTAL VALUES: (Continued)

Composition of saturated solutions

t/°C	KCl mass %	KCl mol % (compiler)	$KClO_3$ mass %	$KClO_3$ mol % (compiler)	Density g cm^{-3}	Nature of the solid phase[a]
30	25.81	7.948	2.29	0.429	1.198	A+B
	25.86	7.967	2.29	0.429	1.197	"
	26.64	8.143	0.87	0.16	1.190	B
	27.30	8.319	0.00	0.00	1.182	"
40	0.00	0.00	11.65[b]	1.902	1.074	A
	4.60	1.25	9.20	1.53	1.084	"
	7.66	2.11	7.64	1.28	1.092	"
	8.71	2.41	7.25	1.22	1.098	"
	10.76	3.010	6.39	1.09	1.106	"
	13.16	3.725	5.44	0.937	1.116	"
	18.43	5.408	4.30	0.768	1.148	"
	20.66	6.162	3.88	0.704	1.165	"
	26.49	8.289	3.15	0.600	1.206	A+B
	26.45	8.273	3.14	0.597	1.206	"
	27.74	8.633	1.54	0.292	1.196	B
	28.75	8.884	0.00	0.00	1.188	"
50	0.00	0.00	14.76[b]	2.482	1.088	A
	8.51	2.41	9.66	1.66	1.105	"
	17.55	5.199	6.01	1.08	1.147	"
	18.53	5.526	5.76	1.04	1.155	"
	27.45	8.800	4.46	0.870	1.214	A+B
	28.24	9.003	3.27	0.634	1.207	B
	30.18	9.457	0.00	0.00	1.194	B

[a] $A = KClO_3$ $B = KCl$

[b] For the binary system the compiler computes the following:

soly of $KClO_3$ = 0.257 mol kg^{-1} at 0°C

= 0.593 mol kg^{-1} at 20°C

= 0.831 mol kg^{-1} at 30°C

= 1.076 mol kg^{-1} at 40°C

= 1.413 mol kg^{-1} at 50°C

COMPONENTS:	ORIGINAL MEASUREMENTS:
(1) Potassium chloride; KCl; [7447-40-7] (2) Potassium chlorate; KClO$_3$; [3811-04-9] (3) Water; H$_2$O; [7732-18-5]	Donald, M.B. *J. Chem. Soc.* 1937, 1325-6.

VARIABLES:	PREPARED BY:
T/K = 293, 323 and 348	Hiroshi Miyamoto

EXPERIMENTAL VALUES:

Composition of saturated solutions at invariant points[a]

| t/°C | KCl | | KClO$_3$ | | Density |
	mass %	mol % (compiler)	mass %	mol % (compiler)	g cm^{-3}
20	25.0	7.58	1.56	0.288	1.177
50	28.3	9.14	4.32	0.648	1.211
75	29.3	9.99	8.08	1.68	1.244

[a] Nature of the solid phases not specified.

AUXILIARY INFORMATION

METHOD/APPARATUS/PROCEDURE:	SOURCE AND PURITY OF MATERIALS:
The objective of this study was to determine the solubilities at the invariant points. The chloride was estimated by Mohr's method, the chlorate by Rupp's method (1), and the water by direct weighing. No other information is given in the original paper.	Nothing specified.
	ESTIMATED ERROR: Nothing specified.
	REFERENCES: 1. Rupp, E. Z. *Anal. Chem.* 1917, 56, 580.

COMPONENTS:	ORIGINAL MEASUREMENTS:
(1) Potassium chloride; KCl; [7447-40-7]	Benrath, A.; Braun, A.
(2) Potassium chlorate; KClO$_3$; [3811-04-9]	Z. Anorg. Allg. Chem. 1940, 244, 348-58.
(3) Water; H$_2$O; [7732-18-5]	

VARIABLES:	PREPARED BY:
T/K = 423, 448 and 473 Composition	Hiroshi Miyamoto and Mark Salomon

EXPERIMENTAL VALUES: Composition of saturated solutions

t/°C	Potassium Chloride mass %	mol % (compiler)	Potassium Chlorate mass %	mol % (compiler)	Nature of the solid phase
150	34.7	12.9	10.40	2.359	KCl
	34.4	12.9	10.88	2.475	"
	31.7	12.6	17.80	4.305	"
	28.5	11.8	23.7	5.99	"
	26.0	11.4	29.5	7.87	"
	18.22	8.175	37.9	10.3	KClO$_3$
	16.90	7.562	38.9	10.6	"
	4.97	2.26	50.6	14.0	"
175	35.5	13.9	13.14	3.122	KCl
	32.3	13.3	19.95	5.014	"
	26.6	12.2	32.0	8.95	"
	20.5	10.7	44.9	14.3	KCl + KClO$_3$
	20.4	10.9	46.0	14.9	"
	20.0	10.7	46.3	15.0	"
	9.28	4.95	55.9	18.1	KClO$_3$
200	38.5	15.4	12.3	3.00	KCl
	35.6	14.7	16.8	4.21	"
	25.2	12.3	36.8	10.9	"
	21.6	11.7	45.9	15.2	"
	14.55	8.903	58.0	21.6	"
	5.76	3.59	66.6	25.2	KCl + KClO$_3$
	3.29	2.08	69.4	26.6	KClO$_3$

AUXILIARY INFORMATION

METHOD/APPARATUS/PROCEDURE:	SOURCE AND PURITY OF MATERIALS:
a 40-50 mg mixture of KCl + KClO$_3$ and water was placed in a 5 cm long glass tube (inner diam = 1.5 mm) and sealed. The tube was heated in a vertical position at the desired temperature.: the method of ascertaining equilibrium not specified. The tube was then rotated permitting the solution to flow to one end and due to the small diameter of the tube, the solids remained in the other end of the tube. The tube was cooled and broken just below the solid residues. Each part of the tube was weighed and dried by heating to dryness. The dried products were weighed and dissolved in nitric acid, and the chloride content determined gravimetrically by precipitation with silver nitrate.	Nothing specified.
	ESTIMATED ERROR:
	Nothing specified.
	COMMENTS AND/OR ADDITIONAL DATA:
	Using literature values for solubilities and melting points of 2 component and 1 component systems, the author prepared Janecke phase diagrams and polytherms. From these diagrams the azeotrope points were determined. The phase diagrams are reproduced on the following page. In these diagrams, m is the moles of water per mole of KCl + KClO$_3$, and x is the mole fraction of KCl in the total KCl + KClO$_3$ content.

continued.....

COMPONENTS:	ORIGINAL MEASUREMENTS:
(1) Potassium chloride; KCl; [7447-40-7]	Benrath, A.; Braun, A.
(2) Potassium chlorate; KClO₃; [3811-04-9]	Z. Anorg. Allg. Chem. 1940, 244, 348-58.
(3) Water; H₂O; [7732-18-5]	

COMMENTS AND/OR ADDITIONAL DATA: (Continued)

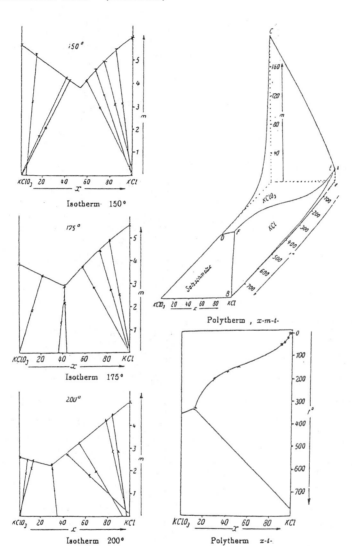

Isotherm 150°

Isotherm 175°

Isotherm 200°

Polytherm, x-m-t-

Polytherm x-t-

point on polytherm	t/°C	mass % KCl	mass % KClO₃	x	m	phases present
E	-11.8	19.8	0.42	98.6	6.9	sln, ice, vapor, KCl, KClO₃
A	-10.7	19.82	--	100	17.02	sln, ice, vapor, KCl
C	-0.8	--	2.97	0	220.8	sln, ice, vapor, KClO₃
D	356	--	--	0	0	m.p. of KClO₃
F	336	--	--	12.0	0	m.p. of KCl-KClO₃ eutectic
B	771	--	--	100	0	m.p. of KCl

COMPONENTS:	ORIGINAL MEASUREMENTS:
(1) Potassium chloride; KCl; [7447-40-7]	Nallet, A.; Paris, R.A.
(2) Potassium chlorate; $KClO_3$; [3811-04-9]	*Bull. Soc. Chim. Fr.* 1956, 488-94.
(3) Water; H_2O; [7732-18-5]	

VARIABLES:	PREPARED BY:
Composition	Hiroshi Miyamoto
T/K = 262.30 to 373	

EXPERIMENTAL VALUES:

Composition of saturated solutions

t/°C	Potassium chloride g/100gH2O	Potassium chloride mass %	Potassium chloride mol % (compiler)	Potassium chlorate g/100gH2O	Potassium chlorate mass %	Potassium chlorate mol % (compiler)	Density g cm^{-3}	Nature of the solid phase
-10.85	24.2	19.5	5.55	0.53	0.53	0.11	1.139	A+B+I
- 9.8	24.7	19.8	5.66	0.56	0.56	0.11	1.43	A+B
- 9.8	22.05	18.07	5.086	0.565	0.562	0.111	1.1295	B+I
- 4	8.93	8.20	2.13	0.968	0.959	0.175	1.062	"
+ 10	31.23	23.80	7.120	1.44	1.42	0.297	1.1709	A+B
+ 30	37.0	27.0	8.48	3.21	3.11	0.634	1.1937	"
+ 50	42.3	29.7	9.91	6.35	5.97	1.39	1.215	"
+ 70	46.5	31.7	11.4	11.6	10.4	2.614	1.240	"
+100	51.8	34.1	14.5	25.9	20.6	6.105	1.280	"

a A = KCl;   B = $KClO_3$;   I = Ice.

METHOD/APPARATUS/PROCEDURE:

Mixtures of salts and water were placed in bottles and shaken in a thermostat for 2 hours at 100°C and for 2 hours or more at lower temperatures. Equilibrium was approached from supersaturation.
The chloride ion concentration was determined by a potentiometric method using silver nitrate solution. After the determination of chloride, the chlorate was reduced with Mohr's salt in mineral acids, and the excess Fe(II) titrated with potassium dichromate solution. The analyses of cations were performed in duplicate. The potassium content was determined by flame photometry and gravimetry.
The nature of the solid phase was determined by Schreinemakers' residues method.
The densities of the saturated solution were also measured.

SOURCE AND PURITY OF MATERIALS:

Potassium chlorate and chloride were recrystallized twice. The purity of the salts was 99.9 %.

ESTIMATED ERROR:
Soly: precision 0.5 % (compiler).
Temp: nothing specified.

REFERENCES:

COMPONENTS:	ORIGINAL MEASUREMENTS:
(1) Potassium chloride; KCl; [7447-40-7]	Munter, P.A.; Brown, R.L.
(2) Potassium chlorate; KClO$_3$; [3811-04-9]	J. Am. Chem. Soc. 1943, 65, 2456-7.
(3) Water; H$_2$O; [7732-18-5]	

VARIABLES:	PREPARED BY:
Composition T/K = 273.2	Hiroshi Miyamoto

EXPERIMENTAL VALUES:

Composition at the isothermally invariant point at 0.0°C

Potassium chloride		Potassium chlorate		Water	
mass %	mol % (compiler)	mass %	mol % (compiler)	mass %	mol % (compiler)
21.36	6.204	0.71	0.13	77.93	93.67

AUXILIARY INFORMATION

METHOD/APPARATUS/PROCEDURE:
Mixtures of the solid salts and water sealed in Pyrex bottles were fastened to a rotor suspended in a constant temperature bath. An aq glycerol solution was used as the bath liquid. A preliminary experiment was carried out in which an original mixt of the solid salts and water was gradually augmented by small additions of salt until the density and composition of the resultant solution became constant. From these data, mixtures of the solid salts and water known to result in satd solutions were prepd for the final test. The equilibrated solutions were sampled by withdrawing clear supernatant solution through cotton filtering plugs directly into a density pipet. After determining the density, the samples were diluted. The chloride content was detd by the Volhard method as modified by Caldwell and Moyer (ref 1). The chlorate content was detd by the method of Dietz as described by Kolthoff and Furman (ref 2). The water content was found by difference.

SOURCE AND PURITY OF MATERIALS:
All the salts used were of c.p. grade and were used without further purification. The chlorate was found to be average 99.9 % pure.

ESTIMATED ERROR:
Soly: nothing specified.
Temp: precision ± 0.1 K (authors).

REFERENCES:
1. Caldwell, J.R.; Moyer, H.V. Ind. Eng. Chem. Anal. Ed. 1935, 7, 38.
2. Kolthoff, I.M.; Furman, N.H. Volumetric Analysis Vol. II, 1929, 388.

COMPONENTS:	ORIGINAL MEASUREMENTS:
(1) Potassium chloride; KCl; [7447-40-7] (2) Potassium chlorate; KClO₃; [3811-04-9] (3) Water; H₂0; [7732-18-5]	Turnetskaya, A.F.; Lepeshkov, I.N. *Zh. Neorg. Khim.* 1965, *10*, 2163-6; *Russ. J. Inorg. Chem. (Engl. Transl.)* 1965, *10*, 1176-8.

VARIABLES:	PREPARED BY:
T/K = 298 and 323 Composition	Hiroshi Miyamoto

EXPERIMENTAL VALUES: Composition of saturated solutions

t/°C	Potassium chloride mass %	mol % (compiler)	Potassium clorate mass %	mol % (compiler)	Density g cm⁻³	Nature of the solid phase[a]
25	26.33	7.950	0.00	0.00	1.1798	A
	26.26	7.949	0.30	0.055	1.1877	"
	25.91	7.935	1.7	0.32	1.189	"
	25.85	7.968	2.35	0.441	–	A+B
	25.84	7.962	2.32	0.435	–	"
	25.59	7.871	2.38	0.445	–	"
	25.63	7.891	2.43	0.455	–	"
	25.56	7.861	2.40	0.449	1.189	"
	25.78	7.950	2.45	0.460	–	"
	25.58	7.871	2.42	0.453	–	"
	25.54	7.853	2.39	0.447	1.1732	"
	21.11	6.228	2.44	0.438	1.1567	B
	17.27	4.933	2.7	0.47	1.1348	"
	11.95	3.280	3.4	0.57	–	"
	5.17	1.36	4.9	0.78	1.0645	"
	1.34	0.344	5.83	0.912	1.058	"
	0.00	0.00	7.99[b]	1.26	1.0568	"

continued.....

AUXILIARY INFORMATION

METHOD/APPARATUS/PROCEDURE:	SOURCE AND PURITY OF MATERIALS:
The isothermal method was used. At 25°C equilibrium in the system was reached after 4-5 days. The potassium was determined gravimetrically with sodium tetraphenyl-borate. At high concentrations, chloride was determined volumetrically by mercuri-metric method, and at low concentrations chloride was detd gravimetrically. ClO₃⁻ was determined volumetrically after reduction to chloride with zinc dust.	KClO₃ and KCl were recrystallized twice and had a purity of 99.7 - 99.8 %.
	ESTIMATED ERROR: Nothing specified.
	REFERENCES:

COMPONENTS:	ORIGINAL MEASUREMENTS:
(1) Potassium chloride; KCl; [7447-40-7] (2) Potassium chlorate; KClO$_3$; [3811-04-9] (3) Water; H$_2$O; [7732-18-5]	Turnetskaya, A.F.; Lepeshkov, I.N. *Zh. Neorg. Khim.* <u>1965</u>, *10*, 2163-6; *Russ. J. Inorg. Chem. (Engl. Transl.)* <u>1965</u>, *10*, 1176-8.

EXPERIMENTAL VALUES: (Continued)

Composition of saturated solutions

t/°C	Potassium chloride		Potassium Chlorate		Density g cm^{-3}	Nature of the solid phase[a]
	mass %	mol % (compiler)	mass %	mol % (compiler)		
50	30.06	9.409	0.00	0.00	1.194	A
	29.67	9.359	1.05	0.201	1.196	"
	28.30	9.204	4.97	0.983	1.216	A+B
	19.84	6.098	7.42	1.39	1.166	B
	7.2	2.1	12.04	2.100	1.1067	"
	2.88	0.814	14.41	2.477	1.097	"
	0.00	0.00	15.84[b]	2.692	1.091	"

[a] A = KCl;   B = KClO$_3$

[b] For the binary system the compiler computed the following:

soly of KClO$_3$ = 0.709 mol kg^{-1} at 25°C

= 1.536 mol kg^{-1} at 50°C

COMPONENTS:	ORIGINAL MEASUREMENTS:
(1) Potassium chloride; KCl; [7447-40-7]	Arkhipov, S.M.; Kashina, N.I.; Kuzina, V.A.
(2) Potassium chlorate; KClO$_3$; [3811-04-9]	Zh. Neorg. Khim. 1969, 14, 567-70; Russ. J. Inorg. Chem. (Engl. Transl.) 1969, 14, 294-6.
(3) Water; H$_2$O; [7732-18-5]	

VARIABLES:	PREPARED BY:
Composition T/K = 298.2	Hiroshi Miyamoto

EXPERIMENTAL VALUES: Composition of saturated solutions at 25.0°C

Potassium chlorate		Potassium chloride		Nature of the
mass %	mol % (compiler)	mass %	mol % (compiler)	solid phase[a]
7.93[b]	1.25	--	--	A
6.21	0.979	1.95	0.505	"
4.65	0.741	5.04	1.32	"
4.07	0.664	8.54	2.29	"
3.11	0.518	11.89	3.253	"
2.80	0.478	15.06	4.222	"
2.53	0.443	18.40	5.300	"
2.00	0.362	22.47	6.683	"
1.93	0.360	25.91	7.955	"
1.94	0.363	26.10	8.029	A+B
1.92	0.359	26.10	8.027	"
1.33	0.248	26.29	8.049	B
--	--	26.72	8.098	"

[a] A = KClO$_3$;   B = KCl

[b] For the binary system the compiler computes the following:

soly of KClO$_3$ = 0.703 mol kg^{-1}

AUXILIARY INFORMATION

METHOD/APPARATUS/PROCEDURE:	SOURCE AND PURITY OF MATERIALS:
The isothermal method was used. The solids (KClO$_3$ and KCl) and water were placed into glass test-tubes held in a thermostat. The rate of rotation of the test-tubes was 45 rev min^{-1}, and equilibrium was reached in 30 hours. Potassium in the liquid phase was analyzed by flame photometry. Chloride was determined by titration of a specimen of the solution with silver nitrate by using potassium chromate as an indicator. The chlorate ion concentration was determined volumetrically by addition of an excess of iron(II) sulfate solution and back-titration of the latter with potassium permanganate solution. The solid phases were identified by the method of residues, crystal optics, and by X-ray diffraction.	The purity of KClO$_3$ and KCl was within 99.9 %.
	ESTIMATED ERROR:
	Soly: the relative error in potassium determination by flame photometry did not exceed 3-5 %. Temp: precision ± 0.1 K (authors).
	REFERENCES:

COMPONENTS:	ORIGINAL MEASUREMENTS:
(1) Potassium chloride; KCl; [7447-40-7] (2) Potassium chlorate; KClO₃; [3811-04-9] (3) Potassium perchlorate; KClO₄; [7778-74-7] (4) Water; H₂O; [7732-18-5]	Karnaukhov, A.S.; Lepeshkov, I.N.; Fursova, A.F. *Zh. Neorg. Khim.* <u>1969</u>, *14*, 2211-3; *Russ. J. Inorg. Chem. (Engl. Transl.)* <u>1969</u>, *14*, 1160-1.

VARIABLES:	PREPARED BY:
T/K = 323 Composition	Hiroshi Miyamoto

COMMENTS AND/OR ADDITIONAL DATA:

The phase diagram of the $KClO_3$-$KClO_4$-H_2O system at 50°C is given below (based on mass % units).

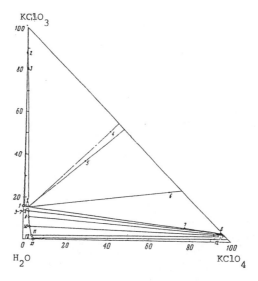

continued.....

AUXILIARY INFORMATION

METHOD/APPARATUS/PROCEDURE:	SOURCE AND PURITY OF MATERIALS:
The isothermal method was used. At 25°C equilibrium was reached in 4-5 days. The potassium content was determined gravimetrically with tetraphenylborate. The chlorate ion concentration at high concentrations was determined volumetrically by the mercurimetric method, and at lower concentrations, gravimetrically. The chlorate ion concentration was determined volumetrically after reduction to chloride with zinc dust. The compositions and the nature of the solid phases were found by chemical analysis with Schreinemakers' method of residues and checked by X-ray diffraction.	Potassium chloride and chlorate were recrystallized twice. The resulting purity was 99.7 - 99.8 %.
	ESTIMATED ERROR: Nothing specified.
	REFERENCES:

COMPONENTS:	ORIGINAL MEASUREMENTS:
(1) Potassium chloride; KC1; [7447-40-7] (2) Potassium chlorate; KC1O₃; [3811-04-9] (3) Potassium perchlorate; KC1O₄; [7778-74-7] (4) Water; H₂0; [7732-18-5]	Karnaukhov, A.S.; Lepeshkov, I.N.; Fursova, A.F. *Zh. Neorg. Khim.* 1969, *14*, 2211-3; *Russ. J. Inorg. Chem. (Engl. Transl.)* 1969, *14*, 1160-1.

EXPERIMENTAL VALUES: (Continued)

Composition of saturated solutions at 50°C

Potassium Chloride mass %	Potassium Chloride mol % (compiler)	Potassium Chlorate mass %	Potassium Chlorate mol % (compiler)	Potassium Perchlorate mass %	Potassium Perchlorate mol % (compiler)	Density g cm⁻³	Nature of the solid phase[a]
29.67	9.385	--	--	1.27	0.216	1.197	D+C
29.65	9.457	1.00	0.194	1.03	0.177	1.197	D+E
29.24	9.450	2.56	0.503	1.00	0.174	--	"
28.17	9.213	4.64	0.923	0.91	0.160	1.218	D+E+B
28.30	9.204	4.97	0.983	--	--	1.216	A+B
28.04	9.286	5.68	1.144	1.06	0.189	--	E+B
17.06	5.177	7.21	1.331	1.47	0.240	1.163	"
10.66	3.145	10.02	1.798	1.68	0.267	--	"
8.33	2.44	11.35	2.022	1.70	0.268	1.120	"
6.25	1.82	12.36	2.187	1.87	0.293	--	"
4.17	1.20	13.28	2.331	2.04	0.317	--	"
2.08	0.596	14.25	2.484	2.22	0.342	1.1002	"
--	--	15.21	2.632	2.39	0.366	1.0982	"
--	--	0.49	0.075	4.68	0.637	1.0179	B+E
0.45	0.11	0.42	0.064	3.97	0.539	--	E+C
0.90	0.23	0.36	0.055	3.27	0.442	--	"
1.67	0.423	0.355	0.0547	3.51	0.479	1.0257	"
2.44	0.623	0.35	0.054	3.75	0.515	--	"
3.72	0.957	0.34	0.053	3.43	0.475	--	"
4.99	1.29	0.32	0.050	3.10	0.432	--	"
7.53	1.99	0.29	0.047	2.85	0.405	--	"
12.61	3.441	0.23	0.038	1.94	0.285	--	"
21.14	6.195	0.12	0.021	1.61	0.254	--	"
25.41	7.731	0.06	0.011	1.44	0.236	--	E+D+C
29.99	9.415	--	--	0.33	0.056	--	A+D

a A = KC1;   B = KC1O₃;   C = KC1O₄;   D - NKC1.mKC1O₄;   E - nKC1O₄.mKC1O₃

COMPONENTS:	ORIGINAL MEASUREMENTS:
(1) Potassium chloride; KCl; [7447-40-7]	Arkhipov, S.M.; Kashina, N.I.;
(2) ¯Potassium chlorate; KClO$_3$; [3811-04-9]	Kuzina, V.A.
(3) Rubidium chloride; RbCl; [7791-11-9]	Zh. Neorg. Khim. 1969, 14, 567-70;
(4) Rubidium chlorate; RbClO$_3$; [13446-71-4]	Russ. J. Inorg. Chem. (Engl. Transl.)
(5) Water; H$_2$O; [7732-18-5]	1969, 14, 294-6.

VARIABLES:	PREPARED BY:
Composition at 298.2 K	Hiroshi Miyamoto

EXPERIMENTAL VALUES:

Experimental data are given on the following page.

AUXILIARY INFORMATION

METHOD/APPARATUS/PROCEDURE:
The isothermal method was used. KClO$_3$,
RbClO$_3$, KCl, RbCl and water were mixed in
glass test tubes held in a thermostat.   The
rate of rotation of the test tubes was 45 rev
min^{-1}, and equilibrium was reached in 30 days.
Specimens of the liquid phases were analyzed
for potassium and rubidium by flame photometry. The chloride content was determined
by titration of a specimen of the solution
with silver nitrate and potassium chromate
as indicator. The chlorate ion concentration was determined volumetrically by addition of excess iron(II) sulfate solution and
back titration of the latter with potassium
permanganate.
The solid phases were identified by the
method of residues, crystal optics, and by
X-ray diffraction.

SOURCE AND PURITY OF MATERIALS:
The purity of the salts used was 99.9 % or
better.

ESTIMATED ERROR:
The relative error in potassium and rubidium determinations by flame photometry did
not exceed 3-5 %.
Temp: precision $\pm$ 0.1 K (authors).

REFERENCES:

COMPONENTS:	ORIGINAL MEASUREMENTS:
(1) Potassium chloride; KCl; [7447-40-7] (2) Potassium chlorate; $KClO_3$; [3811-04-9] (3) Rubidium chloride; RbCl; [7791-11-9] (4) Rubidium chlorate; $RbClO_3$; [13446-71-4] (5) Water; H_2O; [7732-18-5]	Arkhipov, S.M.; Kashina, N.I.; Kuzina, V.A. *Zh. Neorg. Khim.* 1969, *14*, 567-70; *Russ. J. Inorg. Chem. (Engl. Transl.)* 1969, *14*, 294-6.

EXPERIMENTAL VALUES: (Continued)

Composition of saturated solutions at 25.0°C

Potassium Chloride mass %	Potassium Chloride mol % (compiler)	Potassium Chlorate mass %	Potassium Chlorate mol % (compiler)	Rubidium Chloride mass %	Rubidium Chloride mol % (compiler)	Rubidium Chlorate mass %	Rubidium Chlorate mol % (compiler)	Nature of the solid phase[a]
--	--	6.52	1.05	--	--	3.58	0.418	A+B
0.44	0.12	6.25	1.01	--	--	3.55	0.415	"
1.07	0.285	6.09	0.986	--	--	3.58	0.421	"
1.99	0.529	4.96	0.801	--	--	3.67	0.430	"
3.09	0.824	4.51	0.732	--	--	3.63	0.427	"
3.62	0.969	4.08	0.643	--	--	3.91	0.462	"
5.23	1.40	2.77	0.452	--	--	4.05	0.480	"
7.84	2.11	0.56	0.092	--	--	4.22	0.501	"
11.47	3.216	0.61	0.10	--	--	5.14	0.636	"
15.23	4.431	--	--	1.83	0.328	4.29	0.551	"
18.88	5.672	--	--	2.42	0.448	3.56	0.472	"
19.94	6.033	--	--	2.42	0.451	3.31	0.442	"
--	--	--	--	48.20	12.33	0.82	0.150	C+B
2.09	0.882	--	--	47.39	12.33	0.90	0.168	B+E
4.29	1.704	--	--	41.19	10.08	0.94	0.16	"
6.55	2.611	--	--	39.42	9.687	0.96	0.17	"
7.78	3.057	--	--	37.28	9.030	0.97	0.17	"
8.20	3.205	--	--	36.31	8.750	1.18	0.204	"
13.79	5.017	--	--	25.37	5.690	1.71	0.275	"
26.10	8.028	1.93	0.361	--	--	--	--	A+D
24.66	7.710	2.03	0.386	2.68	0.517	--	--	A+E
23.02	7.245	2.11	0.404	4.66	0.904	--	--	"
22.51	7.098	2.00	0.384	5.39	1.05	--	--	"
21.53	7.073	--	--	9.19	1.86	2.57	0.373	B+E
17.15	6.034	--	--	19.18	4.161	2.23	0.346	"
19.17	6.701	--	--	16.92	3.646	2.16	0.333	"
17.82	6.209	--	--	17.93	3.852	2.10	0.323	"
11.48	4.321	--	--	30.47	7.071	1.30	0.216	"
11.63	4.392	--	--	30.56	7.115	1.32	0.220	"
19.01	6.496	--	--	14.79	3.116	2.55	0.385	"
19.90	6.738	--	--	13.04	2.722	2.74	0.409	"
25.30	7.903	1.96	0.372	2.09	0.402	--	--	A+E
23.80	7.626	4.85	0.945	2.82	0.557	--	--	"

a A = $KClO_3$;   B = $RbClO_3$;   C = RbCl;   D = KCl;   E = Solid solution (K,Rb)Cl

COMPONENTS:	ORIGINAL MEASUREMENTS:
(1) Potassium chlorate; KClO₃; [3811-04-9] (2) Potassium bromide; KBr; [7758-02-3] (3) Water; H₂0; [7732-18-5]	Ricci, J.E. *J. Am. Chem. Soc.* <u>1937</u>, 59, 866-7.

VARIABLES:	PREPARED BY:
Composition at 298.15 K	Hiroshi Miyamoto

EXPERIMENTAL VALUES: Composition of saturated solutions at 25.00°C

KClO₃		KBr		Density	Nature of the
mass %	mol % (compiler)	mass %	mol % (compiler)	g cm⁻³	solid phase[a]
7.905[b]	1.246	0.00	0.00	1.047	A
4.59	0.765	9.30	1.596	1.100	"
3.21	0.570	16.99	3.105	1.160	"
2.41	0.458	24.20	4.733	1.216	"
1.87	0.384	31.66	6.700	1.292	"
1.42	0.320	39.47	9.151	1.376	"
1.43	0.324	40.00	9.340	1.385	A+B
1.42	0.322	40.01	9.342	1.386	"
1.37	0.311	40.06	9.354	1.387	"
1.42(Av ± .5)	0.322	40.01	9.342	1.385	"
0.00	0.000	40.63	9.388	1.380	B

[a] A = KClO₃;   B = KBr

[b] For the binary system the compiler computes the following:

$$\text{soly of KClO}_3 = 0.7004 \text{ mol kg}^{-1}$$

AUXILIARY INFORMATION

METHOD/APPARATUS/PROCEDURE:	COMMENTS AND/OR ADDITIONAL DATA:
Mixtures of known composition were stirred in a bath thermostatically controlled at 25°C for at least two days. Potassium bromide was determined by titration with standard silver nitrate solution using Mohr's method. The total solid was determined by evaporation at 100°C followed by heating to 250°C. Potassium chlorate was calculated by difference.	The phase diagram is given below (based on mass %). 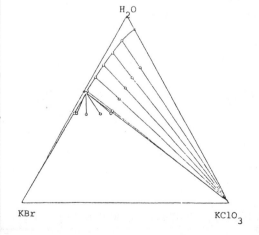

SOURCE AND PURITY OF MATERIALS:
Nothing specified.

ESTIMATED ERROR:
Soly: nothing specified.   Temp: precision ± 0.02 K.

COMPONENTS:	ORIGINAL MEASUREMENTS:
(1) Potassium chlorate; $KClO_3$; [3811-04-9]	Swenson, T.; Ricci, J.E.
(2) Potassium bromate; $KBrO_3$; [7758-01-2]	*J. Am. Chem. Soc.* 1939, *61*, 1974-7.
(3) Water; H_2O; [7732-18-5]	

VARIABLES:	PREPARED BY:
Composition at 298 K	Hiroshi Miyamto

EXPERIMENTAL VALUES: Composition of saturated solutions at 25°C

$KBrO_3$ mass %	$KBrO_3$ mol % (compiler)	$KClO_3$ mass %	$KClO_3$ mol % (compiler)	Density g cm^{-3}	Nature of the solid phase[a]
7.533[b]	0.8712	0	0	1.054	A
6.46	0.755	2.26	0.360	–	SSI
5.75	0.679	4.08	0.656	1.067	"
5.63	0.665	4.29	0.691	–	"
4.936	0.5916	6.546	1.069	1.078	SSI +SSII
4.945	0.5951	6.531	1.067	1.078	"
4.02	0.478	6.75	1.09	1.072	SSII
2.79	0.329	7.08	1.14	1.064	"
2.07	0.243	7.26	1.16	–	"
1.02	0.119	7.60	1.21	1.053	"
0	0	7.895[b]	1.244	1.048	B

[a] $A = KBrO_3$; $B = KClO_3$;

SSI = solid solution containing up to 3 % $KClO_3$ in $KBrO_3$.

SSII = solid solution containing up to 5 % $KBrO_3$ in $KClO_3$.

[b] For binary systems the compiler computes the following:

soly of $KClO_3$ = 0.6995 mol kg^{-1}

soly of $KBrO_3$ = 0.4878 mol kg^{-1}

AUXILIARY INFORMATION

METHOD/APPARATUS/PROCEDURE:

Solubilities were determined according to the usual procedure insofar as method of stirring sampling, filtering, density determination, and temperature control are concerned. Starting with complexes of known composition, and analyzing the saturated solutions at equilibrium, the solid phases were then determined by the methods of graphical or algebraic extrapolation, in addition to occasional analyses of wet and centrifuged residues. The analytical method for the saturated solutions depended on the combined percentage of the $KClO_3$ and $KBrO_3$. For large $KBrO_3$ compositions solutions were analyzed by evaporation, and iodometric titration of the bromate with thiosulfate solution thus allowing the calculation of the percentage of the chlorate by difference. In the presence of a large amount of chlorate, small quantities of bromate were determined as follows. To about 100 ml of solution was added 5 g of sodium iodide (20 ml of 25% solution) giving a concentration of 0.33N after dilution to 100 ml; 1.5 ml of concentrated HCl (0.18 to 0.2N after dilution); titration with 0.2N sodium thiosulfate solution to be started after

continued.....

SOURCE AND PURITY OF MATERIALS:

High grade (99.9 %) potassium bromate was used as received. Potassium chlorate contained small amounts of the corresponding bromate; this bromate content was determined by iodometric titration and the necessary corrections were made when preparing the ternary complexes.

ESTIMATED ERROR:

Soly: precision 2 % (compiler).
Temp: nothing specified.

REFERENCES:

COMPONENTS:	ORIGINAL MEASUREMENTS:
(1) Potassium chlorate; $KClO_3$; [3811-04-9] (2) Potassium bromate; $KBrO_3$; [7758-01-2] (3) Water; H_2O; [7732-18-5]	Swenson, T.; Ricci, J.E. J. Am. Chem. Soc. 1939, 61, 1974-7.

COMMENTS AND/OR ADDITIONAL DATA:

The phase diagram is given below (based on mass %).

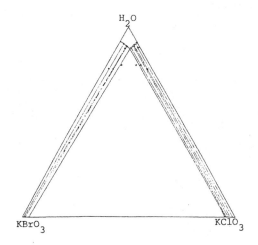

METHOD/APPARATUS/PROCEDURE: (Continued)

1.5 min.

The same conditions, using a 0.02 N
sodium thiosulfate solution for titration and applying the time correction
can be used for the detection of
quantities as small as 0.001 ($\pm$ 0.0005) %
of bromate in chlorate.

COMPONENTS:	ORIGINAL MEASUREMENTS:
(1) Potassium chlorate; $KClO_3$; [3811-04-9] (2) Potassium iodide; KI; [7681-11-0] (3) Water; H_2O; [7732-18-5]	Ricci, J.E. J. Am. Chem. Soc. 1937, 59, 866-7.
VARIABLES: Composition at 298.15 K	PREPARED BY: Hiroshi Miyamoto

EXPERIMENTAL VALUES: Composition of saturated solutions at 25.00°C

	$KClO_3$			KI		Density	Nature of the
mass %	mol % (compiler)		mass %	mol % (compiler)		g cm^{-3}	solid phase[a]
7.905[b]	1.246		0.00	0,00		1.047	A
5.04	0.848		9.33	1.159		1.103	"
3.35	0.612		18.74	2.528		1.178	"
2.30	0.467		28.72	4.303		1.275	"
1.60	0.370		39.26	6.695		1.400	"
1.10	0.296		49.94	9.937		1.555	"
0.82	0.255		58,34	13.39		1.702	"
0.81	0.256		59.27	13.84		1.724	A+B
0.84	0.266		59.30	13.86		1.723	"
0.84	0.266		59.26	13.84		1.725	"
0.83	0.263		59.28	13.85		1.724	"
0.64	0.212		59.36	13.85		1.724	B
0.00	0.000		59.76	13.88		1.718	"

[a] A = $KClO_3$;   B = KI

[b] For the binary system the compiler computes the following:

soly of $KClO_3$ = 0.7004 mol kg^{-1}

AUXILIARY INFORMATION

METHOD/APPARATUS/PROCEDURE:	SOURCE AND PURITY OF MATERIALS:
Mixtures of known composition were stirred in a bath thermostatically controlled at 25°C for at least two days. Potassium iodide was titrated argentometrically by Fajans' method using eocin as an absorption indicator. The total solid was determined by evaporation at 100°C followed by heating to 250°C. Potassium chlorate was calculated by difference.	Nothing specified.
	ESTIMATED ERROR: Soly: nothing specified. Temp: precision ± 0.02 K.
	REFERENCES:

COMPONENTS:	ORIGINAL MEASUREMENTS:
(1) Potassium chlorate; KClO$_3$; [3811-04-9]	Ricci, J.E.
(2) Potassium iodate; KIO$_3$; [7758-05-6]	J. Am. Chem. Soc. 1938, 60, 2040-3.
(3) Water; H$_2$O; [7732-18-5]	
VARIABLES: T/K = 298, 323 Composition	PREPARED BY: Hiroshi Miyamoto

EXPERIMENTAL VALUES: Composition of saturated solutions

t/°C	KIO$_3$ mass %	mol % (compiler)	KClO$_3$ mass %	mol % (compiler)	Density g cm^{-3}	Nature of the solid phase[a]
25	8.45[b]	0.771	0.00	0.000	1.043	A
	7.05	0.648	2.31	0.371	1.070	"
	5.85	0.547	5.31	0.866	1.082	"
	5.43	0.512	6.78	1.117	1.091	A+B
	5.44	0.513	6.79	1.119	1.092	"
	5.44	0.513	6.81	1.122	1.089	"
	5.42	0.511	6.81	1.122	1.086	"
	5.43	0.512	6.80	1.120	1.090(av)	"
	2.92	0.270	7.31	1.180	1.068	B
	0.00	0.000	7.90	1.245	1.048	"
50	13.21[b]	1.265	0.00	0.000		A
	10.87	1.053	3.71	0.628		"
	8.76	0.871	8.58	1.490		"
	7.26	0.749	13.76	2.479		A+B
	7.27	0.750	13.77	2.481		"
	7.27	0.750	13.77	2.481		"
	7.27	0.750	13.77(av)	2.481		"
	5.27	0.535	14.31	2.536		B
	2.41	0.239	15.11	2.616		"
	0.00	0.000	15.78	2.681		"

continued.....

AUXILIARY INFORMATION

METHOD/APPARATUS/PROCEDURE:
Mixtures of KIO$_3$, KClO$_3$ and H$_2$O were stirred for 5-7 days.
The iodate content was determined by treatment with excess potassium iodide and a limited amount of acetic acid, and titration of the liberated iodine with standard thiosulfate solution. The total dissolved solid was determined by evaporation to dryness, and the chlorate salt calculated by difference.

SOURCE AND PURITY OF MATERIALS:
Potassium iodate (c.p. grade) was recrystallized and dried at 100-110°C. Analysis by titration with standard sodium thiosulfate solution showed it to be 100.0 % pure. Potassium chlorate (c.p. grade) was powdered, and then dried at 150-200°C.

ESTIMATED ERROR:
Soly: nothing specified.
Temp: precision ± 0.01 K.

REFERENCES:

COMPONENTS:	ORIGINAL MEASUREMENTS:
(1) Potassium chlorate; $KClO_3$; [3811-04-9]	Ricci, J.E.
(2) Potassium iodate; KIO_3; [7758-05-6]	J. Am. Chem. Soc. 1938, 60, 2040-3.
(3) Water; H_2O; [7732-18-5]	

EXPERIMENTAL VALUES: (Continued)

[a] $A = KIO_3$; $B = KClO_3$

[b] For the binary system the compiler computes the following:

    soly of KIO_3 = 0.431 mol kg^{-1} at 25°C

              = 0.7112 mol kg^{-1} at 50°C

COMMENTS AND/OR ADDITIONAL DATA:

The phase diagram is given below (based on mass % units).

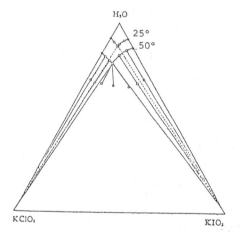

COMPONENTS:	ORIGINAL MEASUREMENTS:
(1) Potassium chlorate; $KClO_3$; [3811-04-9] (2) Potassium hydroxide; KOH; [1310-58-3] (3) Water; H_2O; [7732-18-5]	Bronsted, J.N. *J. Am. Chem. Soc.* 1920, *40*, 1448-54.

VARIABLES:	PREPARED BY:
Concentration of potassium hydroxide T/K = 293	Hiroshi Miyamoto

EXPERIMENTAL VALUES:

Concn of KOH mol/dm^{-3}	Soly of $KClO_3$ $mol\ dm^{-3}$
4.71	0.0924
5.06	0.0882
6.35	0.0609
7.95	0.0445
8.60	0.0410
9.41	0.0351
10.95	0.0287
12.19	0.0254
14.02	0.0215
14.85	0.0195
15.02	0.0191

AUXILIARY INFORMATION

METHOD/APPARATUS/PROCEDURE:	SOURCE AND PURITY OF MATERIALS:
No details given.	Nothing specified.
	ESTIMATED ERROR: Nothing specified.
	REFERENCES:

COMPONENTS:	ORIGINAL MEASUREMENTS:
(1) Potassium chlorate; KClO$_3$; [3811-04-9] (2) Rubidium chlorate; RbClO$_3$; [13446-71-4] (3) Water; H$_2$O; [7732-18-5]	Kirgintsev, A.N.; Kashina, N.I.; Vulikh, A.I.; Korotkevich, B.I. *Zh. Neorg. Khim.* 1965, *10*, 1225-8; *Russ. J. Inorg. Chem. (Engl. Transl.)* 1965, *10*, 662-4.

VARIABLES:	PREPARED BY:
Composition T/K = 298.2	Hiroshi Miyamoto

EXPERIMENTAL VALUES: Composition of saturated solution at 25.0°C

total salts	potassium chlorate			rubidium chlorate	
mol kg^{-1}	y_1^a	g_1/100 g_3	mass %[b]	g_2/100 g_3	mass %[b]
3.94	0.0	0.0	0.0	6.65	6.24
4.76	0.24	1.37	1.35	5.99	5.65
5.11	0.34	2.15	2.10	5.64	5.33
5.18	0.38	2.41	2.35	5.43	5.14
5.11	0.39	2.43	2.37	5.26	5.00
5.85	0.49	3.53	3.41	5.04	4.80
6.07	0.53	3.91	3.76	4.87	4.64
8.04	0.73	6.16	6.68	3.72	3.59
8.40	0.76	7.81	7.24	3.38	3.27
8.41	0.76	7.86	7.29	3.38	3.27
8.40	0.77	7.89	7.31	3.26	3.16
8.25	0.72	7.25	6.76	3.94	3.79
8.23	0.70	7.10	6.63	4.09	3.93
8.28	0.72	7.26	6.77	3.98	3.83
8.35	0.74	7.55	7.02	3.70	3.57
8.22	0.78	7.88	7.30	3.02	2.93
7.59	0.88	8.22	7.60	1.49	1.47
7.21	0.92	8.15	7.54	0.95	0.94
6.98	1.0	8.59	7.91	0.0	0.0
7.02	1.0	8.60	7.92	0.0	0.0

[a] y_1 = mol fraction of KClO$_3$ in mixture of chlorates.

[b] Calculated by the compiler.

AUXILIARY INFORMATION

METHOD/APPARATUS/PROCEDURE:	SOURCE AND PURITY OF MATERIALS:
Solubilities were determined by the method of isothermal relief of supersaturation. Weighed amounts of chlorates were dissolved in water in 50 cm^3 test tubes by heating in a water bath at 65-70°C: the test tubes were then placed in a thermostat at 25°C for 20 m. Supersaturation was removed by stirring at a rate of 60 rev min^{-1} for 10 h. After settling, 2 samples of liquid phase were removed for analysis. The first was evaporated in a drying cupboard at 70-80°C and dried to constant weight at 105°C. The other sample was analyzed for ClO$_3^-$ by adding FeSO$_4$ solution and back-titrating excess iron(II) with permanganate solution. Solid phase compositions were not reported.	The purity of chlorates used was 99.9 % or better.
	ESTIMATED ERROR: Soly: accuracy of y_1 ± 0.01 (authors). Temp: precision ± 0.1 K (authors).
	REFERENCES:

COMPONENTS:	ORIGINAL MEASUREMENTS:
(1) Potassium chlorate; $KClO_3$; [3811-04-9] (2) Cesium chlorate; $CsClO_3$; [13763-67-2] (3) Water; H_2O; [7732-18-5]	Kirgintsev, A.N.; Kashina, N.I.; Vulikh, A.I.; Korotkevich, B.I. *Zh. Neorg. Khim.* <u>1965</u>, *10*, 1225-8; *Russ. J. Inorg. Chem. (Engl. Transl.)* <u>1965</u>, *10*, 662-4.
VARIABLES: Composition T/K = 298.2	PREPARED BY: Hiroshi Miyamoto

EXPERIMENTAL VALUES: Composition of saturated solutions at 25°C

total salts	potassium chlorate			cesium chlorate	
mol kg^{-1}	y_1[a]	g_1/100 g_3	mass %[b]	g_2/100 g_2	mass %[b]
3.59	0.00	0.00	0.00	7.76	7.20
4.39	0.34	1.83	1.80	6.28	5.90
5.49	0.54	3.63	3.50	5.46	5.18
6.60	0.64	5.18	4.92	5.14	4.89
7.85	0.72	6.93	6.48	4.76	4.54
8.38	0.75	7.73	7.17	4.54	4.34
8.35	0.74	7.57	7.03	4.70	4.49
8.29	0.75	7.62	7.08	4.48	4.29
7.73	0.82	7.78	7.22	3.01	2.92
7.01	0.94	8.03	7.43	0.99	0.98
6.98	1.00	8.59	7.91	0.00	0.00

[a] y_1 = mol fraction of $KClO_3$ is mixture of chlorates.

[b] calculated by the compiler.

AUXILIARY INFORMATION

METHOD/APPARATUS/PROCEDURE:	SOURCE AND PURITY OF MATERIALS:
Solubility in this system was studied by the isothermal relief of supersaturation method. Weighed amounts of chlorates were dissolved in water in 50 cm^3 test tubes by heating on a water bath at 65-70°C; the test-tubes were then placed in a thermostat at 25°C for 20 min. Supersaturation was then removed by stirring at 60 rpm for 10 h. After settling two samples of liquid phase were removed for analysis. The first was evaporated in a drying cupboard at 70-80°C and then dried to constant weight at 105°C. The other sample was analyzed for ClO_3^- by adding $FeSO_4$ solution and back-titrating excess iron(II) with permanganate solution. Solid phase compositions not reported.	The purity of chlorates used was 99.9 % or better.
	ESTIMATED ERROR: Soly: accuracy of y_1 ± 0.01 (authors). Temp: precision ± 0.1 K (authors).
	REFERENCES:

COMPONENTS:	ORIGINAL MEASUREMENTS:
(1) Potassium chlorate; KClO₃; [3811-04-9] (2) Calcium chlorate; Ca(ClO₃)₂; [10137-74-3] (3) Water; H₂O; [7732-18-5]	Kirgintsev, A.N.; Kozitskii, V.P. *Zh. Neorg. Khim.* 1968, *13*, 3342-45; *Russ. J. Inorg. Chem. (Engl. Transl.)* 1968, *13*, 1723-5.
VARIABLES: Composition T/K = 298	PREPARED BY: Hiroshi Miyamoto

EXPERIMENTAL VALUES:

Composition of saturated solutions at 25°C[b]

y_1[a]	potassium chlorate		calcium chlorate	
	mass %	mol % (compiler)	mass %	mol % (compiler)
1.000	7.87[c]	1.24	0	0
0.900	7.23	1.15	1.36	0.128
0.816	6.52	1.04	2.49	0.235
0.665	5.60	0.906	4.75	0.455
0.518	4.76	0.785	7.50	0.733
0.438	4.24	0.708	9.20	0.910
0.210	2.73	0.491	17.38	1.85
0.104	1.83	0.363	26.64	3.13
0.103	1.81	0.359	26.63	3.13
0.068	1.47	0.318	33.73	4.32
0.054	1.17	0.254	34.35	4.42
0.052	1.25	0.287	38.31	5.21
0.034	1.14	0.342	54.81	9.74
0.033	1.20	0.387	58.42	11.14

[a] The mole fraction of potassium chlorate based on moles KClO₃ and Ca(ClO₃)₂.

[b] In mol kg⁻¹ units, the authors report the solubility of KClO₃ in terms of the following smoothing equation.

$$\log m_1 = -0.157 - 0.778 \log y_1 - 0.302\,(1-y_1)$$

AUXILIARY INFORMATION

METHOD/APPARATUS/PROCEDURE:	SOURCE AND PURITY OF MATERIALS:
Solubility was measured by the method of isothermal relief of supersaturation. Equilibrium was reached in 6-8 hours. The apparatus for the solubility determination is shown in figure below.	"Analytical reagent" grade calcium chlorate and potassium chlorate were used.
	ESTIMATED ERROR: Soly: nothing specified. Temp: precision ± 0.05 K (authors).

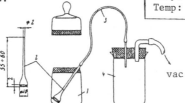

vac

Samples of satd sln to be analyzed were placed in container 1 which had been previously weighed together with the filter stick. Sodium tetraphenylborate solution was added dropwise over a period of 30 min. The precipitate was allowed to settle, and the mother-liquor withdrawn through the filter stick and transferred into beaker 4 through the fine polyvinly chloride tube 3. The precipitate was washed twice with 0.06 % aqueous sodium tetraphenylborate, then four or five times with a few millilitres of distilled water. The container with the precipitate and filter stick was dried for 1.5 hours at 105°C, cooled and weighed.

The calcium content of the solution in beaker 4 was determined by titration with Trilon B.

COMPONENTS:	ORIGINAL MEASUREMENTS:
(1) Potassium chlorate; $KClO_3$; [3811-04-9]	Taylor, A.E.
(2) Ethanol, C_2H_6O; [64-17-5]	J. Phys. Chem. 1897, 1, 718-33.
(3) Water; H_2O; [7732-18-5]	

VARIABLES:	PREPARED BY:
Concentration of ethanol T/K = 303 and 313	Hiroshi Miyamoto

EXPERIMENTAL VALUES:

t/°C	Concn of ethanol mass %	mol % (compiler)	g/g satd soln	Solubility mol % (compiler)	g/g H_2O	mol kg^{-1} (compiler)
30	0	0	0.0923	1.47	0,1017	0.8299
	5	2	0.0772	1.21	0,0880	0.718
	10	4.2	0.0644	1.00	0,0765	0.624
	20	8.9	0.0451	0,690	0,0590	0.481
	30	14	0.0321	0.485	0,0474	0.387
	40	21	0.0235	0.352	0,0400	0.326
	50	28	0.0164	0.245	0,0333	0.272
	60	37	0.0101	0.150	0,0253	0.206
	70	48	0.0054	0.080	0,0182	0.149
	80	61	0.0024	0.035	0,0122	0.100
	90	78	0.0006	0.009	0,0062	0.051
40	0	0	0.1223	2.007	0,1393	1.136
	5	2	0.1048	1.691	0,1233	1.006
	10	4.2	0.0884	1.405	0,1077	0.879
	20	8.9	0.0640	0.995	0,0856	0.698
	30	14	0.0467	0.715	0,0700	0.571
	40	21	0.0341	0.516	0,0588	0.480
	50	28	0.0241	0.362	0,0494	0.403
	60	37	0.0146	0.217	0,0369	0.301
	70	48	0.0078	0.115	0,0263	0.215
	80	61	0.0034	0.050	0,0173	0.141
	90	78	0.0012	0.018	0,0117	0.095

continued.....

AUXILIARY INFORMATION

METHOD/APPARATUS/PROCEDURE:
Small bottles containing the aqueous alcohol and a large excess of powdered salt were placed in an Ostwald thermostat for about half a day at a temperature some ten degrees higher than that at which the solubility was to be determined. During this time, the bottles were shaken frequently and thoroughly. The temperature was lowered and maintained at the desired value for about a day. The solubility was very nearly constant at the end of 3 days, but at least six days were required for many solutions.
About 5 cm^3 of the saturated solution were withdrawn using a pipet and weighed. The solution was evaporated to dryness and weighed.

SOURCE AND PURITY OF MATERIALS:
Potassium chlorate was recrystallized two or three times and dried in an air bath. Ethanol was distilled from lime, stored over dehydrated copper sulfate for one or two days, and finally distilled.

ESTIMATED ERROR:
Soly: accuracy 0.1 % (author).
Temp: nothing specified.

REFERENCES:

COMPONENTS:	ORIGINAL MEASUREMENTS:
(1) Potassium chlorate; $KClO_3$; [3811-04-9] (2) Ethanol; C_2H_6O; [64-17-5] (3) Water; H_2O; [7732-18-5]	Taylor, A.E. J. Phys. Chem. 1897, 1, 718-33.

EXPERIMENTAL VALUES: (Continued)

Fitting equations were given as:

(1)   The concentration of ethanol: 0 - 40 mass %

$$- \log w = (1/1.57) \times 2.260 + \log (x + 0.20) \text{ at } 30°C$$

$$- \log w = (1/1.57) \times 2.000 + \log (x + 0.22) \text{ at } 40°C$$

(2)   The concentration of ethanol: 50 - 90 mass %

$$- \log w = (1/1.2) \times 1.690 + \log (x + 0.20) \text{ at } 30°C$$

$$- \log w = (1/1.2) \times 1.482 + \log (x + 0.22) \text{ at } 40°C$$

where w is the amount of salt in one gram of water,

and x is the amount of alcohol in one gram of water.

COMPONENTS:	ORIGINAL MEASUREMENTS:
(1) Potassium chlorate; KClO$_3$; [3811-04-9] (2) 1,2,3-Propanetriol (glycerol); C$_3$H$_8$O; [56-81-5] (3) Water; H$_2$O; [7732-18-5]	Holm, K. *Pharm. Weekblad* <u>1921</u>, *58*, 1033-7.[1]

VARIABLES:	PREPARED BY:
T/K = 293	T.P. Dirkse

EXPERIMENTAL VALUES:

glycerol composition mass %[a]	solubility KClO$_3$ g/100 g glycerol[b]
86.5	1.32
98.5	1.03

[a] Author only specified % glycerol, and the compiler *assumes* this to mean mass %.

[b] Presumably this refers to grams of KClO$_3$ per 100 grams of the mixed solvent.

AUXILIARY INFORMATION

METHOD/APPARATUS/PROCEDURE:	SOURCE AND PURITY OF MATERIALS:
Isothermal method used with approach from above and below. An excess of dried powdered salt was added to the glycerol-water mixture in a flask, and the cork covered with a layer of paraffin. One set of flasks were agitated in the constant temperature bath while another set of flasks were first heated to 90°C for 1 hour before equilibrating at 20°C. Attainment of equilibrium required weeks to months. Method of analyses not specified.	Two glycerol-water mixtures were prepd with specific gravities of 1.2326 and 1.2645 at 15°C. The compositions of these mixtures were given as 86.5 % and 98.5 % glycerol, respectively (the compiler <u>assumes</u> these are mass % values). No other information given.
	ESTIMATED ERROR: Soly: nothing specified. Temp: precision ± 0.05-0.1 K (author).
	REFERENCES: 1. The data in this paper were also presented at a meeting and published in Pharm. Weekblad before the full paper was published. Holm, K. *Pharm. Weekblad* <u>1921</u>, *58*, 860-2. The paper was read by a Mr. Kok on behalf of Mr. Holm.

COMPONENTS:	ORIGINAL MEASUREMENTS:
(1) Potassium chlorate; $KClO_3$; [3811-04-9]	Taylor, A.E.
(2) 2-Propanone (acetone); C_3H_6O; [67-64-1]	J. Phys. Chem. 1897, 1, 718-33.
(3) Water; H_2O; [7732-18-5]	

VARIABLES:	PREPARED BY:
Concentration of acetone T/K = 303 and 313	Hiroshi Miyamoto

EXPERIMENTAL VALUES:

t/°C	Concn of Acetone mass %	mol % (compiler)	g/g satd soln	Solubility mol % (compiler)	g/g H_2O	mol kg^{-1} (compiler)
30	0	0	0.0923	1.47	0.1017	0.8299
	5	1.6	0.0832	1.32	0.0956	0.7801
	9.09	3.01	0.0763	1.20	0.0909	0.7417
	20	7.2	0.0609	0.944	0.0810	0.6610
	30	12	0.0493	0.757	0.0740	0.6038
	40	17	0.0390	0.593	0.0676	0.5516
	50	24	0.0290	0.437	0.0598	0.4880
	60	32	0.0203	0.304	0.0517	0.4219
	70	42	0.0124	0.184	0.0418	0.3411
	80	55	0.0057	0.084	0.0288	0.2350
	90	74	0.0018	0.027	0.0182	0.1485
40	0	0	0.1223	2.026	0.1393	1.137
	5	1.6	0.1110	1.802	0.1311	1.070
	9.09	3.01	0.1028	1.656	0.1260	1.028
	20	7.2	0.0827	1.31	0.1126	0.9188
	30	12	0.0669	1.04	0.1024	0.8356
	40	17	0.0536	0.826	0.0945	0.7711
	50	24	0.0403	0.614	0.0840	0.6854
	60	32	0.0286	0.431	0.0735	0.5998
	70	47	0.0286	0.251	0.0568	0.4635
	80	55	0.0079	0.117	0.0397	0.3240
	90	74	0.0024	0.035	0.0245	0.1999

continued.....

AUXILIARY INFORMATION

METHOD/APPARATUS/PROCEDURE:	SOURCE AND PURITY OF MATERIALS:
Small bottles containing the aqueous acetone and a large excess of powdered salt were placed in an Ostwald thermostat for about half a day at a temperature some ten degrees higher than that at which the solubility was to be determined. During this time the bottles were shaken frequently and thoroughly. The temperature was lowered and maintained at the desired value for about a day. The solubility was very nearly constant at the end of three days, but at least six days were required for many solutions. About 5 cm^3 of the saturated solution were withdrawn using a pipet and weighed. The solution was evaporated to dryness and weighed.	Potassium chlorate was recrystallized two or three times and dried in an air bath. Acetone was purified by distillation.
	ESTIMATED ERROR: Soly: accuracy 0.1 %. Temp: nothing specified.
	REFERENCES:

COMPONENTS:	ORIGINAL MEASUREMENTS:
(1) Potassium chlorate; $KClO_3$; [3811-04-9] (2) 2-Propanone (acetone); C_3H_6O; [67-64-1] (3) Water; H_2O; [7732-18-5]	Taylor, A.E. *J. Phys. Chem.* 1897, *1*, 718-33.

EXPERIMENTAL VALUES: (Continued)

Fitting equations were given as follows:

(1) The concentration of acetone; 0 - 50 mass %

$$- \log w = (1/3.6) \times 4.273 + \log (x + 0.20) \text{ at } 30°C$$

$$- \log w = (1/3.6) \times 3.640 + \log (x + 0.22) \text{ at } 40°C$$

(2) The concentration of acetone; 60 - 90 mass %

$$- \log w = (1/1.55) \times 1.760 + \log (x + 0.20) \text{ at } 30°C$$

$$- \log w = (1/1.55) \times 1.525 + \log (x + 0.22) \text{ at } 40°C$$

where y is the amount of salt in one gram of water, and x is the amount of acetone in one

gram of water.

COMPONENTS:	ORIGINAL MEASUREMENTS:
(1) Potassium chlorate; KClO$_3$; [3811-04-9] (2) 2-Propanone (acetone); C$_2$H$_6$O; [67-64-1] (3) Water; H$_2$O; [7732-18-5]	Hartley, G.S. *Trans. Faraday Soc.* 1931, *27*, 10-29.
VARIABLES: T/K = 290.8 Concentration of acetone	PREPARED BY: Hiroshi Miyamoto

EXPERIMENTAL VALUES:

		acetone		potassium chlorate	
t/°C	N$_2$[a]	mass % (compiler)	N$_1$[b]	mass % (compiler)	mol/kg^{-1} (compiler)
17.6	0.0000	0.00	0.00953	6.14	0.534
	0.0233	7.14	0.00847	5.24	
	0.0409	12.1	0.00779	4.67	
	0.0513	14.8	0.00740	4.04	

[a] Mole ratio of acetone in the mixture of acetone and water

[b] Mole fraction of solute in saturated solution.

AUXILIARY INFORMATION

METHOD/APPARATUS/PROCEDURE:	SOURCE AND PURITY OF MATERIALS:
Excess recrystallized potassium chlorate was placed in tubes with weighed amounts of water and acetone. The tube was placed in a thermostat and agitated from time to time for several days. Samples of the liquid were quickly transferred to weighed flasks fitted with stoppers. The potassium chlorate samples were carefully evaporated to dryness and weighed.	Potassium chlorate was recrystallized. No other information given.
	ESTIMATED ERROR: Nothing specified.
	REFERENCES:

COMPONENTS:	ORIGINAL MEASUREMENTS:
(1) Potassium chlorate; $KClO_3$; [3811-04-9] (2) 1,2-Ethandiol (ethylene glycol); $C_2H_6O_2$; [107-21-1]	Isbin, H.S.; Kobe, K.A. J. Am. Chem. Soc. 1945, 67, 464-5.
VARIABLES: T/K = 298.15	PREPARED BY: Hiroshi Miyamoto

EXPERIMENTAL VALUES:

   The solubility of $KClO_3$ in ethylene glycol at 25°C is given:

   1.21 g/100 g solvent   (author).

   9.87×10^{-2} mol kg^{-1} (compiler).

AUXILIARY INFORMATION

METHOD/APPARATUS/PROCEDURE:	SOURCE AND PURITY OF MATERIALS:
The solvent and excess solid were sealed in soft-glass test-tubes and rotated for at least one week in a thermostated water bath. All analyses were made on a weight basis by use of weighing pipets. Both the standard gravimetric determination of chloride and the volumetric method, using dichlorofluorescein as an indicator, were used. The chlorate was reduced to the chloride by boiling with excess sulfurous acid.	Technical grade ethylene glycol (Carbide and Carbon Chem. Co) was purified by fractionation. Analytical grade $KClO_3$ was used.
	ESTIMATED ERROR: Soly: precision within 4 %. Temp: precision $\pm$ 0.08 K
	REFERENCES:

COMPONENTS:	ORIGINAL MEASUREMENTS:
(1) Potassium chlorate; KClO$_3$; [3811-04-9]	Isbin, H.S.; Kobe, K.A.
(2) 2-Aminioethanol (monoethanolamine); C$_2$H$_7$NO; [141-43-5]	J. Am. Chem. Soc. 1945, 67, 464-5.

VARIABLES:	PREPARED BY:
T/K = 298.15	Hiroshi Miyamoto

EXPERIMENTAL VALUES:

The solubility of KClO$_3$ in monoethanolamine at 25°C is

0.30g/100g solvent   (authors)

2.45 x 10^{-2} mol kg^{-1} (compiler)

AUXILIARY INFORMATION

METHOD/APPARATUS/PROCEDURE:	SOURCE AND PURITY OF MATERIALS:
The solvent and excess solid were sealed in a soft-glass test-tube and rotated for at least one week in a thermostated water bath. All analyses were made on a weight basis by use of weighing pipets. Both the saturated gravimetric determination of chloride and the volumetric method using dichloro-fluoresein as an indicator were used. The chlorate was reduced to the chloride by boiling with excess sulfurous acid.	Technical grade monoethanolamine (Carbide and Carbon Chem. Co) was purified by fractionation. Analytical grade KClO$_3$ was used.

	ESTIMATED ERROR:
	Soly: precision within 4 %. Temp: precision ± 0.08 K.

	REFERENCES:

COMPONENTS:	ORIGINAL MEASUREMENTS:
(1) Potassium chlorate; $KClO_3$; [3811-04-9] (2) 1,2-Ethanediamine (ethylenediamine); $C_2H_8N_2$; [107-15-3]	Isbin, H,S.; Kobe, K.A. J. Am. Chem. Soc. 1945, 67, 464-5.
VARIABLES: T/K = 298.15	PREPARED BY: Hiroshi Miyamoto

EXPERIMENTAL VALUES:

The solubility of $KClO_3$ in ethylenediamine at 25°C is

0.145g/100g solvent   (authors)

1.18×10^{-2} mol kg^{-1} (compiler)

AUXILIARY INFORMATION

METHOD/APPARATUS/PROCEDURE:	SOURCE AND PURITY OF MATERIALS:
The solvent and excess solid were sealed in soft-glass test-tubes and rotated for at least one week in a thermostated water bath. Analysis was made on a weight basis by use of weighing pipets. Both the standard gravimetric determination of chloride and the volumetric method, using dichloro-fluoresein as an indicator were used. The chlorate was reduced to the chloride by boiling with excess sulfurous acid.	Ethylenediamine was dehydrated and puri-fied by the method given in ref (1). Analytical grade $KClO_3$ was used.
	ESTIMATED ERROR: Soly: precision within 4 %. Temp: precision ± 0.08 K (authors).

	REFERENCES:
	1. Putnam, G.L.; Kobe, K.A.
	   Trans. Electrochem. Soc. 1938,
	   74, 609.

COMPONENTS:	ORIGINAL MEASUREMENTS:
(1) Potassium chlorate; $KClO_3$; [3811-04-9] (2) Dimethylformamide; C_3H_5NO; [68-12-2]	Paul, R.C.; Sreenathan, B.R. *Indian J. Chem.* 1966, *4*, 382-6.

VARIABLES:	PREPARED BY:
One temperature: 298.2 K	Mark Salomon

EXPERIMENTAL VALUES:

The solubility of $KClO_3$ in $HCON(CH_3)_2$ was reported as 18.1 g/100 g solvent   (1.477 mol kg^{-1}, compiler).

The solid phase is the anhydrous salt.

AUXILIARY INFORMATION

METHOD/APPARATUS/PROCEDURE:	SOURCE AND PURITY OF MATERIALS:
Isothermal method used. Excess salt and 10 ml of solvent were placed in a Pyrex test tube, sealed, and rotated in a constant temperature water-bath for 24-30 hours. The seal was broken in a dry box and the slurry quickly filtered. The authors state that the metal was estimated from a known quantity of the saturated (filtered) solution, but no details were given. The saturated solution was colorless.	Dimethylformamide (Baker "analyzed" grade) was further purified as described in (1). A.R. grade $KClO_3$ was warmed and placed under vacuum for 6-8 hours.

	ESTIMATED ERROR:
	  Soly: nothing specified.
	  Temp: precision ± 0.1 K.

	REFERENCES: 1. Paul, R.C.; Guraya, P.S.; Sreenathan, B.R. *Indian J. Chem.* 1963, *1*, 335.

COMPONENTS:	ORIGINAL MEASUREMENTS:
(1) Potassium chlorate; $KClO_3$; [3811-04-9]	Hunt, H.; Boncyk, L.
(2) Ammonia; NH_3; [7664-41-7]	J. Am. Chem. Soc. 1933, 55, 3528-30.

VARIABLES:	PREPARED BY:
T/K = 298	Hiroshi Miyamoto and Mark Salomon

EXPERIMENTAL VALUES:

The solubility of $KClO_3$ in liquid ammonia at 25°C was reported as

2.52 g/100 g NH_3

0.206 mol kg^{-1} (compilers)

AUXILIARY INFORMATION

METHOD/APPARATUS/PROCEDURE:
Two methods were used as described in (1).

Method I. 25 ml test tubes with a constriction at the middle were employed. About 10-
25 g NH_3 were condensed in the bottom, and
the dry salt contained in a small tube
tightly covered with cotton cloth was added
to the test tube; this small tube remained
in the upper part of the test tube as it
could not pass the constriction in the middle of the test tube. The top of the test
tube was drawn to a tip and sealed, and the
tube inverted and placed in a thermostat at
25°C. Equilibrium between NH_3 and the excess salt in the small covered tube required 1-3 weeks with periodic shaking. The
test tube was then inverted and only the
satd sln drained into the lower end (excess
solid remained in the small tube covered
with the cotton cloth). The sln was
frozen and sealed at the constriction,
and weighed. The seal was then broken and
the NH_3 boiled off, and the residue
weighed.

Method II. Excess NH_3 was condensed on a
weighed amount of salt in a tube fitted
with a stopcock. After thermostating at
25°C, NH_3 was slowly permitted to escape
through the stopcock until a crystal of
solid appeared and remained undissolved
upon prolonged shaking.

Authors state that the error due to the
condensation of gaseous NH_3 was not significant since the dead space was kept to
a minimum of about 30 cm^3. However this
amount of dead space was stated to limit
the precision of the method to 0.5 %.

SOURCE AND PURITY OF MATERIALS:
Reagent grade $KClO_3$ was recrystallized
three times from water and then from "a
suitable" anhydrous solvent. The salt was
dried to constant weight in a vacuum oven.

Purification of NH_3 not specified, but
probably similar to that described in (1).
In (1) commercial anhyd ammonia was stored
over metallic sodium for several weeks
before use.

ESTIMATED ERROR:
Soly: accuracy probably around ± 1-2 %.
     (compilers).
Temp: 25.00 + 0.025°C: accuracy established
     by NBS calibration (see ref (1)).

REFERENCES:
1. Hunt, H.; J. Am. Chem. Soc. 1932, 54,
   3509.

COMPONENTS:	EVALUATOR:
(1) Rubidium chlorate; $RbClO_3$; [13446-71-4]	H. Miyamoto Chemistry Department Niigata University, Niigata, Japan
(2) Water; H_2O; [7732-18-5]	and M. Salomon US Army ET & DL Fort Monmouth, NJ, USA August 1984

CRITICAL EVALUATION:

THE BINARY SYSTEM

Solubility data in the binary $RbClO_3$ - H_2O system have been reported in seven publications (1-7). Three publications (1,2,4) report data only for the binary system whereas the remaining publications (3, 5-7) concern isothermal studies on ternary systems which include data for the binary system.

Although some investigators (1-4) did not report the nature of the solid phase, the evaluators assume it to be the anhydrous salt by analogy to the results found in the studies on ternary systems (see below).

A number of different analytical techniques were used to measure the solubilities. In (1-3) the total salt content was determined gravimetrically, and in (5-6) the chlorate content was determined by back titration of iron (II) sulfate with potassium permanganate. In (6) the rubidium content was also determined gravimetrically with sodium tetraphenylborate and by flame photometry. In (7) the chlorate content was determined by argentimetric titration with silver nitrate after reduction of chlorate to chloride.

The experimental solubility data for the binary system are summarized in Table 1 where the evaluators have made appropriate conversions to both mol kg^{-1} and mole fraction units (χ).

Table 1 Experimental solubility data for $RbClO_3$ in water.[a]

T/K	mol kg^{-1}	χ	(ref)	T/K	mol kg^{-1}	χ	(ref)
273.2	0.1265	0.002274	(1)	303.2	0.4779	0.008536	(4)
273.2	0.1282	0.002305	(4)	313.2	0.6827	0.01215	(4)
273.2	0.1282	0.002305	(5)	315.4	0.739	0.01314	(1)
281.2	0.182	0.003268	(1)	323.2	0.9430	0.01670	(4)
283.2	0.2109	0.003785	(4)	323.2	0.9430	0.01670	(5)
293.0	0.317	0.005678	(1)	323.2	0.946	0.01676	(1)
293.2	0.32	0.00573	(2)	333.2	1.3083	0.02303	(4)
293.2	0.3129	0.005605	(4)	343.2	1.7197	0.03005	(4)
298.2	0.3940	0.007048	(3)	349.2	2.020	0.03511	(1)
298.2	0.3940	0.007048	(5)	353.2	2.2704	0.03929	(4)
298.2	0.3920	0.007012	(7)	363.2	2.9040	0.04972	(4)
298.2	0.3893	0.006964	(4)	372.2	3.72	0.06281	(1)
298.2	0.406	0.007261	(6)	373.2	3.720	0.06281	(4)
303.2	0.474	0.008467	(1)				

[a] All data were converted by the evaluators to molality or mole fraction based on original experimental data reported in refs (1-7).

In evaluating the solubility data in Table 1, we used the two smoothing equations based on mole fractions (see the PREFACE, this volume),

$$Y_x = A/(T/K) + B\ln(T/K) + C + D(T/K) \qquad [1]$$

and based on molality (see the INTRODUCTION TO THE SOLUBILITY OF SOLIDS IN LIQUIDS and the PREFACE in Volume 13):

$$Y_m = A/(T/K) + B\ln(T/K) + C \qquad [2]$$

where

$$Y_m = \ln(m/m_o) - nM_2(m - m_o) \qquad [3]$$

COMPONENTS:	EVALUATOR:
(1) Rubidium chlorate; RbClO$_3$; [13446-71-4] (2) Water; H$_2$O; [7732-18-5]	H. Miyamoto Chemistry Department Niigata University, Niigata, Japan and M. Salomon US Army ET & DL Fort Monmouth, NJ USA August, 1984

CRITICAL EVALUATION:

In eq. [3], m is the molality, m_0 is a reference molality (the 298.2 K value was used for the present analyses), n is the hyration number in the solid phase (n = 0 for the present system), and M_2 is the molar mass of the solvent. Additional details on eqs. [2-3] can be found in the PREFACE to volume 13 of the Solubility Data Series (1). By giving all data equal weights, we found that three data points from Table 1 could be rejected on the basis that their differences in the experimental and calculated mole fraction solubilities exceeded $2\sigma_x$ (σ_x is the standard error of estimate as defined in eq. [3] of the LiClO$_3$-H$_2$O critical evaluation). The three data points rejected are from (2) for 293.2 K, from (6) at 298.2 K, and from (1) at 273.2 K. Fitting the remaining 24 data points to eqs. [1] and [2] above resulted in the following:

$$Y_x = -27167.445/(T/K) - 111.3525\ell n(T/K) + 673.495 + 0.145962(T/K) \qquad [4]$$

$$\sigma_y = 0.033 \qquad\qquad \sigma_x = 0.00014$$

and

$$Y_m = -5904.5/(T/K) - 7.8407\ell n(T/K) + 64.466 \qquad [5]$$

$$\sigma_y = 0.015 \qquad\qquad \sigma_m = 0.029$$

The solubilities calculated from eqs. [4] and [5] are designated as *recommended* solubilities, and values at rounded temperatures are given in Table 2 below.

Table 2.   Recommended solubilities in the binary system calculated from the smoothing equations [4] and [5]

T/K	m/mol kg^{-1}	X
273.2	0.1267	0.002286
283.2	0.2050	0.003714
293.2	0.3181	0.005743
298.2	0.3905	0.007025
303.2	0.4751	0.008510
313.2	0.6860	0.012164
323.2	0.9607	0.016860
333.2	1.3091	0.022769
343.2	1.7398	0.030078
353.2	2.2608	0.039009
363.2	2.8779	0.049818
373.2	3.5956	0.06283

COMPONENTS:	EVALUATOR:
(1) Rubidium chlorate; $RbClO_3$; [13446-71-4]	Hiroshi Miyamoto Department of Chemistry Niigata University Niigata, Japan
(2) Water; H_2O; [7732-18-5]	
	December 1984

CRITICAL EVALUATION:

TERNARY SYSTEMS

Data for the solubilities in ternary systems have been reported in 4 publications (3,5-7). A summary of these studies is given in Table 3.

Table 3. Summary of solubility studies in ternary systems

System	T/K	Solid Phase	Reference
$RbClO_3$ - $CsClO_3$ - H_2O	298	Not given	3
$RbClO_3$ - $KClO_3$ - H_2O	298	$RbClO_3$; $KClO_3$	3
$RbClO_3$ - $RbCl$ - H_2O	273, 298, 323	$RbClO_3$; $RbCl$	5
$RbClO_3$ - $NaClO_3$ - H_2O	298	$RbClO_3$; $NaClO_3$	6
$RbClO_3$ - $RbNO_3$ - H_2O	298	$RbClO_3$; $RbNO_3$	7
		Solid solution	

The phase diagrams of the ternary systems, $RbClO_3$-$RbCl$-H_2O(5) and $RbClO_3$-$NaClO_3$-H_2O (6) are simple eutonic types, and no double salts are formed. Although the diagrams of the ternary $RbClO_3$-$CsClO_3$-H_2O(3) and $RbClO_3$-$KClO_3$-H_2O(3) were not reported by Kirgintsev, Kashina, Vulikh and Korotkevich in the original paper, the authors reported that rubidium and cesium chlorate form a continuous series of solid solutions, but potassium and rubidium chlorate do not form solid solutions. The solubility in the ternary $RbClO_3$-$RbNO_3$-H_2O system was studied by Shklovskaya, Arkhipov, Kuzina and Tsibulevskaya (7). The crystallization branch of rubidium nitrate and the solid solution based on rubidium chlorate were determined. In the lower concentration range of rubidium nitrate, a solid solution with rubidium chlorate was formed, and the distribution coefficient (see compilation sheet) of rubidium nitrate in the range of crystallization of solid solution is constant.

OTHER MULTICOMPONENT SYSTEMS

The solubility data for the $RbClO_3$-$RbCl$-$NaClO_3$-$NaCl$-H_2O system was reported by Arkhipov, Kashina and Kuzina (6). They found four crystallization regions in the system. Two ternary points were obtained corresponding to solutions saturated with: (1) $NaCl$+$NaClO_3$+$RbClO_3$; (2) $NaCl$+$RbCl$+$RbClO_3$ (see compilation sheet).
Solubilities in the $RbClO_3$-$RbCl$-$KClO_3$-KCl-H_2O system were studied by Arkhipov, Kashina and Kuzina (8). They found three crystallization fields in the systems: $KClO_3$, $RbClO_3$ and solid solutions of potassium and rubidium chlorides. The main part of the diagram is occupied by the crystallization field of rubidium chlorate (77%) followed by the field of potassium chlorate (20%), and of the solid solutions of potassium and rubidium chlorides (3%).

The $RbClO_3$-$RbCl$-$CsClO_3$-$CsCl$-H_2O system was also studied by Arkhipov, Kashina and Kuzina (9). Three crystallization regions were defined in the system: the field of $CsClO_3$-$RbClO_3$ continuous solid solutions which occupies 99 % of entire area of the diagram, and the fields of $Cs(Rb)Cl$ and $Rb(Cs)Cl$ solid solutions.

COMPONENTS:	EVALUATOR:
(1) Rubidium chlorate; $RbClO_3$; [13446-71-4] (2) Water; H_2O; [7732-18-5]	Hiroshi Miyamoto Department of Chemistry Niigata University Niigata, Japan
	December, 1984

CRITICAL EVALUATION:

REFERENCES:

1. Calzolari, F. *Gazz. Chim. Ital.* 1912, *42*, 85.

2. Treadwell, W. D.; Ammann, A. *Helv. Chim. Acta.* 1938, *21*, 1249.

3. Kirgintsev, A. N.; Kashina, N. I.; Vulikh, A. I.; Korotkevich, B. I. *Zh. Neorg. Khim.* 1965, *10*, 1225; *Russ, J. Inorg. Chem. (Engl. Transl.)* 1965, *10*, 662.

4. Breusov, O. N.; Kahina, N. I.; Revzina, T. V.; Sobolevskaya, N. G. *Zh. Neorg. Khim.* 1967, *12*, 2240; *Russ, J. Inorg. Chem. (Engl. Transl.)* 1967, *12*, 1179.

5. Arkhipov, S. M.; Kashina, N. I.; Revzina, T. V. *Zh. Neorg. Khim.* 1968, *13*, 587; *Russ. J. Inorg. Chem. (Engl. Transl.)* 1968, *13*, 304.

6. Arkhipov, S. M.; Kashina, N. I.; Kuzina, V.A. *Zh. Neorg. Khim.* 1968, *13*, 2872; *Russ. J. Inorg. Chem. (Engl. Transl.)* 1968, *13*, 1476.

7. Shklovskaya, R. M.; Arkhipov, S. M.; Kuzina, V. A.; Tsibulevskaya, T. A. *Zh. Neorg. Khim.* 1976, *21*, 2868; *Russ. J. Inorg. Chem. (Engl. Transl.)* 1976, *21*, 1584.

8. Arkhipov, S. M.; Kashina, N. I.; Kuzina, V. A. *Zh. Neorg. Khim.* 1969, *14*, 567; *Russ. J. Inorg. Chem. (Engl. Transl.)* 1969, *14*, 294.

9. Arkhipov, S. M.; Kashina, N. I.; Kuzina, V. A. *Zh. Neorg. Khim.* 1970, *15*, 1640; *Russ. J. Inorg. Chem. (Engl. Transl.)* 1970, *15*, 840.

10. S. Siekierski, T. Mioduski and M. Salomon, Eds. IUPAC *Solubility Data Series* "*Scandium Yttrium, Lanthanum, and Lanthanide Nitrates.*" Volume 13. Pergamon Press, Oxford, 1983.

COMPONENTS:	ORIGINAL MEASUREMENTS:
(1) Rubidium chlorate; $RbClO_3$; [13446-71-4] (2) Water; H_2O; [7732-18-5]	Calzolari, F. *Gazz. Chim. Ital.* 1912, *42*, 85-92.

VARIABLES:	PREPARED BY:
T/K = 273 to 372	B. Scrosati

EXPERIMENTAL VALUES:

Solubility

t/°C	g/100g H_2O	mol kg^{-1} (compiler)
0	2.138	0.1265
8	3.07	0.182
19.8	5.36	0.317
30	8.00	0.474
42.2	12.48	0.739
50	15.98	0.946
76	34.12	2.020
99	62.8	3.72

AUXILIARY INFORMATION

METHOD/APPARATUS/PROCEDURE:	SOURCE AND PURITY OF MATERIALS:
Method of equilibration not specified, but probably the isothermal method was employed. Aliquots of saturated solution for analysis were withdrawn with a pipet. The aliquots were placed in platinum dishes and the water evaporated. The residues were dried at 120°C to constant weight.	Rubidium chlorate was prepared by treating rubidium sulfate with barium chlorate. The product was repeatedly recrystallized until no trace of sulfate and barium were detected. The purity of the salt was checked by volumetrically determining chlorine in the anhydrous chloride dried at 150-160°C.
	ESTIMATED ERROR: Not possible to estimate due to insufficient data.
	REFERENCES:

COMPONENTS:	ORIGINAL MEASUREMENTS:
(1) Rubidium chlorate; $RbClO_3$; [13446-71-4] (2) Water; H_2O; [7732-18-5]	Treadwell, W.D.; Ammann, A. *Helv. Chim. Acta.* <u>1938</u>, *21*, 1249-56.

VARIABLES:	PREPARED BY:
One temperature; 293 K	Hiroshi Miyamoto

EXPERIMENTAL VALUES:

The solubility of rubidium chlorate in water at 20°C was given as:

$$0.32 \text{ mol kg}^{-1}$$

The concentration solubility product was also given simply as the square of the solubility:

$$1.02 \times 10^{-1} \text{ mol}^2 \text{ kg}^{-2}$$

AUXILIARY INFORMATION

METHOD/APPARATUS/PROCEDURE:	SOURCE AND PURITY OF MATERIALS:
No information was given.	No information was given.
	ESTIMATED ERROR: Nothing specified.
	REFERENCES:

COMPONENTS:	ORIGINAL MEASUREMENTS:
(1) Rubidium chlorate; $RbClO_3$; [13446-71-4] (2) Water; H_2O; [7732-18-5]	Breusov, O.N.; Kashina, N.I.; Revzina, T.V.; Sobolevskaya, N.G. *Zh. Neorg. Khim.* 1967, *12*, 2240-3; *Russ. J. Inorg. Chem.* (*Engl. Transl.*) 1967, *12*, 1179-81.
VARIABLES: T/K = 273.2 to 373.2	PREPARED BY: Hiroshi Miyamoto

EXPERIMENTAL VALUES:

Solubility of $RbClO_3$

t/°C	mass %	mol %	mol kg^{-1} (compiler)
0	2.12	0.230	0.128
10	3.44	0.378	0.211
20	5.02	0.561	0.313
25	6.17	0.696	0.389
30	7.47	0.841	0.478
40	10.34	1.228	0.683
50	13.74	1.670	0.943
60	18.10	2.303	1.308
70	22.51	3.005	1.720
80	27.72	3.929	2.270
90	32.91	4.972	2.904
100	38.59	6.281	3.720

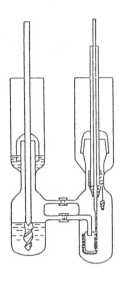

High temp. apparatus

AUXILIARY INFORMATION

METHOD/APPARATUS/PROCEDURE:	SOURCE AND PURITY OF MATERIALS:
Isothermal method. Equilibrium reached in 4-5 h. From 90-100°C, soly detd in apparatus shown in figure. At equilibrium, the apparatus was tilted to allow satd sln to filter through connecting tube into weighed test tubes. The test tube was closed with a stopper, withdrawn, and weighed. Condensation on the walls of the apparatus and loss of water by evaporation was thus prevented. At the lower temperatures, ordinary soly vessels were used, and pipets with glass filters were used for sampling (no other details given). Above 50°C, the pipets were preheated in the thermostat. Saturated solutions analyzed for chlorate by addition of excess ammonium iron(II) sulfate and back-titration of the excess Fe(II) with potassium permanganate.	Results of analysis of $RbClO_3$; Content of $RbClO_3$ 100.2 %. Impurities, %, K <0.05 %; Cs 0.05; Na <0.05.
	ESTIMATED ERROR: Soly: nothing specified. Temp: precision $\pm$ 0.1 K.
	REFERENCES:

COMPONENTS:	ORIGINAL MEASUREMENTS:
(1) Rubidium nitrate; $RbNO_3$; [13126-12-0] (2) Rubidium chlorate; $RbClO_3$; [13446-71-4] (3) Water; H_2O; [7732-18-5]	Shklovskaya, R.M.; Arkhipov, S.M.; Kuzina, V.A.; Tsibulevskaya, T.A. *Zh. Neorg. Khim.* 1976, *21*, 2868-70; *Russ. J. Inorg. Chem. (Engl. Transl.)* 1976, *21*, 1583-4.

VARIABLES:	PREPARED BY:
T/K = 298.2 Composition	Hiroshi Miyamoto

EXPERIMENTAL VALUES: Composition of saturated solutions at 25.0°C

Rubidium Chlorate		Rubidium Nitrate			Nature of the
mass %	mol % (compiler)	mass %	mol % (compiler)	$RbNO_3/RbClO_3$ distrib coeff	solid phase[a]
6.21[b]	0.701	–	–	–	A
5.44	0.629	3.33	0.441	0.008	C
5.05	0.600	6.58	0.896	0.008	"
4.63	0.555	7.92	1.088	0.009	"
3.94	0.479	9.91	1.379	0.008	"
3.88	0.476	11.04	1.553	0.008	"
3.84	0.485	13.78	1.993	0.007	"
3.75	0.482	15.63	2.302	0.008	"
3.45	0.452	17.62	2.643	0.009	"
3.33	0.449	20.36	3.142	0.008	"
3.01	0.419	23.66	3.776	0.008	"
2.85	0.414	27.36	4.551	0.008	"
2.79	0.420	30.46	5.258	0.008	"
2.55	0.409	35.50	6.516	0.009	"
2.57	0.431	38.82	7.453	–	D
2.57	0.431	38.82	7.453	–	"
1.76	0.293	39.19	7.477	–	B
–	–	40.21	7.592	–	"

[a] A = $RbClO_3$;   B = $RbNO_3$;   C = solid solution based on $RbClO_3$;

   D = solid solution based on $RbClO_3$ + $RbNO_3$

[b] For the binary system the compiler computes the following:
      soly of $RbClO_3$ = mol kg^{-1}

AUXILIARY INFORMATION

COMMENTS AND/OR ADDITIONAL DATA:

The distribution coefficients of rubidium nitrate in the chlorate in the range of
crystallization of the solid solution were calculated from the equation

$$D_{RbNO_3/RbClO_3} = (x_1/y_1)\ (1 - y_1/(1-x_1))$$

where x_1 is the mole fraction of rubidium nitrate in the solid phase, and y_1 the mole
fraction of this component in the liquid phase. The results are given in the above
table.

continued...

COMPONENTS:	ORIGINAL MEASUREMENTS:
(1) Rubidium nitrate; RbNO₃; [13126-12-0] (2) Rubidium chlorate; RbClO₃; [13446-71-4] (3) Water; H₂O; [7732-18-5]	Shklovskaya, R.M.; Arkhipov, S.M.; Kuzina, V.A.; Tsibulevskaya, T.A. *Zh. Neorg. Khim.* 1976, *21*, 2868-70; *Russ. J. Inorg. Chem. (Engl. Transl.)* 1976, *21*, 1583-4.

COMMENTS AND/OR ADDITIONAL DATA: (Continued)

The phase diagram is given below (based on mass % units).

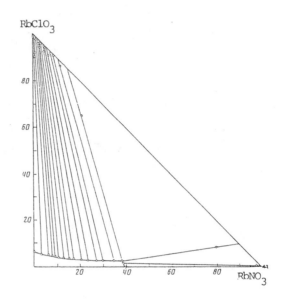

AUXILIARY INFORMATION

METHOD/APPARATUS/PROCEDURE:	SOURCE AND PURITY OF MATERIALS:
Isothermal method. Equilibrium was reached in 20-30 days. Total anion concentration in the liquid phase detd by ion exchange, and chlorate detd by argentometric titrn after reduction to chloride. Nitrate was detd by difference. Specimens of the solid phase were analyzed for chlorate as described above, and for nitrate by reduction to ammonia using Devarda's alloy, volatilization, and colorimetric determination using Nessler's reagent. Solid phase compositions detd by the method of residues, and confirmed from X-ray diffraction patterns.	Highly pure grade RbNO₃ and RbClO₃ were used. No other information given.
	ESTIMATED ERROR: Soly: nothing specified. Temp: precision ± 0.1 K.
	REFERENCES:

COMPONENTS:	ORIGINAL MEASUREMENTS:
(1) Rubidium chloride; RbCl; [7791-11-9] (2) Rubidium chlorate; $RbClO_3$; [13446-71-4] (3) Water; H_2O; [7732-18-5]	Arkhipov, S.M.; Kashina, N.I.; Revezina, T.V. *Zh. Neorg. Khim.* 1968, *13*, 587-8; *Russ. J. Inorg. Chem. (Engl. Transl.)* 1968, *13*, 304.

VARIABLES:	PREPARED BY:
Composition T/K = 273, 298, 323	Hiroshi Miyamoto

EXPERIMENTAL VALUES: Composition of saturated solutions

	$RbClO_3$		RbCl		Nature of the
t/°C	mass %	mol % (compiler)	mass %	mol % (compiler)	solid phase[a]
0	2.12[b]	0.230	0.0	0.0	A
	1.41	0.155	1.95	0.299	"
	0.37	0.051	26.52	5.125	"
	0.29	0.048	42.12	9.821	"
	0.27	0.046	43.16	10.20	A+B
	0.28	0.047	43.11	10.18	"
	0.00	0.000	43.48	10.28	B
25	6.24[b]	0.705	0.0	0.0	A
	5.76	0.651	0.66	0.10	"
	5.36	0.608	1.40	0.222	"
	4.99	0.568	2.18	0.347	"
	4.63	0.529	3.00	0.479	"
	3.83	0.443	5.25	0.849	"
	3.27	0.385	7.70	1.27	"
	2.60	0.316	11.58	1.964	"
	1.91	0.247	18.43	3.324	"
	1.13	0.172	33.90	7.201	"
	0.83	0.15	48.18	12.32	A+B
	0.82	0.15	48.22	12.34	"
	0.34	0.062	48.50	12.37	B
	0.0	0.0	48.60	12.35	"

continued.....

AUXILIARY INFORMATION

METHOD/APPARATUS/PROCEDURE:	SOURCE AND PURITY OF MATERIALS:
The isothermal method was used. At 0°C, glass vessels with an oil seal were immersed in melting ice. At 25 and 50°C, test tubes were mounted in a thermostat with a special device for mixing. The test tubes were rotated at 60 rev min^{-1}, and equilibrium was reached in 10 hours. The liquid and solid phases were analyzed for ClO_3^- by adding an excess of $FeSO_4$ and back-titrating with potassium permanganate. The chloride content was determined by titration with silver nitrate solution with potassium chromate indicator. The composition of the solid phases was found by Schreinemakers' method of residues.	Rubidium chlorate and chloride used had a purity of 99.9 %.
	ESTIMATED ERROR:
	Nothing specified.
	REFERENCES:

COMPONENTS:	ORIGINAL MEASUREMENTS:
(1) Rubidium chloride; RbCl; [7791-11-9]	Arkhipov, S.M.; Kashina, N.I.; Revezina, T.V.
(2) Rubidium chlorate; RbClO$_3$; [13446-71-4]	Zh. Neorg. Khim. 1968, 13, 587-8;
(3) Water; H$_2$O; [7732-18-5]	Russ. J. Inorg. Chem. (Engl. Transl.) 1968, 13, 304.

EXPERIMENTAL VALUES: (Continued)

Composition of saturated solutions

t/°C	RbClO$_3$		RbCl		Nature of the solid phase[a]
	mass %	mol % (compiler)	mass %	mol % (compiler)	
50	13.74[b]	1.670	0.0	0.0	A
	12.21	1.506	3.1	0.53	"
	5.22	0.734	22.92	4.503	"
	1.98	0.387	51.35	14.03	A+B
	1.99	0.390	51.40	14.06	"
	0.0	0.0	52.30	14.04	B

[a] A = RbClO$_3$;   B = RbCl

[b] For the binary system the compiler computes the following:

soly of RbClO$_3$ = 0.128 mol kg^{-1} at 0°C

= 0.394 mol kg^{-1} at 25°C

= 0.9430 mol kg^{-1} at 50°C

COMMENTS AND/OR ADDITIONAL DATA:

The phase diagram for 25°C is given below (based on mass units).

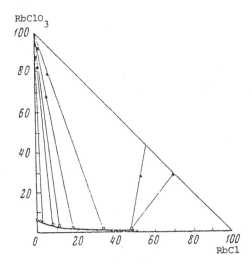

COMPONENTS:	ORIGINAL MEASUREMENTS:
(1) Rubidium chlorate; $RbClO_3$; [13446-71-4] (2) Cesium chlorate; $CsClO_3$; [13763-67-2] (3) Water; H_2O; [7732-18-5]	Kirgintsev, A.N.; Kashina, N.I.; Vulikh, A.I.; Korotkevich, B.I. *Zh. Neorg. Khim.* 1965, *10*, 1225-8; *Russ. J. Inorg. Chem. (Engl. Transl.)* 1965, *10*, 662-4.

VARIABLES:	PREPARED BY:
Composition T/K = 298.2	Hiroshi Miyamoto

EXPERIMENTAL VALUES: Composition of the saturated solutions at 25°C

total salts mol kg^{-1}	rubidium chlorate			cesium chlorate	
	y_1^a	$g_1/100\ g_3$	mass %^{b}	$g_2/100\ g_3$	mass %^{b}
3.62	0.00	0.00	0.00	7.79	7.23^c
3.85	0.12	0.79	0.78	7.31	6.81
4.12	0.23	1.58	1.56	6.89	6.45
4.42	0.33	2.49	2.43	6.37	5.99
4.86	0.46	3.74	3.61	5.73	5.41
5.04	0.54	4.80	4.40	5.02	4.78
4.92	0.62	5.13	4.88	4.08	3.92
4.77	0.67	5.37	5.10	3.44	3.33
4.46	0.78	5.86	5.54	2.15	2.10
4.17	0.86	6.10	5.75	1.23	1.22
3.94	1.00	6.65	6.24^c	0.00	0.00

[a] y_1 = mole fraction of $RbClO_3$ in mixture of chlorates.

[b] calculated by the compiler.

[c] For the binary systems at 25°C the compiler computes the following:

$\qquad$ soly $RbClO_3$ = 0.394 mol kg^{-1} and 0.705 mol %

$\qquad$ soly $CsClO_3$ = 0.360 mol kg^{-1} and 0.645 mol %.

AUXILIARY INFORMATION

METHOD/APPARATUS/PROCEDURE:	SOURCE AND PURITY OF MATERIALS:
Solubility in this system was studied by the isothermal relief of supersaturation method. Weighed amounts of chlorates were dissolved in water in 50 cm^3 test-tubes by heating on a water bath at 65-70°C; the test-tubes were then placed in a thermostat at 25°C for 20 min. Supersaturation was then removed by stirring at 60 rpm for 10 h. After settling two samples of liquid phase were removed for analysis. The first was evaporated in a drying cupboard at 70-80°C and then dried to constant weight at 105°C. The other sample was analyzed for ClO_3^- by adding $FeSO_4$ sln and back-titrating excess iron(II) with permanganate solution.	The purity of chlorates used was 99.9 % or better.
	ESTIMATED ERROR: Soly: accuracy of $y_1 \pm 0.01$ (authors). Temp: precision $\pm$ 0.1 K (authors).
	REFERENCES:

COMPONENTS:	ORIGINAL MEASUREMENTS:
(1) Rubidium chloride; RbCl; [7791-11-9] (2) Rubidium chlorate; RbClO$_3$; [13446-71-4] (3) Cesium chloride; CsCl; [7647-17-8] (4) Cesium chlorate; CsClO$_3$; [13763-67-2] (5) Water; H$_2$O; [7732-18-5]	Arkhipov, S.M.; Kashina, N.I.; Kuzina, V.A. *Zh. Neorg. Khim.* 1970, *15*, 1640-2; *Russ. J. Inorg. Chem. (Engl. Transl.)* 1970, *15*, 840-2.
VARIABLES: Composition; T/K = 298	PREPARED BY: Hiroshi Miyamoto

EXPERIMENTAL VALUES:

Composition of saturated solutions

Rubidium Chloride mass %	Rubidium Chloride mol % (compiler)	Rubidium Chlorate mass %	Rubidium Chlorate mol % (compiler)	Cesium Chloride mass %	Cesium Chloride mol % (compiler)	Cesium Chlorate mass %	Cesium Chlorate mol % (compiler)	Density g cm^{-3}	Nature of the solid phase[a]
-	-	-	-	65.83	17.30	0.54	0.11	1.868	A+B
0.33	0.12	0.23	0.060	65.47	17.14	0.20	0.041	-	E+G
1.43	0.521	0.24	0.063	64.35	16.83	0.21	0.043	-	"
2.42	0.891	0.26	0.069	63.86	16.88	0.22	0.045	-	"
3.98	1.46	0.28	0.073	62.07	16.31	0.24	0.049	-	"
7.38	2.62	0.33	0.084	57.34	14.62	0.27	0.054	1.897	"
14.07	5.107	0.37	0.096	51.91	13.53	0.32	0.065	1.897	F+G
15.85	5.719	0.38	0.098	49.93	12.94	0.32	0.065	1.906	"
16.60	6.037	0.39	0.10	49.56	12.95	0.33	0.067	1.886	"
17.44	6.265	0.40	0.10	48.17	12.43	0.34	0.068	1.831	"
19.36	6.750	0.41	0.10	44.92	11.25	0.34	0.066	1.794	"
19.49	6.588	0.47	0.11	43.17	10.48	0.40	0.076	1.755	"
23.56	7.527	0.51	0.12	36.38	8.348	0.41	0.073	1.613	"
29.48	8.901	0.51	0.11	27.69	6.005	0.42	0.071	1.532	"
35.04	10.11	0.52	0.11	19.80	4.103	0.43	0.069	-	"
40.36	11.08	0.52	0.10	11.75	2.316	0.44	0.067	-	"
40.89	11.15	0.54	0.11	10.84	2.123	0.45	0.069	-	"
45.08	11.83	0.56	0.11	4.36	0.821	0.49	0.072	-	"
48.20	12.33	0.82	0.15	-	-	-	-	1.466	C+D

[a] A = CsCl; B = CsClO$_3$; C = RbCl; D = RbClO$_3$; E = Cs(Rb)Cl solid soln;

F = Rb(Cs)Cl solid soln; G = (Rb, Cs)ClO$_3$ solid soln.

continued.....

COMPONENTS:

(1) Rubidium chloride; RbCl; [7791-11-9]

(2) Rubidium chlorate; $RbClO_3$; [13446-71-4]

(3) Cesium chloride; CsCl; [7647-17-8]

(4) Cesium chlorate; $CsClO_3$; [13763-67-2]

(5) Water; H_2O; [7732-18-5]

ORIGINAL MEASUREMENTS:

Arkhipov, S.M.; Kashina, N.I.;
Kuzina, V.A.

Zh. Neorg. Khim. 1970, *15*, 1640-2;
Russ. J. Inorg. Chem. (Engl. Transl.)
1970, *15*, 840-2.

EXPERIMENTAL VALUES:

AUXILIARY INFORMATION

METHOD/APPARATUS/PROCEDURE:

The method of the solubility measurement was
similar to that described in ref 1.
The isothermal method was used. Equilibrium
was reached in 30 hours. Samples of the
solid and liquid phases were analyzed.
Rubidium and cesium content were determined
by flame photometry.
Chlorate was determined by adding an excess
of iron(II) sulfate to a solution of the
specimen and back-titrating with potassium
permanganate.
The densities of the saturated solutions
were also measured.

SOURCE AND PURITY OF MATERIALS:

"Chemically pure" grade salts with 99.8% or
more purity were used.

ESTIMATED ERROR:

Nothing specified.

REFERENCES:

1. Arkhipov, S.M.; Kashina, N.I.;
Kuzina, V.A. *Zh. Neorg. Khim.*
1968, *13*, 2872; *Russ. J. Inorg. Chem.*
(Engl. Transl.) 1968, *13*.

COMPONENTS:	EVALUATOR:
(1) Cesium chlorate; $CsClO_3$; [13763-67-2] (2) Water; H_2O; [7732-18-5]	H. Miyamoto Department of Chemistry Niigata University Niigata, Japan and M. Salomon US Army ET & DL Fort Monmouth, NJ, USA December, 1984

CRITICAL EVALUATION: THE BINARY SYSTEM

Data for the solubility of $CsClO_3$ in water have been reported in 6 publications (1-6). The publications (1,4) report solubilities as a function of temperature (273-373 K), Treadwell and Ammann (2) report the solubility at 293.2 K, and in (3, 5, 6) solubilities in ternary systems are reported for 298.2 K. In (3 and 6), the detailed phase studies showed that at 298.2 K, the solid phase in the binary system is the anhydrous salt. Although no other study reports the nature of the solid phase over the experimental temperature range of 273-373 K, the evaluators assume it to be the anhydrous salt since when the solubility is plotted as a function of temperature, all data lie on a smooth monotonic curve indicating a single solid phase. Except for three rejected data points (see below), all data could be easily fitted to a single smoothing equation again indicating a single solid phase.

Table 1 summarizes the solubility data for $CsClO_3$ in water as a function of temperature. With the exception of Treadwell and Ammann's (2) use of mol kg^{-1} units, all other original data were reported in mass units. In Table 1 the evaluators have converted from the original units to mole fraction units, and both the original units as well as conversions to mol kg^{-1} units can be found in the compilations.

Table 1. Summary of Solubilities in the Binary System from 273-373 K[a]

T/K	χ	(ref)	T/K	χ	(ref)
273.2	0.002044	(1)	303.2	0.007873	(1)
273.2	0.002061	(4)	303.2	0.007803	(4)
281.2	0.002906	(1)	313.2	0.01149	(4)
283.2	0.003440	(4)	315.4	0.01229	(1)
293.0	0.005202	(1)	323.2	0.01590	(1)
293.2	0.005212	(4)	323.2	0.01610	(4)
293.2	0.0052	(2)	333.2	0.02177	(4)
298.2	0.006486	(4)	343.2	0.02896	(4)
298.2	0.006467	(5)	350.2	0.03352	(1)[e]
298.2	0.006448	(6)	353.2	0.03727	(4)[e]
298.2	0.006457	(6)[b]	363.2	0.04838	(4)
298.2	0.006419	(3)[c]	372.2	0.05988	(1)[e]
298.2	0.006448	(3)[d]	373.2	0.06073	(4)

[a]Mole fraction units converted from original mass units by the evaluators.

[b]Compilation of this work given in the $NaClO_3$ chapter.

[c]Compilation of this work given in the $KClO_3$ chapter.

[d]Compilation of this work given in the $RbClO_3$ chapter.

[e]Rejected data points. See text for discussion.

Giving all mole fraction solubilities in Table 1 equal weights of unity, three data points at 350.2, 353.2 and 372.2 K had to be rejected as their differences between calculated and observed solubilities exceeded twice the standard error of estimate. The remaining 23 data points were fitted to the smoothing equation with the following results:

$$Y_x = -15469.648/(T/K) - 34.5278 \ln(T/K) + 233.128 + 2.2681 \times 10^{-3}(T/K) \qquad [1]$$

$$\sigma_y = 0.026 \qquad\qquad\qquad \sigma_x = 6.7 \times 10^{-5}$$

COMPONENTS:	EVALUATOR:
(1) Cesium chlorate; $CsClO_3$; [13763-67-2] (2) Water; H_2O; [7732-18-5]	H. Miyamoto Department of Chemistry Niigata University Niigata, Japan and M. Salomon US Army ET & DL Fort Monmouth, N.I., USA December, 1984

CRITICAL EVALUATION:

The 23 acceptable data points were also fitted to the smoothing equation based on mol kg^{-1} units, and the resulting equation is:

$$Y_m = -5780.517/(T/K) - 7.1403 \ln(T/K) + 59.864 \qquad [2]$$

$$\sigma_y = 0.013 \qquad\qquad \sigma_m = 0.008$$

The solubilities calculated from the two smoothing equations are designated as *recommended* solubilities, and values at rounded temperatures are given in Table 2.

Table 2. Recommended Solubilities Calculated from the Smoothing Equations [1] and [2]

T/K	m/mol kg^{-1}	χ
273.2	0.114	0.002040
278.2	0.146	0.002627
283.2	0.186	0.003342
288.2	0.234	0.004204
293.2	0.291	0.005230
298.2	0.359	0.006441
303.2	0.439	0.007858
308.2	0.532	0.009500
313.2	0.640	0.01139
318.2	0.764	0.01355
323.2	0.906	0.01599
328.2	1.066	0.01875
333.2	1.246	0.02184
338.2	1.448	0.02527
343.2	1.672	0.02908
348.2	1.921	0.03327
353.2	2.195	0.03787
358.2	2.495	0.04289
363.2	2.822	0.04835
368.2	3.177	0.05427
373.2	3.561	0.06006

TERNARY SYSTEMS

Data for solubilities in ternary systems have been reported in 3 publications (3, 5, 6). A summary of these studies is given in Table 3. Two ternary systems $CsClO_3-KClO_3-H_2O$ and $CsClO_3-RbClO_3-H_2O$ were studied by Kirgintsev, Kashina, Vulikh and Korotkevich (3). Solid solutions were not formed in the former system, but in the latter, rubidium and cesium chlorate form a continuous series of solid solutions.

Kirgintsev and Kizitskii (5) studied solubilities in the $CsClO_3-Ca(ClO_3)_2-H_2O$ system at 298 K, but did not study the system at high concentrations of calcium chlorate owing to high solution viscosity. Compositions of the solid phases were not reported.

Two ternary systems $CsClO_3-CsCl-H_2O$ and $CsClO_3-NaClO_3-H_2O$ were studied by Arkhipov and Kashina (6) using the isothermal method. The solid phases were CsCl and $CsClO_3$ in the former system, and $CsClO_3$ and $NaClO_3$ in the latter. Neither double salts nor solid solutions were reported.

COMPONENTS:	EVALUATOR:
(1) Cesium chlorate; $CsClO_3$; [13763-67-2] (2) Water; H_2O; [7732-18-5]	H. Miyamoto Department of Chemistry Niigata University Niigata, Japan December, 1984

CRITICAL EVALUATION:

Table 3. Summary of Solubility Studies in the Ternary Systems

Ternary system	T/K	Solid Phase	Ref
$CsClO_3$ - $KClO_3$ - H_2O	298	$CsClO_3$; $KClO_3$	3[a]
$CsClO_3$ - $RbClO_3$ - H_2O	298	$CsClO_3$; $RbClO_3$; Solid Sln	3[b]
$CsClO_3$ - $Ca(ClO_3)_2$ - H_2O	298	Not Given	5
$CsClO_3$ - $CsCl$ - H_2O	298	$CsClO_3$; $CsCl$	6
$CsClO_3$ - $NaClO_3$ - H_2O	298	$CsClO_3$; $NaClO_3$	6[c]

[a]See the $KClO_3$ chapter for this compilation.

[b]See the $RbClO_3$ chapter for this compilation

[c]See the $NaClO_3$ chapter for this compilation.

OTHER MULTICOMPONENT SYSTEMS

The $CsClO_3$-$CsCl$-$NaClO_3$-$NaCl$-H_2O system was studied by Arkhipov and Kashina (6) at 298, 323 and 348 K. Solubilities in the quaternary systems, $CsClO_3$-$CsCl$-$NaCl$-H_2O and $CsClO_3$-$NaClO_3$-$NaCl$-H_2O, have been reported (6). Solid phases found in this study were the four pure components ($CsCl$, $CsClO_3$, $NaCl$, $NaClO_3$), and solid solutions formed from $NaCl$ + $CsCl$.

The $RbClO_3$-$RbCl$-$CsClO_3$-$CsCl$-H_2O system was studied by Arkhipov, Kashina and Kuzina (7) at 298 K. Rubidium and cesium chloride formed a restricted series of solid solutions, and cesium and rubidium chlorate solid solutions were also reported.

REFERENCES

1. Calzolari, F. *Gazz. Chim. Ital.* 1912, *42*, 85.

2. Treadwell, W. D.; Ammann, A. *Helv. Chim. Acta* 1938, *21*, 1249.

3. Kirgintsev, A. N.; Kashina, N. I.; Vulikh, A. I.; Korotkevich, B. I. *Zh. Neorg. Khim.* 1965, *10*, 1225; *Russ. J. Inorg. Chem. (Engl. Transl.)* 1965, *10*, 662.

4. Breusov, O. N.; Kashina, N. I.; Revzina, T. V.; Sobolevskaya, N. G. *Zh. Neorg. Khim.* 1976, *12*, 2240; *Russ. J. Inorg. Chem. (Engl. Trans.)* 1967, *12*, 1179.

5. Kirgintsev, A. N.; Kozitskii, V. P. *Zh. Neorg. Khim.* 1968, *13*, 3342; *Russ. J. Inorg. Chem. (Engl. Transl.)* 1968, *13*, 1723.

6. Arkhipov, S. M.; Kashina, N. I. *Zh. Neorg. Khim.* 1970, *15*, 760; *Russ. J. Inorg. Chem. (Engl. Transl.)* 1970, *15*, 391.

7. Arkhipov, S. M.; Kashina, N. I.; Kuzina, V. A. *Zh. Neorg. Khim.* 1970, *15*, 1640; *Russ. J. Inorg. Chem. (Engl. Transl.)* 1970, *15*, 840.

COMPONENTS:	ORIGINAL MEASUREMENTS:
(1) Cesium chlorate; $CsClO_3$; [13763-67-2] (2) Water; H_2O; [7732-18-5]	Calzolari, F. *Gazz. Chim. Ital.* 1912, 42, 85-92.
VARIABLES: T/K = 273 to 372	PREPARED BY: B. Scrosati

EXPERIMENTAL VALUES:

Solubility

t/°C	g/100g H_2O	mol kg^{-1} (compiler)
0	2.46	0.114
8	3.50	0.162
19.8	6.28	0.290
30	9.53	0.440
42.2	14.94	0.667
50	19.40	0.897
77	41.65	1.925
99	76.5	3.54

AUXILIARY INFORMATION

METHOD/APPARATUS/PROCEDURE:	SOURCE AND PURITY OF MATERIALS:
Method of equilibration not specified, but probably the isothermal method was employed. Aliquots of saturated solution for analysis were withdrawn with a pipet. The aliquots were placed in platinum dishes and the water evaporated. The residues were dried at 120°C to constant weight.	Cesium chlorate was prepared by treating cesium sulfate with barium chlorate. The product was repeatedly recrystallized until no trace of sulfate and barium were detected. The purity of the salt obtained was checked by volumetrically determining chlorine in the anhydrous chloride dried at 150-160°C. The result was not given.
	ESTIMATED ERROR: Not possible to estimate due to insufficient data.
	REFERENCES:

COMPONENTS:	ORIGINAL MEASUREMENTS:
(1) Cesium chlorate; $CsClO_3$; [13763-67-2] (2) Water; H_2O; [7732-18-5]	Treadwell, W.D.; Ammann, A. *Helv. Chim. Acta.* <u>1938</u>, *21*, 1249-56.

VARIABLES:	PREPARED BY:
One temperature: 293 K	Hiroshi Miyamoto

EXPERIMENTAL VALUES:

The solubility of cesium chlorate in water at 20°C was given as:

$$0.29 \text{ mol kg}^{-1}$$

The concentration solubility product was also given simply as the square of the solubility:

$$8.41 \times 10^{-2} \text{ mol}^2 \text{ kg}^{-2}$$

AUXILIARY INFORMATION

METHOD/APPARATUS/PROCEDURE:	SOURCE AND PURITY OF MATERIALS:
No information was given.	No information was given.
	ESTIMATED ERROR: Nothing specified.
	REFERENCES:

COMPONENTS:	ORIGINAL MEASUREMENTS:
(1) Cesium chlorate; $CsClO_3$; [13763-67-2] (2) Water; H_2O; [7732-18-5]	Breusov, O.N.; Kashina, N.I.; Revzina, T.V.; Sovolevskaya, N.G. *Zh. Neorg. Khim.* <u>1967</u>, *12*, 2240-3; *Russ. J. Inorg. Chem. (Engl. Transl.)* <u>1967</u>, *12*, 1179-81.
VARIABLES: T/K = 273.2 to 373.2	PREPARED BY: Hiroshi Miyamoto

EXPERIMENTAL VALUES:

	Solubility of $CsClO_3$		
t/°C	mass %	mol %	mol kg^{-1} (compiler)
0	2.42	0.206	0.115
10	3.98	0.344	0.192
20	5.92	0.521	0.291
25	7.27	0.649	0.362
30	8.63	0.780	0.437
40	12.25	1.149	0.645
50	16.42	1.609	0.908
60	21.09	2.177	1.235
70	26.37	2.896	1.655
80	31.74	3.872	2.149
90	37.91	4.838	2.822
100	43.71	6.073	3.589

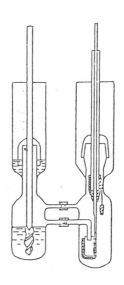

High temperature apparatus

AUXILIARY INFORMATION

METHOD/APPARATUS/PROCEDURE:	SOURCE AND PURITY OF MATERIALS:
Isothermal method. Equilibrium reached in 4-5 h. From 90-100°C, soly detd in apparatus shown in figure. At equilibrium, the apparatus was tilted to allow satd sln to filter through connecting tube into weighed test tubes. The test tube was closed with a stopper, withdrawn, and weighed. Condensation on the walls of the apparatus and loss of water by evaporation was thus prevented. At the lower temperatures, ordinary soly vessels were used, and pipets with glass filters were used for sampling (no other details given). Above 50°C, the pipets were preheated in the thermostat. Satd slns analyzed for chlorate by addition of excess ammonium iron (II) sulfate and back-titration of the excess Fe(II) with potassium permanganate.	Results of analysis of $CsClO_3$; Content of $CsClO_3$ 100.0 % Impurities, %, K <0.05 %; Rb <0.25; Na <0.05.
	ESTIMATED ERROR: Soly: nothing specified. Temp: precision ± 0.1 K.
	REFERENCES:

COMPONENTS:	ORIGINAL MEASUREMENTS:
(1) Cesium chloride; CsCl; [7647-17-8]	Arkhipov, S.M.; Kashina, N.I.
(2) Cesium chlorate; CsClO$_3$; [13763-67-2]	Zh. Neorg. Khim. 1970, 15, 760-4; Russ. J. Inorg. Chem. (Engl. Transl.) 1970, 15, 391-2.
(3) Water; H$_2$0; [7732-18-5]	

VARIABLES:	PREPARED BY:
T/K = 298.2	Hiroshi Miyamoto
Composition	

EXPERIMENTAL VALUES: Composition of saturated solutions

Cesium Chlorate		Cesium Chloride		Nature of the solid phase[a]
mass %	mol % (compiler)	mass %	mol % (compiler)	
7.23[b]	0.645	--	--	A
5.28	0.479	3.75	0.437	"
3.50	0.336	11.35	1.402	"
2.18	0.233	22.41	3.075	"
1.50	0.182	33.62	5.244	"
0.99	0.15	48.30	9.236	"
0.72	0.13	59.03	13.55	"
0.56	0.11	65.39	17.03	"
0.54	0.11	65.92	17.36	A+B
0.53	0.11	65.78	17.26	"
--	--	66.05	17.23	B

[a] A = CsClO$_3$;   B = CsCl

[b] For the binary system the compiler computes the following:

soly of CsClO$_3$ = 0.360 mol kg^{-1}

AUXILIARY INFORMATION

METHOD/APPARATUS/PROCEDURE:	SOURCE AND PURITY OF MATERIALS:
Solubilities were determined by the isothermal method by mixing solid and liquid phases in glass test-tubes in a water thermostat. Specimens of the liquid and solid phases were analyzed for the anions and cesium. Chloride was titrated with silver nitrate solution using potassium chromate as an indicator. Chlorate ion concentration was determined volumetrically by adding an excess of iron(II) sulfate solution and titrating the excess Fe(II) with potassium permanganate solution. Cesium was determined gravimetrically as cesium tetraphenylborate. The solid phases were identified by the method of residues, and by X-ray diffraction.	C.p. grade CsClO$_3$ and CsCl with a purity of 99.9 % or more were used.

ESTIMATED ERROR:
Soly: nothing specified.
Temp: precision ± 0.1 K.

COMMENTS AND/OR ADDITIONAL DATA:

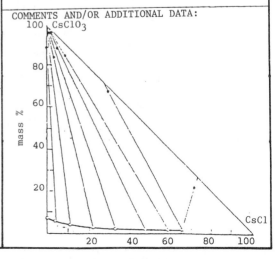

COMPONENTS:	ORIGINAL MEASUREMENTS:
(1) Cesium chlorate; $CsClO_3$; [13763-67-3]	Kirgintsev, A.N.; Kozitskii, V.P.
(2) Calcium chlorate; $Ca(ClO_3)_2$; [10137-79-3]	$Zh.\ Neorg.\ Khim.$ 1968, 13, 3342-5; $Russ.\ J.\ Inorg.\ Chem.\ (Engl.\ Transl.)$ 1968, 13, 1723-5.
(3) Water; H_2O; [7732-18-5]	

VARIABLES:	PREPARED BY:
Composition at 298.2 K	Hiroshi Miyamoto

EXPERIMENTAL VALUES: Composition of saturated solutions

	Cesium Chlorate		Calcium Chlorate	
mass %	mol % (compiler)	mole fraction[a] (y_1)	mass %	mol % (compiler)
7.25[b]	0.673	1	0	0
6.37	0.568	0.867	0.94	0.088
5.79	0.518	0.760	1.75	0.164
5.08	0.457	0.613	3.06	0.288
4.16	0.380	0.417	5.58	0.533
3.15	0.294	0.260	8.59	0.838
2.04	0.216	0.084	21.29	2.355
1.77	0.215	0.049	32.70	4.153
2.05	0.318	0.039	48.64	7.882
2.36	0.403	0.040	53.72	9.582

[a] The mole fraction of cesium chlorate calculated without allowance for the water.

[b] The solubility of $CsClO_3$ in aqueous $Ca(ClO_3)_2$ solutions was given by the following smoothing equation in the original paper.

$$\log m_1 = -0.443 - 0.991 \log y_1 - 0.394\ (1-y_1)$$

where m_1 is the solubility in units of mol kg^{-1}

AUXILIARY INFORMATION

METHOD/APPARATUS/PROCEDURE:
The solubility was measured by the method of isothermal relief of supersaturation. Equilibrium was reached in 6-8 hours. An Apparatus used for analysis of cesium is shown in the figure below

SOURCE AND PURITY OF MATERIALS:
Analytical reagent grade cesium and calcium chlorate were used.

ESTIMATED ERROR:

Soly: nothing specified.
Temp: precision $\pm$ 0.05°C

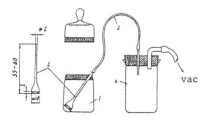

Samples of satd sln to be analyzed were placed in container 1 which had been previously weighed together with the filter stick. The precipitant (1 % aqueous sln of sodium tetraphenylborate) was added dropwise to the sample solution over a period of 30 min, the first portions were added especially slowly. The precipitate was allowed to settle, and the mother-liquor withdrawn through the filter stick and transferred into beaker 4 through the fine polyvinyl chloride tube 3. The precipitate was washed twice with 0.06 % aqueous sodium tetraphenylborate solution, then four or five times with distilled water. The container with the precipitate and filter stick was dried for 1.5 hours at 105°C, cooled and weighed.
The calcium content of the solution in beaker 4 was determined by complexometric titration with Trilon B.

COMPONENTS:	EVALUATOR:
(1) Lithium Bromate; LiBrO₃; [13550-28-2] (2) Water; H₂0; [7732-18-5]	H. Miyamoto Niigata University Niigata, Japan and M. Salomon US Army ET & DL Fort Monmouth, NJ, USA September, 1984

CRITICAL EVALUATION: THE BINARY SYSTEM

Data for the solubility of $LiBrO_3$ in water has been reported in four publications (1-4). Mylius and Funk (1) reported the solubility at 291 K, but a typographical error appears to exist: i.e., they reported the solubility as 60.4 mass % or 153.7 g/100g H_2O, the latter being equivalent to 60.58 mass % (evaluators). While it would appear that the original experimental quantity is the 60.4 mass % value, both values are still too low for 291 K, and were therefore rejected (see below).

Simmons and Waldeck (2) reported solubilities over the temperature range of 278-373 K, and Averko-Antonovich (3) reported results over the wide temperature range of 228-416 K. Chemical analyses of the solid phases showed that above 323 K the solid phase is the anhydrous salt, and below 323 K the solid phase is the monohydrate $LiBrO_3.H_2O$ [55698-66-3]. The existence of the monohydrate as the solid phase in the binary system was confirmed by Campbell et al. (4) who studied ternary systems, and using the Schreinemakers' method of wet residues found the monohydrate at 298.2 K.

A summary of the experimental solubilities are given in Table 1. In this table, the evaluators converted the original mass % units to mole fraction units, and the original units and conversions to mol kg^{-1} units can be found in the compilations.

Table 1. Summary of Experimental Solubilities as a Function of Temperature

T/K	χ	Solid Phase	(ref)	T/K	χ	Solid	(ref)
272.1	0.01511	ice	(3)	298.2	0.2016	$LiBrO_3.H_2O$	(2)
268.4	0.03291	"	(3)	298.2	0.2033	"	(4)
263.4	0.05563	"	(3)	308.2	0.2172	"	(2)
253.0	0.08178	"	(3)	309.1	0.2194	"	(3)
233.2	0.1264	"	(3)	318.2	0.2411		(3)
				323.2	0.2510	"	(2)
228.2	0.1399	$LiBrO_3.H_2O$	(3)	323.2	0.2538	"	(3)
233.2	0.1423	"	(3)				
236.5	0.1463	"	(3)	277.2	0.2119	$LiBrO_3$	(3)
241.7	0.1494	"	(3)	290.7	0.2227	"	(3)
246.7	0.1520	"	(3)	318.2	0.2538	"	(3)
252.2	0.1568	"	(3)	318.2	0.2411[a]	"	(3)
256.7	0.1585	"	(3)	326.2	0.2595	"	(2)
262.4	0.1635	"	(3)	328.2	0.2626	"	(3)
266.4	0.1681	"	(3)	329.2	0.2614	"	(2)
273.2	0.1742	"	(3)	338.2	0.2740	"	(3)
278.2	0.1765	"	(2)	343.7	0.2786	"	(2)
288.2	0.1873	"	(2)	353.2	0.2955	"	(3)
291.2	0.1693[a]	"	(1)	358.2	0.2996	"	(2)
291.2	0.1704[a]	"	(1)	373.2	0.3214	"	(2)
293.2	0.1954	"	(3)	373.7	0.3292	"	(3)
298.1	0.2026	"	(3)	384.2	0.3345[a]	"	(3)
				394.2	0.3659	"	(3)
				416.2	0.4233	"	(3)

[a]Rejected data points (see text for discussion).

All data in the above table were fitted to the smoothing equation.

$$Y_x = A/(T/K) + B\ln(T/K) + C + D(T/K)$$ [1]

COMPONENTS:	EVALUATOR:
(1) Lithium Bromate; LiBrO₃; [13550-28-2] (2) Water; H₂O; [7732-18-5]	H. Miyamoto Niigata University Niigata, Japan and M. Salomon US Army ET & DL Fort Monmouth, NJ, USA September, 1984

CRITICAL EVALUATION:

The complex function Y_x in eq. [1] has been defined previously in the PREFACE and in previous critical evaluations (e.g. see the critical evaluation for the binary LiClO₃–H₂O system). In applying eq. [1] to the mole fraction solubilities in Table 1, we examined each polytherm individually. For each polytherm, all data were initially fitted to eq. [1] and data points rejected when the difference between the experimental and calculated solubilities exceeded two times the standard error of estimate (i.e. for $X_{exptl} - X_{calcd} > 2\sigma_x$). Based on the final tentative and recommended solubilities, the evaluators plotted small portions of the phase diagram in the regions of phase transitions, and our results for the transition temperatures are given below.

Polytherm for Ice as the Solid Phase. The only data reported for this region of the phase diagram are those of Averko-Antonovich (3). Analyses by fitting these data to eq. [1] showed that all data were acceptable. The smoothing equation for these data is:

$$Y_x = -63264.50/(T/K) - 510.3943 \ln(T/K) + 2812.7729 + 1.032927(T/K) \qquad [2]$$

$$\sigma_y = 0.0025 \qquad\qquad \sigma_x = 0.0013$$

The smoothed solubilities calculated from eq. [2] are designated as *tentative*, and values at rounded temperatures are given in Table 2.

Polytherm for LiBrO₃·H₂O as the Solid Phase. All data from references (1-4) were fitted to eq. [1], and only the data of Mylius and Funk (1) had to be rejected. The source of the error in this study cannot be ascertained, but the authors did state that the solid phase was the anhydrous salt. The error in identification of the solid phase and in the low values of the solubility at 291 K indicates a systematic error. The smoothed solubilities based on the data from references (2-4) are designated as *recommended* solubilities, and values at rounded temperatures are given in Table 2. The smoothing equation is given in eq. [3].

$$Y_x = -6721.204/(T/K) - 53.5559 \ln(T/K) + 293.2835 + 0.113089(T/K) \qquad [3]$$

$$\sigma_y = 0.0071 \qquad\qquad \sigma_x = 0.0011$$

The data of Averko-Antonovich (3) indicates a transition from ice as the solid phase to the monohydrate at around 228 K. The evaluators' graphical analysis of the solubility data in this region show this transition to occur at 230.0 K at a mole fraction solubility of $X = 0.1408$.

Polytherm for LiBrO₃ as the Solid Phase. The only solubility data for which the solid phase is the anhydrous salt was reported in (2 and 3). Two data points at 318.2 K and 384.2 K from (3) had to be rejected, but the remaining 14 data points could be fitted to the smoothing equation with the following results:

$$Y_x = -4827.171/(T/K) - 31.1588 \ln(T/K) + 174.786 + 0.056935(T/K) \qquad [4]$$

$$\sigma_y = 0.011 \qquad\qquad \sigma_x = 0.0021$$

The smoothed solubilities calculated from eq. [4] are designated as *recommended* values and are given in Table 2.

Both Simmons and Waldeck (2) and Averko-Antonovich (3) found the temperature for the LiBrO₃·H₂O ⟶ LiBrO₃ transition to be 325 K, and the evaluators find this transition to occur at 325.3 K at a solubility of $X = 0.2587$.

Simmons and Waldeck attempted to measure the melting point of anhydrous LiBrO₃, and although some decomposition was observed, an "average" value of 521 K was reported. Graphical extrapolation (2) yielded a melting point of 533 K, and eq. [4] predicts a value of 502.8 K.

COMPONENTS:	EVALUATOR:
(1) Lithium Bromate; $LiBrO_3$; [13550-28-2] (2) Water; H_2O; [7732-18-5]	H. Miyamoto Niigata University Niigata, Japan and M. Salomon US Army ET & DL Fort Monmouth, NJ,USA September, 1984

CRITICAL EVALUATION:

All tentative and recommended mole fraction solubilities are given in Table 2, and the complete phase diagram based on these data is given in Figure 1.

Table 2.   Tentative and Recommended Solubilities Calculated from
the Smoothing Equations [2], [3] and [4]

T/K	ice[a]	$LiBrO_3 \cdot H_2O$ [b]	$LiBrO_3$ [b]
228.2	0.1491^m	0.1397	
230.0[c]	0.1408	0.1408	
233.2	0.1264	0.1432	
243.2	0.09969	0.1501	
253.2	0.08160	0.1570	
263.2	0.05544	0.1645	
268.2	0.03512	0.1686	
273.2		0.1730	
278.2		0.1777	0.2123^m
283.2		0.1829	0.2170^m
293.2		0.1949	0.2264^m
298.2		0.2019	0.2312^m
303.2		0.2097	0.2360^m
313.2		0.2284	0.2460^m
323.2		0.2527	0.2565^m
325.3[c]		0.2587	0.2587
333.2			0.2677
343.2			0.2800
353.2			0.2934
363.2			0.3082
373.2			0.3255
383.2			0.3431
393.2			0.3639
403.2			0.3875
413.2			0.4144

[a]Tentative solubilities.

[b]Recommended solubilities

[c]Transition temperatures evaluated graphically by the evaluators.

[m]Metastable solubilities.

COMPONENTS:	EVALUATOR:
(1) Lithium Bromate; $LiBrO_3$; [13550-28-2] (2) Water; H_2O; [7732-18-5]	H. Miyamoto Niigata University Niigata, Japan and M. Salomon US Army ET & DL Fort Monmouth, NJ, USA September, 1984

CRITICAL EVALUATION:

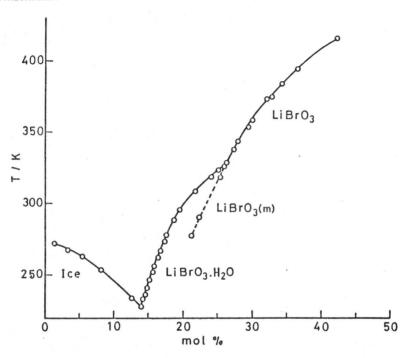

REFERENCES

1. Mylius, F.; Funk, R. *Ber. Dtsch, Chem. Ges.* <u>1897</u>, *30*, 1716.

2. Simmons, J. P.; Waldeck, W. F. *J. Am. Chem. Soc.* <u>1931</u>, *53*, 1725.

3. Averko-Antonovich, I. N. *Zh. Obsch. Khim.* <u>1943</u>, *13*, 272.

4. Campbell, A. N.; Kartzmark, E. M.; Musbally, G. M. *Can. J. Chem.* <u>1967</u>, *45*, 803.

COMPONENTS:	ORIGINAL MEASUREMENTS:
(1) Lithium bromate; $LiBrO_3$; [13550-28-2] (2) Water; H_2O; [7732-18-5]	Mylius, F.; Funk, R. *Ber. Dtsch. Chem. Ges.* <u>1897</u>, *30*, 1716-25.
VARIABLES: T/K = 291	PREPARED BY: Hiroshi Miyamoto

EXPERIMENTAL VALUES:

The solubility of $LiBrO_3$ in water at 18°C was given as follows:

60.4 mass %	(authors)
153.7 g/100g H_2O	(authors)
11.40 mol kg^{-1}	(compiler)

Authors state that the solid phase is the anhydrous salt.

The density of the saturated solution was also given as:

$$1.833 \text{ g cm}^{-3}.$$

AUXILIARY INFORMATION

METHOD/APPARATUS/PROCEDURE:	SOURCE AND PURITY OF MATERIALS:
The salt and water were placed in a bottle. The bottle was agitated in a constant temperature bath for an unspecified time. After the saturated solution settled, an aliquot for analyses was withdrawn with a pipet, and $LiBrO_3$ was determined by evaporation to dryness. The density of the saturated solution was also determined.	The salt was purchased as a "pure chemical", and trace impurities were not present. "The purity sufficed for the solubility determination."
	ESTIMATED ERROR: Soly: precision within 1 %. Temp: nothing specified.
	REFERENCES:

COMPONENTS:	ORIGINAL MEASUREMENTS:
(1) Lithium bromate; $LiBrO_3$; [13550-28-2] (2) Water; H_2O; [7732-18-5]	Simmons, J.P.; Waldeck, W.F. J. Am. Chem. Soc. 1931, 53, 1725-7.
VARIABLES: T/K = 278 - 373	**PREPARED BY:** Hiroshi Miyamoto and Mark Salomon

EXPERIMENTAL VALUES:

Solubility of $LiBrO_3$

t/°C	mass %	mol % (compiler)	mol kg^{-1} (compiler)
5	61.6	17.6	11.9
15	63.3	18.7	12.8
25	65.4	20.2	14.0
35	67.5	21.7	15.4
50	71.5	25.1	18.6
53	72.4	26.0	19.5
56	72.6	26.1	19.6
70.5	74.3	27.9	21.4
85	76.2	30.0	23.7
100	78.0	32.1	26.3

[a]Monodrate → anhydrous salt transition temperature determined graphically is about 52°C, and 50.8°C as determined by cooling studies.

AUXILIARY INFORMATION

METHOD/APPARATUS/PROCEDURE:

Isothermal method used. Water and excess salt were placed in small Pyrex glass-stoppered tubes and agitated until equilibrium was reached (about 3 h). Equilibrium was approached from below because of the tendency to form supersaturated solutions when approaching saturation from above. Samples from 0.5 to 1.5 cm^3 were drawn off by means of pipets into 15 cm^3 weighing bottles. Duplicate samples were evaporated to dryness, and the residues heated to constant mass at 110°C.

Iodometric analyses of the solid phase at "room temperature" showed the solid phase to be the monohydrate. Analyses of the solid phase at 55°C showed it to be the anhydrous salt.

SOURCE AND PURITY OF MATERIALS:

Lithium bromate was prepared by mixing solutions of lithium sulfate and barium bromate by titrating one solution against the other until a drop of either gave no precipitate. The filtrate from the barium sulfate was concentrated, and upon cooling lithium bromate crystallized out. Duplicate iodometric analyses of the dried salt gave results of 99.50 % and 100 % lithium bromate.

ESTIMATED ERROR:

Soly: authors state experimental inaccuracies are negligible. Compilers estimate a precision of ± 0.1 mass % units.

Temp: precision ± 0.02 K to ± 0.1K

REFERENCES:

COMPONENTS:	ORIGINAL MEASUREMENTS:
(1) Lithium bromate; $LiBrO_3$; [13550-28-2] (2) Water; H_2O; [7732-18-5]	Averko-Antonovich, I.N. *Zh. Obshch. Khim.* 1943, 13, 272-8.
VARIABLES: Temperature: 228-416 K	PREPARED BY: Hiroshi Miyamoto

EXPERIMENTAL VALUES:

t/°C	$LiBrO_3$ Solubility mass %	mol % (compiler)	mol kg^{-1} (compiler)	Nature of the solid phase
- 1.05	10.3	1.51	0.852	Ice
- 4.8	20.3	3.29	1.89	"
- 9.8	30.6	5.56	3.27	"
- 20.2	40.0	8.18	4.94	"
- 40.0	52.0	12.6	8.03	"
- 45.0	54.9	14.0	9.03	$LiBrO_3 \cdot H_2O$
- 40.0	55.4	14.2	9.21	"
- 36.7	56.2	14.6	9.52	"
- 31.5	56.8	14.9	9.75	"
- 26.5	57.3	15.2	9.95	"
- 21.0	58.2	15.7	10.3	"
- 16.5	58.5	15.8	10.5	"
- 10.8	59.4	16.4	10.9	"
- 6.8	60.2	16.8	11.2	"
0	61.23	17.42	11.71	$LiBrO_3 \cdot H_2O$
20.1	64.51	19.54	13.48	"
24.9	65.54	20.26	14.10	"
35.9	67.78	21.94	15.60	"
45.0	70.4	24.1	17.6	"
50.0	71.8	25.4	18.9	"
4	66.8	21.2	14.9	$LiBrO_3$ (m)
17.5	68.2	22.3	15.9	"
45	71.8	25.4	18.9	"

continued.....

AUXILIARY INFORMATION

METHOD/APPARATUS/PROCEDURE:	SOURCE AND PURITY OF MATERIALS:
Solubilities above 0°C were studied isothermally. Aliquots of satd sln were withdrawn with a pipet and $LiBrO_3$ detd iodometrically. The satd sln in equilibrium with metastable salt was prepd as follows: the sln satd at 52°C or above was slowly cooled, stirred for 3-5 h at 45°C, and then allowed to settle for 4 h at 45°C. A satd sln at the boiling point (143°C) was prepd by gently heating an unsaturated sln at about 143°C, and excess salt was added to the sln. The resulting satd sln was allowed to settle at the boiling point, and clear sln was withdrawn into a capillary glass tube and allowed to solidify. The tube was cut into three pieces, weighed, and the $LiBrO_3$ content detd iodometrically. Below 0°C a mixture of $LiBrO_3$ and water was placed in a tube equipped with a stirrer, and the tube cooled in a Dewar flask (acetone and solid CO_2). The satd sln was allowed to settle for a few hours, and aliquots withdrawn with a glass tube equipped with a glass-wool or asbestos filter. A water jet-pump was used to filter off the sln, and the slns were analyzed iodometrically. (contd)	No information given. ESTIMATED ERROR: Nothing specified. REFERENCES:

COMPONENTS:	ORIGINAL MEASUREMENTS:
(1) Lithium bromate; LiBrO$_3$; [13550-28-2]	Averko-Antonovich, I.N.
(2) Water; H$_2$O; [7732-18-5]	*Zh. Obshch. Khim.* 1943, *13*, 272-8.

EXPERIMENTAL VALUES: (Continued)

t/°C	KBrO$_3$ Solubility mass %	mol % (compiler)	mol kg^{-1} (compiler)	Nature of the solid phase
45	70.4	24.1	17.6	LiBrO$_3$
55	72.72	26.26	19.77	"
65	73.86	27.40	20.95	"
80	75.84	29.55	23.28	"
100.5	78.6	32.9	27.2	"
111	79.6	34.3	28.9	"
121	81.2	36.6	32.0	"
143	84.6	42.3	40.8	"

METHOD/APPARATUS/PROCEDURE (Continued)

The synthetic method was also used with visual
observation of temperatures of crystallization. The content of LiBrO$_3$ in solution was
previously determined by iodometry.

COMPONENTS:	ORIGINAL MEASUREMENTS:
(1) Lithium bromate; $LiBrO_3$; [13550-28-2]	Campbell, A.N.; Kartzmark, E.M.; Musbally, G.M.
(2) Sodium bromate; $NaBrO_3$; [7789-38-0]	Can. J. Chem. 1967, 45, 803-6.
(3) Water; H_2O; [7732-18-5]	

VARIABLES:	PREPARED BY:
Composition at 298.15 K	Hiroshi Miyamoto

EXPERIMENTAL VALUES: Composition of Saturated Solutions at 25.00°C

Lithium Bromate		Sodium Bromate		Nature of
mass %	mol % (compiler)	mass %	mol % (compiler)	the solid phase
0	0	28.43[a]	4.528	$NaBrO_3$
65.64[a]	20.33	–	–	$LiBrO_3 \cdot H_2O$
49.97	12.35	3.02	0.667	$LiBrO_3 \cdot H_2O + NaBrO_3$

[a]For the binary systems the compiler computes the following:

soly of $LiBrO_3$: 14.17 mol kg^{-1}

soly of $NaBrO_3$: 2.632 mol kg^{-1}

COMMENTS AND/OR ADDITIONAL DATA:

The phase diagram is given in units of mass %.

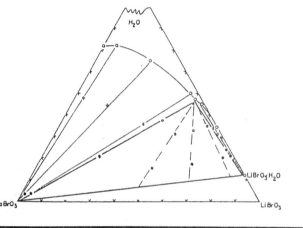

AUXILIARY INFORMATION	
METHOD/APPARATUS/PROCEDURE: Ternary mixtures were stirred for 48 hours. The bromate content was determined iodometrically, and the alkali metal by flame photometry. To determine the nature of the solid phase, the wet residue method of Schreinemakers was used.	SOURCE AND PURITY OF MATERIALS: $LiBrO_3$ and $NaBrO_3$ were certified reagents, and they were dried at 100° C.
	ESTIMATED ERROR: Soly: precision within 1 % (compiler). Temp: precision ± 0.01° K (authors).
	REFERENCES:

COMPONENTS:	ORIGINAL MEASUREMENTS:
(1) Lithium sulfate; Li_2SO_4; [13453-87-7] (2) Lithium bromate; $LiBrO_3$; [13550-28-2] (3) Water; H_2O; [7732-18-5]	Campbell, A.N.; Kartzmark, E.M.; Musbally, G.M. *Can. J. Chem.* 1967, *45*, 803-6.

VARIABLES:	PREPARED BY:
Composition at 298.15 K	Hiroshi Miyamoto

EXPERIMENTAL VALUES: Composition of Saturated Solutions at 25.00°C

Lithium Sulfate		Lithium Bromate		Nature of
mass %	mol % (compiler)	mass %	mol % (compiler)	the solid phase
25.50	5.311	–		$Li_2SO_4.H_2O$
–	–	65.64[b]	20.33	$LiBrO_3.H_2O$
2.82	0.810	46.87	10.98	a

[a] $LiBrO_3.H_2O$ and the solid solution of $LiBrO_3.H_2O$ in $Li_2SO_4.H_2O$ containing 42 mass % Li_2SO_4, 45 mass % $LiBrO_3$ and 13 mass % H_2O.

[b] For the binary system, the compiler computed the following:

soly of $LiBrO_3$ = 14.17 mol kg^{-1}

COMMENTS AND/OR ADDITIONAL DATA:

The phase diagram is given to the right is based on mass % units.

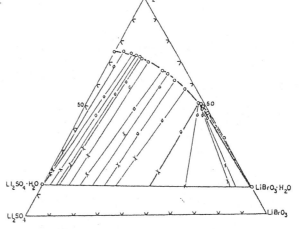

AUXILIARY INFORMATION

METHOD/APPARATUS/PROCEDURE:	SOURCE AND PURITY OF MATERIALS:
Ternary complexes were stirred for 48 hours. The bromate content was determined iodometrically, and the alkali metal by flame photometry. To determine the nature of the solid phase, the wet residue method of Schreinemakers was used.	$LiBrO_3$ and $Li_2SO_4.H_2O$ were certified reagents. $LiBrO_3$ was dried at 100°C and $Li_2SO_4.H_2O$ was used without further purification.
	ESTIMATED ERROR: Soly: precision within 1 % (compiler). Temp: precision $\pm$ 0.01 K (authors).
	REFERENCES:

COMPONENTS:	ORIGINAL MEASUREMENTS:
(1) Lithium bromate; LiBrO$_3$; [13550-28-2] (2) 2-Propanone (acetone); C$_3$H$_6$O; [67-64-1]	Miravitlles, Mille L. Ann. Fís. Quim. (Madrid) 1945, 41, 120-37.

VARIABLES:	PREPARED BY:
T/K = 288, 293 and 298	R. Herrera

EXPERIMENTAL VALUES:

Solubility[a]

t/°C	mass%	mol kg^{-1}
15	0.1013	0.007520
20	0.0897	0.006658
25	0.0803	0.005960

[a]Molalities calculated by the compiler.

AUXILIARY INFORMATION

METHOD/APPARATUS/PROCEDURE:

Saturated solutions were prepared in an Erlenmeyer flask by mixing the dried acetone with an excess of halate for two hours. The solution was constantly stirred by bubbling dry air (air was dried by passing it through CaCl$_2$ while pumping it into the solution). Air going out from the flask after bubbling in the solution carried some acetone vapor during this operation. The solution temperature was kept constant by immersing the flask in a constant temperature water bath. After two hours, the air exit was closed. The resulting pressure forced the saturated solution from the Erlenmeyer through a tube filled with cotton which acted as a filter, and was collected in a small flask. This flask was stoppered and weighed. The halate contained in the sample was weighed after complete evaporation of acetone. In all cases, weights were reported to the fourth decimal figure.

SOURCE AND PURITY OF MATERIALS:

Commercial redistilled acetone. This acetone was then dehydrated three times by leaving it in contact with calcium chloride for forty eight hours each time. Fresh CaCl$_2$ was used in each operation. Finally, the dehydrated acetone was distilled at 56.3°C.

Source and purity of LiBrO$_3$ not specified.

ESTIMATED ERROR:

Nothing specified.

REFERENCES:

COMPONENTS:	EVALUATOR:
(1) Sodium Bromate; NaBrO₃; [7789-38-0]	H. Miyamoto Niigata University Niigata, Japan
(2) Water; H₂O; [7732-18-5]	and M. Salomon US Army ET & DL Fort Monmouth, NJ, USA March, 1984

CRITICAL EVALUATION:

THE BINARY SYSTEM

The evaluators have examined seven publications (1-7) which report the solubility of
NaBrO₃ in water. Linke and Seidell (8) cite three publications by Kremers (9), but the
evaluators were unable to obtain copies of Kremers' papers. Upon detailed comparisons
(see below), the solubilities reported by Kremers are all too large and were rejected.

A summary of the experimental solubilities is given in Table 1. In all cases the equilibrated solid phase is the anhydrous salt. It is noted that Linke and Seidell incorrectly
attribute the experimental solubility at 373 K to Ricci (1) when in fact Ricci did not
report any solubilities above 325 K, and the 373 K result must be from (9). Solubilities
in mole kg^{-1} and mole dm^{-3} units are given in the compilations.

Table 1. Summary of Experimental Solubilities[a]

T/K	mass %	χ	(ref)	T/K	mass %	χ	(ref)
278.15	21.42	0.031519	(1)	308.15	31.35	0.051703	(1)
278.15	21.41	0.031501	(5)	308.2	31.95[b]	0.053080	(4)
283.15	23.24	0.034886	(1)	310.65	32.08[c]	0.053381	(2)
283.15	23.24	0.034886	(2)	313.15	32.80	0.055066	(1)
288.15	24.94	0.038157	(1)	318.15	34.22	0.058478	(1)
293.15	26.69	0.041657	(1)	318.15	34.22	0.058478	(2)
298.2	28.14[b]	0.044665	(4)	323.15	35.50	0.061660	(3)
298.15	28.26	0.044919	(5)	323.15	35.55	0.061787	(1)
298.15	28.29	0.044982	(1)	323.15	35.64	0.062015	(5)
298.15	28.29	0.044982	(3)	325.15	36.09	0.063162	(2)
298.15	28.29	0.044982	(6)	333.2	38.5[b,d]	0.06954	(9)
298.15	28.43[b]	0.045279	(7)	353.2	42.51	0.081121	(4)
303.15	29.85	0.048347	(1)	353.2	43.1[b,d]	0.08294	(9)
303.15	29.85	0.048347	(2)	373.2	47.6[b,d]	0.09784	(9)

[a]Original units are mass %. Mole fractions calculated by evaluators.

[b]Rejected data points. See text for discussion.

[c]Result obtained by graphical extrapolation.

[d]Results quoted from Ref. (8).

It is important to note that there are a number of entries in Table 1 from references
(1-3, 5, 6) which are identical, and since the authors do not indicate that these identical solubilities are the result of the original study (1), the evaluators assume that
these are all independent measurements which must be given equal weights. This is not a
trivial point since by making the assumption that all data reported in (1-3, 5, 6) represent independent measurements, it is obvious that these data have greater weight in our
least squares analyses (see below), and the consequence is that some data must be rejected.
For example, the result at 298.15 K reported by Campbell et al. (7) had to be rejected
which is somewhat of a surprise since Campbell's work is generally of high precision.

All solubility data in Table 1 were fitted to the smoothing equation

$$Y_x = A/(T/K) + B\ln(T/K) + C + D(T/K)$$ [1]

Data were rejected when the difference in the calculated and experimental solubilities
exceeded twice the standard error of estimate.

COMPONENTS:	EVALUATOR:
(1) Sodium Bromate; NaBrO₃; [7789-38-0]	H. Miyamoto Niigata University Niigata, Japan and
(2) Water; H₂0; [7732-18-5]	M. Salomon US Army ET & DL Fort Monmouth, NJ, USA March, 1984

CRITICAL EVALUATION:

Based on the criterion for exceptable data points, two data points from (4) were rejected, the singular data point at 298.15 K from (7) was rejected, and all data from (9) were rejected. Fitting the remaining 22 data points to the smoothing eq. [1] gives the following result for the mole fraction solubilities:

$$Y_x = -24576.69/(T/K) - 124.4405 ln(T/K) + 735.8949 + 0.1698543(T/K) \qquad [2]$$

$$\sigma_y = 0.0026 \qquad\qquad \sigma_x = 7.3 \times 10^{-5}$$

The solubilities calculated from this smoothing equation are designated as *recommended* values, and values at rounded temperatures are given in Table 2. It is interesting to note that we can also obtain a satisfactory fit to eq. [1] by including the melting point of 281 K for NaBrO₃ (10). Thus for 23 data points including the melting point of the solid, the following is obtained:

$$Y_x = -18558.18/(T/K) - 85.7656 ln(T/K) + 513.8164 - 0.107927(T/K) \qquad [3]$$

$$\sigma_y = 0.0033 \qquad\qquad \sigma_x = 8.9 \times 10^{-5}$$

For the 22 acceptable data points, the solubilities in mol kg⁻¹ units were fitted to the following smoothing equation:

$$Y_m = -3893.57/(T/K) - 8.30776 ln(T/K) + 60.39155(T/K) \qquad [4]$$

$$\sigma_y = 0.0029 \qquad\qquad \sigma_m = 0.0089$$

The solubilities calculated from this smoothing equation are designated as *recommended* values, and values at rounded temperatures are given in Table 2.

Finally, for those publications which give density data (1, 2, 5, 6), we were able to calculate solubilities in mol dm⁻³ units (see the compilations). These solubilities were fitted to the following smoothing equation:

$$ln(c/mol\ dm^{-3}) = -5813.91/(T/K) - 15.3545 ln(T/K) + 107.8469(T/K) \qquad [5]$$

$$\sigma_c = 0.0070$$

Solubilities calculated from eq. [5] are designated as *recommended* solubilities, and values at rounded temperatures are given in Table 2.

MULTICOMPONENT SYSTEMS

Ternary systems of two saturating components of NaBrO₃ and an alkali metal halide are all of the simple eutonic type (1, 4). The NaBrO₃-NaBr-H₂0 system was studied by both Ricci (1) and Klebanov and Basova (4), but direct comparisons cannot be made since the temperatures used differ in these studies.

The compilations should be consulted for other ternary systems involving sulfates (1, 2), halates (3, 5, 7) or sodium molybdate (6). A number of quaternary systems were reported in (4).

Note that the compilations of (3) and (7) can be found in the chapters on NaClO₃ and LiBrO₃, respectively.

COMPONENTS:	EVALUATOR:
(1) Sodium Bromate; $NaBrO_3$; [7789-38-0] (2) Water; H_2O; [7732-18-5]	H. Miyamoto Niigata University Niigata, Japan and M. Salomon US Army ET & DL Fort Monmouth, NJ, USA March, 1984

CRITICAL EVALUATION:

Table 2. Recommended Solubilities Calculated from Eqs. [2], [4] and [5]. Solid Phase in all Cases is the Anhydrous Salt.

T/K	χ	$m/\text{mol kg}^{-1}$	$c/\text{mol dm}^{-3}$
278.15	0.03154	1.814	1.694
283.15	0.03484	2.003	1.864
288.15	0.03819	2.199	2.034
293.15	0.04157	2.400	2.204
298.15	0.04497	2.606	2.370
303.15	0.04837	2.815	2.533
308.15	0.05176	3.027	2.689
313.15	0.05513	3.235	2.839
318.15	0.05847	3.453	
323.15	0.06179	3.666	
328.15	0.06507	3.877	
333.15	0.06833	4.086	
338.15	0.07155	4.292	
343.15	0.07477	4.493	
348.15	0.07796	4.689	
353.15	0.08115	4.880	

REFERENCES

1. Ricci, J. E. *J. Am. Chem. Soc.* 1934, *56*, 299.

2. Ricci, J. E. *J. Am. Chem. Soc.* 1935, *57*, 805.

3. Swenson, T.; Ricci, J. E. *J. Am. Chem. Soc.* 1935, *61*, 1974.

4. Klebanov, G. S.; Basova, E. P. *Zh. Prikl. Khim.* 1939, *12*, 1601.

5. Ricci, J. E.; Aleshnick, J. J. *J. Am. Chem. Soc.* 1944, *66*, 980.

6. Ricci, J. E.; Linke, W. F. *J. Am. Chem. Soc.* 1947, *69*, 1080.

7. Campbell, A. N.; Kartzmark, E. M.; Musbally, G. M. *Can. J. Chem.* 1967, *45*, 803.

8. Linke, W. F.; Seidell, A. *Solubilities of Inorganic and Metal-Organic Compounds.* Vol II, 4th Edition. Am. Chem. Soc. Washington, DC. 1965.

9. Kremers. *Pogg. Ann.* 1855, *94*, 271; 1855, *95*, 468; 1856, *97*, 5.

10. Dean, J. A., Ed. *Lange's Handbook of Chemistry: Twelfth Edition.* McGraw-Hill, NY. 1979.

COMPONENTS:	ORIGINAL MEASUREMENTS:
(1) Sodium bromate; $NaBrO_3$; [7789-38-0] (2) Water; H_2O; [7732-18-5]	Ricci, J.E. J. Am. Chem. Soc. 1934, 56, 299-303.

VARIABLES:	PREPARED BY:
T/K = 278 to 323	Hiroshi Miyamoto

EXPERIMENTAL VALUES: Solubility of $NaBrO_3$

t/°C	mass %	mol % (compiler)	mol kg^{-1} (compiler)	Density g cm^{-3}	Nature of the solid phase
5	21.42	3.152	1.807	1.194	$NaBrO_3$
10	23.24	3.489	2.006	1.211	"
15	24.94	3.816	2.202	1.232	"
20	26.69	4.166	2.413	1.248	"
25	28.29	4.498	2.614	1.257	"
30	29.85	4.835	2.820	1.284	"
35	31.35	5.170	3.026	1.288	"
40	32.80	5.507	3.235	1.310	"
45	34.22	5.848	3.448	–	"
50	35.55	6.179	3.656	–	"

AUXILIARY INFORMATION

METHOD/APPARATUS/PROCEDURE:	SOURCE AND PURITY OF MATERIALS:
Mixtures of $NaBrO_3$ and water were placed in a bottle, and rotated in a large water thermostat for two days which was found to be sufficient for attainment of equilibrium. Samples of the saturated solution were withdrawn by means of a calibrated pipet provided with a folded filter paper at the tip. The bromate content was determined by titration with standard sodium thiosulfate solution.	C.p. grade $NaBrO_3$ was recrystallized, dried to the anhydrous state, and then kept constantly in a 100°C oven.

ESTIMATED ERROR:
Soly: accuracy within 0.2 %. Temp: precision ± 0.01 K. Densities: precision about 0.1 %.

REFERENCES:

COMPONENTS:	ORIGINAL MEASUREMENTS:
(1) Sodium bromate; $NaBrO_3$; [7789-38-0] (2) Water-d_2; D_2O; [7789-20-0] (3) Water; H_2O; [7732-18-5]	Noonan, E.C. J. Am. Chem. Soc. <u>1948</u>, 70, 2915-8.
VARIABLES: T/K = 278.15	PREPARED BY: W.A. Van Hook

EXPERIMENTAL VALUES:

t/°C	water-d_2	Soly $NaBrO_3$ moles/100 moles solvent
5	0	3.253[a]
	91.59	2.899
	100.0	2.867[b]

[a] Solubility in H_2O taken from ref (1).

[b] Extrapolated by the author assuming a linear dependence between solubility and mass % D_2O.

AUXILIARY INFORMATION

METHOD/APPARATUS/PROCEDURE:	SOURCE AND PURITY OF MATERIALS:
Appropriate excess of purified salts were placed in ampoules, and heavy water was distilled in under vacuum and the ampules sealed. Equilibrium was approached from the high temperature side only by rotating the ampules for 12 to 48 hours in a water-bath. After settling one hour, 2-5 ml samples of solution were removed with pipets fitted with glass wool filters. The pipets were kept at the same temperature as the solutions. Samples of the solution were transferred to tared 30 ml platinum crucibles contained in suitable weighing bottles, and evaporated to dryness. All solubility determinations were performed in duplicate.	Commercial reagent grade salt was recrystallized at least twice. Heavy water was treated by distillation from alkaline permanganate and then from crystals of potassium dichromate or chromic anhydride. The product was found to have a conductivity of 2×10^{-6} S cm^{-1} or better.
	ESTIMATED ERROR: Soly: precision 0.5 % or better (author). Temp: precision $\pm$ 0.01 K (author).
	REFERENCES: 1. Ricci, A. J. Am. Chem. Soc. <u>1934</u>, 56, 230.

COMPONENTS:	ORIGINAL MEASUREMENTS:
(1) Sodium bromate; $NaBrO_3$; [7789-38-0]	Ricci, J.E.; Linke, W.F.
(2) Disodium (I-4)-tetroxomolybdate (2-) (sodium molybdate); Na_2MoO_4; [7631-95-0]	J. Am. Chem. Soc. 1947, 69, 1080-3.
(3) Water; H_2O; [7732-18-5]	

VARIABLES:	PREPARED BY:
Composition at 298.15° K	Hiroshi Miyamoto

EXPERIMENTAL VALUES: Composition of saturated solutions at 25.0°C

| Sodium Molybdate | | Sodium Bromate | | Density | Nature of |
mass %	mol % (compiler)	mass %	mol % (compiler)	g cm^{-3}	the solid phase[a]
39.38	5.378	0.00	0.00	1.432	A
38.30	5.280	1.80	0.339	1.442	"
37.09	5.171	3.86	0.734	1.453	"
35.57	5.022	6.33	1.22	1.466	A+B
35.58	5.021	6.29	1.21	1.468	"
35.60	5.025	6.28	1.21	1.470	"
(Av) 35.58	5.022	6.30	1.21	1.468	"
32.64	4.489	7.49	1.41	1.440	B
27.53	3.639	9.86	1.78	1.398	"
22.44	2.868	12.56	2.190	1.363	"
16.18	1.998	16.35	2.756	1.326	"
11.47	1.385	19.40	3.197	1.304	"
4.85	0.573	24.42	3.936	1.278	"
0.00	0.000	28.29[b]	4.498	1.264	

[a] A = $Na_2MoO_4 \cdot 2H_2O$;   B = $NaBrO_3$

[b] For the binary system the compiler computes the following:

soly of $NaBrO_3$ = 2.614 mol kg^{-1}

AUXILIARY INFORMATION

METHOD/APPARATUS/PROCEDURE:	SOURCE AND PURITY OF MATERIALS:
Isothermal method. Saturated solutions were prepared by stirring complexes of known compositions. Aliquots of saturated solution for analyses were withdrawn with calibrated pipets fitted with filters at the tips.	C.p. grade sodium molybdate dihydrate was used. The salt was completely dehydrated at 180°C, and stored at 150°C. The purity of this anhydrous salt was found to be 100.0 %.

Bromate content in the saturated solutions was determined iodometrically. In the presence of molybdate, a slight excess of aqueous HCl solution was required to obtain the correct end-point within the short titration time.

The total salt content of liquid and solid samples was determined by evaporation and drying to constant weight.

C.p. grade sodium bromate used and was found to be pure within 1/1000.

COMMENTS AND/OR ADDITIONAL DATA:

The phase diagram for this ternary system is given (superimposed) on the phase diagram for the Na_2MoO_4-$NaIO_3$-H_2O system (see the compilation for this latter system).

ESTIMATED ERROR:

Soly: the accuracy of titration was within 0.1 %.
Temp: precision ± 0.04 K.

REFERENCES:

COMPONENTS:	ORIGINAL MEASUREMENTS:
(1) Sodium carbonate; Na_2CO_3; [497-19-8] (2) Sodium bromate; $NaBrO_3$; [7789-38-0] (3) Water; H_2O; [7732-18-5]	Klebanov, G.S.; Basova, E.P. *Zh. Prikl. Khim.* <u>1939</u>, *12*, 1601-9.

VARIABLES:	PREPARED BY:
Composition at 353 K	Hiroshi Miyamoto

EXPERIMENTAL VALUES: Composition of saturated solutions at 90°C

Sodium Bromate		Sodium Carbonate		Nature of the
mass %	mol % (compiler)	mass %	mol % (compiler)	solid phase[a]
42.51[b]	8.112	–	–	A
36.58	6.895	5.36	1.44	"
28.10	5.171	12.50	3.275	"
21.88	4.013	18.84	4.919	A+B
15.51	2.740	22.60	5.684	B
8.65	1.45	25.75	6.164	"
–	–	30.95	7.079	"

[a] A = $NaBrO_3$;   B = $Na_2CO_3.H_2O$

[b] For the binary system the compiler computes the following:

soly of $NaBrO_3$ = 4.900 mol kg^{-1}

AUXILIARY INFORMATION

METHOD/APPARATUS/PROCEDURE:

The isothermal method was used. Prior to the experiment the carbon dioxide content in solution was checked by phenolphthalein. The salt and water were placed into a tube equipped with a stirrer, and the tube placed in a water thermostat. A layer of paraffin placed on the surface of water in the thermostat at 80°C. Equilibrium was reached in a day. The sodium bromate content was determined iodometrically by titration with 0.1 mol dm^{-3} thiosulfate solution. The sodium carbonate was titrated with 0.1 mol dm^{-3} HCl.
The composition of the solid phase was identified by Schreinemakers' method, and by crystal optics.

SOURCE AND PURITY OF MATERIALS:

Chemically pure grade sodium carbonate was used without further purification. Sodium bromate was prepared as follows: (1) The salt was synthesized by the following reaction: $Br_2 + 5Cl_2 + 12NaOH = 2NaBrO_3 + 10NaCI + 6H_2O$. (2) $KBrO_3$ was reacted with $BaCl_2$. The $Ba(BrO_3)_2$ obtained was treated with Na_2SO_4. The pptd $BaSO_4$ was removed by filtration and $NaBrO_3$ crystallized from the filtrate. The product was recryst to remove foreign ions.

ESTIMATED ERROR:

Nothing specified.

REFERENCES:

COMPONENTS:	ORIGINAL MEASUREMENTS:
(1) Sodium bromate; $NaBrO_3$; [7789-38-0] (2) Sodium hydrogen carbonate; $NaHCO_3$; [144-55-8] (3) Water; H_2O; [7732-18-5]	Klebanov, G.S.; Basova, F.P. *Zh. Prikl. Khim.* 1939, *12*, 1601-9.

VARIABLES:	PREPARED BY:
T/K = 298 and 308 Composition	Hiroshi Miyamoto

EXPERIMENTAL VALUES: Composition of saturated solutions

t/°C	Sodium Bromate		Sodium Hydrogen Carbonate		Nature of the
	mass %	mol % (compiler)	mass %	mol % (compiler)	solid phase[a]
25	28.14[b]	4.466	–	–	A
	25.94	4.089	1.80	0.510	"
	24.34	3.884	4.76	1.36	A+B
	18.47	2.760	4.90	1.32	B
	12.24	1.732	6.18	1.57	"
	6.98	0.948	7.55	1.84	"
	–	–	9.34	2.16	"
35	31.95[b]	5.308	–	–	A
	29.00	4.757	2.13	0.628	"
	28.02	4.679	4.88	1.46	A+B
	23.25	3.673	5.02	1.42	B
	17.88	2.683	5.98	1.61	"
	11.90	1.697	7.40	1.90	"
	7.86	1.08	8.20	2.03	"
	6.26	0.855	8.96	2.20	"
	–	–	10.55	2.467	"

[a] A = $NaBrO_3$;   B = $NaHCO_3$

[b] For the binary system the compiler computes the following:

soly of $NaBrO_3$ = 2.595 mol kg^{-1} at 25°C
= 3.112 mol kg^{-1} at 35°C

AUXILIARY INFORMATION

METHOD/APPARATUS/PROCEDURE:	SOURCE AND PURITY OF MATERIALS:
The isothermal method was used. The salt and water were placed into a tube equipped with a stirrer, and the tube placed in a water thermostat at 25 or 35°C. Equilibrium was reached in a day. The sodium bromate content was determined iodometrically by titration with 0.1 mol dm^{-3} thiosulfate solution. The sodium hydrogen carbonate content was determined by titration with 0.1 mol dm^{-3} hydrochloric acid using methyl orange indicator. The composition of the solid phase was determined by Schreinemakers' method, and by crystal-optics.	Chemically pure grade $NaHCO_3$ was used without further purification. Sodium bromate was prepd as follows: (1) The salt was synthesized by the following reaction: Br_2 + $5Cl_2$ + 12NaOH = $2NaBrO_3$ + 10NaCI + $6H_2O$. (2) $KBrO_3$ was reacted with $BaCl_2$, and the $Ba(BrO_3)_2$ obtnd was treated with Na_2SO_4. The pptd $BaSO_4$ was removed by filtration and $NaBrO_3$ crystallized from the filtrate. The product was recryst to remove foreign ions.
	ESTIMATED ERROR: Nothing specified.
	REFERENCES:

COMPONENTS:	ORIGINAL MEASUREMENTS:
(1) Sodium nitrate; $NaNO_3$; [7631-99-4]	Ricci, J.E.
(2) Sodium bromate; $NaBrO_3$; [7789-38-0]	J. Am. Chem. Soc. 1934, 56, 299-303.
(3) Water; H_2O; [7732-18-5]	
VARIABLES:	**PREPARED BY:**
Composition at 298.15 K	Hiroshi Miyamoto

EXPERIMENTAL VALUES: Composition of saturated solutions at 25°C

Sodium Nitrate		Sodium Bromate		Density $g\ cm^{-3}$	Nature of the solid phase[a]
mass %	mol % (compiler)	mass %	mol % (compiler)		
47.87	16.29	0.00	0.00	1.384	A
46.50	16.10	2.43	0.474	1.405	"
44.46	15.80	6.04	1.21	1.432	"
42.57	15.51	9.39	1.93	1.455	A+B
42.60	15.52	9.37	1.92	1.455	"
(Av) 42.59	15.52	9.38	1.92	1.455	"
39.57	14.02	10.23	2.042	1.441	B
32.54	10.87	12.41	2.336	1.387	"
25.54	8.114	14.94	2.674	1.353	"
18.48	5.614	17.79	3.044	1.314	"
11.33	3.319	21.25	3.506	1.288	"
5.00	1.43	24.92	4.014	1.270	"
0.00	0.00	28.29[b]	4.498	1.257	"

[a] A = $NaNO_3$;   B = $NaBrO_3$

[b] For the binary system the compiler computes the following:

$$\text{soly of } NaBrO_3 = 2.614 \text{ mol kg}^{-1}$$

AUXILIARY INFORMATION

METHOD/APPARATUS/PROCEDURE:	SOURCE AND PURITY OF MATERIALS:
The ternary complexes were prepared from weighed amounts of water and the two anhydrous salts: these complexes were rotated in a large thermostat for about two days, a time found to be sufficient for attaining equilibrium. Samples of the saturated solution were withdrawn by means of a calibrated pipet provided with a folded filter paper at the tip. The bromate was determined by titration with standard sodium thiosulfate solution, and the total solids by evaporation at 100°C and drying at 250°C. Sodium nitrate was found by difference. For the determination of solid phase compositions, the method of algebraic extrapolation of tie-lines was used.	C.p. grade salts were recrystallized, dried to the anhydrous state, and kept constantly in a 100°C oven.
	ESTIMATED ERROR: Soly: accuracy within 0.2 %. Temp: precision ± 0.01 K. Densities: precision about 0.1 %.
	REFERENCES:

COMPONENTS:	ORIGINAL MEASUREMENTS:
(1) Sodium sulfate; Na_2SO_4; [7757-82-6] (2) Sodium bromate; $NaBrO_3$; [7789-38-0] (3) Water; H_2O; [7732-18-5]	Ricci, J.E. J. Am. Chem. Soc. <u>1934</u>, 56, 299-303.
VARIABLES: Composition at 298.15 K	PREPARED BY: Hiroshi Miyamoto

EXPERIMENTAL VALUES: Composition of saturated solutions at 25°C

Sodium sulfate		Sodium bromate		Density	Nature of the
mass %	mol % (compiler)	mass %	mol % (compiler)	g cm^{-3}	solid phase[a]
21.90	3.434	0.00	0.00	1.205	A
20.48	3.281	3.34	0.504	1.225	"
19.17	3.156	7.06	1.094	1.254	"
17.94	3.041	10.79	1.721	1.275	"
16.94	2.952	14.11	2.315	1.303	"
16.46	2.914	15.87	2.644	1.319	A+B
16.43	2.907	15.86	2.641	1.316	"
16.47	2.914	15.83	2.637	1.320	"
16.45	2.911	15.85	2.640	1.312	"
(Av)16.45	2.911	15.85	2.640	1.317	"
12.35	2.153	18.71	3.071	1.303	B
8.28	1.436	21.72	3.521	1.288	"
4.09	0.697	24.96	4.003	1.284	"
0.00	0.00	28.29[b]	4.498	1.257	"

[a] A = $Na_2SO_4 \cdot 10H_2O$;   B = $NaBrO_3$

[b] For the binary system the compiler computes the following:

soly of $NaBrO_3$ = 2.614 mol kg^{-1}

AUXILIARY INFORMATION

METHOD/APPARATUS/PROCEDURE:	SOURCE AND PURITY OF MATERIALS:
The ternary complexes were prepared from weighed amounts of water and the two anhydrous salts: these complexes were rotated in a large thermostat for about two days, a time found to be sufficient for attaining equilibrium. Samples of the saturated solution were withdrawn by means of a calibrated pipet provided with a folded filter paper at the tip. The bromate was determined by titration with standard sodium thiosulfate solution, and the total solids by evaporation at 100°C and drying at 250°C. Potassium sulfate was found by difference. For the determination of solid phase compositions, the method of algebraic extrapolation of tie-line was used.	C.p. grade salts were recrystallized, dried to the anhydrous state, and kept constantly in a 100°C oven.
	ESTIMATED ERROR: Soly: accuracy within 0.2 %. Temp: precision ± 0.01 K. Densities: precision about 0.1 %.
	REFERENCES:

COMPONENTS:	ORIGINAL MEASUREMENTS:
(1) Sodium sulfate; Na_2SO_4; [7757-82-6] (2) Sodium bromate; $NaBrO_3$; [7789-38-0] (3) Water; H_2O; [7732-18-5]	Ricci, J.E. *J. Am. Chem. Soc.* 1935, 57, 805-10.
VARIABLES: T/K = 283 - 325 Composition	PREPARED BY: Hiroshi Miyamoto

EXPERIMENTAL VALUES: Composition of saturated solutions

| t/°C | Sodium Sulfate | | Sodium Bromate | | Density | Nature of the |
	mass %	mol % (compiler)	mass %	mol % (compiler)	g cm^{-3}	solid phase[a]
10	8.26	1.13	0.00	0.00	1.079	A
	6.96	0.990	5.40	0.723	1.112	"
	5.20	0.795	14.21	2.045	1.175	"
	4.41	0.712	19.93	3.027	1.230	"
	4.41	0.713	20.10	3.059	1.228	A+C
	4.37	0.706	20.12	3.061	1.228	"
	(Av) 4.40	0.711	20.11	3.061	1.227	"
	3.61	0.582	20.67	3.138	1.226	C
	1.83	0.294	21.96	3.316	1.217	"
	0.00	0.00	23.24	3.489	1.211	"
30	29.14	4.958	0.00	0.00	1.286	A
	26.92	4.747	5.18	0.860	1.312	"
	26.02	4.690	7.85	1.33	1.333	"
	25.28	4.665	10.43	1.812	1.351	"
	24.95	4.646	11.46	2.008	1.361	A+C
	25.03	4.659	11.36	1.990	1.364	"
	(Av) 25.02	4.658	11.38	1.994	1.362	"
	21.04	3.844	13.86	2.384	1.343	C
	12.43	2.201	19.89	3.315	1.311	"
	0.00	0.00	29.85[b]	4.835	1.284	"

continued....

AUXILIARY INFORMATION

METHOD/APPARATUS/PROCEDURE:

The ternary complexes were prepared from weighed amounts of water and the two anhydrous salts; these complexes were rotated in a large thermostat. Two weeks of stirring were required for attaining equilibrium. Samples of the saturated solution were withdrawn by means of a calibrated pipet provided with a folded filter paper at the tip. The bromate content was determined by titration with standard sodium thiosulfate solution, and the total solids by evaporation at 100°C and drying at 250°C. Na_2SO_4 was found by difference.
For the determination of solid phase compositions, the method of algebraic extrapolation of tie-lines was used.

SOURCE AND PURITY OF MATERIALS:

C.p. grade sodium sulfate and sodium bromate were recrystallized and dried to the anhydrous state, and then kept constantly in a 100°C oven.

ESTIMATED ERROR:

Soly: accuracy probably about 0.2% as in (1). Temp: precision probably ± 0.01 K as in (1).

REFERENCES:

1. Ricci, J.E. *J. Am. Chem. Soc.*
   1934, 56, 249.

COMPONENTS:	ORIGINAL MEASUREMENTS:
(1) Sodium sulfate; Na_2SO_4; [7757-82-6]	Ricci, J.E.
(2) Sodium bromate; $NaBrO_3$; [7789-38-0]	*J. Am. Chem. Soc.* 1935, 57, 805-10.
(3) Water; H_2O; [7732-18-5]	

EXPERIMENTAL VALUES: (Continued)

Composition of saturated solutions

t/°C	Sodium Sulfate		Sodium Bromate		Density g cm^{-3}	Nature of the solid phase[a]
	mass %	mol % (compiler)	mass %	mol % (compiler)		
37.5	32.70	5.805	0.00	0.00	–	B
	31.20	5.643	2.99	0.509	–	"
	30.68	5.639	4.77	0.825	–	"
	30.53	5.658	5.57	0.972	–	"
	(30.4)	5.63	(5.7)	0.99	–	B+C
	30.36	5.631	5.80	1.01	–	S
	29.56	5.464	6.33	1.10	–	"
	28.04	5.184	7.85	1.37	–	"
	26.45	4.916	9.83	1.72	–	"
	25.11	4.704	11.78	2.077	–	"
	24.08	4.546	13.38	2.377	–	S+C
	24.18	4.566	13.31	2.366	–	"
	(Av)24.14	4.559	13.35	2.373	–	"
	25.67	4.883	12.39	2.219		C(m)
	25.01	4.739	12.75	2.274	–	"
	24.71	4.677	12.95	2.307	–	"
	23.01	4.325	14.11	2.496	–	C
	15.28	2.793	19.61	3.374	–	"
	0.00	0.00	(32.08)[c]	5.338	–	"
45	32.07	5.650	0.00	0.00	–	B
	30.35	5.441	3.12	0.526	–	"
	29.18	• 5.321	5.64	0.968	–	"
	28.82	5.275	6.29	1.08	–	B+S
	28.79	5.268	6.30	1.09	–	"
	28.74	5.261	6.37	1.10	–	"
	(Av)28.78	5.267	6.32	1.09	–	"
	30.44	5.559	4.50	0.774	–	S(m)
	29.95	5.471	5.01	0.861	–	"
	29.52	5.393	5.45	0.937	–	"
	29.21	5.343	5.85	1.01	–	"
	27.76	5.093	7.53	1.30	–	S
	26.56	4.905	9.23	1.60	–	"
	25.85	4.785	10.12	1.763	–	"
	24.18	4.522	12.56	2.211	–	"
	22.92	4.318	14.38	2.550	–	"
	22.77	4.297	14.65	2.602	–	"
	21.58	4.079	15.96	2.840	–	"
	20.94	3.979	17.00	3.041	–	"
	20.76	3.947	17.22	3.082	–	S+C
	20.90	3.978	17.15	3.072	–	"
	(Av)20.86	3.969	17.17	3.075	–	"
	22.47	4.299	16.00	2.882	–	C(m)
	22.19	4.243	16.23	2.921	–	"
	21.59	4.117	16.62	2.983	–	"
	21.09	4.017	17.03	3.053	–	"
	19.81	3.755	17.93	3.199	–	C
	16.54	3.103	20.38	3.599	–	"
	8.10	1.49	27.14	4.694	–	"
	0.00	0.00	34.22[b]	5.848	–	"

continued.....

[c]Extrapolated value.

COMPONENTS:	ORIGINAL MEASUREMENTS:
(1) Sodium sulfate; Na_2SO_4; [7757-82-6]	Ricci, J.E.
(2) Sodium bromate; $NaBrO_3$; [7789-38-0]	J. Am. Chem. Soc. 1935, 57, 805-10.
(3) Water; H_2O; [7732-18-5]	

EXPERIMENTAL VALUES: (Continued)

Composition of Saturated Solutions

t/°C	Sodium Sulfate		Sodium Bromate		Density g cm^{-3}	Nature of the solid phase[a]
	mass %	mol %	mass %	mol %		
52	31.47	5.504	0.00	0.00	–	B
	29.71	5.279	3.03	0.507	–	"
	28.17	5.062	5.47	0.925	–	"
	27.73	5.064	7.19	1.24	–	B+S
	27.64	5.059	7.45	1.28	–	"
	(Av)27.7	5.06	7.3	1.3	–	"
	26.03	4.776	9.24	1.60	–	S
	23.17	4.315	13.24	2.321	–	"
	21.39	4.033	15.96	2.833	–	"
	19.26	3.675	18.98	3.409	–	"
	18.12	3.490	20.80	3.771	–	S+C
	18.15	3.492	20.69	3.747	–	"
	(Av)18.13	3.490	20.77	3.764	–	"
	19.60	3.789	19.61	3.568	–	C(m)
	19.03	3.673	20.06	3.644	–	"
	16.27	3.115	22.19	3.999	–	C
	8.62	1.623	28.49	5.048	–	"
	0.00	0.00	36.09 [b]	6.316	–	"

[a] $A = Na_2SO_4.10H_2O$; $B = Na_2SO_4$; $C = NaBrO_3$; S = solid solution

[b] For the binary system the compiler computes the following:

$$\text{soly of } NaBrO_3 = 2.006 \text{ mol kg}^{-1} \text{ at } 10°C$$
$$= 2.820 \text{ mol kg}^{-1} \text{ at } 30°C$$
$$= 3.130 \text{ mol kg}^{-1} \text{ at } 37.5°C$$
$$= 3.448 \text{ mol kg}^{-1} \text{ at } 45°C$$
$$= 3.742 \text{ mol kg}^{-1} \text{ at } 52°C$$

COMMENTS AND/OR ADDITIONAL DATA:

The phase diagrams are given below (based on mass % units).

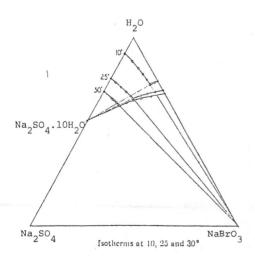

Isotherms at 10, 25 and 30°

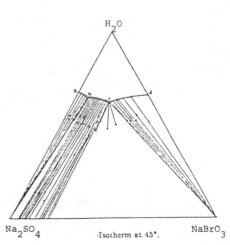

Isotherm at 45°.

COMPONENTS:	ORIGINAL MEASUREMENTS:
(1) Sodium chloride; NaCl; [7647-14-1] (2) Sodium bromate; NaBrO$_3$; [7789-38-0] (3) Water; H$_2$0; [7732-18-5]	Ricci, J.E. J. Am. Chem. Soc. <u>1934</u>, 56, 299-303.

VARIABLES:	PREPARED BY:
Composition at 283.15 and 298.15 K	Hiroshi Miyamoto

EXPERIMENTAL VALUES: Composition of saturated solutions

t/°C	Sodium chloride mass %	mol % (compiler)	Sodium bromate mass %	mol % (compiler)	Density g cm^{-3}	Nature of the solid phase[a]
10	26.32	9.919	0.00	0.00	–	A
	24.53	9.619	5.02	0.762	1.236	A+B
	24.53	9.619	5.02	0.762	1.233	"
	24.51	9.608	5.01	0.761	1.235	"
	(Av) 24.52	9.614	5.02	0.762	1.235	"
	23.61	9.214	5.32	0.804	1.229	B
	20.75	7.995	6.41	0.957	1.213	"
	16.15	6.125	8.58	1.26	1.199	"
	9.84	3.70	12.75	1.857	1.192	"
	4.85	1.84	17.28	2.534	1.193	"
	0.00	0.00	23.24[b]	3.489	1.211	"
25	26.46	9.984	0.00	0.00	1.195	A
	25.55	9.827	2.48	0.369	1.215	"
	24.35	9.598	5.62	0.858	1.236	"
	23.93	9.536	6.92	1.07	1.247	A+B
	23.95	9.546	6.92	1.07	1.248	"
	23.93	9.536	6.92	1.07	1.246	"
	23.95	9.545	6.91	1.07	1.247	"
	23.92	9.530	6.91	1.07	1.249	"
	(Av) 23.94	9.541	6.92	1.07	1.247	"

continued.....

AUXILIARY INFORMATION

METHOD/APPARATUS/PROCEDURE:	SOURCE AND PURITY OF MATERIALS:
The ternary complexes were prepared from weighed amounts of water and the two anhydrous salts: these complexes were rotated in a large thermostat for about two days, a time found to be sufficient for attaining equilibrium. Samples of saturated solution were withdrawn by means of a calibrated pipet provided with a folded filter paper at the tip. The bromate was determined by titration with standard sodium thiosulfate solution, and the total solids by evaporation at 100°C and drying at 250°C. Sodium chloride was found by difference. For the determination of solid phase compositions, the method of algebraic extrapolation of tie-lines was used.	C.p. grade salts were recrystallized, dried to the anhydrous state, and kept constantly in a 100°C oven.

ESTIMATED ERROR:
Soly: accuracy within 0.2 %. Temp: precision ± 0.01 K. Densities: precision about 0.1 %.

REFERENCES:

COMPONENTS:	ORIGINAL MEASUREMENTS:
(1) Sodium chloride; NaCl; [7647-14-1] (2) Sodium bromate; $NaBrO_3$; [7789-38-0] (3) Water; H_2O; [7732-18-5]	Ricci, J.E. *J. Am. Chem. Soc.* <u>1934</u>, *56*, 299-303.

EXPERIMENTAL VALUES: (Continued)

Composition of saturated solutions

t/°C	Sodium chloride		Sodium bromate		Density g cm^{-3}	Nature of the solid phase[a]
	mass %	mol % (compiler)	mass %	mol % (compiler)		
25	20.99	8.279	8.32	1.27	1.234	B
	17.55	6.869	10.34	1.568	1.234	"
	12.95	5.053	13.67	2.066	1.228	"
	9.98	3.91	16.31	2.473	1.225	"
	8.27	3.25	17.98	2.736	1.228	"
	6.17	2.44	20.27	3.107	1.229	"
	3.76	1.50	23.13	3.585	1.241	"
	0.00	0.00	28.29[b]	4.498	1.257	"

[a] A = NaCl;   B = $NaBrO_3$

[b] For the binary system the compiler computes the following:

$$\text{soly of } NaBrO_3 = 2.006 \text{ mol kg}^{-1} \text{ at } 10°C$$
$$= 2.614 \text{ mol kg}^{-1} \text{ at } 25°C$$

COMMENTS AND/OR ADDITIONAL DATA:

The phase diagram is given below (based on mass % units).

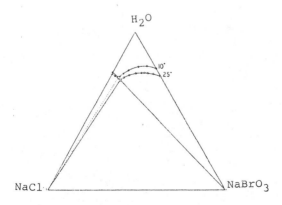

COMPONENTS:	ORIGINAL MEASUREMENTS:
(1) Sodium bromide; NaBr; [7647-15-6]	Ricci, J.E.
(2) Sodium bromate; $NaBrO_3$; [7789-38-0]	J. Am. Chem. Soc. 1934, 56, 299-303.
(3) Water; H_2O; [7732-18-5]	

VARIABLES:	PREPARED BY:
Composition at 283, 298 and 318 K	Hiroshi Miyamoto

EXPERIMENTAL VALUES: Composition of saturated solutions

t/°C	Sodium Bromide mass %	mol %	Sodium Bromate mass %	mol %	Density g cm^{-3}	Nature of the solid phase[a]
10	45.89	12.93	0.00	0.00	1.492	A
	44.50	12.77	2.58	0.505	1.519	A+B
	44.54	12.79	2.58	0.505	1.516	"
	44.49	12.76	2.58	0.505	1.515	"
(Av)	44.51	12.77	2.58	0.505	1.517	"
	43.09	12.18	2.83	0.545	1.498	B
	39.40	10.72	3.55	0.658	1.452	"
	11.10	2.488	14.46	2.210	1.240	"
	5.33	1.18	18.73	2.827	1.220	"
	0.00	0.00	23.24[b]	3.489	1.211	"
25	48.41	14.11	0.00	0.00	1.530	A
	47.37	14.00	1.90	0.383	1.546	"
	46.84	13.95	2.93	0.595	1.555	A+B
	46.81	13.94	2.94	0.597	1.558	"
	46.82	13.94	2.94	0.597	1.555	"
	46.81	13.94	2.94	0.597	1.553	"
(Av)	46.82	13.94	2.94	0.597	1.555	"

continued.....

AUXILIARY INFORMATION

METHOD/APPARATUS/PROCEDURE:

The ternary complexes were prepared from weighed amounts of water and the two anhydrous salts: these complexes were rotated in a large thermostat for about two days, a time found to be sufficient for attaining equilibrium.
Samples of the saturated solution were withdrawn by means of a calibrated pipet provided with a folded filter paper at the tip. The bromate was determined by titration with standard sodium thiosulfate solution, and the total solids by evaporation at 100°C and drying at 250°C. NaBr was found by difference. For the determination of solid phase compositions, the method of algebraic extrapolation of tie-line was used.

SOURCE AND PURITY OF MATERIALS:

C.p. grade salts were recrystallized, dried to the anhydrous state, and kept constantly in a 100°C oven.

ESTIMATED ERROR:
Soly: accuracy within 0.2 %
Temp: precision ± 0.01 K.
Densities: precision about 0.1 %.

REFERENCES:

COMPONENTS:	ORIGINAL MEASUREMENTS:
(1) Sodium bromide; NaBr; [7647-15-6]	Ricci, J.E.
(2) Sodium bromate; $NaBrO_3$; [7789-38-0]	J. Am. Chem. Soc. 1934, 56, 299-303.
(3) Water; H_2O; [7732-18-5]	

EXPERIMENTAL VALUES: (Continued)

Composition of saturated solutions

	Sodium Bromide		Sodium bromate		Density	Nature of the
$t/°C$	mass %	mol % (compiler)	mass %	mol % (compiler)	g cm^{-3}	sold phase[a]
25	45.62	13.40	3.15	0.631	1.542	B
	39.24	10.81	4.61	0.866	1.462	"
	38.66	10.59	4.78	0.893	1.457	"
	29.83	7.628	7.86	1.37	1.377	"
	21.27	5.183	12.04	2.001	1.320	"
	13.82	3.275	16.72	2.702	1.282	"
	6.46	1.51	22.38	3.564	1.270	"
	0.00	0.00	28.29[b]	4.498	1.257	"
45	52.55	16.24	0.00	0.00	–	A
	50.66	16.09	3.51	0.760	–	A+B
	50.70	16.11	3.50	0.758	–	"
(Av)	50.68	16.10	3.51	0.761	–	"
	49.39	15.45	3.72	0.793	–	B
	28.69	7.554	11.17	2.006	–	"
	7.91	1.98	26.65	4.545	–	"
	0.00	0.00	34.22[b]	5.848	–	"

[a] A = $NaBr.2H_2O$;   B = $NaBrO_3$

[b] For the binary system the compiler computes the following:

$$\text{soly of } NaBrO_3 = 2.006 \text{ mol kg}^{-1} \text{ at } 10°C$$

$$= 2.614 \text{ mol kg}^{-1} \text{ at } 25°C$$

$$= 3.448 \text{ mol kg}^{-1} \text{ at } 45°C$$

COMMENTS AND/OR ADDITIONAL DATA:

The phase diagram is given below (based on mass % units).

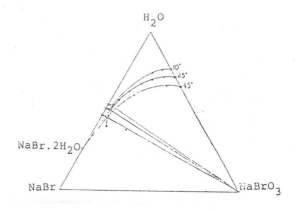

COMPONENTS:	ORIGINAL MEASUREMENTS:
(1) Sodium bromide; NaBr; [7647-15-6] (2) Sodium bromate; NaBrO$_3$; [7789-38-0] (3) Water; H$_2$O; [7732-18-5]	Klebanov, G.S.; Basova, E.P. *Zh. Prikl. Khim.* <u>1939</u>, *12*, 1601-9.

VARIABLES:	PREPARED BY:
Composition at 303 K	Hiroshi Miyamoto

EXPERIMENTAL VALUES: Composition of saturated solutions at 30°C

Sodium Bromide		Sodium Bromate		Nature of the
mass %	mol % (compiler)	mass %	mol % (compiler)	solid phase[a]
50.57	15.19	-	-	A
47.92	14.58	3.31	0.687	A+B
39.69	11.05	4.94	0.938	B
27.94	7.150	9.69	1.69	"
18.18	4.450	15.30	2.554	"
8.73	2.100	22.59	3.699	"
-	-	31.95[b]	5.308	"

[a] A = NaBr.2H$_2$O;   B = NaBrO$_3$

[b] For the binary system the compiler computes the following:

soly of KBrO$_3$ = 3.112 mol kg^{-1}

AUXILIARY INFORMATION

METHOD/APPARATUS/PROCEDURE:	SOURCE AND PURITY OF MATERIALS:
The isothermal method was used. Prior to the experiment, the carbon dioxide content in solution was checked by phenolphthalein. The salt and water were placed into a tube equipped with a stirrer. The tube was kept in a water thermostat. Equilibrium was reached in a day. The sodium bromate content was determined iodometrically. The sodium bromide content was determined as follows: sulfurous acid solution was added to the sample solution containing sodium bromate and bromide, and the solution boiled to remove excess SO$_2$. Bromide was determined by Volhard's method using standardized silver nitrate solution. The sodium bromide content was calculated by difference. The composition of the solid phase was identified by Schreinemakers' method, and by crystallography.	Analytical grade sodium bromide was used. The sodium bromide contained 0.2% NaCl or less. Sodium bromate was prepd as follows: (1) Barium chloride was added to barium bromate solution, and the resulting solution was treated with sodium sulfate solution. The sodium bromate obtained was recrystallized. (2) The reaction Br$_2$ + 5Cl$_2$ + 12NaOH = 2NaBrO$_3$ + 10NaCl + 6H$_2$O was used to prepare the sodium bromate.
	ESTIMATED ERROR: Nothing specified.
	REFERENCES:

COMPONENTS:	ORIGINAL MEASUREMENTS:
(1) Sodium bromate; $NaBrO_3$; [7789-38-0]	Ricci, J.E.
(2) Sodium iodide; NaI; [7681-82-5]	J. Am. Chem. Soc. 1934, 56, 299-301.
(3) Water; H_2O; [7732-18-5]	

VARIABLES:	PREPARED BY:
Composition at 298.15 K	Hiroshi Miyamoto

EXPERIMENTAL VALUES: Composition of saturated solutions at 25°C

| Sodium iodide | | Sodium bromate | | Density | Nature of the |
mass %	mol % (compiler)	mass %	mol % (compiler)	g cm^{-3}	solid phase[a]
64.71	18.06	0.00	0.00	1.904	A
63.98	18.02	1.17	0.327	1.911	A+B
64.00	18.03	1.17	0.327	1.913	"
63.93	17.98	1.16	0.324	1.920	"
64.00	18.03	1.17	0.327	1.916	"
(Av)63.95	18.00	1.17	0.327	1.914	"
62.13	16.90	1.30	0.351	1.874	B
60.65	16.07	1.44	0.379	1.836	"
54.89	13.26	2.23	0.535	1.727	"
48.11	10.61	3.62	0.793	1.619	"
40.76	8.296	5.78	1.17	1.521	"
32.21	6.067	8.92	1.67	1.438	"
17.32	2.967	16.57	2.819	1.332	"
0.00	0.00	28.29[b]	4.498	1.257	"

[a] A = $NaI \cdot 2H_2O$;   B = $NaBrO_3$

[b] For the binary system the compiler computes the following:

soly of $NaBrO_3$ = 2.614 mol kg^{-1}

AUXILIARY INFORMATION

METHOD/APPARATUS/PROCEDURE:	SOURCE AND PURITY OF MATERIALS:
The ternary complexes were prepared from weighed amounts of water and the two anhydrous salts: these complexes were rotated in a large thermostat for about two days, a time found to be sufficient for attaining equilibrium. Samples of the saturated solution were withdrawn by means of a calibrated pipet provided with a folded filter paper at the tip. The bromate was determined by titration with standard sodium thiosulfate solution, and the total solids by evaporation at 100°C and drying at 250°C. Sodium iodide was found by difference. For the determination of solid phase compositions, the method of algebraic extrapolation of tie-lines was used.	C.p. grade salts were recrystallized, dried to the anhydrous state, and kept constantly in a 100°C oven.
	ESTIMATED ERROR:
	Soly: accuracy within 0.2 %. Temp: precision ± 0.01 K. Densities: precision about 1 %.
	REFERENCES:

COMPONENTS:	ORIGINAL MEASUREMENTS:
(1) Sodium Bromate; $NaBrO_3$; [7789-38-0]	Ricci, J. E.; Aleshnick, J. J.
(2) Silver Bromate; $AgBrO_3$; [7783-89-3]	J. Am. Chem. Soc. 1944, 66, 980-3.
(3) Water; H_2O; [7732-18-5]	

VARIABLES:	PREPARED BY:
T/K = 278, 298, 323 Composition	H. Miyamoto

EXPERIMENTAL VALUES:

Composition of Saturated Aqueous Solutions:

t/°C	NaBrO₃ (mass %)	NaBrO₃ (mol kg⁻¹) (compiler)	AgBrO₃ (mass %)	AgBrO₃ (mol kg⁻¹) (compiler)	Density (kg dm⁻³)	Nature of solid phase[b]
5	21.41	1.805	–	–	1.192	N
	21.32	1.796	–	–	1.190	SSI + AN
	--	--	0.0905[a]	0.000384	0.9998	A
25	28.26	2.611	–	–	1.264	N
	28.26	2.611	–	–	1.264	SSI
	28.24	2.608	–	–	1.261	SSI
	28.21	2.604	–	–	1.262	SSI
	28.16	2.598	–	–	1.261	SSI
	28.14	2.595	–	–	1.260	SSI
	28.08	2.587	–	–	1.260	SSI
	27.97[a]	2.573	–	–	1.260	SSI + AN
	27.84	2.557	–	–	1.257	AN
	27.78	2.549	–	–	1.257	AN
	27.66	2.534	–	–	1.256	AN
	27.41	2.502	–	–	1.252	AN
	27.35	2.495	–	–	1.251	AN
	27.05	2.457	–	–	1.248	AN
	26.42	2.380	–	–	1.241	AN
	25.36	2.252	–	–	1.232	AN
	24.32	2.130	–	–	1.220	AN
	22.71	1.947	–	–	1.203	AN
	21.28	1.791	–	–	1.185	SSII

continued.....

AUXILIARY INFORMATION

METHOD/APPARATUS/PROCEDURE:	SOURCE AND PURITY OF MATERIALS:
The ternary complexes were prepared by weight, using distilled water, c.p. NaBrO3 and c.p. AgBrO3. Attainment of equilibrium was proved in almost all cases by repeated analysis of the solution after further stirring. The complexes were stirred for periods of 2 or 8 weeks. In the last one or two mixtures at 25°C in above table (next page), AgBrO3 content was determined gravimetrically as AgBr after reduction with NaNO2. NaBrO3 was determined by evaporation of the solution to dryness.	C.p. AgBrO3 and c.p. NaBrO3 were used. The purity of the c.p. NaBrO3 was found to be (100 ± 0.1%). The purity of the c.p. AgBrO3 was determined to be 98.2% silver bromate and 1.8% sodium bromate.
	ESTIMATED ERROR: Solubility errors in solubility of AgBrO3 in water and of NaBrO3 in water are ±0.004 mass % and ±0.02 mass %, respectively. Temperature: nothing specified.
	REFERENCES:

COMPONENTS:	ORIGINAL MEASUREMENTS:
(1) Sodium Bromate; NaBrO3; [7789-38-0]	Ricci, J. E.; Aleshnick, J. J.
(2) Silver Bromate; AgBrO3; [7783-89-3]	J. Am. Chem. Soc. 1944, 66, 980-3.
(3) Water; H2O; [7732-18-5]	

EXPERIMENTAL VALUES: (Continued)

Composition of Saturated Aqueous Solutions:

t/°C	NaBrO3 (mass %)	NaBrO3 (mol kg⁻¹) (compiler)	AgBrO3 (mass %)	AgBrO3 (mol kg⁻¹) (compiler)	Density (kg dm⁻³)	Nature of solid phase[b]
25	16.99	1.356	–	–	1.143	SSII
	13.04	0.994	–	–	1.108	SSII
	10.28	0.759	–	–	1.079	SSII
	8.39	0.607	–	–	1.062	SSII
	7.17	0.512	0.01	0.0004	1.051	SSII
	3.92	0.270	0.03	0.0013	1.025	SSII
	--	--	0.204[c]	0.00867	0.9985	A
50	35.64	3.670	–	–	1.341	N
	35.24[d]	5.606	–	–	1.334	SSI + AN
	35.05	3.576	–	–	1.334	AN
	34.73	3.526	–	–	1.331	AN
	34.57	3.501	–	–	–	AN
	28.77	2.677	–	–	1.258	SSII
	23.32	2.015	–	–	1.196	SSII
	--	--	0.430[c]	0.0183	0.9934	A

[a]Average of 16 determinations.

[b]N = NaBrO3; A = AgBrO3; SSI = solid solution containing up to 2.5-3.0 mass % AgBrO3
SSII = solid solution containing AgBrO3 from 61 to 95 mass %

[c]The solubilities of pure AgBrO3 were determined on samples of c.p. AgBrO3 which were
repeatedly washed with considerable quantities of water. The purity of about 99.7%
was finally thus obtained, but the author stated that great accuracy cannot be claimed
for these solubilities.

[d]Average of 3 determinations.

The phase diagram is given below for this system at 25°C.

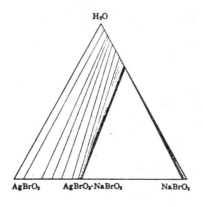

COMPONENTS:	ORIGINAL MEASUREMENTS:
(1) Sodium carbonate; Na_2CO_3; [497-19-8]	Klebanov, G.S.; Basova, E.P.
(2) Sodium bromide; NaBr; [7647-15-6]	*Zh. Prikl. Khim.* 1939, *12*, 1601-9.
(3) Sodium bromate; $NaBrO_3$; [7789-38-0]	
(4) Water; H_2O; [7732-18-5]	

VARIABLES:	PREPARED BY:
Composition at 353 K	Hiroshi Miyamoto

EXPERIMENTAL VALUES: Composition of saturated solutions at 80°C

Sodium Bromide		Sodium Bromate		Sodium Carbonate		Nature of the
mass %	mol % (compiler)	mass %	mol % (compiler)	mass %	mol % (compiler)	solid phase[a]
52.67	16.56	-	-	1.05	0.321	A+B
48.20	15.79	6.60	1.47	1.18	0.375	A+B+C
52.20	17.77	6.25	1.45	-	-	A+C
29.82	8.596	12.26	2.410	4.66	1.30	B+C
17.05	4.728	17.09	3.232	9.34	2.51	"
10.81	2.939	19.18	3.556	11.80	3.115	"
-	-	21.88	4.013	18.84	4.919	"

[a] A = NaBr;   B = $Na_2CO_3 \cdot H_2O$;   C = $NaBrO_3$

AUXILIARY INFORMATION

METHOD/APPARATUS/PROCEDURE:	SOURCE AND PURITY OF MATERIALS:
Isothermal method. Prior to the experiment, the CO_2 content in water was checked with phenolphthalein. The salts and water were placed into a tube equipped with a stirrer, and the tube placed in a water thermostat. A layer of paraffin was placed on the surface of the water in the thermostat at 80°C. Equilibrium was reached in a day. The sodium bromate content was detd iodometrically by titrn with 0.1 mol dm^{-3} thiosulfate solution. The sodium bromide content was detd as follows: sulfurous acid was added to an aliquot of saturated sln, and the sln boiled to remove excess SO_2. The bromide was detd by Volhard's method using standard silver nitrate sln, and the sodium bromide content was obtained by difference. The sodium carbonate was detd by titrn with 0.1 mol dm^{-3}. Solid phase compositions detd by Schreinemakers' method, and by crystal optics.	Analytical grade NaBr was used. The NaBr contained less than 0.2% NaCl. Chemically pure grade Na_2CO_3 was used. Sodium bromate was prepd as follows: (1) The reaction, $Br_2 + 5Cl_2 + 12NaOH = 2NaBrO_3 + 10NaCl + 6H_2O$, was used to prepare the sodium bromate. (2) Barium chloride was reacted with $KBrO_3$. The $Ba(BrO_3)_2$ obtained was treated with Na_2SO_4. The pptd $BaSO_4$ was removed by filtration and $NaBrO_3$ crystallized from the filtrate. The product was recryst to remove foreign ions.
	ESTIMATED ERROR:
	Nothing specified.
	REFERENCES:

COMPONENTS:	ORIGINAL MEASUREMENTS:
(1) Sodium bromide; NaBr; [7647-15-6]	Klebanov, G.S.; Basova, E.P.
(2) Sodium bromate; NaBrO$_3$; [7789-38-0]	Zh. Prikl. Khim. 1939, 12, 1601-9.
(3) Sodium hydrogen carbonate; NaHCO$_3$; [144-55-8]	
(4) Water; H$_2$O; [7732-18-5]	

VARIABLES:	PREPARED BY:
Composition at 298 and 308 K	Hiroshi Miyamoto

EXPERIMENTAL VALUES: Composition of saturated solutions

t/°C	Sodium Bromide mass %	mol % (compiler)	Sodium Bromate mass %	mol % (compiler)	NaHCO$_3$ mass %	mol % (compiler)	Nature of the solid phase[a]
25	47.91	13.95	-	-	0.44	0.16	A+B
	46.92	13.85	1.92	0.387	0.38	0.14	"
	46.20	13.74	3.06	0.621	0.42	0.15	A+B+C
	46.82	13.94	2.94	0.597	-	-	A+C
	39.07	10.79	4.44	0.836	0.62	0.21	B+C
	31.25	8.100	6.95	1.23	0.70	0.22	"
	27.96	7.091	8.09	1.40	0.98	0.30	"
	18.97	4.585	12.18	2.007	1.50	0.444	"
	10.08	2.387	17.44	2.816	3.04	0.882	"
	4.34	1.02	21.52	3.455	3.94	1.14	"
	-	-	24.34	3.884	4.76	1.36	"
35	49.20	14.59	-	-	0.48	0.17	A+B
	49.60	14.99	0.90	0.19	0.46	0.17	"
	48.82	14.79	1.78	0.368	0.46	0.17	"
	47.94	14.62	2.98	0.620	0.52	0.19	"
	47.79	14.59	3.20	0.666	0.50	0.19	A+B+C
	47.92	14.58	3.31	0.0687	-	-	A+C
	41.05	11.67	4.66	0.904	0.60	0.21	B+C
	33.85	9.066	6.92	1.264	0.78	0.26	"
	28.85	7.464	8.76	1.546	1.04	0.330	"
	17.48	4.310	15.12	2.542	1.60	0.483	"
	9.68	2.34	19.72	3.255	2.94	0.872	"
	-	-	28.02	4.679	4.88	1.464	"

[a] A = NaBr.2H$_2$O;   B = NaHCO$_3$;   C = NaBrO$_3$.

AUXILIARY INFORMATION

METHOD/APPARATUS/PROCEDURE:	SOURCE AND PURITY OF MATERIALS:
Isothermal method. Prior to the experiment the CO$_2$ content in water was checked with phenolphthalein. The salt and water were placed in a tube equipped with a stirrer, and the tube thermostated at 25 or 35°C Equilibrium was reached in a day. The sodium bromate content was detd iodometrically by titrn with 0.1 mol dm^{-3} thiosulfate solution. The sodium bromide content was detd by adding sulfurous acid solution to an aliquot of saturated solution, and the solution boiled to remove excess SO$_2$. Bromide was detd by Volhard's method using standard AgNO$_3$ solution. The sodium bromide content was calcd by difference. The sodium hydrogen carbonate content was detd by titrn with 0.1 mol dm^{-3} HCl using methyl orange indicator. Solid phase compositions detd by Schreinemakers' method, and by crystal optics.	Analytical grade sodium bromide and chemically pure grade NaHCO$_3$ were used. Sodium bromate was prepd as follows: (1) The reaction, Br$_2$ + 5Cl$_2$ + 12NaOH = 2NaBrO$_3$ + 10NaCl + 6H$_2$O, was used to prepare the sodium bromate. (2) Barium chloride was reacted with KBrO$_3$. The Ba(BrO$_3$)$_2$ obtained was treated with Na$_2$SO$_4$. The pptd BaSO$_4$ was removed by filtration and NaBrO$_4$ crystallized from the filtrate. The product was recrystallized to remove foreign ions.
	ESTIMATED ERROR:
	Nothing specified.
	REFERENCES:

COMPONENTS:	ORIGINAL MEASUREMENTS:
(1) Sodium Bromate; $NaBrO_3$; [7789-38-0] (2) N-Methylacetamide; C_3H_7NO; [79-16-3]	Dawson, L.R.; Berger, J.E.; Vaughn, J.W.; Eckstrom, H.C. *J. Phys. Chem.* 1963, *67*, 281-3.
VARIABLES: T/K = 313	PREPARED BY: Hiroshi Miyamoto and Mark Salomon

EXPERIMENTAL VALUES:

The solubility of $NaBrO_3$ in n-methylacetamide, $CH_3CONHCH_3$, at 40°C was given as

$$0.10 \text{ mol dm}^{-3}$$

AUXILIARY INFORMATION

METHOD/APPARATUS/PROCEDURE:

An "approximate" solubility was determined by the conductivity method. About 0.5 to 1.0 gram of salt and 10 ml of solvent were placed in a large test tube, stoppered and covered with aluminum foil, and heated to 60°C. Upon cooling to 40°C, the occurence of precipitation was assumed to indicate the existence of a saturated solution. Conductivities were measured in duplicate in "the usual manner." Experiental details and the measured electrolytic conductivities were not given.

The concentration of the salt in the saturated solution was determined from the experimental electrolytic conductivities, but details on the calculation were not given. Presumably the limiting law was used as in (1).

SOURCE AND PURITY OF MATERIALS:

N-Methylacetamide was prepared by reacting monoethylamine with glacial acetic acid and subsequent heating to distill off the water. The product was purified by fractional distillation followed by five or more fractional freezing cycles. The electrolytic conductance of the purified solvent ranged from 0.5×10^{-5} to 2×10^{-5} S cm^{-1}.

Reagent grade $NaBrO_3$ was dried in a vacuum desiccator over anhydrous magnesium perchlorate without further treatment.

ESTIMATED ERROR:

Soly: authors "believe" the solubility to be accurate to within 5 %.

Temp: not specified.

REFERENCES:

1. Dawson, L.R.; Wilhoit, E.D.; Holmes, R.R.; Sears, P.G. *J. Am. Chem. Soc.* 1957, *79*, 3004 ($\Lambda\infty$ values are given in this paper).

COMPONENTS:	ORIGINAL MEASUREMENTS:
(1) Sodium bromate; NaBrO$_3$; [7789-38-0] (2) Hydrazine; N$_2$H$_4$; [302-01-2]	Welsh, T.W.B.; Broderson, H.J. J. Am. Chem. Soc. 1915, 37, 816-24.
VARIABLES: Room temperature (compiler's assumption)	PREPARED BY: Mark Salomon and Hiroshi Miyamoto

EXPERIMENTAL VALUES:

The solubility of NaBrO$_3$ in hydrazine at room temperature was given as

$$0.01 \text{ g/ cm}^3 \text{ N}_2\text{H}_4$$

The authors stated that the chief object of the research was to obtain
qualitative and approximate quantitative data, and the temperature was not kept constant.

AUXILIARY INFORMATION

METHOD/APPARATUS/PROCEDURE:	SOURCE AND PURITY OF MATERIALS:
The solubility vessel has a glass tube to which a U-shaped capillary tube was attached to the bottom. A stopcock at the end of the capillary permitted the adjustment of the rate of flow of dry nitrogen. About 1 cc of anhydrous hydrazine was placed in the tube, and small amounts of NaBrO3 added from a weighing bottle. After each addition of NaBrO3, a loosely fitting cork was placed in the top of the solubility tube. Nitrogen was bubbled through solution until the salt dissolved. The process was repeated until no more salt would dissolve. Temperature was not kept constant. The accuracy of this method is very poor. In addition, the authors stated that it was difficult to prevent the oxidation of hydrazine.	Anhydrous hydrazine was prepared by first partially dehydrating commercial hydrazine with sodium hydroxide according to the method of Rasching (1). Further removal of water by distn from barium oxide after the method of de Bruyn (2). The type of distillation apparatus employed and the procedure followed in the respective distillations were those described by Welsh (3). The product was found on analysis to contain 99.7 % hydrazine. The hydrazine was stored in 50 cm^3 sealed tubes. Sodium bromate was the ordinary pure chemical of standard manufacture.

ESTIMATED ERROR:

Soly: accuracy ± 50 % at best (compilers).

REFERENCES:
1. Raschig, F. Ber. Dtsch. Chem. Ges. 1927,
   43, 1927.: Hale, C.F.; Shetterly, F.F.
   J. Am. Chem. Soc. 1911, 33, 1071.
2. de Bruyn, L. Trav. Chim. Pays-Bas.
   1895, 14, 458.
3. Welsh, T.W. J. Am. Chem. Soc. 1915,
   37, 497.

COMPONENTS:	EVALUATOR:
(1) Potassium Bromate; $KBrO_3$; [7758-01-2] (2) Water; H_2O; [7732-18-5]	H. Miyamoto Niigata University Niigata, Japan and M. Salomon U.S. ARMY ET & DL Fort Monmouth, NJ, USA June, 1985

CRITICAL EVALUATION:

THE BINARY SYSTEM

Solubility data for the binary $KBrO_3$-H_2O system have been reported in 13 publications (1-13). Breusov et al. (13) also cite data from a handbook (14) which the evaluators cannot trace. For those solubility values from (14) which we cannot locate, the majority are much higher than those reported in (1-13), and they must be rejected in any case. Note that the compilation of reference (6) is given in the $KClO_3$ chapter.

No hydrates of $KBrO_3$ have been reported, and the solid phase in all studies is the anhydrous salt. All studies employed the isothermal method with the exception of the study by Benrath et al. (5) who used the synthetic method over the temperature range of 407-585 K. In analyzing the solubility data, we have treated the low and high temperature data separately: i.e. from the isothermal studies over the temperature range of 273-373 K, and from the synthetic study over the temperature range of 407-585 K. Summaries of these data are given in Tables 1 and 2, respectively.

Table 1.   Experimental Solubilities from 273-373 K[a]

T/K	mass %	χ	(ref)	T/K	mass %	χ	(ref)
273.2	2.98	0.003303	(13)	303.15	8.785	0.01028	(3)
273.2	2.96	0.003280	(4)	303.2	8.84	0.01035	(13)
278.15	3.642	0.004061	(3)	308.15	10.13	0.01201	(3)
283.15	4.510	0.005069	(3)	313.15	11.58	0.01393	(3)
283.2	4.54	0.005104	(13)	313.2	11.70	0.01409[b]	(4)
288.15	5.397	0.006117	(3)	313.2	11.67	0.01405	(13)
293.15	6.460	0.007395	(3)	318.15	13.08	0.01597	(3)
293.2	6.43	0.007359	(4)	323.15	14.69	0.01824	(3)
293.2	6.42	0.007346	(13)	323.2	14.82	0.01842	(13)
298.15	7.733	0.008712	(3)	333.2	18.21	0.02345	(4)
298.15	7.533	0.008712	(6)	333.2	18.08	0.02325	(13)
298.15	7.635[b]	0.008839	(7)	343.2	21.76	0.02913	(13)
298.2	7.49	0.008658	(9)	353.2	25.53	0.03566	(4)
298.2	7.53	0.008708	(10)	353.2	25.35	0.03534[b]	(13)
298.2	7.7[b]	0.00892	(11)	363.2	29.40	0.04299	(13)
298.2	7.49	0.008658	(12)	373.2	33.31	0.05113	(13)
298.2	7.52	0.008696	(8)				
298.2	7.55	0.008733	(13)				

[a]Mole fraction solubilities calculated by evaluators.

[b]Rejected data points (see text for discussion).

Table 2.   High Temperature Solubility Data from ref. (5)[a]

T/K	mass %	χ	T/K	mass %	χ
407.2	43.6	0.07697	499.2	70.6	0.2057
422.2	48.4	0.09189	503.2	72.6	0.2223
433.2	51.1	0.1013	522.2	72.6	0.2223
440.2	53.5	0.1104	527.2	74.4	0.2387
443.2	54.1	0.1128	538.2	77.2	0.2675
445.2	57.3	0.1265	547.2	79.1	0.2899
449.2	59.9	0.1388	552.2	81.1	0.3164
466.2	63.2	0.1563	559.2	81.4	0.3207
477.2	64.2	0.1621	570.2	83.1	0.3466
484.2	67.4	0.1824	585.2	86.4	0.4066

[a]Mole fraction solubilities calculated by evaluators.

COMPONENTS:	EVALUATOR:
(1) Potassium Bromate; $KBrO_3$; [7758-01-2] (2) Water; H_2O; [7732-18-5]	H. Miyamoto Niigata University Niigata, Japan and M. Salomon U.S. Army ET & DL Fort Monmouth, NJ, USA June, 1985

CRITICAL EVALUATION:

Solubility data in mol dm^{-3} units were reported in (1,2), and while Ricci et al. (3,6) reported data in mass % units, densities for the saturated solutions were also given enabling the evaluators to convert these mass units into volume units. Table 3 lists the solubility data in volume units from 278-313 K.

Table 3. Experimental Solubilities Based on Volume Units[a]

T/K	density/g cm^{-3}	c/mol dm^{-3}	reference
278.15	1.024	0.2233	3
283.15	1.035	0.2795	3
288.15	1.042	0.3367	3
293.15	1.048	0.4054	3
298.15	1.054	0.4754	3,6
298.2	----	0.4715[b]	1
298.2	----	0.478	2
303.15	1.062	0.5587	3
308.15	1.074	0.6515	3
313.15	1.083	0.7510	3

[a]Original data from (3,6) are in mass %, and conversions to mol dm^{-3} calculated by evaluators using experimental densities of saturated solutions given in the table.

[b]Rejected data point.

Over the temperature range of 273-373 K, the mole fraction solubility data were fitted to the smoothing equation, and four data points were rejected on the basis that the difference between calculated and observed solubilities exceeded $2\sigma_m$ (see Table 1 for rejected data points). The remaining 30 data points were used to obtain the following smoothing equation:

$$Y_x = -22549.21/(T/K) - 83.2220 \ln(T/K) + 514.161 + 0.092301(T/K) \qquad [1]$$

$$\sigma_y = 0.011 \qquad\qquad \sigma_x = 6.3 \times 10^{-5}$$

The smoothed solubilities calculated from eq. [1] are designated as *recommended* values, and are given in Table 4.

Over the temperature range of 407-585 K, the data of Benrath et al. (5) were used to obtain the following smoothing equation:

$$Y_x = -45971.8/(T/K) - 169.820 \ln(T/K) + 1060.642 + 0.16898(T/K) \qquad [2]$$

$$\sigma_y = 0.079 \qquad\qquad \sigma_x = 0.0086$$

The standard error in the solubilities is quite high, and we designate the calculated (smoothed) solubilities as *tentative* values. Using eq. [2] to calculate solubilities outside the experimental temperature range is not recommended. The melting point of $KBrO_3$ calculated from eq. [2] is 660.1 K which is not in good agreement with the hand-book value of 623 K (15): according to (15) the pure salt decomposes at 643 K.

For the solubilities in mol kg^{-1} units over the temperature range of 273-373 K, the 30 acceptable data used to derive the smoothing equation [1] were fitted to the following smoothing equation:

$$Y_m = -6137.40/(T/K) - 10.4118 \ln(T/K) + 79.904 \qquad [3]$$

$$\sigma_y = 0.0070 \qquad\qquad \sigma_m = 0.0088$$

COMPONENTS:	EVALUATOR:
(1) Potassium Bromate; $KBrO_3$; [7758-01-2] (2) Water; H_2O; [7732-18-5]	H. Miyamoto Niigata University Niigata, Japan and M. Salomon U.S. Army ET & DL Fort Monmouth, NJ, USA June, 1985

CRITICAL EVALUATION:

The mol kg^{-1} solubilities calculated from eq.[3] are designated as *recommended* values, and values at rounded temperatures are given in Table 4.

The solubilities in units of mol dm^{-3} were fitted to the simple smoothing equation:

$$ln(c/mol\ dm^{-3}) = -7507.54/(T/K) - 15.3273ln(T/K) + 111.769 \qquad [4]$$

$$\sigma_c = 0.0018$$

We designate the smoothed solubilities as *tentative*, and values from 278-313 K are given in Table 4.

> Table 4. Recommended and Tentative Solubilities Calculated from the
>   Smoothing Equations [1], [3], [4]. Solid Phase is the
>   Anhydrous Salt.

T/K	χ[a]	m/mol kg^{-1}[a]	c/mol dm^{-3}[b]
273.2	0.003294	0.1848	
278.2	0.004098	0.2286	0.2266
283.2	0.005036	0.2809	0.2777
288.2	0.006116	0.3404	0.3364
293.2	0.007347	0.4099	0.4030
298.2	0.008737	0.4883	0.4777
303.2	0.01029	0.5766	0.5608
308.2	0.01202	0.6743	0.6522
313.2	0.01392	0.7849	0.7519
318.2	0.01601	0.9043	
323.2	0.01827	1.038	
333.2	0.02335	1.336	
343.2	0.02918	1.679	
353.2	0.03574	2.066	
363.2	0.04303	2.493	
373.2	0.05105	2.955	

[a]Recommended values.

[b]Tentative values.

TERNARY SYSTEMS

The solubility of $KBrO_3$ in ternary systems have been reported in a number of publications (3, 4, 6, 8-12), and detailed information can be found in the compilations. Since no two studies are exactly alike, comparisons cannot be made and therefore critical evaluation is not possible at this time.

COMPONENTS:	EVALUATOR:
(1) Potassium Bromate; $KBrO_3$; [7758-01-2] (2) Water; H_2O; [7732-18-5]	H. Miyamoto Niigata University Niigata, Japan and M. Salomon U.S. Army ET & DL Fort Monmouth, NJ, USA June, 1985

CRITICAL EVALUATION:

REFERENCES

1. Geffcken, G. Z. *Physik. Chem.* <u>1904</u>, *49*, 257.

2. Rothmund, V. Z. *Physik. Chem.* <u>1909</u>, *69*, 523.

3. Ricci, J. E. *J. Am. Chem. Soc.* <u>1934</u>, *56*, 299.

4. Gerasimov, Ya. I. *Zh. Obshch. Khim.* <u>1934</u>, *4*, 223.

5. Benrath, A.; Gjedebo, F.; Schiffers, B.; Wunerlich, H. Z. *Anorg. Allgem. Chem.*
    <u>1937</u>, *231*, 285.

6. Swenson, T.; Ricci, J. E. *J. Am. Chem. Soc.* <u>1939</u>, *61*, 1974.

7. Chang, T. L.; Hsieh, Y. Y. *Sci. Repts. Natl. Tsing Hua Univ.* <u>1948</u>, **A5,** 252.

8. Ricci, J. E.; Offenbach, J. A. *J. Am. Chem. Soc.* <u>1951</u>, *73*, 1597.

9. Kirgintsev, A.; Yakobi, N. Y. *Zh. Neorg. Khim.* <u>1968</u>, *13*, 2851; *Russ. J. Inorg.*
    Chem. (Engl. Transl.) <u>1968</u>, *13*, 1468.

10. Kirgintsev, A. N.; Shklovskaya, R. M.; Arkhipov, S. M. *Izv. Akad. Nauk SSSR Ser.*
    Khim. <u>1971</u>, 2631; *Bull. Acad. Sci. USSR Div. Chem. Sci.* <u>1971</u>, 2501.

11. Serbrennikov, V. V.; Balyreva, V. A.; Larionova, I. S. *Zh. Neorg. Khim.* <u>1982</u>,
    27, 2959; *Russ. J. Inorg. Chem. (Engl. Transl)* <u>1982</u>, *27*, 1677.

12. Azarova, L. A.; Vinogradov, E. E. *Zh. Neorg. Khim.* <u>1982</u>, *27*, 2967; *Russ. J. Inorg.*
    Chem. (Engl. Transl.) <u>1982</u>, *27*, 1681.

13. Breusov, O. N.; Kashina, N. I.; Revzina, T. V.; Sobolevskaya, N.G.
    Zh. Neorg. Khim. <u>1967</u>, *12*, 2240.

14. *Spravochnik po Rastvorimosti Solevykh Sistem (Handbook on Solubility in Salt Sys).*
    Goshkhimizdat, Moscow, 1961. Vol. 3.

15. Dean, J. A., Ed. *Lange's Handbook of Chemistry: Twelfth Edition.*
    McGraw-Hill, NY. 1979.

COMPONENTS:	ORIGINAL MEASUREMENTS:
(1) Potassium bromate; $KBrO_3$; [7758-01-2] (2) Water; H_2O; [7732-18-5]	Ricci, J.E. J. Am. Chem. Soc. 1934, 56, 299-303.

VARIABLES:	PREPARED BY:
T/K = 278 - 323	Hiroshi Miyamoto

EXPERIMENTAL VALUES: Solubility of $KBrO_3$

t/°C	mass %	mol % (compiler)	mol kg^{-1} (compiler)	Density g cm^{-3}	Solid phase
5	3.642	0.4061	0.2263	1.024	$KBrO_3$
10	4.510	0.5069	0.2828	1.035	"
15	5.397	0.6117	0.3416	1.042	"
20	6.460	0.7395	0.4135	1.048	"
25	7.533	0.8712	0.4878	1.054	"
30	8.785	1.028	0.5767	1.062	"
35	10.13	1.201	0.6750	1.074	"
40	11.58	1.393	0.7842	1.083	"
45	13.08	1.597	0.9011	-	"
50	14.69	1.824	1.031	-	"

AUXILIARY INFORMATION

METHOD/APPARATUS/PROCEDURE:	SOURCE AND PURITY OF MATERIALS:
Mixtures of $KBrO_3$ and water were placed in bottles and rotated in a large water thermo-stat for two days, a time found to be sufficient for attaining equilibrium. Samples of the saturated solution were withdrawn by means of a calibrated pipet provided with a folded filter paper at the tip. The bromate was determined by titration with standard sodium thiosulfate solution.	C.p. grade $KBrO_3$ was recrystallized, dried to the anhydrous state, and stored in a 100°C oven.
	ESTIMATED ERROR: Soly: accuracy within 0.2 %. Temp: precision $\pm$ 0.01 K. Densities: precision about 0.1 %.
	REFERENCES:

COMPONENTS:	ORIGINAL MEASUREMENTS:
(1) Potassium bromate; KBrO$_3$; [7758-01-2] (2) Water; H$_2$O; [7732-18-5]	Benrath, A.; Gjedebo, F.; Schiffers, B.; Wunderlich, H. Z. Anorg. Allg. Chem. 1937, 231, 285-97.
VARIABLES: T/K = 407 to 585	PREPARED BY: Hiroshi Miyamoto

EXPERIMENTAL VALUES:

		Solubility
t/°C	mass %	mol kg^{-1} (compiler)
134	43.6	4.63
149	48.4	5.62
160	51.1	6.26
167	53.5	6.89
170	54.1	7.06
172	57.3	8.04
186	59.9	8.94
193	63.2	10.3
204	64.2	10.7
211	67.4	12.4
226	70.6	14.4
230	72.6	15.9
249	72.6	15.9
254	74.4	17.4
265	77.2	20.3
274	79.1	22.7
279	81.1	25.7
286	81.4	26.2
297	83.1	29.4
312	86.4	38.0

AUXILIARY INFORMATION

METHOD/APPARATUS/PROCEDURE:	SOURCE AND PURITY OF MATERIALS:
Synthetic method used with visual observation of temperature of crystallization and solubilization (ref 1). The weighed salt and water were placed in a small tube. The tubes were set in an oven equipped with a mica window. A thermometer was immersed in the oven.	No information was given.
	ESTIMATED ERROR: Nothing specified.
	REFERENCES: 1. Jaenecke, E. Z. Physik. Chem. 1936, A177, 7.

COMPONENTS:	ORIGINAL MEASUREMENTS:

COMPONENTS:

(1) Potassium bromate; $KBrO_3$; [7758-01-2]

(2) Water; H_2O; [7732-18-5]

ORIGINAL MEASUREMENTS:

Breusov, O. N.; Kashina, N. I.; Revzina, T. V.; Sobolevskaya, N. G.

Zh. Neorg. Khim. <u>1967</u>, *12*, 2240-3; *Russ. J. Inorg. Chem. (Engl. Transl.)* <u>1967</u>, *12*, 1179-81.

VARIABLES:

T/K = 273 to 373

PREPARED BY:

M. Salomon and H. Miyamoto

EXPERIMENTAL VALUES:

Solubility of $KBrO_3$[a]

t/°C	mass %	mol %	mol kg^{-1}
0	2.98	0.3303	0.1839
10	4.54	0.5104	0.2848
20	6.42	0.7346	0.4108
25	7.55	0.8733	0.4890
30	8.84	1.035	0.5806
40	11.67	1.405	0.7911
50	14.82	1.842	1.042
60	18.08	2.325	1.322
70	21.76	2.913	1.665
80	25.35	3.534	2.033
90	29.40	4.299	2.494
100	33.31	5.113	2.991

[a] Mol % and mol kg^{-1} solubilities calcd by compilers.

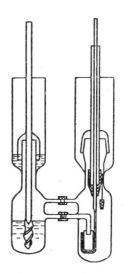

AUXILIARY INFORMATION

METHOD/APPARATUS/PROCEDURE:

Isothermal method. Equilibrium reached in 4-5 h. From 90-100°C, the solubility was determined in the apparatus shown in the figure. At equilibrium the apparatus was tilted to allow saturated solution to filter through connecting tube into weighed test tubes. The test tubes were closed with a stopper, withdrawn and weighed. Condensation on the walls of the apparatus and loss of water by evaporation was thus prevented. At the lower temperatures, ordinary solubility vessels were used, and pipets with glass filters were used for sampling (no other details given). Above 50°C, the pipets were preheated in the thermostat.

Bromate was determined iodometrically.

SOURCE AND PURITY OF MATERIALS:

Results of analysis of $KBrO_3$:

content of $KBrO_3$ = 99.3 %, and impurities are Rb (0.1 %), Cs (0.01 %), Na (0.016 %), SO_4 (0.005 %), and Fe (0.0001 %).

The alkali metal impurities were dtd by flame photometry, the iron colorimetrically, and sulfate nephelometrically.

ESTIMATED ERROR:

Soly: nothing specified.

Temp: precision ± 0.1 K.

REFERENCES:

COMPONENTS:	ORIGINAL MEASUREMENTS:
(1) Potassium bromate; $KBrO_3$; [7758-01-2] (2) Water-d_2; D_2O; [7789-20-0] (3) Water; H_2O; [7732-18-5]	Chang, T.L.; Hsieh, Y.Y. *Sci. Repts. Natl. Tsing Hua Univ.* <u>1948</u>, A5, 252-9.

VARIABLES:	PREPARED BY:
T/K = 298.15	G. Jansco and H. Miyamoto

EXPERIMENTAL VALUES:

t/°C	Water-d_2 mol %	Potassium Bromate mole/55.51 moles of H_2O-D_2O mixture
25	0	0.494 0.495 (Av)0.495[a]
	32.5	0.479 0.479 (Av)0.479
	66.2	0.460 0.458 (Av)0.459
	100	0.443[b]

[a] Average values calculated by compiler.

[b] The solubility in 100 % D_2O was obtained from the solubilities in the H_2O-D_2O mixtures by linear extrapolation.

AUXILIARY INFORMATION

METHOD/APPARATUS/PROCEDURE:	SOURCE AND PURITY OF MATERIALS:
Saturated solutions of $KBrO_3$ in the H_2O-D_2O mixtures were prepared by the method of supersaturation. The supersaturated solutions were prepared by agitating excess salt with the water mixture for one hour at 60°C; the time of agitation in the 25°C bath was 3 hours. A sample of the clear solution was delivered into a weighing bottle, the solvent evaporated and the residual pure salt was dried in vacuum at 100°C and weighed.	Potassium bromate was purified by recrystallization from conductivity water and found to be free from bromide. The salt was dried over calcium chloride in a desiccator for several days before use. D_2O content of the water mixture was determined by pycnometer both before and after each measurement. The mole percentage was calculated from the specific gravity at 25°C (ref 1).

	ESTIMATED ERROR:
	Soly: accuracy about 1 % (authors). Temp: precision $\pm$ 0.03 K.

	REFERENCES:
	1. Swift, E. Jr. *J. Am. Chem. Soc.* <u>1939</u>, 61, 198.

COMPONENTS:	ORIGINAL MEASUREMENTS:
(1) Potassium bromate; $KBrO_3$; [7758-01-2] (2) Sodium nitrate; $NaNO_3$; [7631-99-4] (3) Water; H_2O; [7732-18-5]	Geffcken, G. Z. Physik. Chem. 1904, 49, 257-302.
VARIABLES: T/K = 298 Concentration of $NaNO_3$	PREPARED BY: Hiroshi Miyamoto

EXPERIMENTAL VALUES:

Concn of $NaNO_3$ c_2/mol dm^{-3}	Soly of $KBrO_3$ c_1/mol dm^{-3}
0	0.4715
0.5	0.5745
1	0.6497
2	0.7680
3	0.9026
4	1.031

AUXILIARY INFORMATION

METHOD/APPARATUS/PROCEDURE:	SOURCE AND PURITY OF MATERIALS:
Mixtures of aqueous $NaNO_3$ solution and solid $KBrO_3$ were placed in bottles, and the bottles rotated in a thermostat. After equilibrium was established, the saturated solutions were allowed to settle in the thermostat. Samples were withdrawn with a pipet equipped with a cotton-wool filter. The determination of $KBrO_3$ was rapidly performed by iodometric titration.	No information given.
	ESTIMATED ERROR: Nothing specified.
	REFERENCES:

COMPONENTS:	ORIGINAL MEASUREMENTS:
(1) Potassium bromate; $KBrO_3$; [7758-01-2]	Geffcken, G.
(2) Sodium chloride; NaCl; [7647-14-5]	Z. *Physik. Chem.* 1904, 49, 257-302.
(3) Water; H_2O; [7732-18-5]	

VARIABLES:	PREPARED BY:
$T/K = 298$ Concentration of NaCl	Hiroshi Miyamoto

EXPERIMENTAL VALUES:

Concn of NaCl $c_2/\text{mol dm}^{-3}$	Soly of $KBrO_3$ $c_1/\text{mol dm}^{-3}$
0	0.4715
0.5	0.5220
1	0.5616
2	0.6042
3	0.6244
4	0.6400

AUXILIARY INFORMATION

METHOD/APPARATUS/PROCEDURE:	SOURCE AND PURITY OF MATERIALS:
Mixtures of aqueous NaCl solution and solid $KBrO_3$ were placed in bottles, and the bottles rotated in a thermostat. After equilibrium was established, the saturated solutions were allowed to settle in the thermostat. Samples were withdrawn with a pipet equipped with a cotton-wool filter. The determination of $KBrO_3$ was rapidly performed by iodometric titration.	No information given.
	ESTIMATED ERROR: Nothing specified.
	REFERENCES:

COMPONENTS:	ORIGINAL MEASUREMENTS:
(1) Potassium nitrate; KNO_3; [7757-79-1]	Ricci, J.E.
(2) Potassium bromate; $KBrO_3$; [7758-01-2]	J. Am. Chem. Soc. 1934, 56, 299-303.
(3) Water; H_2O; [7732-18-5]	

VARIABLES:	PREPARED BY:
Composition at 298.15 K	Hiroshi Miyamoto

EXPERIMENTAL VALUES: Composition of saturated solutions at 25.00°C

Potassium Nitrate		Potassium Bromate		Density	Nature of the
mass %	mol % (compiler)	mass %	mol % (compiler)	g cm^{-3}	solid phase[a]
27.71	6.393	0.00	0.00	1.193	A
27.27	6.459	2.64	0.379	1.211	"
27.01	6.475	3.90	0.566	1.228	A+B
27.01	6.475	3.90	0.566	1.225	"
27.01	6.475	3.90	0.566	1.223	"
(Av) 27.01	6.475	3.90	0.566	1.225	"
23.17	5.335	4.00	0.558	1.193	B
16.98	3.678	4.23	0.555	1.148	"
11.10	2.280	4.64	0.577	1.110	"
5.05	0.991	5.61	0.666	1.074	"
0.00	0.00	7.533[b]	0.871	1.054	"

[a] $A = KNO_3$; $B = KBrO_3$

[b] For the binary system the compiler computes the following:

soly of $KBrO_3 = 0.4878$ mol kg^{-1}

AUXILIARY INFORMATION

METHOD/APPARATUS/PROCEDURE:	SOURCE AND PURITY OF MATERIALS:
The ternary complexes were prepared from weighed amounts of water and the two anhydrous salts: these complexes were rotated in a large thermostat for about two days, a time found to be sufficient to reach equilibrium. Samples of the saturated solution were withdrawn with a calibrated pipet provided with a folded filter paper at the tip. The bromate was determined by titration with standard sodium thiosulfate solution, and the total solid by evaporation at 100°C and drying at 250°C. Potassium nitrate was found by difference. For the determination of solid phases, the method of algebraic extrapolation of tie-lines was used.	C.p. grade salts were recrystallized, dried to the anhydrous state, and stored in a 100°C oven.
	ESTIMATED ERROR: Soly: accuracy within 0.2 %. Temp: precision ± 0.01 K. Densities: precision about 0.1 %.
	REFERENCES:

COMPONENTS:	ORIGINAL MEASUREMENTS:
(1) Potassium sulfate; K_2SO_4; [7778-80-5] (2) Potassium bromate; $KBrO_3$; [7758-01-2] (3) Water; H_2O; [7732-18-5]	Ricci, J.E. J. Am. Chem. Soc. <u>1934</u>, 56, 299-303.

VARIABLES:	PREPARED BY:
Composition at 298.15 K	Hiroshi Miyamoto

EXPERIMENTAL VALUES: Composition of Saturated Solutions at 25.00°C

Potassium Sulfate		Potassium Bromate		Density	Nature of the
mass %	mol % (compiler)	mass %	mol % (compiler)	g cm^{-3}	solid phase[a]
10.76	1.231	0.00	0.00	1.083	A
10.12	1.170	1.69	0.204	1.094	"
9.45	1.10	3.40	0.414	1.103	"
9.34	1.10	4.00	0.490	1.108	A+B
9.36	1.10	4.00	0.490	1.108	"
9.35	1.10	4.01	0.491	1.108	"
(Av)9.35	1.10	4.00	0.490	1.108	"
8.20	0.954	4.27	0.519	1.100	B
5.44	0.620	5.02	0.597	1.083	"
2.67	0.299	6.08	0.712	1.066	"
0.00	0.00	7.53[b]	0.871	1.054	"

[a] A = K_2SO_4;   B = $KBrO_3$

[b] For the binary system the compiler computes the following:

soly of $KBrO_3$ = 0.4876 mol kg^{-1}

AUXILIARY INFORMATION

METHOD/APPARATUS/PROCEDURE:	SOURCE AND PURITY OF MATERIALS:
The ternary complexes were prepared from weighed amounts of water and the two anhydrous salts: these complexes were rotated in a large thermostat for about two days, a time found to be sufficient for attaining equilibrium. Samples of the saturated solution were withdrawn by means of a calibrated pipet provided with a folded filter paper at the tip. The bromate was determined by titration with standard sodium thiosulfate solution, and the total solid by evaporation at 100°C and drying at 250°C. Potassium sulfate was found by difference. For the determination of solid phases, the method of algebraic extrapolation of tielines was used.	C.p. grade salts were recrystallized, dried to the anhydrous state, and stored in a 100°C oven.
	ESTIMATED ERROR: Soly: accuracy within 0.2 %. Temp: precision $\pm$ 0.01 K. Densities: precision about 0.1 %.
	REFERENCES:

COMPONENTS:	ORIGINAL MEASUREMENTS:
(1) Potassium chloride; KCl; [7447-40-7]	Ricci, J.E.
(2) Potassium bromate; KBrO$_3$; [7758-01-2]	J. Am. Chem. Soc. 1934, 56, 299-303.
(3) Water; H$_2$O; [7732-18-5]	

VARIABLES:	PREPARED BY:
Composition at 298.15 K	Hiroshi Miyamoto

EXPERIMENTAL VALUES: Composition of saturated solutions at 25.00°C

| Potassium Chloride | | Potassium Bromate | | Density | Nature of the |
mass %	mol % (compiler)	mass %	mol % (compiler)	g cm^{-3}	solid phase[a]
26.36	7.961	0.00	0.00	1.179	A
25.93	7.930	1.48	0.202	1.187	"
25.90	7.930	1.61	0.220	1.197	A+B
25.89	7.926	1.60	0.219	1.189	"
25.88	7.923	1.61	0.220	1.190	"
(Av) 25.89	7.926	1.61	0.220	1.192	"
24.87	7.544	1.65	0.223	1.183	B
19.71	5.718	1.97	0.255	1.147	"
14.45	4.020	2.44	0.303	1.112	"
9.03	2.418	3.24	0.387	1.082	"
4.33	1.130	4.63	0.539	1.058	"
0.00	0.00	7.533[b]	0.8712	1.054	"

[a] A = KCl;   B = KBrO$_3$

[b] For the binary system the compiler computes the following:

soly of KBrO$_3$ = 0.4878 mol kg^{-1}

AUXILIARY INFORMATION

METHOD/APPARATUS/PROCEDURE:	SOURCE AND PURITY OF MATERIALS:
The ternary complexes were prepared from weighed amounts of water and the two anhydrous salts: these complexes were rotated in a large thermostat for about two days, a time found to be sufficient for attaining equilibrium. Samples of the saturated solution were withdrawn by means of a calibrated pipet provided with a folded filter paper at the tip. The bromate was determined by titration with standard sodium thiosulfate solution, and the total solid by evaporation at 100°C and drying at 250°C. Potassium chloride was found by difference. For the determination of solid phases, the method of algebraic extrapolation of tie-lines was used.	C.p. grade salts were recrystallized, dried to the anhydrous state, and stored in a 100°C oven.
	ESTIMATED ERROR:
	Soly: accuracy within 0.2 %. Temp: precision ± 0.01 K. Densities: precision about 0.1 %.
	REFERENCES:

COMPONENTS:	ORIGINAL MEASUREMENTS:
(1) Potassium bromide; KBr; [7758-02-3]	Gerasimov, Ya. I.
(2) Potassium bromate; KBrO₃; [7758-01-2]	*Zh. Obshch. Khim.* 1934, 4, 223-7.
(3) Water; H₂0; [7732-18-5]	

Corrected subscripts:

COMPONENTS:	ORIGINAL MEASUREMENTS:
(1) Potassium bromide; KBr; [7758-02-3]	Gerasimov, Ya. I.
(2) Potassium bromate; $KBrO_3$; [7758-01-2]	*Zh. Obshch. Khim.* 1934, 4, 223-7.
(3) Water; H_2O; [7732-18-5]	

VARIABLES:	PREPARED BY:
T/K = 273, 293, 313, 333 and 353 Composition	Hiroshi Miyamoto

EXPERIMENTAL VALUES: Composition of saturated solutions

t/°C	Potassium Bromate mass %	mol % (compiler)	Potassium Bromide mass %	mol % (compiler)	Nature of the solid phase[a]
0	2.96[b]	0.328	–	–	
	0.646	0.100	35.16	7.649	
	0.665	0.103	35.16	7.651	
	0.57	0.088	35.08	7.617	
	–	–	35.30	7.629	
20	6.43[b]	0.736	–	–	
	2.73	0.336	11.80	2.040	
	1.85	0.251	22.20	4.227	
	1.22	0.201	39.26	9.061	
	–	–	39.4	8.96	
40	11.70[b]	1.409	–	–	A
	7.32	0.915	8.50	1.49	"
	4.32	0.594	20.845	4.022	"
	2.60	0.418	35.97	8.109	"
	2.19	0.380	42.34	10.32	A+B
	1.28	0.220	42.44	10.22	B
	0.34	0.058	42.97	10.29	"
	–	–	43.54	10.45	"

continued.....

AUXILIARY INFORMATION

METHOD/APPARATUS/PROCEDURE:	SOURCE AND PURITY OF MATERIALS:
The isothermal method was used. After complexes of salts and water were vigorously stirred for 1.5 to 2.0 hours, equilibrium was established. The bromate content was determined iodometrically. The bromide content was determined by Volhard's method: the solution containing bromide was treated with excess standard silver nitrate solution and the residual silver nitrate determined by titration with standard ammonium thiocynate solution. The determination of the composition of solid phases was not described in the original paper.	No information was given.
	ESTIMATED ERROR: Nothing specified.
	REFERENCES:

COMPONENTS:	ORIGINAL MEASUREMENTS:
(1) Potassium bromide; KBr; [7758-02-3]	Gerasimov, Ya. I.
(2) Potassium bromate; KBrO₃; [7758-01-2]	*Zh. Obshch. Khim.* 1934, 4, 223-7.
(3) Water; H₂0; [7732-18-5]	

EXPERIMENTAL VALUES: (Continued)

Composition of saturated solutions

t/°C	Potassium Bromate mass %	mol % (compiler)	Potassium Bromide mass %	mol % (compiler)	Nature of the solid phase[a]
60	18.21[b]	2.345	–	–	A
	16.43	2.161	3.94	0.727	"
	15.58	2.039	4.35	0.799	"
	14.92	1.942	4.48	0.818	"
	13.14	1.765	9.37	1.766	"
	12.83	1.720	9.52	1.791	"
	12.74	1.716	10.07	1.903	"
	12.09	1.663	12.69	2.449	"
	5.71	0.908	31.90	7.119	"
	3.73	0.684	44.56	11.46	"
	3.70	0.681	44.92	11.61	"
	3.75	0.745	49.93	13.92	"
	3.70	0.684	45.17	11.72	A+B
	–	–	46.2	11.50	B
80	25.53[b]	3.566	–	–	A
	20.74	2.976	7.45	1.50	"
	5.66	1.08	45.57	12.26	"
	5.62	1.09	46.5	12.7	A+B
	–	–	49.72	13.02	B

[a] A = KBrO₃;   B = KBr

[b] For the binary system the compiler computes the following:

$$\text{soly of KBrO}_3 = 0.183 \text{ mol kg}^{-1} \text{ at } 0°C$$

$$= 0.411 \text{ mol kg}^{-1} \text{ at } 20°C$$

$$= 0.7934 \text{ mol kg}^{-1} \text{ at } 40°C$$

$$= 1.333 \text{ mol kg}^{-1} \text{ at } 60°C$$

$$= 2.053 \text{ mol kg}^{-1} \text{ at } 80°C$$

COMPONENTS:	ORIGINAL MEASUREMENTS:
(1) Potassium bromide; KBr; [7758-02-3] (2) Potassium bromate; $KBrO_3$; [7758-01-2] (3) Water; H_2O; [7732-18-5]	Ricci, J.E. J. Am. Chem. Soc. 1934, 56, 299-303.

VARIABLES:	PREPARED BY:
Composition at 298.15 K	Hiroshi Miyamoto

EXPERIMENTAL VALUES: Composition of saturated solutions at 25.00°C

Potassium Bromide		Potassium Bromate		Density	Nature of the
mass %	mol % (compiler)	mass %	mol % (compiler)	g cm^{-3}	solid phase[a]
40.62	9.384	0.00	0.00	1.381	A
40.08	9.347	1.20	0.199	1.389	"
40.00	9.348	1.43	0.238	1.392	A+B
39.99	9.344	1.43	0.238	1.393	"
39.99	9.344	1.43	0.238	1.392	"
(Av) 39.99	9.344	1.43	0.238	1.392	"
34.82	7.639	1.62	0.253	1.328	B
26.05	5.185	2.06	0.292	1.237	"
17.48	3.199	2.73	0.356	1.161	"
7.82	1.32	4.29	0.517	1.089	"
0.00	0.00	7.533[b]	0.8712	1.054	"

[a] A = KBr; B = $KBrO_3$

[b] For the binary system the compiler computes the following:

soly of $KBrO_3$ = 0.4878 mol kg^{-1}

AUXILIARY INFORMATION

METHOD/APPARATUS/PROCEDURE:	SOURCE AND PURITY OF MATERIALS:
The ternary complexes were prepared from weighed amounts of water and the two anhydrous salts: these complexes were rotated in a large thermostat for about two days, a time found to be sufficient for attaining equilibrium. Samples of the saturated solution were withdrawn by means of a calibrated pipet provided with a folded filter paper at the tip. The bromate was determined by titration with standard sodium thiosulfate solution, and the total solid by evaporation at 100°C and drying at 250°C. Potassium bromide was found by difference. For the determination of solid phases, the method of algebraic extrapolation of tie-lines was used.	C.p. grade salts were recrystallized, dried to the anhydrous state, and stored in a 100°C oven.
	ESTIMATED ERROR: Soly: accuracy within 0.2 %. Temp: precision ± 0.01 K. Densities: precision about 0.1 %.
	REFERENCES:

COMPONENTS:	ORIGINAL MEASUREMENTS:
(1) Potassium bromate; $KBrO_3$; [7758-01-2]	Ricci, J.E.
(2) Potassium iodide; KI; [7681-11-0]	*J. Am. Chem. Soc.* 1934, 56, 299-303.
(3) Water; H_2O; [7732-18-5]	

VARIABLES:	PREPARED BY:
Composition at 298.15 K	Hiroshi Miyamoto

EXPERIMENTAL VALUES: Composition of Saturated solutions at 25.00°C

Potassium Iodide mass %	mol % (compiler)	Potassium Bromate mass %	mol % (compiler)	Density g cm^{-3}	Nature of the solid phase[a]
59.76	13.88	0.00	0.00	1.718	A
59.15	13.83	0.96	0.22	1.728	A+B
59.22	13.87	0.96	0.22	1.727	"
59.22	13.87	0.96	0.22	1.730	"
(Av) 59.20	13.86	0.96	0.22	1.729	"
58.14	13.34	0.99	0.23	1.707	B
50.06	10.01	1.21	0.240	1.565	"
38.99	6.634	1.63	0.276	1.402	"
28.60	4.277	2.17	0.323	1.278	"
18.85	2.539	2.96	0.396	1.182	"
8.77	1.080	4.54	0.556	1.103	"
0.00	0.00	7.533[b]	0.871	1.054	"

[a] A = KI;   B = $KBrO_3$

[b] For the binary system the compiler computes the following:

   soly of $KBrO_3$ = 0.4878 mol kg^{-1}.

AUXILIARY INFORMATION

METHOD/APPARATUS/PROCEDURE:	SOURCE AND PURITY OF MATERIALS:
The ternary complexes were prepared from weighed amounts of water and the two anhydrous salts: these complexes were rotated in a large thermostat for about two days, a time found to be sufficient to reach equilibrium. Samples of the saturated solution were withdrawn with a calibrated pipet provided with a folded filter paper at the tip. The bromate was determined by titration with standard sodium thiosulfate solution, and the total solid by evaporation at 100°C and drying at 250°C. Potassium iodide was found by difference. For the determination of solid phases, the method of algebraic extrapolation of tie-lines was used.	C.p. grade salts were recrystallized, dried to the anhydrous state, and stored in a 100°C oven.
	ESTIMATED ERROR: Soly: accuracy within 0.2 %. Temp: precision ± 0.01 K. Densities: precision about 0.1 %.
	REFERENCES:

COMPONENTS:	ORIGINAL MEASUREMENTS:
(1) Potassium bromate; $KBrO_3$; [7758-01-2] (2) Rubidium bromate; $RbBrO_3$; [13446-70-3] (3) Water; H_2O; [7732-18-5]	Kirgintsev, A.N.; Shklovskaya, R.M.; Arkhipov, S.M. *Izv. Akad. Nauk SSSR Ser. Khim.* 1971, 2631-4; *Bull. Acad. Sci. USSR Div. Chem. Sci. (Engl. Transl.)* 1971, 2501-4.

VARIABLES:	PREPARED BY:
Composition at 298.2 K	Hiroshi Miyamoto

EXPERIMENTAL VALUES: Composition of saturated solutions

t/°C	Potassium Bromate mass %	mol % (compiler)	Rubidium Bromate mass %	mol % (compiler)	m^a mol kg^{-1}	$y_1{}^b$
25	7.53[c]	0.871	0.00	0.00	0.488	1.00
	6.68	0.769	0.43	0.039	0.452	0.95
	5.74	0.657	0.68	0.061	0.412	0.91
	4.94	0.563	0.97	0.086	0.362	0.87
	3.94	0.446	1.25	0.111	0.313	0.80
	2.88	0.324	1.69	0.149	0.262	0.69
	2.25	0.252	1.80	0.158	0.227	0.61
	1.54	0.172	1.99	0.174	0.198	0.50
	1.01	0.112	2.23	0.194	0.171	0.37
	0.49	0.054	2.46	0.213	0.151	0.20
	0.0	0.0	2.83[c]	0.245	0.136	0.0

[a] m = the total molality of the salts in liquid phase.

[b] y_1 = the mole fraction of $KBrO_3$ based on total bromate content.

[c] For the binary systems the compiler computes the following:

   soly of $KBrO_3$ = 0.488 mol kg^{-1}

   soly of $RbBrO_3$ = 0.136 mol kg^{-1}

AUXILIARY INFORMATION

METHOD/APPARATUS/PROCEDURE:	SOURCE AND PURITY OF MATERIALS:
The solubility was studied by the isothermal relief of supersaturation followed by mixing of the solid phase and mother liquor for 24 hours at 25°C. To verify the solubility of the method used to establish equilibrium, the solubilities for several points were detd by the method of isothermal saturation with mixing for 30 days. The number of moles of the anion (n_1) were detd by iodometric titrn. Alkali metal contents were determined in the same samples by flame photometry from three parallel analyses. In each analysis the authors calculated the sum of the moles of the cations (n_2). The composition of the solid phase was detd by the Schreinemakers' method of residues. A phase diagram indicating the existence of a hydrate was not given in the original paper. Densities of the saturated solutions at 25°C were determined, but the data were not given.	C.p. grade $KBrO_3$ and $RbBrO_3$ were recrystallized from double distilled water.
	ESTIMATED ERROR:
	REFERENCES:

COMPONENTS:	ORIGINAL MEASUREMENTS:
(1) Potassium bromate; $KBrO_3$; [7758-01-2] (2) Cesium bromate; $CsBrO_3$; [13454-75-6] (3) Water; H_2O; [7732-18-5]	Kirgintsev, A.I.; Yakobi, N.Y. *Zh. Neorg. Khim.* <u>1968</u>, *13*, 2851-3; *Russ. J. Inorg. Chem. (Engl. Transl.)* <u>1968</u>, *13*, 1467-8.

VARIABLES:	PREPARED BY:
Composition at 298.2 K	Hiroshi Miyamoto

EXPERIMENTAL VALUES: Composition of saturated solutions

t/°C	Potassium Bromate mass %	mol % (compiler)	Cesium Bromate mass %	mol % (compiler)	y_1[a]	m[b]	Nature of the solid phase[c]
25	0.00	0.0	3.66[d]	0.262	0.0	0.146	A
	0.33	0.037	3.44	0.246	0.13	0.153	"
	0.68	0.076	3.21	0.230	0.25	0.170	"
	1.19	0.135	3.92	0.284	0.39	0.191	"
	1.70	0.191	2.64	0.190	0.50	0.212	"
	2.36	0.266	2.37	0.171	0.61	0.244	"
	4.04	0.461	1.98	0.145	0.76	0.338	"
	5.28	0.608	1.80	0.133	0.82	0.414	"
	6.04	0.700	1.71	0.127	0.85	0.464	"
	7.03	0.822	1.60	0.120	0.87	0.528	"
	7.12	0.834	1.61	0.121	0.87	0.535	"
	7.22	0.845	1.53	0.115	0.88	0.538	A+B
	7.38	0.861	1.02	0.0762	0.92	0.525	B
	7.49[d]	0.866	0.0	0.0	1.00	0.485	"

[a] y_1 = the mole fraction of $KBrO_3$ based only on total bromate composition.

[b] m = the total molality of the salts in liquid phase.

[c] A = $CsBrO_3$;   B = $KBrO_3$

[d] For binary systems the compiler computes the following:

soly of $KBrO_3$ = 0.485 mol kg^{-1};   soly of $CsBrO_3$ = 0.146 mol kg^{-1}.

AUXILIARY INFORMATION

METHOD/APPARATUS/PROCEDURE:	SOURCE AND PURITY OF MATERIALS:
The isothermal relief of supersaturation method was employed. The supersaturated solutions were stirred for 7-8 hours. The composition of the coexisting phases was determined by the method of indirect analyses (ref 1 and 2), the parameters measured being the sum of the salts and the total number of moles of salt determined by an ion-exchange method.	Analytical reagent grade $KBrO_3$ and $CsBrO_3$ were recrystallized from double-distilled water.
	ESTIMATED ERROR: Soly: the accuracy in determining y_1 was within 5 %. Temp: precision $\pm$ 0.1 K.

	REFERENCES:
	1. Kirgintsev, A.I.; Kashina, N.I.;
	   Vulikh, A.I.; Korotkevich, B.I.
	   Zh. Neorg. Khim. <u>1965</u>, *10*, 1225;
	   Russ. J. Inorg. Chem. (Engl. Transl.)
	   <u>1965</u>, *10*, 662.
	2. Kirgintsev, A.I.; Trushnikova, L.N.
	   Zh. Neorg. Khim. <u>1963</u>, *13*, 2843;
	   Russ. J. Inorg. Chem. (Engl. Transl.)
	   <u>1963</u>, *13*, 1591.

COMPONENTS:	ORIGINAL MEASUREMENTS:
(1) Potassium bromate; $KBrO_3$; [7758-01-2]	Azarova, L.A.; Vinogradov, E.E.
(2) Barium bromate; $Ba(BrO_3)_2$; [13967-90-3]	*Zh. Neorg. Khim.* 1982, *27*, 2967-70; *Russ. J. Inorg. Chem. (Engl. Transl.)*
(3) Water; H_2O; [7732-18-5]	1982, *27*, 1681-3;

VARIABLES:	PREPARED BY:
Composition at 298 K	Hiroshi Miyamoto

EXPERIMENTAL VALUES:

Composition of saturated solutions

Barium Bromate		Potassium Bromate		Nature of the
mass %	mol % (compiler)	mass %	mol % (compiler)	solid phase[a]
0.79[b]	0.036	--	--	A
0.098	0.0046	1.52	0.166	"
0.33	0.016	3.88	0.435	"
trace	–	4.90	–	"
trace	–	7.09	–	"
0.57	0.028	6.96	0.805	A+B
0.37	0.018	7.44	0.863	"
0.023	0.0011	7.62	0.882	"
0.47	0.023	7.49	0.870	"
--	--	7.49[b]	0.866	B

[a] A = $Ba(BrO_3)_2 \cdot H_2O$;   B = $KBrO_3$

[b] For binary systems the compiler computes the following:

soly of $KBrO_3$ = 0.485 mol kg^{-1}

soly of $Ba(BrO_3)_2$ = 0.020 mol kg^{-1}

AUXILIARY INFORMATION

METHOD/APPARATUS/PROCEDURE:	COMMENTS AND/OR ADDITIONAL DATA:
Probably the isothermal method was used. Equilibrium was reached in 10-12 days. The potassium content was detd gravimetrically with sodium tetraphenylborate. The bromate concentration was detd by iodometric titrn using sodium thiosulfate. The barium content was detd gravimetrically as the sulfate. The compositions of the solid phases were determined by Schreinemakers' method of residues, and by X-ray diffraction.	The phase diagram is given below (based on mass % units). 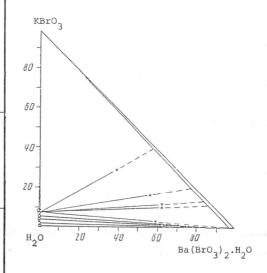
SOURCE AND PURITY OF MATERIALS: "Analytical grade" potassium bromate was used. Barium bromate monohydrate was prepd by mixing solns of $KBrO_3$ and $BaCl_2$. The formula of the salt obtained was determined by chemical analysis and checked by X-ray diffraction.	
ESTIMATED ERROR: Nothing specified.	

COMPONENTS:	ORIGINAL MEASUREMENTS:
(1) Potassium bromate; KBrO₃; [7758-01-2]	Serebrennikov, V.V.; Batyreva, V.A.; Larionova, I.S.
(2) Yttrium bromate; Y(BrO₃)₃; [15162-95-5]	*Zh. Neorg. Khim.* <u>1982</u>, *27*, 2959-61; *Russ. J. Inorg. Chem. (Engl. Transl.)* <u>1982</u>, *27*, 1677-9.
(3) Water; H₂O; [7732-18-5]	

VARIABLES:	PREPARED BY:
Composition at 298 K	Hiroshi Miyamoto

EXPERIMENTAL VALUES: Composition of saturated solutions

Yttrium Bromate		Potassium Bromate		Nature of the
mass %	mol % (compiler)	mass %	mol % (compiler)	solid phase[a]
45.5[b]	3.08	0.0	0.00	A
48.6	3.52	0.7	0.14	A+B
49.3	3.62	0.7	0.15	"
47.3	3.36	0.9	0.18	"
48.7	3.55	1.0	0.21	"
48.6	3.54	1.0	0.21	"
48.6	3.53	0.9	0.19	"
47.3	3.37	1.1	0.22	"
45.1	3.09	1.2	0.23	B
42.4	2.88	3.4	0.65	"
37.7	2.33	2.3	0.40	"
23.4	1.27	7.9	1.21	"
2.3	0.097	8.1	0.96	"
0.0	0.00	7.7[b]	0.89	"

[a] A = Y(BrO₃)₃·9H₂O;   B = KBrO₃

[b] For binary systems the compiler computes the following:

$$\text{soly of KBrO}_3 = 1.77 \text{ mol kg}^{-1}$$

$$\text{soly of Y(BrO}_3)_3 = 0.50 \text{ mol kg}^{-1}$$

AUXILIARY INFORMATION

METHOD/APPARATUS/PROCEDURE:	COMMENTS AND/OR ADDITIONAL DATA:
The solubility was probably studied by the isothermal method. Mixtures of the salts and water were continuously stirred in glass bottles for seven days. The potassium bromate content in the liquid phase was determined by flame photometry, and yttrium bromate was determined complexometrically. The composition of the solid phase was determined by X-ray analysis.	The phase diagram is given below (based on mass % units). 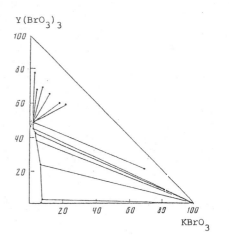

SOURCE AND PURITY OF MATERIALS:
Nothing specified.

ESTIMATED ERROR:
Nothing specified.

COMPONENTS:	ORIGINAL MEASUREMENTS:
(1) Potassium bromate; KBrO₃; [7758-01-2]	Serebrennikov, V.V.; Batyreva, V.A.; Larionova, I.S.
(2) Samarium bromate; Sm(BrO₃)₃; [28958-26-1]	Zh. Neong. Khim. 1982 27, 2959-61; Russ. J. Inong. Chem. (Engl. Transl.) 1982, 27, 1677-9.
(3) Water; H₂0; [7732-18-5]	

VARIABLES:	PREPARED BY:
Composition at 298 K	Hiroshi Miyamoto

EXPERIMENTAL VALUES: Composition of saturated solutions at 25°C

Samarium Bromate		Potassium Bromate		Nature of the
mass %	mol % (compiler)	mass %	mol % (compiler)	solid phase[a]
41.4[b]	2.33	0.0	0.00	A
40.3	2.25	0.6	0.11	A+B
40.0	2.23	1.1	0.20	"
39.9	2.24	1.4	0.25	"
39.9	2.24	1.6	0.29	"
40.3	2.29	1.9	0.35	"
38.3	2.10	1.7	0.30	"
38.9	2.15	1.7	0.30	"
42.6	2.62	4.4	0.86	"
39.6	2.22	1.8	0.32	"
39.4	2.20	1.7	0.30	"
40.3	2.29	1.8	0.33	"
44.8	2.75	2.0	0.39	"
40.0	2.26	1.9	0.34	B
38.3	2.11	2.1	0.37	"
28.7	1.39	2.7	0.42	"
19.3	0.83	3.6	0.50	"
12.5	0.50	4.1	0.52	"
4.7	0.18	5.7	0.68	"
0.0	0.00	7.7[b]	0.89	"

[a] A = Sm(BrO₃)₃·9H₂0;   B = KBrO₃

[b] For binary systems the compiler computes the following:

$$\text{soly of } KBrO_3 = 1.77 \text{ mol kg}^{-1}$$
$$\text{soly of } Sm(BrO_3)_3 = 1.32 \text{ mol kg}^{-1}$$

METHOD/APPARATUS/PROCEDURE:	COMMENTS AND/OR ADDITIONAL DATA:
The solubility was probably studied by the isothermal method. Mixtures of the salts and water were continuously stirred in glass bottles for seven days. The potassium bromate content in the liquid phase was determined by flame photometry, and samarium bromate was determined spectro-photometrically. The composition of the solid phases were determined by X-ray analysis.	The phase diagram is given below (based on mass % units). 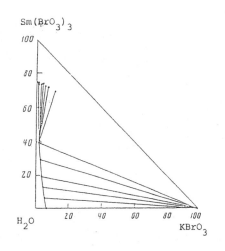

SOURCE AND PURITY OF MATERIALS:

Nothing specified.

ESTIMATED ERROR:

Nothing specified.

COMPONENTS:

(1) Potassium bromate; KBrO$_3$ [7758-01-2]

(2) Silver bromate; AgBrO$_3$; [7783-89-3]

(3) Water; H$_2$O; [7732-18-5]

ORIGINAL MEASUREMENTS:

Ricci, J. E.; Offenbach, J. A.

J. Am. Chem. Soc. 1951, 73, 1597-9.

VARIABLES:

Composition
T/K = 298

PREPARED BY:

H. Miyamoto and M. Salomon

EXPERIMENTAL VALUES:

Composition of Saturation Solutions at 25°C[a]

KBrO$_3$		AgBrO$_3$		Density	Solid
mass %	mol %	mass %	mole %	g/cm^3	Phase
0	0	0.193[c]	0.01477	0.9983	AgBrO$_3$
3.21		Not given	Not given	1.022	"
5.70		Not given	Not given	1.049	"
7.03		Not given	Not given	1.050	"
7.52		Not given	Not given	1.054	AgBrO$_3$ + KBrO$_3$
7.57		Not given	Not given	1.052	"
7.56		Not given	Not given	1.053	"
7.55		Not given	Not given	1.054	"
7.52[b]	0.8696			1.053	KBrO$_3$

[a]Mole % calculated by compilers.

[b]Solubility of KBrO$_3$ = 0.4869 mol kg^{-1}

[c]Solubility of AgBrO$_3$ = 0.008202 mol kg^{-1}

AUXILIARY INFORMATION

METHOD/APPARATUS/PROCEDURE:

Ternary mixtures, AgBrO$_3$-KBrO$_3$-H$_2$O, of known composition were made to come to come to equilibrium at 25°C. The mixture was stirred for 2 weeks. The solution was simply evaporated to dryness for its KBrO$_3$ content since the solubility of AgBrO$_3$ in the presence of KBrO$_3$ was found to be negligible. The determination method of AgBrO$_3$ in solution was not given, but it was probably by Volhard titration with KSCN since this was the method used for AgClO$_3$-NaClO$_3$-H$_2$O system reported in the same paper.

SOURCE AND PURITY OF MATERIALS:

AgBrO$_3$ was prepared by adding a dilute sln of bromic acid to Ag$_2$CO$_3$ in the presence of HNO$_3$. After some heating and digestion, the solid was washed with water, and finally dried at 110°C. The two batches prepared gave 99.93 and 99.75% AgBrO$_3$ by Br analysis. Ag$_2$CO$_3$ was made by addition of Na$_2$CO$_3$ to an excess of an aqueous AgNO$_3$ solution. Bromic acid solution was made from dilute H$_2$SO$_4$ and solid Ba(BrO$_3$)$_2$.H$_2$O. KBrO$_3$ purity was 99.97%.

ESTIMATED ERROR:

Nothing specified in original article.
Soly: $\pm$ .01 mass % (compiler)
Temp: precision probably better than 0.1 K
      (compiler).

REFERENCES:

COMPONENTS:	ORIGINAL MEASUREMENTS:
(1) Potassium bromate; KBrO$_3$; [7758-01-2] (2) Alcohols (3) Water; H$_2$O; [7732-18-5]	Rothmund, V. Z. Physik. Chem. 1909, 69, 523-46.
VARIABLES: T/K = 298 Composition	PREPARED BY: Hiroshi Miyamoto and Mark Salomon

EXPERIMENTAL VALUES:

Composition of Solvent			soly of KBrO$_3$/mol dm^{-3}
pure water			0.478

binary mixtures containing 0.5 mol dm^{-3} of the following:

methanol;	CH$_4$O;	[67-56-1]	0.444
ethanol;	C$_2$H$_6$O;	[64-17-5]	0.421
1,2-ethanediol; (ethylene glycol)	C$_2$H$_6$O$_2$;	[107-21-1]	0.448
1-propanol;	C$_3$H$_8$O;	[71-23-8]	0.409
1,2,3-propanetriol; (glycerol)	C$_3$H$_8$O$_3$;	[56-81-5]	0.451
2-methyl-2-butanol;	C$_5$H$_{12}$O;	[75-85-4]	0.383

AUXILIARY INFORMATION

METHOD/APPARATUS/PROCEDURE:	SOURCE AND PURITY OF MATERIALS:
The salt and solvent were placed in a bottle sealed with a rubber stopper, and rotated in a thermostat for at least 14 hours. After the saturated solutions were allowed to settle, aliquots of saturated solution were withdrawn with a pipet fitted with a glass-wool or cotton-wool filter. The bromate content was determined by iodometric titration.	Potassium bromate was repeatedly recrystallized. No information of the source and purity of the solvents was given.
	ESTIMATED ERROR: Nothing specified.
	REFERENCES:

COMPONENTS:	ORIGINAL MEASUREMENTS:
(1) Potassium bromate; $KBrO_3$; [7758-01-2]	Rothmund, V.
(2) Ethyl ether; $C_4H_{10}O$; [60-29-7]	Z. Physik. Chem. 1909, 69, 523-46
(3) Water; H_2O; [7732-18-5]	

VARIABLES:	PREPARED BY:
T/K = 298	Hiroshi Miyamoto
Concentration of ethyl ether	

EXPERIMENTAL VALUES:

Concn ethyl ether/mol dm^{-3}	soly of $KBrO_3$/mol dm^{-3}
0	0.478
0.5	0.395

AUXILIARY INFORMATION

METHOD/APPARATUS/PROCEDURE:	SOURCE AND PURITY OF MATERIALS:
The salt and solvent were placed in a bottle, sealed with a rubber stopper, and rotated in a thermostat for at least 14 hours. After the saturated solution was allowed to settle, aliquots were withdrawn with a pipet fitted with a glass-wool or cotton-wool filter. The bromate content was determined by iodometric titration.	Potassium bromate was repeatedly recrystallized. No information of the source and purity of the solvents was given.
	ESTIMATED ERROR:
	Nothing specified.
	REFERENCES:

COMPONENTS:	ORIGINAL MEASUREMENTS:
(1) Potassium bromate; $KBrO_3$; [7758-01-2] (2) Sugars (3) Water; H_2O; [7732-18-5]	Rothmund, V. Z. Physik. Chem. 1909, 69, 523-46.

VARIABLES:	PREPARED BY:
T/K = 298 Composition	Hiroshi Miyamoto and Mark Salomon

EXPERIMENTAL VALUES:

Composition of Solvent			Soly of $KBrO_3$/ mol dm^{-3}
pure water			0.478

binary mixtures containing 0.5 mol dm^{-3}
of the following:

D-glucose;	$C_6H_{12}O_6$;	[50-99-7]	0.463
D-mannitol;	$C_6H_{14}O_6$;	[69-65-8]	0.451

AUXILIARY INFORMATION

METHOD/APPARATUS/PROCEDURE:	SOURCE AND PURITY OF MATERIALS:
The salt and solvent were placed in a bottle, sealed with a rubber stopper, and rotated in a thermostat for at least 14 hours. After the saturated solutions were allowed to settle, aliquots were withdrawn with a pipet fitted with a glass-wool or cotton-wool filter. The bromate content was determined by iodometric titration.	Potassium bromate was repeatedly recrystallized. No information of the source and purity of the solvents was given.
	ESTIMATED ERROR: Nothing specified.
	REFERENCES:

COMPONENTS:	ORIGINAL MEASUREMENTS:
(1) Potassium bromate; $KBrO_3$; [7758-01-2]	Rothmund, V.
(2) Formaldehyde; CH_2O; [50-00-0]	Z. *Physik. Chem.* 1909, 69, 523-46.
(3) Water; H_2O; [7732-18-5]	

VARIABLES:	PREPARED BY:
T/K = 298	Hiroshi Miyamoto
Concentration of formaldehyde	

EXPERIMENTAL VALUES:

Formaldehyde concn/mol dm^{-3}	soly of $KBrO_3$/mol dm^{-3}
0	0.478
0.5	0.397

AUXILIARY INFORMATION

METHOD/APPARATUS/PROCEDURE:	SOURCE AND PURITY OF MATERIALS:
The salt and solvent were placed in a bottle, sealed with a rubber stopper, and rotated in a thermostat for at least 14 hours. After the saturated solution was allowed to settle, aliquots of the saturated solution were withdrawn with a pipet fitted with a glass-wool or cotton-wool filter. The bromate content was determined by iodometric titration.	Potassium bromate was repeatedly recrystallized. No information of the source and purity of the solvents was given.
	ESTIMATED ERROR:
	Nothing specified.
	REFERENCES:

COMPONENTS:	ORIGINAL MEASUREMENTS:
(1) Potassium bromate; $KBrO_4$; [7758-01-2] (2) 2-Propanone (acetone); C_3H_6O; [67-64-1] (3) Water; H_2O; [7732-18-5]	Rothmund, V. Z. *Physik. Chem.* 1909, 69, 523-46.

VARIABLES:	PREPARED BY:
T/K = 298 Concentration of acetone	Hiroshi Miyamoto

EXPERIMENTAL VALUES:

Concn acetone/mol dm^{-3}	soly of $KBrO_3$/mol dm^{-3}
0	0.478
0.5 mol dm^{-3} soln	0.425

AUXILIARY INFORMATION

METHOD/APPARATUS/PROCEDURE:	SOURCE AND PURITY OF MATERIALS:
The salt and solvent were placed in a bottle, sealed with a rubber stopper, and rotated in a thermostat for at least 14 hours. After the saturated solution was allowed to settle, aliquots were withdrawn with a pipet fitted with a glass-wool or cotton-wool filter. The bromate content was determined by iodometric titration.	Potassium bromate was repeatedly recrystallized. No information of the source and purity of the solvents was given.
	ESTIMATED ERROR: Nothing specified.
	REFERENCES:

COMPONENTS:	ORIGINAL MEASUREMENTS:
(1) Potassium bromate; KBrO$_3$; [7758-01-2]	Rothmund, V.
(2) Acids	Z. Physik. Chem. 1909, 69, 523-46.
(3) Water; H$_2$O; [7732-18-5]	

VARIABLES:	PREPARED BY:
T/K = 298 Composition	Hiroshi Miyamoto and Mark Salomon

EXPERIMENTAL VALUES:

Composition of Solvent			Soly of KBrO$_3$/mol dm^{-3}
pure water			0.478

binary mixtures containing 0.5 mol dm^{-3} of the following:

acetic acid;	C$_2$H$_2$O;	[64-19-7]	0.456
glycine ;	C$_2$H$_5$NO$_2$;	[56-40-6]	0.501

(aminoacetic acid)

AUXILIARY INFORMATION

METHOD/APPARATUS/PROCEDURE:	SOURCE AND PURITY OF MATERIALS:
The salt and solvent were placed in a bottle, sealed with a rubber stopper, and rotated in a thermostat for at least 14 hours. After the saturated solutions were allowed to settle, aliquots were withdrawn with a pipet fitted with a glass-wool or cotton-wool filter. The bromate content was determined by iodometric titration.	Potassium bromate was repeatedly recrystallized. No information of the source and purity of the solvents was given.
	ESTIMATED ERROR: Nothing specified.
	REFERENCES:

COMPONENTS:	ORIGINAL MEASUREMENTS:
(1) Potassium bromate; $KBrO_3$; [7758-01-2]	Rothmund, V.
(2) Phenol; C_6H_6O; [108-95-2]	Z. Physik. Chem. 1909, 69, 523-46.
(3) Water; H_2O; [7732-18-5]	

VARIABLES:	PREPARED BY:
One temperature: 298 K	Hiroshi Miyamoto
Concentration of phenol	

EXPERIMENTAL VALUES:

Concn of phenol/mol dm^{-3}	soly of $KBrO_3$/mol dm^{-3}
0	0.478
0.5 mol dm^{-3}	0.426

AUXILIARY INFORMATION

METHOD/APPARATUS/PROCEDURE:	SOURCE AND PURITY OF MATERIALS:
The salt and solvent were placed in a bottle, sealed with a rubber stopper, and rotated in a thermostat for at least 14 hours. After the saturated solution was allowed to settle, aliquots of saturated solution were withdrawn with a pipet fitted with a glass-wool or cotton-wool filter. The bromate content was determined by iodometric titration.	Potassium bromate was repeatedly recrystallized. No information of the source and purity of the solvent was given.
	ESTIMATED ERROR:
	Nothing specified.
	REFERENCES:

COMPONENTS:	ORIGINAL MEASUREMENTS:
(1) Potassium bromate; KBrO$_3$; [7758-01-2] (2) Acetic acid, methyl ester (methyl acetate); C$_3$H$_6$O$_2$; [79-20-9] (3) Water; H$_2$O; [7732-18-5]	Rothmund, V. Z. *Physik. Chem.* 1909, *69*, 523-46.

VARIABLES:	PREPARED BY:
T/K = 298 Concentration of methyl acetate	Hiroshi Miyamoto

EXPERIMENTAL VALUES:

Concn methyl acetate/mol dm^{-3}	soly of KBrO$_3$/mol dm^{-3}
0	0.478
0.5	0.420

AUXILIARY INFORMATION

METHOD/APPARATUS/PROCEDURE:	SOURCE AND PURITY OF MATERIALS:
The salt and solvent were placed in a bottle, sealed with a rubber stopper, and rotated in a thermostat for at least 14 hours. After the saturated solution was allowed to settle, aliquots were withdrawn with a pipet fitted with a glass-wool or cotton-wool filter. The bromate content was determined by iodometric titration.	Potassium bromate was repeatedly recrystallized. No information of the source and purity of the solvents was given.
	ESTIMATED ERROR:
	REFERENCES:

COMPONENTS:	ORIGINAL MEASUREMENTS:
(1) Potassium bromate; $KBrO_3$; [7758-01-2]	Rothmund, V.
(2) Amines	Z. Physik. Chem. 1909, 69, 523-46.
(3) Water; H_2O; [7732-18-5]	

VARIABLES:	PREPARED BY:
T/K = 298 Composition	Hiroshi Miyamoto and Mark Salomon

EXPERIMENTAL VALUES:

Composition of Solvent			soly of $KBrO_3$/mol dm^{-3}
pure water			0.478
binary mixtures containing 0.5 mol dm^{-3} of the following:			
diethylamine;	C_4H_4N;	[109-89-7]	0.384
pyridine;	C_5H_5N;	[110-86-1]	0.415
piperidine;	$C_5H_{11}N$;	[110-89-4]	0.396

AUXILIARY INFORMATION

METHOD/APPARATUS/PROCEDURE:	SOURCE AND PURITY OF MATERIALS:
The salt and solvent were placed in a bottle, sealed with a rubber stopper, and rotated in a thermostat for at least 14 hours. After the saturated solutions were allowed to settle, aliquots were withdrawn with a pipet fitted with a glass-wool or cotton-wool filter. The bromate content was determined by iodometric titration.	Potassium bromate was repeatedly recrystallized. No information of the source and purity of the solvents was given.
	ESTIMATED ERROR:
	Nothing specified.
	REFERENCES:

COMPONENTS:	ORIGINAL MEASUREMENTS:
(1) Potassium bromate; $KBrO_3$; [7758-01-2]	Rothmund, V.
(2) Dimethylpyrone; $C_7H_8O_2$; [?][a]	Z. Physik. Chem. 1909, 69, 523-46.
(3) Water; H_2O; [7732-18-5]	

VARIABLES:	PREPARED BY:
T/K = 298	Hiroshi Miyamoto
Concentration of dimethylpyrone	

EXPERIMENTAL VALUES:

concn of dimethylpyrone mol dm^{-3}	soly of $KBrO_3$/mol dm^{-3}
0	0.478
0.5	0.478

[a] There are nine isomers of dimethylpyrone, and the author did not specify which isomer was used. The isomer listed in the Aldrich Catalog is 2,6-dimethyl-γ-pyrone (2,6-dimethyl- 4H -pyran-4-one): [1004-36-0]. Other isomers are:

2,3-dimethyl-4H-pyran-4-one [73761-48-5]	3,5-dimethyl-2H-pyran-2-one [63233-31-8]
2,5-dimethyl-4H-pyran-4-one [?]	3,6-dimethyl-2H-pyrane-2-one [53034-20-1]
3,5-dimethyl-4H-pyran-4-one [19083-61-5]	4,5-dimethyl-2H-pyran-2-one [61906-92-1]
3,4-dimethyl-2H-pyran-2-one [62968-83-6]	4,6-dimethyl-2H-pyran-2-one [645-09-2]

AUXILIARY INFORMATION

METHOD/APPARATUS/PROCEDURE:	SOURCE AND PURITY OF MATERIALS:
The salt and solvent were placed in a bottle, sealed with a rubber stopper, and rotated in a thermostat for at least 14 hours. After the saturated solution was allowed to settle, aliquots were withdrawn with a pipet fitted with a glass-wool or cotton-wool filter. The bromate content was determined by iodometric titration.	Potassium bromate was repeatedly recrystallized. No information of the source and purity of the solvents was given.
	ESTIMATED ERROR:
	Nothing specified.
	REFERENCES:

COMPONENTS:	ORIGINAL MEASUREMENTS:
(1) Potassium bromate; $KBrO_3$; [7758-01-2] (2) Ammonia and amides (3) Water; H_2O; [7732-18-5]	Rothmund, V. Z. Physik. Chem. 1909, 69, 523-46.

VARIABLES:	PREPARED BY:
T/K = 298 Composition	Hiroshi Miyamoto and Mark Salomon

EXPERIMENTAL VALUES:

Composition of Solvent			Soly of $KBrO_3$/mol dm^{-3}
pure water			0.478

Binary mixtures containing 0.5 mol dm^{-3} of the following:

ammonia;	NH_3;	[7664-41-7]	0.445
formamide;	CH_3NO;	[75-12-7]	0.473
acetamide;	C_2H_5NO;	[60-35-5]	0.445
urea;	CH_4N_2O;	[57-13-6]	0.477
ethyl carbamate;	$C_3H_7NO_2$;	[51-79-6]	0.433
(urethane)			

AUXILIARY INFORMATION

METHOD/APPARATUS/PROCEDURE:	SOURCE AND PURITY OF MATERIALS:
The salt and solvent were placed in a bottle sealed with a rubber stopper, and rotated in a thermostat for at least 14 hours. After the saturated solutions were allowed to settle, aliquots of saturated solution were withdrawn with a pipet fitted with a glass-wool or cotton-wool filter. The bromate content was determined by iodometric titration.	Potassium bromate was repeatedly recrystallized. No information of the source and purity of the solvents was given.
	ESTIMATED ERROR: Nothing specified.
	REFERENCES:

COMPONENTS:	ORIGINAL MEASUREMENTS:
(1) Potassium bromate; $KBrO_3$; [7758-01-2] (2) Lithium chloride; LiCl; [7447-41-8] (3) Ethanol; C_2H_6O; [64-17-5]	Gross, P.; Kuzmany, P.; Wald, M. *J. Am. Chem. Soc.* <u>1937</u>, 59, 2692-4.

VARIABLES:	PREPARED BY:
Concentration of LiCl at 288.15 K	Hiroshi Miyamoto

EXPERIMENTAL VALUES:

The solubility and solubility product of $KBrO_3$ in 100 % ethanol[a] containing LiCl are:

Concn LiCl $10^4\ c_1/mol\ dm^{-3}$[b]	Soly $KBrO_3$ $10^4\ c_2/mol\ dm^{-3}$	$-\log K_{s0}$	$-\log K_{s0}^\circ$
0.000	0.8088	8.184	8.251
1.191	0.8583	8.133	8.240
2.942	0.8947	8.097	8.244
5.769	0.9465	8.048	8.242
11.07	1.015	7.987	8.248
15.87	1.076	7.936	8.246
20.34	1.139	7.895	8.243
24.52	1.169	7.864	8.246

[a] The authors gave d_4^{15} = 0.79359 and ε = 25.7_6 D.

[b] Initial LiCl concentration (see below).

[c] $\log K_{s0}^\circ = \log K_{s0} - 7.52 I^{\frac{1}{2}}$, where I is the ionic strength.

AUXILIARY INFORMATION

METHOD/APPARATUS/PROCEDURE:

Solid $KBrO_3$ and ethanol containing LiCl were placed in a glass flask, and the flask rotated in a thermostat at 15.000°C for about 12 hours. The authors state that 12 hours are sufficient to insure equilibrium. For the analysis 25 cm^3 of the filtered alcoholic solution was placed in a small flask and evaporated in a boiling water bath while N_2 gas was forced through the liquid. The precipitated $KBrO_3$ was titrated iodometrically.
Although not specifically stated, it appears that the LiCl concentrations in the above table are *initial* concentrations.

SOURCE AND PURITY OF MATERIALS:

Analytical reagent grade $KBrO_3$ was recrystallized twice. Ethanol was treated first with lime, then with KOH, with silver oxide and with aluminum amalgam. Finally, it was distilled from water-free sulfanilic acid in a stream of N_2. The boiling point was 78.03°C. LiCl prepared from purified carbonate and HCl, and dried in a stream of HCl. LiCl concentration determined gravimetrically.

ESTIMATED ERROR:
Soly: precision within 0.6 % (compiler).
Temp: precision ± 0.0025 K.

REFERENCES:

COMPONENTS:	ORIGINAL MEASUREMENTS:
(1) Potassium bromate; $KBrO_3$; [7758-01-2] (2) 2-Flurancarboxaldehyde (furfural); $C_5H_4O_2$; [98-01-1]	Trimble, F. *Ind. Eng. Chem.* <u>1941</u>, *33*, 660-2.

VARIABLES:	PREPARED BY:
T/K = 298	Hiroshi Miyamoto

EXPERIMENTAL VALUES:

The solubility of $KBrO_3$ in furfural at 25°C was given as:

$$0.01 \text{ mass \%} \qquad \text{(author)}$$

$$6 \times 10^{-4} \text{ mol kg}^{-1} \qquad \text{(compiler)}$$

AUXILIARY INFORMATION

METHOD/APPARATUS/PROCEDURE:

Furfural and an excess of solute were agitated for 24 hours in a constant-temperature bath. After agitation the mixture was filtered at the same temperature as that employed for saturation.

For the determination of the amount of the salt in furfural, about 40 grams of the solution were accurately weighed in a beaker, and 100 cm³ of water added; the resulting solution was evaporated until the total volume was reduced to about 10 cm³. The solution was transferred to a weighed crucible, and then the evaporation continued to dryness. The residue contained in the crucible was moistened with a few drops of concentrated H_2SO_4 and ignited at about 700°C, and then weighed as the sulfate. All determinations were made in duplicate.

SOURCE AND PURITY OF MATERIALS:

Furfural used was purified by carefully fractionating the technical grade twice under about 12 mm pressure (boiling point 54-55°C) through a well insulated 7-foot (2.13 meter) Hempel column packed with 7 mm Rasching rings.

C.p. grade $KBrO_3$ was used.

ESTIMATED ERROR:

Soly: duplicates checked within 25 % of each other.

Temp: nothing specified.

REFERENCES:

COMPONENTS:	ORIGINAL MEASUREMENTS:
(1) Potassium bromate; $KBrO_3$; [7758-01-2] (2) N-Methylacetamide; C_3H_7NO; [79-16-3]	Dawson, L.R.; Berger, J.E.; Vaughn, J.W.; Eckstrom, H.C. J. Phys. Chem. 1963, 67, 281-3.
VARIABLES: T/K = 313	PREPARED BY: Hiroshi Miyamoto and Mark Salomon

EXPERIMENTAL VALUES:

The solubility of $KBrO_3$ in n-methylacetamide, $CH_3CONHCH_3$, at 40°C was given as

$$0.03 \ mol \ dm^{-3}$$

AUXILIARY INFORMATION

METHOD/APPARATUS/PROCEDURE:

An "approximate" solubility was determined by the conductivity method. About 0.5 to 1.0 gram of salt and 10 ml of solvent were placed in a large test tube, stoppered and covered with aluminum foil, and heated to 60°C.

Upon cooling to 40°C, the occurence of precipitation was assumed to indicate the existence of a saturated solution. Conductivities were measured in duplicate in "the usual manner." Experimental details and the measured electrolytic conductivities were not given.

The concentration of the salt in the saturated solution was determined from the experimental electrolytic conductivities, but details on the calculation were not given. Presumably the limiting law was used as in (1).

SOURCE AND PURITY OF MATERIALS:

N-Methylacetamide was prepared by reacting monoethylamine with glacial acetic acid and subsequent heating to distill off the water. The product was purified by fractional distillation followed by five or more fractional freezing cycles. The electrolytic conductance of the purified solvent ranged from 0.5×10^{-5} to 2×10^{-5} cm^{-1}.

Reagent grade $KBrO_3$ was dried in a vacuum desiccator over anhydrous magnesium perchlorate without further treatment.

ESTIMATED ERROR:

Soly: authors "believe" the solubility to be accurate to within 5 %.

Temp: not specified.

REFERENCES:

1. Dawson, L.R.; Wilhoit, E.D.; Holmes, R.R.; Sears, P.G. J. Am. Chem. Soc. 1957, 79, 3004 (Λ^{∞} values are given in this paper).

COMPONENTS:	ORIGINAL MEASUREMENTS:
(1) Potassium bromate; KBrO₃; [7789-38-0] (2) Ammonia; NH₃; [7664-41-7]	Hunt, H.; Boncyk, L. *J. Am. Chem. Soc.* 1933, 55, 3528-30.

VARIABLES:	PREPARED BY:
T/K = 298.15	Mark Salomon and Hiroshi Miyamoto

EXPERIMENTAL VALUES:

The solubility of $KBrO_3$ in liquid ammonia at 25°C was reported as

$$0.002 \text{ g/100g } NH_3.$$

AUXILIARY INFORMATION

METHOD/APPARATUS/PROCEDURE:
Two methods were used as described in (1).

Method I. 25 ml test tubes with a constriction at the middle were employed. About 10-25 g NH_3 were condensed in the bottom, and the dry salt contained in a small tube tightly covered with cotton cloth was added to the test tube: this small tube remained in the upper part of the test tube as it could not pass the construction in the middle of the test tube. The top of the test tube was drawn to a tip and sealed, and the tube inverted and placed in a thermostat at 25°C. Equilibrium between NH_3 and the excess salt in the small covered tube required 1-3 weeks with periodic shaking. The test tube was then inverted and only the satd sln drained into the lower end (excess solid remained in the small tube covered with the cotton cloth). The sln was frozen and sealed at the constriction, and weighed. The seal was then broken and the NH_3 boiled off, and the residue weighed.

Method II. Excess NH_3 was condensed on a weighed amount of salt in a tube fitted with a stopcock. After thermostating at 25°C, NH_3 was slowly permitted to escape through the stopcock until a crystal of solid appeared and remained undissolved upon prolonged shaking.

Authors state that the error due to the condensation of gaseous NH_3 was not significant since the dead space was kept to a minimum of about 30 cm³. However this amount of dead space was stated to limit the precision of the method to 0.5 %.

SOURCE AND PURITY OF MATERIALS:
Reagent grade $KBrO_3$ was recrystallized three times from water and then from "a suitable" anhydrous solvent. The salt was dried to a constant weight in a vacuum oven.

Purification of NH_3 not specified, but probably similar to that described in (1). In (1) commercial anhyd ammonia was stored over metallic sodium for several weeks before use.

ESTIMATED ERROR:

Soly: accuracy probably around ± 1-2 % (compilers).
Temp: 25 + 0.025°C accuracy established by NBS calibration as described in (1).

REFERENCES:

1. Hunt, H.; *J. Am. Chem. Soc.* 1932, 54, 3509.

COMPONENTS:	EVALUATOR:
(1) Rubidium Bromate; RbBrO₃; [13446-70-3] (2) Water; H₂0; [7732-18-5]	H. Miyamoto Niigata University Niigata, Japan and M. Salomon US Army ET & DL Fort Monmouth, NJ, USA February 1986

CRITICAL EVALUATION:

THE BINARY SYSTEM

The solubility of rubidium bromate in water has been reported in 3 publications (1-3): note that the compilation for Ref. (3) is given in the KBrO₃ chapter. The solid phase in all cases was reported (2,3) to be the anhydrous salt. Breusov et al. (2) report a small break at 311.2 K when log (χ) was plotted as a function of T/K. Since the solid phase is the anhydrous salt and since both dilatometric and X-ray studies showed no signs of polymorphism, these authors concluded that this phenomenon may be due to large changes in hydration of the ions in solution. When the evaluators plotted the simple function of solubility as a function of T/K, a smooth monotonous curve was obtained with no indication of any breaks.

Buell and McCrosky (1) determined a melting point of 603 K for RbBrO₃.

In all cases the solubilities of Bruesov et al. (2) are slightly higher than those of Buell and McCrosky (1), the maximum difference being close to 3% at 298.2 K. Since it is not possible to determine any sources of error in either study (1-3), we have uncritically fitted all data points to the smoothing equations (see the compilations for the experimental results and compilers' conversions). For mole fraction solubilities we derive the following smoothing equation:

$$Y_x = -14463.4/(T/K) - 32.984 ln(T/K) + 218.108 + 0.025875(T/K) \qquad [1]$$

$$\sigma_y = 0.024 \qquad\qquad \sigma_x = 1.1 \times 10^{-4}$$

For solubilities in mol kg^{-1} units we obtain the following equation:

$$Y_m = -5626.5/(T/K) - 7.185 ln(T/K) + 63.842 \qquad [2]$$

$$\sigma_y = 0.012 \qquad\qquad \sigma_m = 0.0071$$

Solubilities calculated from eqs. [1] and [2] are designated as *tentative* solubilities, and values at rounded temperatures are given in the Table following the references.

REFERENCES

1. Buell, H. D.; McCrosky, C. R. *J. Am. Chem. Soc.* 1921, *43*, 2031.

2. Breusov, O. N.; Kashina, N. I.; Revzina, T. V.; Sobolevskaya, N. G. *Zh. Neorg. Khim.* 1967, *12*, 2240; *Russ. J. Inorg. Chem. (Engl. Transl.)* 1967, *12*, 1179.

3. Kirgintsev, A. N.; Shklovskaya, R. M.; Arkhipov, S. M. *Izv. Akad. Nauk SSSR Ser. Khim.* 1971, 2631; *Bull. Acad. Sci. USSR Div. Chem. Sci.* 1971, *2501*.

COMPONENTS:	EVALUATOR:
(1) Rubidium Bromate; $RbBrO_3$; [13446-70-3] (2) Water; H_2O; [7732-18-5]	H. Miyamoto Niigata University Niigata, Japan and M. Salomon US Army ET & DL Fort Monmouth, NJ, USA February 1986

CRITICAL EVALUATION:

Table 1. Tentative Solubilities Calculated from the Smoothing Equations
[1] and [2]. In all cases the Solid Phase is Anhydrous $RbBrO_3$.

T/K	X	mol kg^{-1}
273.2	0.000828	0.0462
283.2	0.001327	0.0738
293.2	0.002038	0.1132
298.2	0.002489	0.1383
303.2	0.003012	0.1675
308.2	0.003615	0.2013
313.2	0.004305	0.2400
323.2	0.005970	0.3338
333.2	0.008063	0.4522
343.2	0.01063	0.5980
353.2	0.01373	0.7739
363.2	0.01738	0.9818
373.2	0.02164	1.223

COMPONENTS:	ORIGINAL MEASUREMENTS:
(1) Rubidium bromate; RbBrO$_3$; [13446-70-3] (2) Water; H$_2$O; [7732-18-5]	Buell, H.D.; McCrosky, C.R. J. Am. Chem. Soc. 1923, 43, 2031-4.
VARIABLES: T/K = 298, 303, 308 and 313	PREPARED BY: Hiroshi Miyamoto and Mark Salomon

EXPERIMENTAL VALUES:

Solubility of RbBrO$_3$

t/°C	g/100g H$_2$O	mol kg^{-1} (compiler)
25	2.994	0.1403
	2.895	0.1357
	2.917	0.1367
	2.917	0.1367
	(Av)2.93 (σ = 0.04)	0.137
30	3.584	0.1680
	3.578	0.1677
	3.509	0.1645
	3.559	0.1667
	(Av)3.56 (σ = 0.03)	0.166
35	4.310	0.2020
	4.247	0.1990
	4.295	0.2013
	4.269	0.2001
	(Av)4.28 (σ = 0.03)	0.201
40	5.104	0.2392
	5.116	0.2398
	5.021	0.2353
	5.092	0.2386
	(Av)5.08 (σ = 0.02)	0.238

AUXILIARY INFORMATION

METHOD/APPARATUS/PROCEDURE:
The method for determining the solubility is similar to that described in ref 1. Mixtures of rubidium bromate and water were shaken in a thermostat. About 5 hours were required to attain equilibrium. Two methods of analysis were used. In the first method, aliquots of the saturated solutions were weighed, carefully evaporated to dryness, and dried at 115°C to constant weight.
In the second method, the iodometric method was used to determine the bromate concentration. Both methods were of equal precision.

SOURCE AND PURITY OF MATERIALS:
RbCl of "doubtful purity" was converted to the alum, recrystallized, and digested with excess BaCO$_3$ on a hot plate. The sln was filtered, treated with Ba(OH)$_2$ and CO$_2$, and filtered again. The salt was then treated with excess "pure" bromic acid and allowed to crystallize. The resulting RbBrO$_3$ was recrystallized three times.

Source and purity of water not specified.

ESTIMATED ERROR:
Soly: precision in analyses about ± 0.3 %
      (compilers), standard deviations for
      solubility measurements given in table
      calculated by the compilers.
Temp: nothing specified.

REFERENCES:

1. McCrosky, C.R.; Buell, H.D.
   J. Am. Chem. Soc. 1920, 42, 1786.

COMPONENTS:	ORIGINAL MEASUREMENTS:
(1) Rubidium bromate; $RbBrO_3$; [13446-70-3] (2) Water; H_2O; [7732-18-5]	Breusov, O.N.; Kashina, N.I.; Revzina, T.V.; Sobolevskaya, N.G. *Zh. Neorg. Khim.* 1967, *12*, 2240-3; *Russ. J. Inorg. Chem. (Engl. Transl.)* 1967, *12*, 1179-81.
VARIABLES: T/K = 273 to 373	PREPARED BY: Hiroshi Miyamoto

EXPERIMENTAL VALUES:

Solubility of $RbBrO_3$[a]

t/°C	mass %	mol %	mol kg^{-1} (compiler)
0	0.98	0.0835	0.0464
10	1.53	0.131	0.0728
20	2.37	0.205	0.1138
25	2.93	0.254	0.1415
30	3.45	0.301	0.1675
40	4.92	0.435	0.2425
50	6.72	0.608	0.3376
60	8.90	0.818	0.4579
70	11.17	1.051	0.5893
80	14.06	1.367	0.7667
90	17.15	1.718	0.9701
100	20.96	2.177	1.243

[a] The nature of the solid phase was not specified.

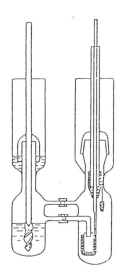

High temp. apparatus

AUXILIARY INFORMATION

METHOD/APPARATUS/PROCEDURE:	SOURCE AND PURITY OF MATERIALS:
Isothermal method. Equilibrium reached in 4-5 h. From 90-100°C, soly detd in apparatus shown in figure. At equilibrium, the apparatus was tilted to allow saturated solution to filter through connecting tube into weighed test tubes. The test tube was closed with a stopper, withdrawn, and weighed. Condensation on the walls of the apparatus and loss of water by evaporation was thus prevented. At the lower temperatures, ordinary soly vessels were used, and pipets with glass filters were used for sampling (no other details given). Above 50°C, the pipets were preheated in the thermostat. Bromate was determined iodometrically.	Results of analysis of $RbBrO_3$; Content of $RbBrO_3$ = 98.6 %. Impurities (mass %): K 0.12; Cs 0.1; Na 0.014; SO_4 0.1; Fe < 0.0025.
	ESTIMATED ERROR: Soly: nothing specified. Temp: precision $\pm$ 0.1 K.
	REFERENCES:

COMPONENTS:	ORIGINAL MEASUREMENTS:
(1) Rubidium bromate; $RbBrO_3$; [13446-70-3] (2) Cesium bromate; $CsBrO_3$; [13454-75-6] (3) Water; H_2O; [7732-18-5]	Kirgintsev, A.N.; Shklovskaya, R.M.; Arkhipov, S.M. *Izv. Acad. Nauk SSSR Ser. Khim.* 1971, 2631-4; *Bull. Acad. Sci. USSR Div. Chem. Sci. (Engl. Transl.)* 1971, 2501-4.
VARIABLES: Composition at 298.2 K	PREPARED BY: Hiroshi Miyamoto

EXPERIMENTAL VALUES: Composition of saturated solutions at 25.0°C

Rubidium Bromate		Cesium Bromate		m^a	y_1^b
mass %	mol % (compiler)	mass %	mol % (compiler)	mol kg^{-1}	
2.83[c]	0.245	0.00	0.00	0.136	1.00
2.56	0.222	0.50	0.036	0.148	0.86
2.52	0.220	1.06	0.0757	0.166	0.74
2.18	0.191	1.49	0.107	0.169	0.64
2.01	0.176	2.02	0.145	0.176	0.55
1.83	0.160	2.13	0.153	0.182	0.51
1.63	0.143	2.47	0.177	0.179	0.45
1.23	0.108	2.81	0.202	0.172	0.35
0.81	0.071	3.09	0.221	0.164	0.24
0.44	0.038	3.23	0.231	0.152	0.14
0.00	0.000	3.71[c]	0.265	0.148	0.00

[a] m = the total molality of the salts in liquid phase.

[b] y_1 = the mole fraction of $RbBrO_3$ based on total salts.

[c] For binary systems the compiler computes the following:

soly of $RbBrO_3$ = 0.136 mol kg^{-1}

soly of $CsBrO_3$ = 0.148 mol kg^{-1}

AUXILIARY INFORMATION

METHOD/APPARATUS/PROCEDURE:	SOURCE AND PURITY OF MATERIALS:
Isothermal relief of supersaturation method. Super saturated solutions were prepared, and the solid and liquid phases separated. The mother liquor was equilibrated at 25°C for 24 hours. The number of moles of the anion was determined by iodometric titration. Alkali metal contents were determined in the same sample by the method of flame photometry from three parallel analyses. In each analysis the authors calculated the sum of cations. The composition of the solid phases was established by the Schreinemakers' method of residues. The authors did not give a phase diagram.	C.p. grade $RbBrO_3$ and $CsBrO_3$ were recrystallized from double distilled water.
	ESTIMATED ERROR: Soly: precision within 2 %. Temp: precision ± 0.1 K.
	REFERENCES:

COMPONENTS:	EVALUATOR:
(1) Cesium Bromate; $CsBrO_3$; [13454-75-6] (2) Water; H_2O; [7732-18-5]	H. Miyamoto Niigata University Niigata, Japan and M. Salomon US Army ET & DL Fort Monmouth, NJ, USA February 1986

CRITICAL EVALUATION:

THE BINARY SYSTEM

Data for the solubility of cesium bromate in water have been reported in five publications (1-5). The compilations based on references (4, 5) are given in the chapters on $KBrO_3$ and $RbBrO_3$, respectively. The isothermal method was used in all studies, and in (3-5) the solid phase was determined to be the anhydrous salt. Buell and McCrosky (2) reported a melting point of 693 K for the pure salt. A summary of the experimental solubility data is given in Table 1. Solubilities in mol kg^{-1} units are given in the compilations.

Table 1.   Summary of Experimental Solubilities[a]

T/K	mass %	χ	(ref)	T/K	mass %	χ	(ref)
273.2	1.17	0.000817	(3)	308.2	5.06[b,c]	0.003667	(2)
283.2	1.90	0.001336	(3)	313.2	6.28	0.004607	(3)
293.2	2.09[b]	0.001472	(3)	323.2	8.56	0.006425	(3)
298.2	3.75	0.002684	(3)	333.2	11.32	0.008740	(3)
298.2	3.54[b,c]	0.002527	(2)	343.2	14.48	0.011560	(3)
298.2	3.66	0.002617	(4)	353.2	17.99	0.014926	(3)
298.2	3.71	0.002654	(5)	363.2	22.01[b]	0.019121	(3)
303.2	4.34[b,c]	0.003121	(1,2)	373.2	25.96	0.023647	(3)
303.2	4.46	0.003214	(3)				

[a]Conversions to mole fraction units by evaluators.

[b]Rejected data points (see text for discussion).

[c]Original units are g/100 g H_2O, and conversion to mass % by evaluators.

Inspection of Table 1 shows that all data reported by Buell and McCrosky (1,2) are significantly lower than corresponding data from (3,4,5) by as much as 6%. All data reported in (1,2) were rejected. Breusov et al. (3) find that when log (χ) is plotted as a function of T/K, a break in the curve is observed at 300 K, and which is attributed to a change in hydration of ions in solution. However the evaluators plotted the simple function of mole fraction vs T/K, and we find a smooth monotonous curve in which only the data point at 293.2 K falls off the curve. In fitting the data to the smoothing equation, again it is found that the point at 293.2 K from (3) deviates significantly from the calculated value (for all data points in Table 1 except those from (1,2), we find that $\chi_{exptl} = 0.001472$ compared to the calculated value of $\chi = 0.002008$). Using our criteria that all acceptable data points should not be greater or less than $2\sigma_x$ from the smoothed calculated values, one more data point from (3) at 363.2 K was rejected. The final smoothing equation based on the 12 remaining data points based on mole fraction units is:

$$Y_x = -32233.93/(T/K) - 137.0375 \ell n(T/K) + 825.2493 + 0.178260(T/K) \qquad [1]$$

$$\sigma_y = 0.018 \qquad\qquad \sigma_x = 2.8 \times 10^{-5}$$

Based on the 12 acceptable data points, the derived smoothing equation based on mol kg^{-1} units is:

$$Y_m = -6754.46/(T/K) - 10.4410 \ell n(T/K) + 82.116 \qquad [2]$$

$$\sigma_y = 0.013 \qquad\qquad \sigma_m = 0.0067$$

Solubilities calculated from eqs. [1] and [2] are designated as *tentative* values, and calculated solubilities at rounded temperatures are given in Table 2.

COMPONENTS:	EVALUATOR:
(1) Cesium Bromate; $CsBrO_3$; [13454-75-6] (2) Water; H_2O; [7732-18-5]	H. Miyamoto Niigata University Niigata, Japan and M. Salomon US Army ET & DL Fort Monmouth, NJ, USA February 1986

CRITICAL EVALUATION:

Table 2. Tentative Solubilities Calculated from Eqs. [1] and [2].
In all Cases the Solid Phase is Anhydrous $CsBrO_3$.

T/K	X	m/mol kg^{-1}
273.2	0.000813	0.0457
283.2	0.001356	0.0752
293.2	0.002138	0.1181
298.2	0.002636	0.1457
303.2	0.003213	0.1779
308.2	0.003878	0.2153
313.2	0.004636	0.2582
323.2	0.006460	0.3624
333.2	0.008740	0.4937
343.2	0.01153	0.6545
353.2	0.01490	0.8466
363.2	0.01892	1.0709
373.2	0.02367	1.3274

REFERENCES

1. McCrosky, C. R.; Buell, H. D. *J. Am. Chem. Soc.* 1920, *42*, 1786.

2. Buell, H. D.; McCrosky, C. R. *J. Am. Chem. Soc.* 1921, *43*, 2031.

3. Breusov, O. N.; Kashina, N. I.; Revizina, T. V.; Sobolevskaha, N. G. *Zh. Neorg. Khim.* 1967, *12*, 2240; *Russ. J. Inorg. Khim. (Engl. Transl.)* 1967, *12*, 1179.

4. Kirgintsev, A. I.; Yakobi, N. Y. *Zh. Neorg. Khim.* 1968, *13*, 2851; *Russ. J. Inorg. Chem (Engl. Transl.)* 1968, *13*, 1467.

5. Kirgintsev, A. N.; Shklovskaya, R. M.; Arkhipov, S. M. *Izv. Acad. Nauk SSSR Ser. Khim.* 1971, 2631; *Bull. Acad. Sci. USSR Div. Chem. Sci.* 1971, 2501.

COMPONENTS:	ORIGINAL MEASUREMENTS:
(1) Cesium bromate; $CsBrO_3$; [13454-75-6] (2) Water; H_2O; [7732-18-5]	McCrosky, C.R.; Buell, H.D. J. Am. Chem. Soc. 1920, 42, 1786-9.

VARIABLES:	PREPARED BY:
T/K = 303.2	Hiroshi Miyamoto

EXPERIMENTAL VALUES:

Solubility of cesium bromate in water at 30°C[a]

g/100g H_2O	mol kg^{-1}
4.484	0.1800
4.573	0.1837
4.525	0.1817
4.549	0.1827
4.483	0.1800
4.577	0.1837
(Av)4.53	0.182

[a] Molalities calculated by the compiler.

AUXILIARY INFORMATION

METHOD/APPARATUS/PROCEDURE:	SOURCE AND PURITY OF MATERIALS:
Mixtures of cesium bromate and water were shaken in a thermostat. About 5 hours were allowed for the salt to come into equilibrium with the solvent before the saturated solution was withdrawn for analysis. Aliquots of the saturated solution were weighed and then carefully evaporated to dryness until constant in weight.	Cesium bromate was prepared by neutralization of CsOH with bromic acid followed by addition of excess bromic acid. The solution was evaporated somewhat and allowed to crystallize. The product was recrystallized from water and then dried.

	ESTIMATED ERROR: Soly: standard deviation(σ) 0.04 for g/100g H_2O units. Temp: precision $\pm$ 0.3 K.
	REFERENCES:

COMPONENTS:	ORIGINAL MEASUREMENTS:
(1) Cesium bromate; $CsBrO_3$; [13454-75-6] (2) Water; H_2O; [7732-18-5]	Buell, H.D.; McCrosky, C.R. J. Am. Chem. Soc. 1921, 43, 2031-4.
VARIABLES: T/K = 298, 303 and 308	PREPARED BY: Hiroshi Miyamoto and Mark Salomon

EXPERIMENTAL VALUES:

Solubility of $CsBrO_3$

t/°C	g/100g H_2O	mol kg^{-1} (compiler)
25	3.627	0.1444
	3.664	0.1458
	3.710	0.1477
	(Av) 3.68 (σ = 0.04)	0.146
30	4.484	0.1800
	4.573	0.1837
	4.525	0.1817
	(Av) 4.53 (σ = 0.04)	0.182
35	5.357	0.2170
	5.410	0.2193
	5.215	0.2110
	(Av) 5.32 (σ = 0.10)	0.216

AUXILIARY INFORMATION

METHOD/APPARATUS/PROCEDURE:	SOURCE AND PURITY OF MATERIALS:
The method for determining the solubility is similar to that described in ref 1. Mixtures of cesium bromate and water were agitated in a thermostat. About 5 hours were required to attain equilibrium. Two methods of analysis were used. In the first method, aliquots of the saturated solutions were weighed, carefully evaporated to dryness, and dried at 115°C to constant weight. In the second method, the iodometric method was used to determine the bromate concentration. Both methods were of equal precision.	Nothing specified, but the compiler assumes that the preparation of cesium bromate was similar to that described in ref 1.
	ESTIMATED ERROR: Soly: precision in analyses about ± 0.3 % (compilers). Standard deviations for solubility measurements given in table calculated by compilers. Temp: nothing specified.
	REFERENCES: 1. McCrosky, C.R.; Buell, H.D. J. Am. Chem. Soc. 1920, 42, 1786.

COMPONENTS:	ORIGINAL MEASUREMENTS:
(1) Cesium bromate; CsBrO$_3$; [13454-75-6] (2) Water; H$_2$O; [7732-18-5]	Breusov, O.N.; Kashina, N.I.; Revzina, T.V.; Sobolevskaya, N.G. Zh. Neorg. Khim. 1967, 12, 2240-3; Russ. J. Inorg. Chem. (Engl. Transl.) 1967, 12, 1179-81.

VARIABLES:	PREPARED BY:
T/K = 273 to 373	Hiroshi Miyamoto

EXPERIMENTAL VALUES:

Solubility of CsBrO$_3$

t/°C	mass %	mol %	mol kg^{-1} (compiler)
0	1.17	0.0817	0.0454
10	1.90	0.134	0.0743
20	2.09	0.212	0.0818
25	3.75	0.268	0.149
30	4.46	0.321	0.179
40	6.28	0.461	0.257
50	8.56	0.642	0.359
60	11.32	0.874	0.489
70	14.48	1.156	0.649
80	17.99	1.493	0.841
90	22.01	1.912	1.082
100	25.96	2.365	1.344

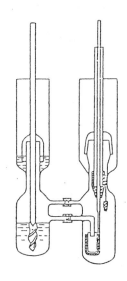

High Temp. Apparatus

AUXILIARY INFORMATION

METHOD/APPARATUS/PROCEDURE:	SOURCE AND PURITY OF MATERIALS:
Isothermal method. Equilibrium reached in 4-5 h. From 90-100°C, soly detd in apparatus shown in figure. At equilibrium, the apparatus was tilted to allow satd sln to filter through connecting tube into weighed test tubes. The test tube was closed with a stopper, withdrawn, and weighed. Condensation on the walls of the apparatus and loss of water by evaporation was thus prevented. At the lower temperatures, ordinary soly vessels were used, and pipets with glass filters were used for sampling (no other details given). Above 50°C, the pipets were preheated in the thermostat. Bromate was determined iodometrically.	Results of analysis of CsBrO$_3$; Content of CsBrO$_3$ = 99.3 % Impurities(mass %): K <0.002; Rb 0.09; Na 0.0025; SO$_4$ 0.05; Fe 0.005.
	ESTIMATED ERROR: Soly: nothing specified. Temp: precision ± 0.1 K.
	REFERENCES:

COMPONENTS:	EVALUATOR:
(1) Lithium Iodate; LiIO3; [13765-03-2] (2) Water; H2O; [7732-18-5]	H. Miyamoto Niigata University Niigata, Japan and M. Salomon US Army ET & DL Fort Monmouth, NJ, USA February, 1986

CRITICAL EVALUATION:

THE BINARY SYSTEM

Solubility data for LiIO3 in water have been reported in 37 publications (1-34, 44-46).
At 298.2 K Shklovskaya et al. (5,7,8,11,14-20,22,23,25,27,28, 44)   reported the identical
solubility of 43.82 mass %, and although the work spans a period of 10 years (1974-1983),
it is not possible to determine the number of independent measurements in these 18 publications. The solubility of 43.30 mass % at 298.2 K reported in (24) is distinctly lower
than all other findings and is therefore rejected. Unezawa et al. (33) reported the
solubility to be about 76 g/100 g H2O (43.2 mass %) over the temperature range of
278-343 K, and although this is an interesting result its qualitative nature led us to
conclude that it should not be compiled. However the importance of this paper is that
the authors identified the stable solid phase at room temperature as hexagonal LiIO3
(i.e. the α-phase) which is consistent with other quantitative data as discussed below.
A graphical summary of the solubility of LiIO3 in water is given in the polytherm figure
below. In all cases the equilibrated solid phase is the anhydrous salt.

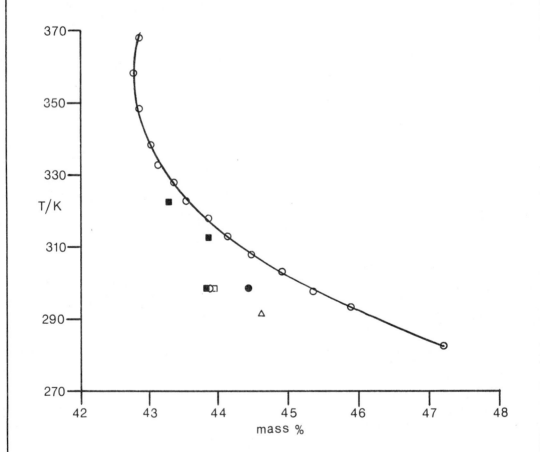

Figure 1. Solubility of LiIO3

○ Ref (2)   ● Ref (34)
□ Ref (9,12)   ■ Ref (3-8, 11, 14-20,22,23,25,27,28,44)
△ Ref (1)   ● Ref (33)

COMPONENTS:	EVALUATOR:
(1) Lithium Iodate; LiIO$_3$; [13765-03-2] (2) Water; H$_2$O; [7732-18-5]	H. Miyamoto Niigata University Niigata, Japan and M. Salomon US Army ET & DL Fort Monmouth, NJ, USA February 1986

CRITICAL EVALUATION:

There are a number of problems in critically evaluating the data in the figure, and this is undoubtedly due to the uncertainties in the form of the solid phase. LiIO$_3$ is polymorphic, and according to (35,36) the stable structure is hexagonal with 2 molecules per unit cell with lithium and iodine both lying within oxygen octahedra. By growing crystals from saturated solutions at 343 K, Unezawa et al. (33) were able to identify the less stable tetragonal form of LiIO$_3$ (the β-phase), and solubilities for the α-phase and β-phase were published in graphical form in (34). In (34) the solubility of stable α-LiIO$_3$ appears to be constant at 80 g/100 g H$_2$O (44.4 mass %) between 298-353 K (see the compilation for this figure). According to (34) the transition between α-LiIO$_3$ and β-LiIO$_3$ occurs at 343 K. The Lithium Iodate Crystal Research Group in The Peoples Republic of China (37) investigated saturated and supersaturated solutions of α-LiIO$_3$ hexagonal) and β-LiIO$_3$ (tetragonal) as a function of temperature, and reported that when crystals co-exist, α-LiIO$_3$ is more stable below 333 K. This paper contains only graphical data and therefore was not compiled.

Desvignes and Romissent (38) used X-ray diffraction to demonstrate that the stable solid at room temperature is α-LiIO$_3$ and that there is an irreversible change to β-LiIO$_3$ when the crystals are heated to 528 K: this is similar to the irreversible transformation temperature of 533 K reported in (33). Liminga and Abrahams (39) also reported that the stable solid at room temperature is α-LiIO$_3$.

To complicate matters, Ricci and Amron (2) reported a third phase, an octahedral form, stable at higher temperatures, metastable down to 283 K, and the transition between stable and metastable solutions for this new solid occurs at around 328 K. This solid phase was identified by microscopic analysis, but X-ray confirmation has not been report-ed. Since the polytherm for Ricci and Amron's octahedral phase lies fairly close to that for the tetragonal form reported by Unezawa and Tatuoka (34), it would appear to the evaluators that both are identical: i.e. that the quantitative solubilities of the octa-hedral form are in fact the solubilities for the less stable tetragonal form: i.e. for β-LiIO$_3$. It should be noted that Ricci and Amron used (after purification) a commercial salt and a product prepared by metathesis, and both salts were dried at 383-456 K before use.

In view of the above uncertainties in the nature of the solid phases and the scatter in the reported solubilities (see figure), we are only able to make limited recommenda-tions on the solubility of LiIO$_3$ as a function of temperature. There is no doubt that at room temperature the stable solid phase is α-LiIO$_3$ and that its solubility at 298.2 K is probably close to 43.82 mass % (i.e. $\chi = 0.0717$ and m = 4.29 mol kg^{-1}). Transition to the β-phase probably occurs between 313-328 K, and the solubility of β-LiIO$_3$ as a function of temperature is probably the same as those reported by Ricci and Amron for the so-called octahedral phase. The reported solubilities at 313 K (13) and 323 K (3,4,6,10) appear imprecise and probably can be attributed to mixed (α- and β-) solid phases. We cannot offer either *tentative* or *recommended* solubilities at this time.

TERNARY SYSTEMS

1.   One Saturating Component.

1.1. The solubility of lithium iodate in aqueous iodic acid solution.

Shklovskaya et al. (13) determined the solubility of lithium iodate in 10 mass % iodic acid solution over a wide temperature range. The solubility was found to exhibit a minimum near 363 K.

COMPONENTS:	EVALUATOR:
(1) Lithium iodate; LiIO$_3$; [13765-03-2] (2) Water; H$_2$O; [7732-18-5]	H. Miyamoto Niigata University Niigata, Japan February, 1986

CRITICAL EVALUATION:

1.2. The solubility of lithium iodate in ethanol-water mixtures.

Arkhipov, Pruntsev and Kidyarov (40) have measured the solubility of lithium iodate in mixtures of ethanol and water, but reported only graphical data. The figure with some comments is included in the compilation section of this chapter.

2.   Two Saturating Components.

Solubility data for the ternary aqueous systems have reported in a number of publications which are summarized in Tables 1-4.

The System With Iodic Acid. The system has been reported in 3 publications (2,3,13) at the three temperatures 298, 313 and 323 K, and phase diagrams at each temperature were also included. The dominant feature in this system is the formation of solid solutions mLiIO$_3$.nHIO$_3$ where m and n were not specified.

Lukasiewicz, Pietaszemska and Zmija (41) measured solubilities in the ternary LiIO$_3$ - HIO$_3$-H$_2$O system at 313, 323, and 328 K. The pH range of the saturated solutions was 1.9 to 3.5. The composition in the solid phase was not reported.

The System With Lithium Hydroxide. Solubilities in the LiIO$_3$-LiOH-H$_2$O system were studied by the isothermal method at 298 K (9). Lithium hydroxide and iodate exists in a restricted series of solid solutions.

Systems With Alkali Metal Iodates. Solubility studies for ternary aqueous systems with alkali metal iodates have been reported in 4 publications (4-7). The summary of these studies is given in Table 1.

The ternary systems with NaIO$_3$ and CsIO$_3$ are simple eutonic, and no double salts were found. The dominant feature of the ternary systems with KIO$_3$ and RbIO$_3$ is the existence of double salts, KIO$_3$.2LiIO$_3$ and RbIO$_3$.2LiIO$_3$, respectively.

Table 1. Summary for Solubility Studies of the Ternary Systems with
Alkali Metal Iodates

Ternary System	T/K	Solid Phase	Reference
LiIO$_3$ + NaIO$_3$ + H$_2$O	298	LiIO$_3$; NaIO$_3$.H$_2$O	5
LiIO$_3$ + KIO$_3$ + H$_2$O	298	LiIO$_3$; KIO$_3$; KIO$_3$.2LiIO$_3$	7
LiIO$_3$ + KIO$_3$ + H$_2$O	323	LiIO$_3$; KIO$_3$; KIO$_3$.2LiIO$_3$	4
LiIO$_3$ + RbIO$_3$ + H$_2$O	298	LiIO$_3$; RbIO$_3$; RbIO$_3$.2LiIO$_3$	7
LiIO$_3$ + RbIO$_3$ + H$_2$O	323	LiIO$_3$; RbIO$_3$; RbIO$_3$.2LiIO$_3$	6
LiIO$_3$ + CsIO$_3$ + H$_2$O	298	LiIO$_3$; CsIO$_3$	5
LiIO$_3$ + CsIO$_3$ + H$_2$O	323	LiIO$_3$; CsIO$_3$	6

Systems With Alkaline Earth Metal Iodates. Solubility studies for these systems have been reported in 4 publications (29-32), and are summarized in Table 2.

The ternary system with Mg(IO$_3$)$_2$ was studied at 298 and 323 K. No double salts were formed at 298 K, and the system is one of the eutonic type. In the study at 323 K, lithium and magnesium iodates form a restricted series of solid solutions. The dominant feature of the ternary systems containing calcium iodate is the absence of double salts: the phase diagrams are of the eutonic type.

COMPONENTS:	EVALUATOR:
(1) Lithium iodate; LiIO$_3$; [13765-03-2]	H. Miyamoto
	Department of Chemistry
(2) Water; H$_2$0; [7732-18-5]	Niigata University
	Niigata, Japan
	February, 1986

CRITICAL EVALUATION:

Table 2. Summary for Solubility Studies of the Ternary Systems with
Alkaline earth metal iodates

Ternary System	T/K	Solid Phase	Reference
LiIO$_3$ + Mg(IO$_3$)$_2$ + H$_2$0	298	α-LiIO$_3$; Mg(IO$_3$)$_2$.4H$_2$0	30
LiIO$_3$ + Mg(IO$_3$)$_2$ + H$_2$0	323	LiIO$_3$; Mg(IO$_3$)$_2$.4H$_2$0;	29
		nMg(IO$_3$)$_2$.4H$_2$0.nLiIO$_3$	
LiIO$_3$ + Ca(IO$_3$)$_2$ + H$_2$0	298	LiIO$_3$; Ca(IO$_3$)$_2$.6H$_2$0	32
LiIO$_3$ + Ca(IO$_3$)$_2$ _ H$_2$0	323	LiIO$_3$; Ca(IO$_3$)$_2$.H$_2$0	31

Systems With Rare Earth Iodates: Recently, two ternary systems (14,24) were studied by the isothermal method at 298 K. The systems LiIO$_3$-Sm(IO$_3$)$_3$-H$_2$0 (14) and LiIO$_3$ - Nd(IO$_3$)$_3$-H$_2$0 (24) were of the eutonic type, and no double salts were reported.

Systems With The Other Iodates. Solubility studies of the ternary aqueous systems containing lithium iodate and other iodates (with the exception of the above systems) have been reported in 9 publications (11,12,19-23,25,44), and summary of these studies is given in Table 3.

Table 3. Summary for Solubility Studies of the Ternary Systems With
the Other Iodates.

Ternary Systems	T/K	Solid Phase	Reference
LiIO$_3$ + Al(IO$_3$)$_3$ + H$_2$0	298	LiIO$_3$; Al(IO$_3$)$_3$.6H$_2$0	11
LiIO$_3$ + NH$_4$IO$_3$ + H$_2$0	298	LiIO$_3$; NH$_4$IO$_3$; NH$_4$IO$_3$.2LiIO$_3$	12
LiIO$_3$ + Ga(IO$_3$)$_3$ + H$_2$0	298	α-LiIO$_3$; Ga(IO$_3$)$_3$.2H$_2$0;	19
		2Ga(IO$_3$)$_3$.LiIO$_3$.4H$_2$0;	
		Ga(IO$_3$)$_3$.LiIO$_3$.2H$_2$0; Solid	
		solution based on Ga(IO$_3$)$_3$.2H$_2$0	
LiIO$_3$ + In(IO$_3$)$_2$ + H$_2$0	298	LiIO$_3$; In(IO$_3$)$_3$.H$_2$0;	20
		LiIO$_3$.In(IO$_3$)$_3$.H$_2$0;	
		2LiIO$_3$.In(IO$_3$)$_3$.H$_2$0	
LiIO$_3$ + TlIO$_3$ + H$_2$0	298	LiIO$_3$; TlIO$_3$; LiIO$_3$.TlIO$_3$	21
LiIO$_3$ + Hf(IO$_3$)$_4$ + H$_2$0	298	LiIO$_3$; Hf(IO$_3$)$_4$; Solid Solution	22
LiIO$_3$ + Zr(IO$_3$)$_4$ + H$_2$0	298	LiIO$_3$; Zr(IO$_3$)$_4$	23
LiIO$_3$ + Ti(IO$_3$)$_4$ + H$_2$0	298	LiIO$_3$; Ti(IO$_3$)$_4$.2H$_2$0	25
		Solid solution	
LiIO$_3$ + AgIO$_3$ + H$_2$0	298	LiIO$_3$; AgIO$_3$; Solid Solution	44

COMPONENTS:	EVALUATOR:
(1) Lithium iodate; $LiIO_3$; [13765-03-2]	Hiroshi Miyamoto
	Department of Chemistry
(2) Water; H_2O; [7732-18-5]	Niigata University
	Niigata, Japan
	February, 1986

CRITICAL EVALUATION:

The ternary systems with $Al(IO_3)_3$ (11) and $Zr(IO_3)_4$ (23) are of the simple eutonic type, and no double salts form. The dominant feature of the ternary systems with NH_4IO_3 (12) and $TlIO_3$ (21) is the existence of the double salts $NH_4IO_3.2LiIO_3$ and $LiIO_3.TlIO_3$. In the ternary $LiIO_3-In(IO_3)_3-H_2O$ system, two double salts $LiIO_3.In(IO_3)_3.H_2O$ and $2LiIO_3.In(IO_3)_3.H_2O$ are formed.

The solubilities in the ternary systems $LiIO_3-Hf(IO_3)_4-H_2O$ (22), $LiIO_3-Ti(IO_3)_4-H_2O$ and $LiIO_3-AgIO_3-H_2O$ (44) have been studied by the isothermal method and are characterized by the formation of solid solutions A continuous series of solid solutions based on hafnium iodate in the $LiIO_3-Hf(IO_3)_4-H_2O$ system and titanium iodate in the $LiIO_3-Ti(IO_3)_4-H_2O$ system are formed.

Systems With Lithium Halides. Solubility studies for these systems have been reported in 3 publications (18,45,46), and are summarized in Table 4.

Solubilities in these systems were studied by the isothermal method at 298 K. Incongruently soluble compounds are formed.

Table 4. Summary for Solubility Studies of the Ternary Systems With Lithium Halides

Ternary System	T/K	Solid Phase	Reference
$LiIO_3$ + $LiCl$ + H_2O	298	$\alpha-LiIO_3$; $LiCl$; $LiIO_3.LiCl.H_2O$	17
		$3LiIO_3.3LiCl.3H_2O$	
$LiIO_3$ + $LiBr$ + H_2O	298	$\alpha-LiIO_3$; $LiBr.2H_2O$;	45
		$LiIO_3.2LiBr.4H_2O$; $LiIO_3.2LiBr.4H_2O$	
$LiIO_3$ + LiI + H_2O	298	$\alpha-LiIO_3$; $LiI.3H_2O$;	46
		$3LiIO_3.2LiI.6H_2O$	

Systems With Lithium Salts. The solubility in the ternary $LiIO_3-LiNO_3-H_2O$ system was studied by the isothermal method at 298 K (8) and 323 K (10). This system at each temperature is of the simple eutonic type, and no double salts were formed.

Solubilities in ternary systems with lithium sulfate (15), perrhenate (18), carbonate (26) and phosphate (27,28) were studied by the isothermal method at 298 K only. These systems are of the simple eutonic type, and no double salts were formed.

OTHER MULTI-COMPONENT SYSTEMS

Quaternary Systems With Iodic Acid And The Other Iodates. Solubility data for the quaternary system $LiIO_3-HIO_3-KIO_3-H_2O$ at 323 K and $LiIO_3-HIO_3Al(IO_3)_3-H_2O$ at 298 K have been reported by Azarova, Vinogradov and Lepeshkov (42) and Shklovskaya, Arkhipov, Kidyarov and Tsibulevskaya (43), respectively. The systems were studied by the isothermal method. The double salts found experimentally are:

$$KiO_3.HIO_3; \quad KIO_3.2LiIO_3; \quad KIO_3.2H_2O \qquad (41)$$

$$Al(IO_3)_3.2HIO_3.6H_2O \qquad (42)$$

Solid solutions of $LiIO_3$ and HIO_3 (41,42) were found in both systems.

COMPONENTS:	EVALUATOR:
(1) Lithium iodate; $LiIO_3$; [13765-03-2] (2) Water; H_2O; [7732-18-5]	Hiroshi Miyamoto Department of Chemistry Niigata University Niigata, Japan February, 1986

CRITICAL EVALUATION:

Five Components System.   Although the five component system $LiIO_3$-KIO_3-$LiOH$-KOH-H_2O has been studied by Vinogradov, Lepeshkov and Tarasova (16), numerical solubility data were reported only for two quaternary systems, $LiIO_3$-KIO_3-$LiOH$-H_2O and KIO_3-$LiOH$-KOH-H_2O. In the quaternary $LiIO_3$-KIO_3-$LiOH$-H_2O system, the double salt $KIO_3 \cdot 2LiIO_3$ was formed and a solid solution of lithium iodate and hydroxide was also found.

REFERENCES

1. Mylius, F.; Funk, R. Ber. Dtsch. Chem. Ges. 1897, 30, 1716.

2. Ricci, J. E.; Amron, I. J. Am. Chem. Soc. 1951, 73, 3613.

3. Azarova, L. A.; Vinogradov, E. E.; Mikhailova, E. M.; Pakhomov, V. I. Zh. Neorg. Khim. 1973, 18, 239; Russ. J. Inorg. Chem. (Engl. Transl.) 1973, 18, 124.

4. Azarova, L. A.; Vinogradov, E. E.; Mikhailova, E. M.; Pakhomov, V. I. Zh. Neorg. Khim. 1973, 18, 2559; Russ. J. Inorg. Chem. (Engl. Transl.) 1973, 18, 1357.

5. Shklovskaya, R. M.; Arkhipov, S. M.; Kidyarov, E. I.; Mitnitskii, P. L. Zh. Neorg. Khim. 1974, 19, 1975; Russ. J. Inorg. Chem. (Engl. Transl.) 1974, 19, 1082.

6. Karataeva, I. M.; Vinogradov, E. E. Zh. Neorg. Khim. 1974, 19, 3156; Russ. J. Inorg. Chem. (Engl. Transl.) 1974, 19, 1726.

7. Shklovskaya, R. M.; Kashina, N. I.; Arkhipov, S. M.; Kuzina, V. A.; Kidyarov, B. I. Zh. Neorg. Khim. 1975, 20, 783; Russ. J. Inorg. Chem. (Engl. Transl.) 1975, 20, 441.

8. Shklovskaya, R. M.; Arkhipov, S. M.; Kidyarov, B. I. Zh. Neorg. Khim. 1975, 20, 1442; Russ. J. Inorg. Chem. (Engl. Transl.) 1975, 20, 811.

9. Tarasova, G. N.; Vinogradov, E. E.; Lepeshkov, I. N. Zh. Neorg. Khim. 1976, 21, 874; Russ. J. Inorg. Chem. (Engl. Transl.) 1976, 21, 478.

10. Vinogradov, E. E.; Karataeva, I. M. Zh. Neorg. Khim 1976, 21, 1664; Russ. J. Inorg. Chem. (Engl. Transl.) 1976, 21, 910.

11. Zhklovskaya, R. M.; Arkhipov, S. M.; Kidyarov, B. I.; Kuzina, V. A.; Poleva, G. V. Zh. Neorg. Khim. 1976, 21, 3116; Russ. J. Inorg. Chem. (Engl. Transl.) 1976, 21, 1718.

12. Tarasova, G. N.; Vinogradov, E. E.; Lepeshkov, I. N. Zh. Neorg. Khim. 1976, 21, 3373; Russ. J. Inorg. Chem. (Engl. Transl.) 1976, 21, 1858.

13. Shklovskaya, R. M.; Arkhipov, S. M.; Kidyarov, B. I.; Mitnitskii, P. L. Izv. Sib. Otd. Akad. Nauk SSSR Ser. Khim. Nauk 1976, (6), 89.

14. Shklovskaya, R. M.; Arkhipov, S. M.; Kidyarov, B. I.; Zherdienko, L. P. Zh. Neorg. Khim 1977, 22, 1139; Russ. J. Inorg. Chem. (Engl. Transl.) 1977, 22, 624.

15. Shklovskaya, R. M.; Arkhipov, S. M.; Kidyarov, B. I.; Tsibulevskaya, K. A. Zh. Neorg. Khim. 1978, 23, 2565; Russ. J. Inorg. Chem. (Engl. Transl.) 1978, 23, 1421.

16. Vinogradov, E. E.; Lepeshkov, I. N.; Tarasova, G. N. Zh. Neorg. Khim. 1978, 23, 3360; Russ. J. Inorg. Chem. (Engl. Transl.) 1978, 23, 1865.

17. Shklovskaya, R. M.; Arkhipov, S. M.; Kidyarov, B. I.; Kuzina, V. A. Zh. Neorg. Khim. 1979, 24, 203; Russ. J. Inorg. Chem. (Engl. Transl.) 1979, 24, 113.

COMPONENTS:	EVALUATOR:
(1) Lithium iodate; $LiIO_3$; [13765-3-2] (2) Water; H_2O; [7732-18-5]	Hiroshi Miyamoto Department of Chemistry Niigata University Niigata, Japan February, 1986

CRITICAL EVALUATION:

REFERENCES: (Continued)

18. Shklovskaya, R. M.; Arkhipov, S. M.; Kidyarov, B. I. *Zh. Neorg. Khim.* 1979, 24, 2287; *Russ. J. Inorg. Chem. (Engl. Transl.)* 1979, 24, 1269.

19. Shklovskaya, R. M.; Arkhipov, S. M.; Kidyarov, B. I.; Tokareva, A. G.; Kuzina, V. A. *Zh. Neorg. Khim.* 1980, 25, 1122; *Russ. J. Inorg. Chem. (Engl. Transl.)* 1980, 25, 618.

20. Shklovskaya, R. M.; Arkhipov, S. M.; Kidyarov, B. I.; Poleva, G. A.; Kuzina, V. A. *Zh. Neorg. Khim.* 1981, 26, 791; *Russ. J. Inorg. Chem. (Engl. Transl.)* 1981, 26, 425.

21. Arkhipov, S. M.; Kashina, N. I.; Kidyarov, B. I.; Kuzina, V. A. *Zh. Neorg. Khim.* 1981, 26, 1447; *Russ. J. Inorg. Chem. (Engl. Transl.)* 1981, 26, 779.

22. Shklovkaya, R. M.; Arkhipov, S. M.; Kidyarov, B. I.; Tokareva, A. G. *Zh. Neorg. Khim.* 1981, 26, 1701; *Russ. J. Inorg. Chem. (Engl. Transl.)* 1981, 26, 919.

23. Shklovskaya, R. M.; Arkhipov, S. M.; Kidyarov, B. I.; Poleva, G. V.; Troitskaya, N. I. *Zh. Neorg. Khim.* 1982, 27, 257; *Russ. J. Inorg. Chem. (Engl. Transl.)* 1982, 27, 145.

24. Vinogradov, E. E.; Tarasova, G. N. *Zh. Neorg. Khim.* 1982, 27, 269; *Russ. J. Inorg. Chem. (Engl. Transl.)* 1982, 27, 153.

25. Shklovskaya, R. M.; Arkhipov, S. M.; Kidyarov, B. I.; Kuzina, V. A.; Vdovkina, T. E. *Zh. Neorg. Khim.* 1982, 27, 513; *Russ. J. Inorg. Chem. (Engl. Transl.)* 1982, 27, 292.

26. Arkhipov, S. M.; Kashina, N. I.; Kidyarov, B. I. *Zh. Neorg. Khim.* 1982, 27, 539; *Russ. J. Inorg. Chem. (Engl. Transl.)* 1982, 27, 306.

27. Shklovskaya, R. M.; Arkhipov, S. M.; Kidyarov, B. I.; Vdovkina, T. E. *Zh. Neorg. Khim.* 1982, 27, 1597; *Russ. J. Inorg. Chem. (Engl. Transl.)* 1982, 27, 902.

28. Shklovskaya, R. M.; Arkhipov, S. M.; Kidyarov, B. I.; Vdovkina, T. E. *Zh. Neorg. Khim.* 1982, 27, 2985; *Russ. J. Inorg. Chem. (Engl. Transl.)* 1982, 27, 1692.

29. Azarova, L. A.; Vinogradov, E. E.; Pakhomov, V. I. *Zh. Neorg. Khim.* 1976, 21, 2801; *Russ. J. Inorg. Chem. (Engl. Transl.)* 1976, 21, 1545.

30. Shklovskaya, R. M.; Arkhipov, S. M.; Kidyarov, B. I. *Izv. Sib. Otd. Akad. Nauk SSSR Ser. Khim. Nauk* 1979, (9), 75.

31. Azarova, L. A.; Vinogradov, E. E. *Zh. Neorg. Khim.* 1977, 22, 273; *Russ. J. Inorg. Chem. (Engl. Transl.)* 1977, 22, 153.

32. Arkhipov, S. M.; Kashina, N. I.; Kidyarov, B. I. *Zh. Neorg. Khim.* 1978, 23, 1422; *Russ. J. Inorg. Chem. (Engl. Transl.)* 1978, 23, 784.

33. Umezawa, T.; Ninomiya, Y.; Tatuoka, S. *J. Appl. Crystallogr.* 1970, 3, 417.

34. Umezawa, T.; Tatuoka, S. *Jpn. J. Appl. Phys.* 1972, 11, 408.

35. Zachariasen, W. H.; Barta, F. A. *Phys. Rev.* 1931, 37, 1626.

36. Barta, F. A.; Zachariasen, W. H. *Phys. Rev.* 1931, 37, 1693.

37. Lithium Iodate Crystal Research Group, Institute of Physics *Wu Li Hsueh Pao* 1975, 24, 91; *C. A.* 1976, 83, 106309m.

38. Desvignes, J. M.; Remoissent, M. *Mater. Res. Bull.* 1971, 6, 705.

COMPONENTS:	EVALUATOR:
(1) Lithium iodate; $LiIO_3$; [13765-03-2] (2) Water; H_2O; [7732-18-5]	Hiroshi Miyamoto Department of Chemistry Niigata University Niigata, Japan February, 1986

CRITICAL EVALUATION:

REFERENCES: (Continued)

39. Liminga, R.; Abrahams, S. C. *J. Apll. Crystallogr.* <u>1976</u>, *9*, 42.

40. Arkhipov, S. M.; Pruntsev, A. E.; Kidyarov, B. I. *Zh. Neorg. Khim.* <u>1977</u>, *22*, 3394;
     Russ. J. Inorg. Chem. (Engl. Transl.) <u>1977</u>, *22*, 1855.

41. Lukasiewicz, T.; Pietaszewska, J.; Smija, J. *Biul. Wojak, Acad. Teck.* <u>1979</u>, *28*, 85.

42. Azarova, L. A.; Vinogradov, E. E.; Lepeshkov, I. M. *Zh. Neorg. Khim.* <u>1978</u>, *23*, 1952;
     Russ. J. Inorg. Chem. (Engl. Transl.) <u>1978</u>, *23*, 1072.

43. Shklovskaya, R. M.; Arkhipov, S. M.; Kidyarov, B. I.; Tsibulevskaya, K. A.
     Zh. Neorg. Khim. <u>1979</u>, *24*, 253; *Russ. J. Inorg. Chem. (Engl. Transl.)* <u>1979</u>, *24*,
     141.

44. Shklovskaya, R. M.; Arkhipov, S. M.; Kidyarov, B. I.; Vdovkina, T. E.; Poleva, G. V.
     Zh. Neorg. Khim. <u>1983</u>, *28*, 2431; *Russ. J. Inorg. Chem. (Engl. Transl.)* <u>1983</u>, *28*,
     1382.

COMPONENTS:	ORIGINAL MEASUREMENTS:
(1) Lithium iodate; $LiIO_3$; [13765-03-2] (2) Water; H_2O; [7732-18-5]	Mylius, F.; Funk, R. *Ber. Dtsch. Chem. Ges.* <u>1897</u>, *80*, 1716-25.

VARIABLES:	PREPARED BY:
T/K = 291	Hiroshi Miyamoto

EXPERIMENTAL VALUES:

The solubility of $(LiIO_3)_2$ in water at 18°C was given as:

44.6 mass %	(authors)
80.3 g/100 g H_2O	(authors)
4.42 mol kg^{-1}	(compiler)

The density of the saturated solution was given as 1.568 gm^{-3}.

AUXILIARY INFORMATION

METHOD/APPARATUS/PROCEDURE:	SOURCE AND PURITY OF MATERIALS:
The salt and water were placed in a bottle and the bottle agitated in a constant temperature bath for a long time (time not specified). After the saturated solution settled, an aliquot for analyses was removed with a pipet. $LiIO_3$ was determined by evaporation to dryness. The density of the saturated solution was also determined.	The salt used was purchased as a "pure chemical" and trace impurities were absent.
	ESTIMATED ERROR: Soly: precision ± 1 %. Temp: nothing specified.
	REFERENCES:

COMPONENTS:	ORIGINAL MEASUREMENTS:
(1) Lithium iodate; $LiIO_3$; [13765-03-2] (2) Water; H_2O; [7732-18-5]	Ricci, J.E.; Amron, I. J. Am. Chem. Soc. <u>1951</u>, 73, 3613-8.
VARIABLES: Temperature: 9.93 to 95.1°C	PREPARED BY: Hiroshi Miyamoto

EXPERIMENTAL VALUES: Solubility of $LiIO_3$

t/°C	mass %	mol % (compiler)	mol kg^{-1} (compiler)	Approach from
9.93	47.19(m)	8.133	4.914	U
20.24	45.96(m)	7.742	4.658	S
24.95	45.33(m)	7.591	4.560	U&S
29.94	44.89(m)	7.467	4.479	U
34.95	44.45(m)	7.345	4.400	U
40.00	44.12(m)	7.255	4.342	U
45.00	43.84(m)	7.178	4.293	U&S
50.06	43.51(m)	7.090	4.236	S
55.1	43.35(?)	7.047	4.208	U
60.2	43.10	6.980	4.165	U
65.3	43.00	6.954	4.149	U
75.5	42.82	6.907	4.118	U
85.5	42.76	6.891	4.108	S
95.1	42.85	6.914	4.123	U

m = metastable

U: undersaturation;   S: supersaturation

AUXILIARY INFORMATION

METHOD/APPARATUS/PROCEDURE:
Isothermal method. Many measurements were
made in an attempt to determine the stable
solubility curve of the forms of $LiIO_3$
from 10 to 95°C. The solubility curve was
determined with some points approached
from undersaturation, some from supersaturation, and a few from both directions.
The values obtained represent measurements
agreeing on repeated analysis with continued
stirring at each temperature.
For each point, the solid phase was examined
microscopically.

ESTIMATED ERROR:

Soly: precision about 0.1 % (compiler).

Temp: precision about ± 0.05 K (compiler).

SOURCE AND PURITY OF MATERIALS:
Some of the lithium iodate was made by purification of two samples of commercial c.p.
material which assayed ∿97% $LiIO_3$. One
sample contained insoluble $Ba(IO_3)_2$ and gave
an acid reaction. Part of it was simply recrystallized twice, and part was neutralized
with Kahlbaum LiOH before the second crystallization. The other sample contained insoluble Li_2CO_3 and gave an alkaline reaction; this was neutralized with iodic
acid and LiOH before two recrystallizations.
The rest of the salt used was made from
Kahlbaum Li_2CO_3 and c.p. iodic acid using
LiOH for final neutralization. The final
product was obtained by slow evaporation
with stirring on a hot-plate. After decantation, the crystals were filtered by
suction and washed with water. Ground and
dried at 110-180°C, the product was found
to be 99.9 to 100.1% pure by determination
of lithium as Li_2SO_4, and iodate by titration with $Na_2S_3O_3$ solution.

COMPONENTS:	ORIGINAL MEASUREMENTS:
(1) Lithium iodate; LiIO$_3$; [13765-03-2] (2) Water; H$_2$O; [7732-18-5]	Umezawa, T.; Tatuoka, S. *Jpn. J. Appl. Phys.* <u>1972</u>, *11*, 408.

VARIABLES:	PREPARED BY:
Two crystal forms: hexagonal and tetragonal Temperature: T/K = 278-253	Hiroshi Miyamoto and Mark Salomon

EXPERIMENTAL VALUES:

The solubilities of solutions in equilibrium with hexagonal and tetragonal solid phases were reported in graphical form. The polytherms are reproduced in the figure below.

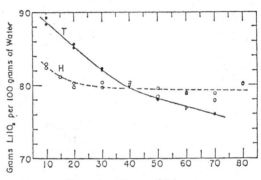

AUXILIARY INFORMATION

METHOD/APPARATUS/PROCEDURE:	SOURCE AND PURITY OF MATERIALS:
Saturated solutions starting with hexagonal or tetragonal crystals prepared at room temperature. Solutions were then stirred for 4-5 hours at the desired temperature. About 10 ml of the saturated solution were placed in a weighing bottle using a pipet with a filter at its tip. The solvent was evaporated in an oven, and the lithium iodate solubility determined gravimetrically. The pH of the saturated solutions varied between 8.3 and 8.7. The hexagonal → tetragonal transition temperature is around 40°C. As reported in an earlier publication (1), the solubility of hexagonal LiIO$_3$ is nearly constant over the temperature range studied. Note that reference (1) has not been rejected, and therefore this paper has not been compiled.	Nothing specified.

ESTIMATED ERROR:
Nothing specified.

REFERENCES:
1. Umezawa, T.; Ninomiya, Y.; Tatuoka, S. *J. Appl. Crystallogr.* <u>1970</u>, *3*, 417.

COMPONENTS:	ORIGINAL MEASUREMENTS:
(1) Lithium iodate; $LiIO_3$; [13765-03-2] (2) Iodic acid; HIO_3; [7782-68-2] (3) Water; H_2O; [7782-18-5]	Shklovskaya, R.M.; Arkhipov, S.M. Kidyarov, B.I.; Mitnitski, P.L. *Izv. Sib. Otd. Akad. Nauk SSSR Ser. Khim. Nauk* <u>1976</u>, (6), 89-91.
VARIABLES: T/K = 273 to 373 K	PREPARED BY: Hiroshi Miyamoto

EXPERIMENTAL VALUES:

The solubility of lithium iodate in aqueous solutions
containing 10 mass % HIO_3[a]

t/°C	solubility of $LiIO_3$	
	mass %	mol kg^{-1} (compiler)
0	45.07	4.512
10	45.30	4.554
20	43.62	4.255
25	43.53	4.239
30	42.91	4.133
40	42.83	4.120
50	42.70	4.098
60	42.52	4.068
70	42.19	4.013
80	42.1	4.00
90	41.1	3.84
100	42.19	4.013

[a]Initial composition of aqueous solution is
10 mass % HIO_3.

AUXILIARY INFORMATION

METHOD/APPARATUS/PROCEDURE:	SOURCE AND PURITY OF MATERIALS:
The compiler assumes that saturated solutions were prepared isothermally. Equilibrium was reached in 8 hours. The iodate concentration of the saturated solutions was determined by titration with thiosulfate solution.	"Chemically pure" grade $LiIO_3$ was used. The total amount of impurities did not exceed 0.001 %. Iodic acid was prepared as described in ref(1).
	ESTIMATED ERROR: Nothing specified.
	REFERENCES: 1. Vulikh, A.I.; Bogatyrev, V.L.; Kaz'minskaya, V.A.; Zherdienko, L.P. *Methody Polucheniya Khimcheskikh Reaktivov i Preparatov IREA, Vyp. 16.M., S.5.*

COMPONENTS:	ORIGINAL MEASUREMENTS:
(1) Lithium iodate; $LiIO_3$; [13765-03-2] (2) Ammonium iodate; NH_4IO_3; [13446-09-8] (3) Water; H_2O; [7732-18-5]	Tarasova, G.N.; Vinogradov, E.E.; Lepeshkov, I.N. Zh. Neorg. Khim. 1976, 21, 3373-6; Russ. J. Inorg. Chem. (Engl. Transl.) 1976, 21, 1858-60.
VARIABLES: Composition at 298.2 K.	PREPARED BY: Hiroshi Miyamoto

EXPERIMENTAL VALUES: Composition of saturated solutions at 25.0°C

LiIO₃		NH₄IO₃		Nature of
mass %	mol % (compiler)	mass %	mol % (compiler)	the solid phase[a]
0.00	0.000	3.72[b]	0.359	A
1.13	0.116	2.79	0.270	"
9.14	1.001	1.48	0.153	A+C
9.35	1.026	1.47	0.152	"
9.28	1.083	1.50	0.155	"
9.17	1.005	1.46	0.151	"
9.34	1.026	1.50	0.155	"
19.48	2.344	0.12	0.014	C
37.21	5.557	0.15	0.021	"
37.93	5.721	0.16	0.023	"
42.27	6.769	0.06	0.009	"
41.92	6.692	0.19	0.029	C+B
42.14	6.735	0.05	0.008	"
43.90[b]	7.195	–	–	B

[a] A = NH_4IO_3;   B = $LiIO_3$;   C = $NH_4IO_3 \cdot 2LiIO_3$

[b] For binary systems the compiler computes the following:

$$\text{soly of } LiIO_3 = 4.303 \text{ mol kg}^{-1}$$

$$\text{soly of } NH_4IO_3 = 0.200 \text{ mol kg}^{-1}$$

AUXILIARY INFORMATION

METHOD/APPARATUS/PROCEDURE:

Isothermal method used. Equilibrium was reached in 5-6 days. The liquid and solid phases were analyzed for Li^+ by the periodate method. Iodate was determined by titration with sodium thiosulfate in the presence of H_2SO_4 and KI and NH_4^+ was determined gravimetrically with sodium tetraphenyborate.

The composition and nature of the solid phases were found by use of Schreinemakers' method of residues, X-ray diffraction, thermography and infrared spectroscopy.

SOURCE AND PURITY OF MATERIALS:

Lithium iodate was prepd from lithium carbonate and iodic acid. NH_4IO_3 was prepd by mixing a slight excess of NH_4OH with HIO_3 in water. The precipitate was then filtered and washed to remove the excess NH_3.

ESTIMATED ERROR:

Soly: nothing specified.
Temp: precision ± 0.1 K.

COMMENTS AND/OR ADDITIONAL DATA:

The phase diagram is given below (based on mass % units).

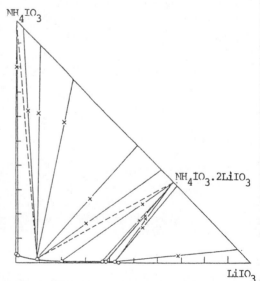

COMPONENTS:	ORIGINAL MEASUREMENTS:
(1) Lithium sulfate; Li_2SO_4; [13453-86-6] (2) Lithium iodate; $LiIO_3$; [13765-03-2] (3) Water; H_2O; [7732-18-5]	Shklovskaya, R.M.; Arkhipov, S.M.; Kidyarov, B.I.; Tsibulevskaya, K.A. *Zh. Neorg. Khim.* 1978, *23*, 2565-6; *Russ. J. Inorg. Chem. (Engl. Transl.)* 1978, *23*, 1421-2.

VARIABLES:	PREPARED BY:
Composition at 298.2 K	Hiroshi Miyamoto

EXPERIMENTAL VALUES: Composition of saturated solutions at 25.0°C

$LiIO_3$		Li_2SO_4		Nature of
mass %	mol % (compiler)	mass %	mol % (compiler)	the solid phase[a]
43.82[b]	7.173	–	–	A
41.53	6.778	2.25	0.607	"
37.76	5.956	3.80	0.991	"
35.99	5.690	5.87	1.535	"
32.40	4.965	7.36	1.865	"
27.35	4.078	10.67	2.632	"
25.44	3.755	11.91	2.908	"
22.30	3.247	14.19	3.417	"
19.51	2.799	16.00	3.797	"
17.66	2.550	18.50	4.418	A+B
16.51	2.355	18.73	4.418	B
14.98	2.112	19.41	4.526	"
12.96	1.789	19.80	4.521	"
9.43	1.268	21.30	4.736	"
7.51	0.945	22.03	4.825	"
5.55	0.727	23.17	5.020	"
2.99	0.385	24.34	5.183	"
–	–	25.84	5.401	"

[a] $A = \alpha\text{-}LiIO_3$; $B = Li_2SO_4 \cdot H_2O$

[b] For the binary system the compiler computes the following:

soly of $LiIO_3 = 4.289$ mol kg^{-1}

AUXILIARY INFORMATION

METHOD/APPARATUS/PROCEDURE:	COMMENTS AND/OR ADDITIONAL DATA:
Isothermal method used. Equilibrium was reached in 15-20 days. Aliquots of liquid phases were analyzed for iodate by iodometric titration and for sulfate gravimetrically as barium sulfate. Before precipitating the sulfate ion, the aliquots were treated with hydroxylamine hydrochloride in acidic medium to reduce IO_3^-, after which iodine was removed by boiling the solution. The solid phases were identified by the method of residues, and X-ray diffraction.	The phase diagram is given below (based on mass % units). 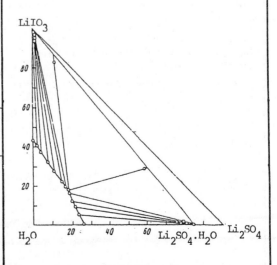

SOURCE AND PURITY OF MATERIALS:

Highly pure grade $\alpha\text{-}LiIO_3$ and lithium sulfate monohydrate were used.

ESTIMATED ERROR:

Soly: nothing specified.
Temp: precision ± 0.1 K.

COMPONENTS:	ORIGINAL MEASUREMENTS:
(1) Lithium chloride; LiCl; [7447-41-8]	Shklovskaya, R.M.; Arkhipov, S.M.; Kidyarov, B.I.; Kuzina, V.A.
(2) Lithium iodate; LiIO$_3$; [13765-03-2]	*Zh. Neorg. Khim.* 1979, *24*, 203-6;
(3) Water; H$_2$O; [7732-18-5]	*Russ. J. Inorg. Chem. (Engl. Transl.)* 1979, *24*, 113-4.

VARIABLES:	PREPARED BY:
Composition at 298.2 K	Hiroshi Miyamoto

EXPERIMENTAL VALUES: Composition of saturated solutions at 25.0°C

| LiIO$_3$ | | LiCl | | Nature of |
mass %	mol % (compiler)	mass %	mol % (compiler)	the solid phase[a]
43.82[b]	7.173	–	–	A
39.58	6.147	0.96	0.64	"
32.34	4.662	3.73	2.31	"
20.08	2.600	9.37	5.203	"
12.58	1.543	13.77	7.247	"
6.96	0.831	18.77	9.617	"
3.37	0.407	25.91	13.42	"
2.04	0.255	32.82	17.59	"
1.97	0.251	35.61	19.46	"
1.90	0.243	35.94	19.68	A+C
1.95	0.250	36.60	20.15	C
1.70	0.218	36.96	20.34	C+D
0.89	0.115	39.11	21.67	D
0.71	0.093	41.03	23.01	"
0.52	0.069	42.99	24.42	"
0.39	0.052	44.68	25.67	"
0.33	0.044	45.16	26.03	D+B
–	–	45.51	26.19	B

[a] A = α-LiIO$_3$;   B = LiCl;   C = LiIO$_3$.LiCl.H$_2$O;   D = 2LiIO$_3$.3LiCl.3H$_2$O

[b] For the binary system the compiler computes the following:
     soly of LiIO$_3$ = 4.289 mol kg^{-1}

AUXILIARY INFORMATION

METHOD/APPARATUS/PROCEDURE:
The ternary system LiIO$_3$-LiCl-H$_2$O was studied by the isothermal method. Equilibrium was established in 20-30 days. Aliquots of the liquid phases were analyzed for lithium by ion exchange, and for iodate by iodometric titration. The chloride was determined by difference. The solid phases were identified by the method of residues, checked by X-ray diffraction, and thermographically.

SOURCE AND PURITY OF MATERIALS:

α-LiIO$_3$ and LiCl.H$_2$O were of special purity grade.

ESTIMATED ERROR:

Soly: nothing specified.
Temp: precision ± 0.1 K.

COMMENTS AND/OR ADDITIONAL DATA:
The phase diagram based on mass % units is given below.

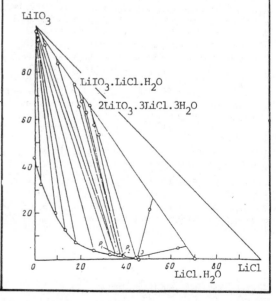

COMPONENTS:	ORIGINAL MEASUREMENTS:
(1) Lithium bromide; LiBr; [7550-35-8] (2) Lithium iodate; $LiIO_3$; [13765-03-2] (3) Water; H_2O; [7732-18-5]	Arkhipov, S.M.; Kashina, N.I.; Kidyarov, B.I.; Kuzina, V.A. *Zh. Neorg. Khim.* 1983, *28*, 2647-9; *Russ. J. Inorg. Chem. (Engl. Transl.)* 1983, *28*, 1503-4.
VARIABLES: Composition at 298.2 K	PREPARED BY: Hiroshi Miyamoto

EXPERIMENTAL VALUES: Composition of saturated solutions at 25.0°C

$LiIO_3$		LiBr		Nature of
mass %	mol % (compiler)	mass %	mol % (compiler)	the solid phase[a]
43.80[b]	7.168	–	–	A
36.40	5.521	2.39	0.759	"
33.51	5.103	6.00	1.913	"
26.40	3.864	10.77	3.301	"
17.10	2.427	18.66	5.545	"
6.39	0.915	31.66	9.497	"
3.34	0.501	39.01	12.25	"
2.13	0.340	45.33	15.13	"
1.86	0.303	47.25	16.10	A+B
1.46	0.239	48.01	16.43	B
0.86	0.145	51.07	18.03	"
0.65	0.113	53.50	19.47	"
0.40	0.071	55.60	20.75	"
0.32	0.058	56.47	21.32	"
0.18	0.034	59.58	23.49	B+C
0.17	0.032	59.90	23.73	C
0.14	0.027	60.40	24.09	"
0.15	0.029	61.04	24.59	C+D
–	–	61.55	24.93	D

[a] A = α-$LiIO_3$;   B = $LiIO_3 \cdot LiBr \cdot 2H_2O$;   C = $LiIO_3 \cdot 2LiBr \cdot 4H_2O$;   D = $LiBr \cdot 2H_2O$

[b] For the binary system the compiler computes the following:
        soly of $LiIO_3$ = 4.286 mol kg^{-1}

AUXILIARY INFORMATION

METHOD/APPARATUS/PROCEDURE:

Isothermal method used. Equilibrium was reached in 20-30 days. The lithium content was determined by ion exchange, and the iodate content was determined iodometrically. The bromide content was obtained by difference.

The bromide content in a sample of the liquid phase containing low concentration of iodate was determined by argentometric titration.

The composition of the solid phase was determined by the method of residues, and the result was checked by X-ray analysis.

SOURCE AND PURITY OF MATERIALS:

"Special purity" grade α-$LiIO_3$ and $LiBr \cdot 2H_2O$ were used.

ESTIMATED ERROR:

Soly: nothing specified.
Temp: precision $\pm$ 0.1 K.

COMMENTS AND/OR ADDITIONAL DATA:

Isotherm based on mass % units is reproduced below.

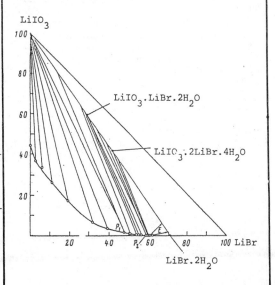

COMPONENTS:	ORIGINAL MEASUREMENTS:
(1) Lithium iodide; LiI; [10377-51-2] (2) Lithium iodate; LiIO₃; [13765-03-2] (3) Water; H₂0; [7732-18-5]	Shklovskaya, R.M.; Arkhipov, S.M. Kidyarov, B.I.; Vdovkina, T.E.; Kuzina, V.A. *Zh. Neorg. Khim.* **1983**, *28*, 2701-3; *Russ, J. Inorg. Chem. (Engl. Transl.)* **1983**, *28*, 1533-4.

VARIABLES:	PREPARED BY:
Composition at 298.2 K	Hiroshi Miyamoto

EXPERIMENTAL VALUES: Composition of saturated solutions

LiIO₃ mass %	mol % (compiler)	LiI mass %	mol % (compiler)	Nature of the solid phase
43.82[b]	7.173	–	–	A
34.31	5.534	8.86	1.942	"
29.52	4.662	12.34	2.648	"
25.89	4.126	16.77	3.631	"
18.97	3.000	23.41	5.029	"
14.24	2.276	29.11	6.322	"
11.96	1.930	32.17	7.054	"
8.11	1.363	38.97	8.895	"
3.95	0.710	47.78	11.67	"
3.12	0.606	53.37	14.08	"
1.77	0.360	57.37	15.84	A+C
1.14	0.235	58.75	16.43	C
0.91	0.188	59.06	16.54	"
0.79	0.168	60.75	17.50	"
0.73	0.159	62.12	18.34	B+C
–	–	62.00	18.01	B

[a] A = α-LiIO₃;   B = LiI.3H₂0;   C = 3LiIO₃.2LiI.6H₂0

[b] For the binary system the compiler computes the following:

soly = 4.289 mol kg⁻¹

AUXILIARY INFORMATION

METHOD/APPARATUS/PROCEDURE:	SOURCE AND PURITY OF MATERIALS:
Isothermal method used. Equilibrium was reached in 20-30 days. The iodate content in samples of coexisting phases was determined by iodometric titration. The iodide content in the liquid phase was determined by argentometric titration. The composition of solid phases was determined by the method of residues, and the result checked by X-ray analysis.	"Special purity" grade α-LiIO₃ and analytical grade LiI were used

METHOD column continues:

| ESTIMATED ERROR:

Soly: nothing specified.
Temp: precision ± 0.1° K. | COMMENT AND/OR ADDITIONAL DATA:
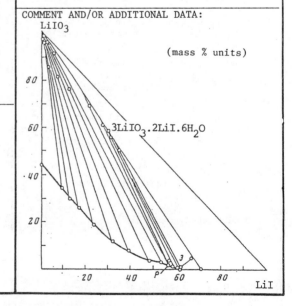 |

COMPONENTS:	ORIGINAL MEASUREMENTS:
(1) Lithium iodate; $LiIO_3$; [13765-03-2] (2) Sodium iodate; $NaIO_3$; [7681-55-2] (3) Water; H_2O; [7732-18-5]	Shklovskaya, R.M.; Arkhipov, S.M.; Kidyarov, E.I.; Mitnitskii, P.L. Zh. Neorg. Khim. 1974, 19, 1975-6; Russ. J. Inorg. Chem. (Engl. Transl.) 1974, 19, 1082-3.
VARIABLES: Composition at 298.2 K	PREPARED BY: Hiroshi Miyamoto

EXPERIMENTAL VALUES: Composition of saturated solutions at 25.0°C

| \ | $LiIO_3$ | | $NaIO_3$ | | Nature of |
mass %	mol % (compiler)	mass %	mol ° (compiler)		the solid phase[a]
43.82[b]	7.173	–	–		A
43.63	7.127	0.045	0.0068		"
41.90	7.008	3.32	0.510		A+B
38.75	6.160	3.05	0.446		B
32.56	4.808	3.94	0.535		"
30.87	4.413	3.17	0.416		"
26.96	3.707	4.03	0.509		"
24.58	3.285	4.11	0.505		"
18.97	2.374	4.11	0.473		"
14.09	1.680	4.63	0.507		"
10.22	1.177	5.27	0.558		"
6.237	0.6926	5.69	0.581		"
–	–	8.66[b]	0.856		"

[a] $A = LiIO_3$; $B = NaIO_3 \cdot H_2O$

[b] For binary systems the compiler computes the following:

soly of $LiIO_3$ = 4.289 mol kg^{-1}

soly of $NaIO_3$ = 0.479 mol kg^{-1}

AUXILIARY INFORMATION

METHOD/APPARATUS/PROCEDURE:	SOURCE AND PURITY OF MATERIALS:
Solubility in the system was studied by the isothermal method. Equilibrium between the liquid and solid phases was established in 30 days. The alkali metal content of the liquid and solid phases was determined by flame photometry, and iodate was estimated by a volumetric method. The compiler assumes that iodate content was determined iodometrically. The solid phases were identified by the method of residues, and by X-ray diffraction analysis.	Chemically pure $LiIO_3$ and $NaIO_3$ were recrystallized twice from aqueous solutions.
	ESTIMATED ERROR: Soly: nothing specified. Temp: precision ± 0.1 K.
	REFERENCES:

COMPONENTS:	ORIGINAL MEASUREMENTS:
(1) Lithium iodate; $LiIO_3$; [13765-03-2] (2) Potassium iodate; KIO_3; [7758-05-6] (3) Water; H_2O; [7732-18-5]	Azarova, L.A.; Vinogradov, E.E.; Mikhailove, E.M.; Pakhomov, V.I. *Zh. Neorg. Khim.* 1973, *18*, 2559-63; *Russ. J. Inorg. Chem. (Engl. Transl.)* 1973, *18*, 1357-60.
VARIABLES: Composition at 323.2 K	PREPARED BY: Hiroshi Miyamoto

EXPERIMENTAL VALUES: Composition of saturated solutions at 50.0°C.

LiIO₃		KIO₃		Nature of
mass %	mol % (compiler)	mass %	mol % (compiler)	the solid phase[a]
0.00	0.000	13.08[b]	1.251	A
3.95	0.455	11.47	1.124	"
1.47	0.168	13.14	1.277	"
3.66	0.428	13.19	1.312	"
9.87	1.19	9.49	0.969	"
13.86	1.733	9.05	0.962	A+C
14.22	1.817	10.56	1.147	"
14.55	1.849	9.76	1.054	"
16.15	2.073	9.04	0.986	"
16.15	2.073	9.04	0.986	"
13.91	1.713	7.66	0.802	"
14.91	1.901	9.69	1.050	"
14.25	1.791	9.09	0.971	C
14.39	1.808	8.92	0.952	"
31.55	4.391	0.42	0.050	"
29.67	4.048	0.71	0.082	"
18.53	2.345	5.49	0.590	"
42.27	6.793	0.30	0.041	C+B
41.68	6.648	0.37	0.050	"
40.00	6.231	0.40	0.053	"
42.15	6.769	0.37	0.050	B
43.28[b]	7.028	0.00	0.000	"

[a] A = KIO_3;   B = $LiIO_3$;   C = $KIO_3 \cdot 2LiIO_3$

[b] For binary systems, the compiler computes the following:

soly of $LiIO_3$ = 4.20 mol kg^{-1},   soly of KIO_3 = 0.703 mol kg^{-1}

METHOD/APPARATUS/PROCEDURE:	SOURCE AND PURITY OF MATERIALS:
Solubilities were determined isothermally at 50°C in a thermostated water bath. Equilibrium in the system was reached after continual stirring for 4-5 days. The total IO_3^- content in the sample was found by iodometric titration. Lithium was determined by flame photometry and the periodate method. Potassium was determined with tetraphenylborate. The composition and nature of the solid phases were determined by Schreinemakers' method, X-ray diffraction, thermography, infrared spectroscopy, and crystallography.	C.p. grade potassium iodate used. Lithium iodate prepared from lithium carbonate and HIO_3. Purities checked by chemical and X-ray diffraction methods, however the results were not given.
	ESTIMATED ERRORS: Nothing specified.
	COMMENTS AND/OR ADDITIONAL DATA: The phase diagram is given below (based on mass % units). 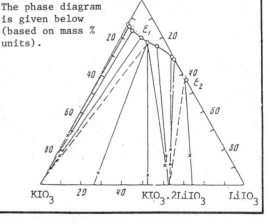

COMPONENTS:	ORIGINAL MEASUREMENTS:
(1) Lithium iodate; LiIO₃; [13765-03-2] (2) Potassium iodate; KIO₃; [7758-05-6] (3) Water; H₂0; [7732-18-5]	Shklovskaya, R.M.; Kashina, N.I.; Arkhipov, S.M.; Kuzina, V.A.; Kidyarov, B.I. *Zh. Neorg. Khim.* <u>1975</u>, *20*, 783–5; *Russ. J. Inorg. Chem. (Engl. Transl.)* <u>1975</u>, *20*, 441–2.
VARIABLES: Composition at 298 K.	PREPARED BY: Hiroshi Miyamoto

EXPERIMENTAL VALUES: Composition of saturated solutions at 25°C

LiIO₃ mass %	LiIO₃ mol % (compiler)	KIO₃ mass %	KIO₃ mol % (compiler)	Nature of the solid phase[a]
0.00	0.000	8.39[b]	0.765	A
6.42	0.722	6.70	0.640	"
13.10	1.560	5.47	0.554	"
14.16	1.700	5.18	0.529	A+C
14.82	1.782	4.64	0.474	C
17.41	2.112	2.91	0.300	"
19.54	2.392	1.60	0.166	"
23.50	2.990	1.06	0.115	"
26.71	3.523	0.89	0.100	"
31.94	4.459	0.28	0.033	"
36.22	5.339	0.18	0.023	"
38.32	5.811	0.16	0.021	"
42.87	6.927	0.07	0.010	B+C
43.82[b]	7.173	0.00	0.000	B

[a] A = KIO₃;   B = LiIO₃;   C = KIO₃·2LiIO₃

[b] For binary systems the compiler computes the following:

$$\text{soly of LiIO}_3 = 4.289 \text{ mol kg}^{-1}$$

$$\text{soly of KIO}_3 = 0.428 \text{ mol kg}^{-1}$$

AUXILIARY INFORMATION

METHOD/APPARATUS/PROCEDURE:	COMMENTS AND/OR ADDITIONAL DATA:
The solubility in KIO₃-LiIO₃-H₂0 system was studied by the isothermal method. Equilibrium was reached in 10 days. The iodate content in the liquid was determined iodometrically, and the potassium content detd gravimetrically as tetraphenylborate. Lithium concentrations were determined by difference, and in several instances by flame photometry. X-ray diffraction patterns were recorded.	The phase diagram is given below (based on mass % units). 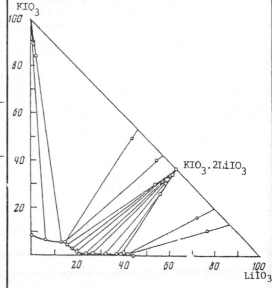
SOURCE AND PURITY OF MATERIALS: Chemically pure grade KIO₃ and LiIO₃ were used.	
ESTIMATED ERROR: Nothing specified.	

COMPONENTS:	ORIGINAL MEASUREMENTS:
(1) Lithium iodate; $LiIO_3$; [13765-03-2]	Karataeva, I.M.; Vinogradov, E.E.
(2) Rubidium iodate; $RbIO_3$; [13446-76-9]	*Zh. Neorg. Khim.* 1974, *19*, 3156-60; *Russ. J. Inorg. Chem. (Engl. Transl.)* 1974, *19*, 1726-9.
(3) Water; H_2O; [7732-18-5]	

VARIABLES:	PREPARED BY:
Composition at 323 K	Hiroshi Miyamoto

EXPERIMENTAL VALUES:　　　Composition of saturated solutions at 50°C

	$LiIO_3$		$RbIO_3$		Nature of
mass %	mol % (compiler)		mass %	mol % (compiler)	the solid phase[a]
43.25[c]	7.020		–	–	A
39.40	6.062		0.12	0.013	"
36.23	5.341		0.17	0.018	A+C
36.27	5.351		0.19	0.020	"
36.17	5.336		0.28	0.029	C
33.48	4.780		0.48	0.048	"
25.90	3.386		0.94	0.086	"
27.20	3.598		0.64	0.059	"
26.11	3.419		0.88	0.080	"
25.05	3.246		1.04	0.0941	"
22.79	2.890		1.43	0.127	C+B[b]
22.71	2.876		1.40	0.124	"
22.72	2.877		1.40	0.124	"
22.64	2.864		1.37	0.121	"
19.10	2.318		1.23	0.104	B
11.98	1.353		1.61	0.127	"
11.72	1.323		1.81	0.143	"
5.11	0.542		2.11	0.156	"
1.74	0.181		3.59	0.261	"
0.57	0.059		4.19	0.303	"
–	–		4.39[b]	0.317	"

continued.....

AUXILIARY INFORMATION

METHOD/APPARATUS/PROCEDURE:
The compiler assumes that the isothermal method was used. Equilibrium between the liquid and solid phases was established in 14 days. The liquid and solid phases were analyzed for ions: Li^+ by the periodate method, Rb^+ gravimetrically with sodium tetraphenylborate, and IO_3^- by iodometric titration in sulfuric acid solution.

To determine the composition and nature of solid phases formed in the systems, the authors used Schreinemakers' method of wet residues, X-ray diffraction, thermography, and infrared spectroscopy.

SOURCE AND PURITY OF MATERIALS:
Lithium iodate was prepared from lithium carbonate and iodic acid. Although the purity of lithium iodate was checked by chemical, thermal and X-ray diffraction analyses, the results were not given. C.p. grade rubidium iodate was used.

ESTIMATED ERROR:
Nothing specified.

REFERENCES:

COMPONENTS:	EVALUATOR:
(1) Lithium iodate; $LiIO_3$; [13765-03-2]	Karataeva, I.M.; Vinogradov, E.E.
(2) Rubidium iodate; $RbIO_3$; [13446-76-9]	*Zh. Neorg. Khim.* 1974, *19*, 3156-60; *Russ. J. Inorg. Chem. (Engl. Transl.)* 1974, *19*, 1726-9.
(3) Water; H_2O; [7732-18-5]	

COMMENTS AND/OR ADDITIONAL DATA:

The phase diagram for 50°C is given below

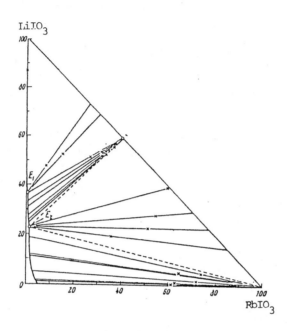

EXPERIMENTAL VALUES (Continued)

[a] $A = LiIO_3$; $B = RbIO_3$; $C = 2LiIO_3 \cdot RbIO_3$

[b] The compiler assumes that $2LiIO_3 \cdot RbIO_3 \cdot RbIO_3$ given in the original paper should read $2LiIO_3 \cdot RbIO_3 + RbIO_3$.

For the binary systems, the compiler computes the following:

   soly of $LiIO_3 = 4.191$ mol kg^{-1}

   soly of $RbIO_3 = 0.176$ mol kg^{-1}

COMPONENTS:	ORIGINAL MEASUREMENTS:
(1) Lithium iodate; $LiIO_3$; [13765-03-2] (2) Rubidium iodate; $RbIO_3$; [13446-76-9] (3) Water; H_2O; [7732-18-5]	Shklovskaya, R.M.; Kashina, N.I.; Arkhipov, S.M.; Kuzina, V.A.; Kidyarov, B.I. *Zh. Neorg. Khim.* <u>1975</u>, *20*, 783-5; *Russ. J. Inorg. Chem. (Engl. Transl.)* <u>1975</u>, *20*, 441-2
VARIABLES:	PREPARED BY:
Composition at 298 K.	Hiroshi Miyamoto

EXPERIMENTAL VALUES: Composition of saturated solutions at 25°C

LiIO_3		RbIO_3		Nature of
mass %	mol % (compiler)	mass %	mol % (compiler)	the solid phase[a]
0.00	0.000	2.36[b]	0.167	A
4.36	0.454	1.09	0.0793	"
10.58	1.168	0.75	0.058	"
15.88	1.849	0.67	0.054	"
21.15	2.603	0.47	0.040	A+C
25.65	3.314	0.23	0.021	C
30.57	4.188	0.148	0.0142	"
34.08	4.879	0.105	0.0105	"
38.37	5.814	0.067	0.0071	"
43.11	6.987	0.037	0.0042	B+C
43.82[b]	7.173	0.000	0.0000	B

[a] A = $RbIO_3$;   B = $LiIO_3$;   C = $RbIO_3 \cdot 2LiIO_3$

[b] For binary systems the compiler computes the following:

soly of $LiIO_3$ = 4.289 mol kg^{-1}

soly of $RbIO_3$ = 0.0928 mol kg^{-1}

AUXILIARY INFORMATION

METHOD/APPARATUS/PROCEDURE:	COMMENTS AND/OR ADDITIONAL DATA:
The solubility in $LiIO_3$-$RbIO_3$-H_2O system was studied by the isothermal method. Equilibrium was reached in 10 days. The iodate content in the liquid was determined iodometrically, and rubidium determined gravimetrically as the tetraphenyl borate. Lithium was determined by difference, and in several instances by flame photometry. X-ray diffraction patterns were recorded. The composition of the solid phases were determined by the method of residues, and was checked by X-ray diffraction.	The phase diagram is given below (based on mass % units). 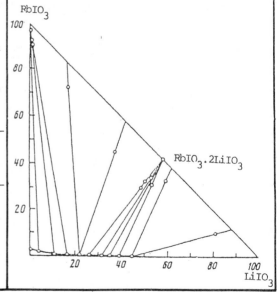
SOURCE AND PURITY OF MATERIALS: Chemical pure grade $LiIO_3$ and $RbIO_3$ were used.	
ESTIMATED ERROR: Nothing specified.	

COMPONENTS:	ORIGINAL MEASUREMENTS:
(1) Lithium iodate; $LiIO_3$; [13765-03-2] (2) Cesium iodate; $CsIO_3$; [13454-81-4] (3) Water; H_2O; [7732-18-5]	Shklovskaya, R.M.; Arkhipov, S.M.; Kidyarov, E.I.; Mitnitskii, P.L. *Zh. Neorg. Khim.* <u>1974</u>, *19*, 1975-6; *Russ. J. Inorg. Chem. (Engl. Transl.)* <u>1974</u>, *19*, 1082-3.
VARIABLES: Composition at 298.2 K	PREPARED BY: Hiroshi Miyamoto

EXPERIMENTAL VALUES: Composition of saturated solutions at 25.0°C

$LiIO_3$ mass %	mol % (compiler)	mass %	mol % (compiler)	Nature of the solid phase[a]
43.82[b]	7.173	–	–	A
42.98	7.005	0.519	0.0500	"
42.52	6.896	0.647	0.0620	A+B
39.70	6.178	0.609	0.0560	B
36.45	5.423	0.610	0.0536	"
34.19	4.936	0.611	0.0521	"
32.36	4.562	0.611	0.0509	"
28.13	3.762	0.612	0.0484	"
24.98	3.217	0.612	0.0466	"
21.16	2.609	0.618	0.0450	"
15.35	1.777	0.640	0.0438	"
12.63	1.422	0.646	0.0430	"
8.103	0.8726	0.744	0.0473	"
4.59	0.479	1.006	0.06203	"
–	–	2.61[b]	0.157	"

[a] $A = LiIO_3$; $B = CsIO_3$

[b] For binary systems the compiler computes the following:

soly of $LiIO_3$ = 4.289 mol kg^{-1}

soly of $CsIO_3$ = 0.0871 mol kg^{-1}

AUXILIARY INFORMATION

METHOD/APPARATUS/PROCEDURE:	SOURCE AND PURITY OF MATERIALS:
Solubility in the system was studied by the isothermal method. Equilibrium between liquid and solid phases was established in 30 days. Lithium content in samples of the liquid and solid phases were determined by flame photometry, and cesium was determined gravimetrically as the tetraphenylborate. The authors report that iodate was determined by volumetric method. The compiler assumes that this is an iodometric titration. The solid phases were identified by the method of residues, and by X-ray diffraction analysis.	C.p. grade $LiIO_3$ and $CsIO_3$ were recrystallized twice from aqueous solutions.
	ESTIMATED ERROR: Soly: nothing specified. Temp: precision ± 0.1 K.
	REFERENCES:

COMPONENTS:	ORIGINAL MEASUREMENTS:
(1) Lithium iodate; $LiIO_3$; [13765-03-2]	Karataeva, I.M.; Vinogradov, E.E.
(2) Cesium iodate; $CsIO_3$; [13454-81-4]	$Zh.$ $Neorg.$ $Khim.$ 1974, 19, 3156-60; $Russ.$ $J.$ $Inorg.$ $Chem.$ $(Engl.$ $Transl.)$ 1974, 19, 1726-9.
(3) Water; H_2O; [7732-18-5]	

VARIABLES:	PREPARED BY:
Composition at 323 K	Hiroshi Miyamoto

EXPERIMENTAL VALUES: Composition of saturated solutions

LiIO$_3$		CsIO$_3$		Nature of
mass %	mol % (compiler)	mass %	mol % (compiler)	the solid phase[a]
43.29[b]	7.031	–	–	A
42.91	7.119	1.73	0.170	"
43.73	7.358	1.83	0.182	A+B
43.62	7.328	1.84	0.183	"
43.78	7.361	1.74	0.173	"
34.24	5.024	1.73	0.150	B
31.32	4.435	1.94	0.162	"
30.07	4.203	2.16	0.178	"
19.12	2.348	2.23	0.162	"
3.68	0.393	4.10	0.259	"
–	–	5.07[b]	0.312	"

[a] $A = LiIO_3$; $B = CsIO_3$

[b] For binary systems the compiler computes the following:

$$\text{soly of } LiIO_3 = 4.198 \text{ mol kg}^{-1}$$

$$\text{soly of } CsIO_3 = 0.174 \text{ mol kg}^{-1}$$

AUXILIARY INFORMATION

METHOD/APPARATUS/PROCEDURE:	SOURCE AND PURITY OF MATERIALS:
The compiler assumes that the isothermal method was used. Equilibrium between the liquid and solid phases was established in 14 days. The liquid and solid phases were analyzed for ions: Li^+ by the periodate method, IO_3^- by iodometric titration in sulfuric acid solution, and Cs^+ by difference. To determine the composition and nature of solid phases formed in this system, the authors used Schreinemakers' method of wet residues, X-ray diffraction, thermography and infrared spectroscopy.	Lithium iodate was prepared from lithium carbonate and iodic acid. Although the purity of the lithium iodate was checked by chemical, thermal and X-ray diffraction analyses, the results were not given. C.p. grade cesium iodate was used.
	ESTIMATED ERROR:
	Nothing specified.
	REFERENCES:

COMPONENTS:	ORIGINAL MEASUREMENTS:
(1) Lithium carbonate; Li_2CO_3; [554-13-2] (2) Lithium iodate; $LiIO_3$; [13765-03-2] (3) Water; H_2O; [7732-18-5]	Arkhipov, S.M.; Kashina, N.I.; Kidyarov, B.I. *Zh. Neorg. Khim.* 1982, *27*, 539; *Russ. J. Inorg. Chem. (Engl. Transl.)* 1982, *27*, 306-7.
VARIABLES: Composition at 298.2 K	PREPARED BY: Hiroshi Miyamoto

EXPERIMENTAL VALUES: Composition of saturated solutions at 25.0°C

Li_2CO_3		$LiIO_3$		Nature of
mass %	mol % (compiler)	mass %	mol % (compiler)	the solid phase[a]
1.21	0.298	–	–	A
1.01	0.251	1.22	0.123	"
0.76	0.193	3.90	0.403	"
0.73	0.186	4.28	0.444	"
0.31	0.083	9.80	1.068	"
0.11	0.033	19.71	2.377	"
0.074	0.025	29.37	3.960	"
0.063	0.022	35.13	5.095	"
0.049	0.019	39.87	6.168	"
0.040	0.016	42.48	6.821	"
0.037	0.015	43.71	7.147	A+B
–	–	43.80[b]	7.168	B

[a] A = Li_2CO_3;   B = $LiIO_3$

[b] For the binary system the compiler computes the following:

$$\text{soly of } LiIO_3 = 4.286 \text{ mol kg}^{-1}$$

AUXILIARY INFORMATION

METHOD/APPARATUS/PROCEDURE:	SOURCE AND PURITY OF MATERIALS:
The isothermal method was used. Equilibrium was reached in 7 days. Samples of the liquid phase were analyzed for iodate by iodometric titration, and carbonate by back-titration using methyl red indicator. The compositions of the solid phases were determined by the method of residues and checked by X-ray diffraction.	"Special purity" grade α-lithium iodate and lithium carbonate were used.
	ESTIMATED ERROR: Soly: nothing specified. Temp: precision ± 0.1 K.
	REFERENCES:

COMPONENTS:	ORIGINAL MEASUREMENTS:
(1) Lithium nitrate; $LiNO_3$; [7790-69-4] (2) Lithium iodate; $LiIO_3$; [13765-03-2] (3) Water; H_2O; [7732-18-5]	Shklovskaya, R.M.; Arkhipov, S.M.; Kidyarov, B.I. *Zh. Neorg. Khim.* <u>1975</u> *20*, 1442-4; *Russ. J. Inorg. Chem. (Engl. Transl.)* <u>1975</u>, *20*, 811-2.
VARIABLES: Composition at 298 K.	PREPARED BY: Hiroshi Miyamoto

EXPERIMENTAL VALUES: Composition of saturated solutions at 25°C

| $LiIO_3$ | | $LiNO_3$ | | Nature of |
mass %	mol % (compiler)	mass %	mol % (compiler)	the solid phase[a]
0.00	0.000	47.10	18.88	A
2.37	0.367	45.90	18.76	A+B
2.37	0.367	45.90	18.76	"
2.48	0.375	43.75	17.47	B
2.96	0.446	42.84	17.04	"
4.49	0.644	36.39	13.77	"
5.51	0.752	30.45	10.97	"
7.63	1.041	27.80	10.01	"
10.16	1.392	25.12	9.08	"
12.04	1.640	22.26	8.00	"
15.67	2.161	19.01	6.92	"
19.35	2.685	15.12	5.53	"
24.93	3.547	10.72	4.02	"
43.82[b]	7.173	0.00	0.00	"

[a] A = $LiNO_3 \cdot 3H_2O$;   B = $LiIO_3$

[b] For the binary system the compiler computes the following:

soly of $LiIO_3$ = 4.289 mol kg^{-1} at 25°C.

AUXILIARY INFORMATION

METHOD/APPARATUS/PROCEDURE:	COMMENTS AND/OR ADDITIONAL DATA:
Equilibrium in the ternary system was reached isothermally after 10-15 days. Specimens of the liquid and solid phases were analyzed volumetrically, presumably (compiler) iodometrically for the iodate, and gravimetrically for the nitrate ion with nitron as a precipitant.	The phase diagram is given below (based on mass % units). 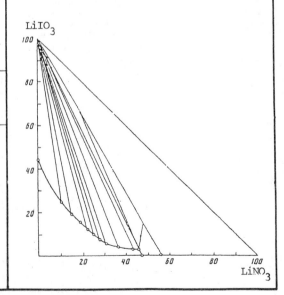

SOURCE AND PURITY OF MATERIALS:

Chemically pure grade $LiIO_3$ and $LiNO_3$ were used.

ESTIMATED ERROR:

Nothing specified.

COMPONENTS:	ORIGINAL MEASUREMENTS:
(1) Lithium nitrate; $LiNO_3$; [7790-69-4] (2) Lithium iodate; $LiIO_3$; [13765-03-2] (3) Water; H_2O; [7732-18-5]	Vinogradov, E.E.; Karataeva, I.M. *Zh. Neorg. Khim.* 1976, *21*, 1664-6; *Russ. J. Inorg. Chem. (Engl. Transl.)* 1976, *21*, 910-1.
VARIABLES:	PREPARED BY:
Composition at 323 K	Hiroshi Miyamoto

EXPERIMENTAL VALUES: Composition of saturated solutions at 50°C

LiIO₃		LiNO₃		Nature of
mass %	mol % (compiler)	mass %	mol % (compiler)	the solid phase[a]
43.25[b]	7.020	0.00	0.00	A
33.86	5.168	6.2	2.50	"
24.05	3.436	12.17	4.587	"
16.99	2.345	17.49	6.369	"
10.12	1.403	26.31	9.624	"
6.88	0.995	34.24	13.06	"
4.02	0.654	48.06	20.63	"
1.36	0.250	60.78	29.48	"
1.25	0.240	63.87	32.29	"
1.46	0.275	62.38	30.99	A+B
1.49	0.284	63.17	31.75	"
1.40	0.263	62.18	30.77	"
1.38	0.258	61.98	30.57	"
1.36	0.257	62.77	31.30	"
1.32	0.248	62.37	30.91	"
1.39	0.262	62.51	31.07	"
0.30	0.056	63.57	31.48	B
0.00	0.000	64.41	32.11	"

[a] A = $LiIO_3$;   B = $LiNO_3$

[b] For the binary system the compiler computes the following:

soly of $LiIO_3$ = 4.191 mol kg^{-1}

AUXILIARY INFORMATION

METHOD/APPARATUS/PROCEDURE:

The method used was similar to that described in (1) (presumably an isothermal method:
compiler). Equilibrium was reached in 14
days. The liquid and solid phases were
analyzed for Li^+ by periodate method, and
IO_3^- by titration with sulfuric acid and KI.
Composition and nature of the solid phases
determined by Schreinemakers' method of
residues, X-ray diffraction, thermography,
and IR spectroscopy.

SOURCE AND PURITY OF MATERIALS:

The compiler assumes that lithium iodate was
prepd from lithium carbonate and iodic acid
as described in (1). The source of $LiNO_3$
was not given.

REFERENCES:
1. Karataeva, I.M.; Vinogradov, E.E.
   Zh. Neorg. Khim. 1974, *19*, 3156.

ESTIMATED ERROR:

Nothing specified.

COMMENTS AND/OR ADDITIONAL DATA:

The phase diagram is given below (based on
mass % units).

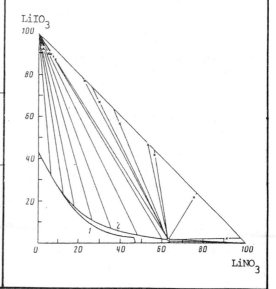

COMPONENTS:	ORIGINAL MEASUREMENTS:
(1) Lithium phosphate; Li_3PO_4; [10377-52-3] (2) Lithium iodate; $LiIO_3$; [13765-03-2] (3) Water; H_2O; [7732-18-5]	Shklovskaya, R.M.; Arkhipov, S.M.; Kidyarov, B.I.; Vdovkina, T.E. *Zh. Neorg. Khim.* 1982, 27, 2985-6; *Russ. J. Inorg. Chem. (Engl. Transl.)* 1982, 27, 1692-3.

VARIABLES:	PREPARED BY:
Composition at 298.2 K	Hiroshi Miyamoto

EXPERIMENTAL VALUES: Composition of saturated solutions

$LiIO_3$		Li_3PO_3		Nature of the solid phase[a]
mass %	mol % (compiler)	mass %	mol % (compiler)	
43.82[b]	7.173	–	–	A
43.67	7.133	0.0024	0.00062	A+B
42.01	6.697	0.0036	0.00090	B
38.96	5.948	0.0038	0.00091	"
35.14	5.094	0.0040	0.00091	"
32.23	4.500	0.0042	0.00092	"
29.53	3.986	0.0045	0.00095	"
27.50	3.622	0.0049	0.0010	"
24.33	3.087	0.0061	0.0012	"
21.02	2.569	0.0067	0.0013	"
16.22	1.882	0.0070	0.0013	"
13.05	1.465	0.0073	0.0013	"
10.08	1.098	0.0081	0.0014	"
5.20	0.541	0.0085	0.0014	"
–	–	0.036	0.056	"

[a] A = $LiIO_3$;   B = Li_3PO_4

[b] For the binary system the compiler computes the following:

soly of $LiIO_3$ = 4.289 mol kg^{-1}

AUXILIARY INFORMATION

METHOD/APPARATUS/PROCEDURE:	COMMENTS AND/OR ADDITIONAL DATA:
Isothermal method used. Equilibrium was established after 15-20 days. The iodate content was determined iodometrically. The phosphate concn in solution was determined colorimetrically as the vanadomolybdo-phosphate complex, and in the residues by titration after dissolving the solid in acid. The composition of the solid phase was determined by the method of residues and checked by X-ray diffraction.	The phase diagram is given below (based on mass units).

SOURCE AND PURITY OF MATERIALS:

"Pure grade" lithium iodate and chemically pure grade lithium phosphate were used.

ESTIMATED ERROR:

Soly: nothing specified.
Temp: precision ± 0.1 K.

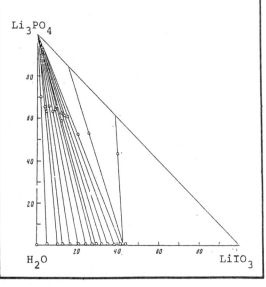

COMPONENTS:	ORIGINAL MEASUREMENTS:
(1) Lithium iodate; $LiIO_3$; [13705-03-2] (2) Silver iodate; $AgIO_3$; [7783-97-3] (3) Water; H_2O; [7732-18-5]	Shklovskaya, R.M.; Arkhipov, S.M.; Kidyarov, B.I.; Vdovkina, T.E.; Poleva, G.V. *Zh. Neorg. Khim.* <u>1983</u>, *28*, 2431-3; *Russ. J. Inorg. Chem. (Engl. Transl.)* <u>1983</u>, *28*, 1382-3.
VARIABLES: Composition at 298.2 K	PREPARED BY: Hiroshi Miyamoto

EXPERIMENTAL VALUES: Composition of saturated solutions at 25.0°C

$LiIO_3$		$AgIO_3$		Nature of
mass %	mol % (compiler)	mass %	mol % (compiler)	the solid phase[a]
–	–	0.0052	3.3×10^{-4}	A
7.48	0.795	0.00012	8.2×10^{-6}	SS
13.62	1.538	0.00067	4.9×10^{-5}	"
19.23	2.304	0.00062	4.8×10^{-5}	"
22.91	2.860	0.00059	4.7×10^{-5}	"
26.61	3.468	0.00056	4.7×10^{-5}	"
30.05	4.082	0.00052	4.5×10^{-5}	"
34.62	4.984	0.00050	4.6×10^{-5}	"
38.78	5.905	0.00048	4.7×10^{-5}	"
43.35	7.047	0.00047	4.9×10^{-5}	SS+B
43.35	7.047	0.00047	4.9×10^{-5}	"
43.82[b]	7.173	–	–	B

[a] A = $AgIO_3$ B = $\alpha\text{-}LiIO_3$;   SS = solid solution.

[b] For binary systems the compiler computes the following:

soly of $LiIO_3$ = 4.289 mol kg^{-1}

soly of $AgIO_3$ = 1.8×10^{-4} mol kg^{-1}

AUXILIARY INFORMATION

METHOD/APPARATUS/PROCEDURE:	REFERENCES:
Isothermal method used. Equilibrium was established after 25-30 days. The iodate content in samples of coexisting phases was determined iodometrically, and the silver content in liquid phase was determined by atomic absorption. The composition of solid phase was determined by the method of residues, and the result was checked by X-ray analysis.	1. Rene, M.; Claude, G.J. *Solid State Chem.* <u>1980</u>, *32*, 177.
	COMMENT AND/OR ADDITIONAL DATA:

1. Rene, M.; Claude, G.J. *Solid State Chem.* <u>1980</u>, *32*, 177.

SOURCE AND PURITY OF MATERIALS:

"Special purity" grade $\alpha\text{-}LiIO_3$ was used. Silver iodate was prepared by mixing aqueous silver nitrate and sodium iodate solutions. The product contained Ag 37.98 mass % and IO_3 61.94 mass %, and the ratio IO_3/Ag = 1.01. The X-ray diffraction pattern of the product was consistent with that in literature (ref 1).

ESTIMATED ERROR:

Soly: 1-3 rel. %.
Temp: precision ± 0.1 K.

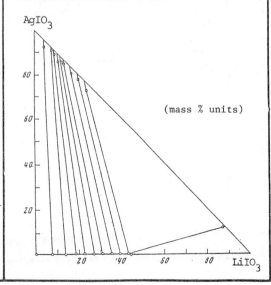

(mass % units)

COMPONENTS:	ORIGINAL MEASUREMENTS:
(1) Lithium iodate; $LiIO_3$; [13765-03-2] (2) Aluminum iodate; $Al(IO_3)_3$; [15123-75-8] (3) Water; H_2O; [7732-18-5]	Shklovskaya, R.M.; Arkhipov, S.M.; Kidyarov, B.I.; Kuzina, V.A; Poeva, G.V. *Zh. Neorg. Khim.* 1976, *21*, 3116-9; *Russ. J. Inorg. Chem.* (*Engl. Transl.* 1976, *21*, 1718-20.
VARIABLES: Composition at 298.2 K	PREPARED BY: Hiroshi Miyamoto

EXPERIMENTAL VALUES: Composition of saturated solutions at 25.0°C

LiIO₃ mass %	mol % (compiler)	Al(IO₃)₃ mass %	mol % (compiler)	Nature of the solid phase[a]
0.00	0.000	5.70[b]	0.197	A
4.91	0.526	3.16	0.112	"
8.18	0.893	1.96	0.0705	"
11.96	1.355	1.85	0.0691	"
15.71	1.841	1.37	0.0529	"
20.20	2.484	1.29	0.0523	"
24.73	3.203	1.26	0.0538	"
28.50	3.874	1.49	0.0668	"
30.48	4.249	1.53	0.0703	"
33.73	4.912	1.64	0.0787	"
39.00	6.150	2.11	0.110	"
40.96	6.678	2.41	0.130	A+B
40.96	6.678	2.41	0.130	"
42.18	6.840	0.94	0.0502	B
43.82[b]	7.173	0.00	0.0000	"

[a] A = $Al(IO_3)_2.6H_2O$;   B = $LiIO_3$

[b] For binary systems the compiler computes the following:

soly of $LiIO_3$ = 4.289 mol kg^{-1}

soly of $Al(IO_3)_3$ = 0.110 mol kg^{-1}

continued.....

AUXILIARY INFORMATION

METHOD/APPARATUS/PROCEDURE:	SOURCE AND PURITY OF MATERIALS:
The isothermal method used. Equilibrium was reached in 15-20 days. Aluminum content was determined by complexometric titration, lithium by flame photometry. The composition of the solid phases was determined by the method of residues, and checked by X-ray diffraction. The X-ray diffraction patterns were recorded on a URS-50-I diffractometer with Cu radiation. The IR spectra and thermogram were also recorded.	Aluminum iodate was prepared at 80-90°C by neutralization of a saturated solution of iodic acid with an equivalent amount of freshly precipitated aluminum hydroxide. The solution was cooled to room temperature, and the solid dried and analyzed. Found, mass %: Al 4.03; IO_3 78.7; H_2O 17.6. Calcd. For $Al(IO_3)_3.6H_2O$, mass %: Al 4.09; IO_3 79.53; H_2O 16.38 (by difference). "Very pure" grade $LiIO_3$ was used.
	ESTIMATED ERROR: Soly: error in flame photometry analysis did not exceed 1-3 rel %. Temp: precision $\pm$ 0.1 K.
	REFERENCES:

COMPONENTS:	ORIGINAL MEASUREMENTS:
(1) Lithium iodate; $LiIO_3$; [13765-03-2] (2) Aluminum iodate; $Al(IO_3)_3$; [15123-75-8] (3) Water; H_2O; [7732-18-5]	Shklovshaya, R.M.; Arkhipov, S.M. Kidyarov, B.I.; Kuzina, V.A.; Poleva, G.V. *Zh. Neorg. Khim.* 1976, *21*, 3116-9; *Russ. J. Inorg. Chem. (Engl. Transl.)* 1976, *21*, 1718-20.

COMMENTS AND/OR ADDITIONAL DATA:

The phase diagram is given below (based on mass % units).

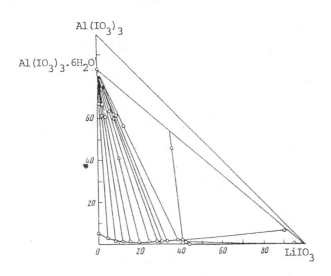

AUXILIARY INFORMATION

METHOD/APPARATUS/PROCEDURE:	SOURCE AND PURITY OF MATERIALS:
	ESTIMATED ERROR:
	REFERENCES:

COMPONENTS:	ORIGINAL MEASUREMENTS:
(1) Lithium iodate; $LiIO_3$; [13765-03-2]	Shklovskaya, R.M.; Arkhipov, S.M.; Kidyarov, B.I.; Tokareva, A.C.; Kuzina, V.A.
(2) Gallium iodate; $Ga(IO_3)_3$; [70504-12-0]	
(3) Water; H_2O; [7732-18-5]	*Zh. Neorg. Khim.* 1980, 25, 1112-6; *Russ. J. Inorg. Chem.* (Engl. Transl.) 1980, 25, 618-9.

VARIABLES:	PREPARED BY:
Composition at 298.2 K	Hiroshi Miyamoto

EXPERIMENTAL VALUES: Composition of saturated solutions

LiIO_3		Ga(IO_3)_3		
mass %	mol % (compiler)	mass %	mol % (compiler)	Nature of the solid phase[a]
		0.85[b]	0.026	A
1.12	0.112	0.24	0.0074	B
2.49	0.253	0.13	0.0040	"
4.8	0.50	0.061	0.0019	"
6.57	0.692	0.040	0.0013	"
7.97	0.851	0.035	0.0011	"
10.99	1.209	0.032	0.0011	"
15.47	1.781	0.028	0.00099	"
17.35	2.038	0.030	0.0011	B+C
17.49	2.058	0.028	0.0010	C
19.78	2.385	0.029	0.0011	"
22.88	2.856	0.032	0.0012	"
25.16	3.225	0.035	0.0014	"
28.51	3.803	0.037	0.0015	"
32.98	4.651	0.041	0.0018	"
34.80	5.025	0.045	0.0020	"
36.37	5.363	0.052	0.0023	"
37.45	5.603	0.049	0.0022	C+D
37.61	5.640	0.051	0.0023	D
39.47	6.073	0.056	0.0026	"
42.57	6.848	0.064	0.0031	"

continued .....

AUXILIARY INFORMATION

METHOD/APPARATUS/PROCEDURE:

The investigation was carried out by the isothermal method. Equilibrium was reached in 25-30 days. Samples of the coexisting phases were analyzed for lithium by flame photometry. The photometry was carried out on solutions in which the lithium concentration did not exceed 10 μg ml^{-1}. The gallium content of liquid phases was determined by atomic absorption. Solutions for which the lithium and iodate ion concentrations range from 20 to 100 μg ml^{-1} do not influence the absorption of gallium. Analysis of these solutions was carried out by the restricted standards method with standard solutions based on gallium iodate. The solid phases were identified by the method of residues and checked by X-ray diffraction.

SOURCE AND PURITY OF MATERIALS:

"Special purity" grade lithium iodate was used.
Gallium iodate was made by the reaction of gallium nitrate with iodic acid.

ESTIMATED ERROR:

Soly: rel. error 1-3 % (flame photometry) and 3-5 % (atomic absorption measurement).
Temp: precision ± 0.1 K.

REFERENCES:

COMPONENTS:	EVALUATOR:
(1) Lithium iodate; $LiIO_3$; [13765-03-2] (2) Gallium iodate; $Ga(IO_3)_3$; [70504-12-0] (3) Water; H_2O; [7732-18-5]	Shklovskaya, R.M.; Arkhipov, S.M.; Kidyarov, B.I.; Tokareva, A.G.; Kuzina, V.A. *Zh. Neorg. Khim.* <u>1980</u>, *25*, 1112-6; *Russ. J. Inorg. Chem.* (*Engl. Transl.*) <u>1980</u>, *25*, 618-9.

EXPERIMENTAL VALUES: (Continued)

Composition of saturated solutions

LiIO₃		Ga(IO₃)₃		Nature of
mass %	mol % (compiler)	mass %	mol % (compiler)	the solid phase[a]
42.87	6.927	0.066	0.0033	D+E
43.82[b]	7.173	–	–	E

[a] Solid phase compositions are:

   $A = Ga(IO_3)_3 \cdot 2H_2O$; B = solid solution based on $Ga(IO_3)_3 \cdot 2H_2O$

   $C = 2Ga(IO_3)_3 \cdot LiIO_3 \cdot 4H_2O$; $D = Ga(IO_3)_3 \cdot LiIO_3 \cdot 2H_2O$; $E = \alpha\text{-}LiIO_3$

[b] For binary systems the compiler computes the following:

   soly of $LiIO_3 = 4.289$ mol kg^{-1}

   soly of $Ga(IO_3)_3 = 0.014$ mol kg^{-1}

COMMENTS AND/OR ADDITIONAL DATA:

The phase diagram is given below.

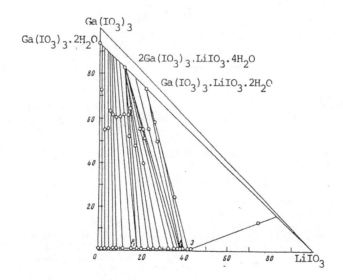

COMPONENTS:	ORIGINAL MEASUREMENTS:
(1) Lithium iodate; LiIO$_3$; [13765-03-2] (2) Indium iodate; In(IO$_3$)$_3$; [65597-32-2] (3) Water; H$_2$O; [7732-18-5]	Shklovskaya, R.M.; Arkhipov, S.M.; Kidyarov, B.I.; Poleva, G.A.; Kuzina, V.A. *Zh. Neorg. Khim.* 1981, *26*, 791-4; *Russ. J. Inorg. Chem. (Engl. Transl.)* 1981, *26*, 425-7.

VARIABLES:	PREPARED BY:
Composition at 298.2 K	Hiroshi Miyamoto

EXPERIMENTAL VALUES: Composition of saturated solutions at 25.0°C

LiIO$_3$		In(IO$_3$)$_3$		Nature of
mass %	mol % (compiler)	mass %	mol % (compiler)	the solid phase[a]
–	–	0.049[b]	0.0014	A
3.51	0.359	0.0080	0.00023	"
5.76	0.602	0.0075	0.00022	"
7.26	0.770	0.0070	0.00021	"
9.75	1.059	0.0065	0.00020	"
11.61	1.285	0.0060	0.00019	"
12.3	1.371	0.0070	0.00022	A+B
12.3	1.371	0.0070	0.00022	"
13.90	1.574	0.0074	0.00024	B
15.36	1.766	0.0087	0.00028	"
17.30	2.031	0.0084	0.00028	"
18.30	2.171	0.009	0.00030	B+C
18.30	2.171	0.009	0.00030	"
19.31	2.316	0.0083	0.00028	C
21.46	2.636	0.0068	0.00024	"
24.36	3.092	0.0038	0.00014	"
26.39	3.430	0.0033	0.00012	"
29.38	3.959	0.0021	0.000080	"
33.67	4.788	0.0020	0.000081	"
36.50	5.388	0.0019	0.000080	"
38.08	5.743	0.0018	0.000077	"
41.28	6.511	0.0017	0.000076	"

continued.....

AUXILIARY INFORMATION

METHOD/APPARATUS/PROCEDURE:	SOURCE AND PURITY OF MATERIALS:
Isothermal method. Equilibrium was estab- lished in 25-30 days. Samples of satd sln and solids were analyzed for lithium by flame emission spectroscopy. The lithium content was detd by comparing the test solution with solutions containing only lithium. Indium in the liquid phase was detd by flame absorption spectrometry. The solid phases were identified by the method of "residues" and checked by X-ray diffraction. The thermographic investiga- tion was carried out on an OD-102 derivato- graph.	"Special purity" grade lithium iodate was used, and indium iodate was made by reaction of indium nitrate with iodic acid.

	ESTIMATED ERROR:
	Soly: precision in lithium analysis $\leq$ 3 %. Temp: precision $\pm$ 0.1 K.

	REFERENCES:

COMPONENTS:	EVALUATOR:
(1) Lithium iodate; $LiIO_3$; [13765-03-2] (2) Indium iodate; $In(IO_3)_3$; [65597-32-2] (3) Water; H_2O; [7732-18-5]	Shklovskaya, R.M.; Arkhipov, S.M.; Kidyarov, B.I.; Poleva, G.A.; Kuzina, V.A. *Zh. Neorg. Khim.* 1981, 26, 791-4; *Russ. J. Inorg. Chem. (Engl. Transl.)* 1981, 26, 425-7.

EXPERIMENTAL VALUES (Continued)

Composition of saturated solutions at 25.0°C

	$LiIO_3$		$In(IO_3)_3$		Nature of
mass %	mol % (compiler)	mass %	mol % (compiler)		the solid phase[a]
42.42	6.802	0.0028	0.00013		C+D
42.42	6.802	0.0028	0.00013		"
43.82[b]	7.173	–	–		D

[a] A = $In(IO_3)_3 \cdot H_2O$;   B = $LiIO_3 \cdot In(IO_3)_3 \cdot H_2O$ C = $2LiIO_3 \cdot In(IO_3)_3 \cdot H_2O$;

  D = $LiIO_3$

[b] For binary systems the compiler computes the following:

     soly of $LiIO_3$ = 4.289 mol kg^{-1}

     soly of $In(IO_3)_3$ = 7.7 x 10^{-4} mol kg

COMMENTS AND/OR ADDITIONAL DATA:

The phase diagram is given below

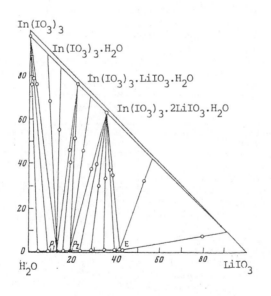

COMPONENTS:	ORIGINAL MEASUREMENTS:
(1) Lithium iodate; $LiIO_3$; [13765-03-2] (2) Thallium iodate; $TlIO_3$; [14767-09-0] (3) Water; H_2O; [7732-18-5]	Arkhipov, S.M.; Kashina, N.I.; Kidyarov, B.I.; Kuzina, V.A. *Zh. Neorg. Khim.* 1981, 26, 1447-9; *Russ. J. Inorg. Chem.* (*Engl. Transl.*) 1981, 26, 779-80.
VARIABLES: Composition at 298.2 K	PREPARED BY: Hiroshi Miyamoto

EXPERIMENTAL VALUES: Composition of saturated solutions at 25.0°C

$TlIO_3$		$LiIO_3$		Nature of
mass %	mol % (compiler)	mass %	mol % (compiler)	the solid phase[a]
0.066[b]	0.0031	–	–	A
0.0030	0.00015	4.85	0.502	"
0.0019	0.000099	9.44	1.02	"
0.0025	0.00014	15.50	1.785	"
0.0029	0.00016	16.80	1.961	"
0.0040	0.00024	22.84	2.849	"
0.0058	0.00036	25.15	3.222	"
0.0070	0.00046	31.36	4.331	"
0.0072	0.00050	34.40	4.939	"
0.0075	0.00055	39.55	6.088	"
0.0094	0.00071	40.82	6.397	A+C
0.0092	0.00069	41.13	6.474	C
0.0090	0.00069	42.12	6.726	"
0.0091	0.00071	43.40	7.061	C+B
0.0063	0.00049	43.66	7.131	B
–	–	43.79[b]	7.165	"

[a] A = $TlIO_3$;   B = $LiIO_3$;   C = $LiIO_3 \cdot TlIO_3$

[b] For binary systems the compiler computes the following:

soly of $LiIO_3$ = 4.284 mol kg^{-1}

soly of $TlIO_3$ = 1.7 x 10^{-3} mol kg^{-1}

AUXILIARY INFORMATION

METHOD/APPARATUS/PROCEDURE:	SOURCE AND PURITY OF MATERIALS:
Isothermal method used. Equilibrium was reached in 20 days. The iodate in samples of the liquid and solid phases with low $TlIO_3$ concentrations was detd by iodometric titration, and the thallium by flame emission photometry. The lithium concentration was obtained by difference. In samples of solid phases at higher $TlIO_3$ concentrations, the thallium was detd gravimetrically as the chromate, and lithium by flame emission photometry. The compositions of the solid phases were detd by the method of residues and X-ray diffraction.	"Special purity" lithium iodate was used. Thallium iodate was made from thallium nitrate and lithium iodate.

SOURCE AND PURITY OF MATERIALS section continues with:

COMMENTS AND/OR ADDITIONAL DATA:

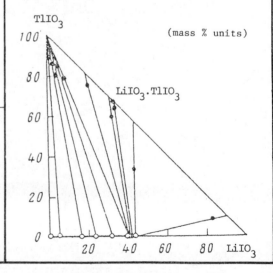

ESTIMATED ERROR:

Soly: 0.3 rel % (samples of higher $TlIO_3$) and nothing specified (samples of lower $TlIO_3$).
Temp: precision ± 0.1 K.

COMPONENTS:	ORIGINAL MEASUREMENTS:
(1) Lithium iodate; $LiIO_3$; [13765-03-2] (2) Titanium iodate; $Ti(IO_3)_4$; [73621-77-9] (3) Water; H_2O; [7732-18-5]	Shklovskaya, R.M.; Arkhipov, S.M.; Kidyarov, B.I.; Kuzina, V.A.; Vdovkina, T.E. Zh. Neorg. Khim. 1982, 27, 513-6; Russ. J. Inorg. Chem. (Engl. Transl.) 1982, 27, 292-4.
VARIABLES: Composition at 298.2 K	PREPARED BY: Hiroshi Miyamoto

EXPERIMENTAL VALUES:

Composition of saturated solutions at 25.0°C

$LiIO_3$		$Ti(IO_3)_3$		Nature of the solid phase[a]
mass %	mol % (compiler)	mass %	mol % (compiler)	
–	–	0.047[b]	0.00011	A
2.47	0.250	0.061	0.0015	B
4.95	0.514	0.22	0.0056	"
9.79	1.07	0.43	0.011	"
11.51	1.278	0.45	0.012	"
13.62	1.547	0.53	0.015	"
17.30	2.044	0.59	0.017	"
21.83	2.712	0.61	0.018	"
26.40	3.454	0.51	0.016	"
29.52	4.009	0.47	0.016	"
31.88	4.458	0.44	0.015	"
34.30	4.946	0.41	0.014	"
39.46	6.107	0.45	0.017	"
42.50	6.870	0.43	0.017	"
42.92	6.975	0.38	0.015	C
42.92	6.975	0.38	0.015	"
43.82[b]	7.173	–	–	D

[a] A = $Ti(IO_3)_4 \cdot 2H_2O$;   B = Solid solution based on $Ti(IO_3)_4 \cdot 2H_2O$;

C = Solid solution + $LiIO_3$;   D = $LiIO_3$

continued...

AUXILIARY INFORMATION

METHOD/APPARATUS/PROCEDURE:	SOURCE AND PURITY OF MATERIALS:
Isothermal method used. Equilibrium was reached in 25-30 days. The iodate concentration in the coexisting phases was de-determined iodometrically. The liquid phases were analyzed for tianium colorimetrically with chromotropic acid after preliminary reduction of the iodate ion with hydroxylammonium sulfate in an acidic medium followed by removal of iodine by evaporation of the solution. The lithium content was determined by difference, and also checked by flame emission spectrometry in the solutions after removal of titanium. The composition of the solid phases were determined by the method of Schreinemakers' resisues and checked by X-ray diffraction.	$Ti(IO_3)_4 \cdot 2H_2O$ prepared by mixing freshly precipitated titanium hydroxide and a stoichiometric amount of 75 % iodic acid solution at 50 to 60°C. Purity of the product was reported as: found: Ti 6.12 %; IO_3 89.63 %; IO_3:Ti = 4.01. calcd for $Ti(IO_3)_4 \cdot 2H_2O$: Ti 6.11 %; IO_3 89.29 %. Special purity grade lithium iodate was used.
	ESTIMATED ERROR: Soly: nothing specified. Temp: precision $\pm$ 0.1 K.
	REFERENCES:

COMPONENTS:	ORIGINAL MEASUREMENTS:
(1) Lithium iodate; $LiIO_3$; [13765-03-2] (2) Titanium iodate; $Ti(IO_3)_4$; [73621-77-9] (3) Water; H_2O; [7732-18-5]	Shklovskaya, R.M.; Arkhipov, S.M.; Kidyarov, B.I.; Kuzina, V.A.; Vdovkina, T.E. *Zh. Neorg. Khim.* 1982, *27*, 513-6; *Russ. J. Inorg. Chem. (Engl. Transl.)* 1982, *27*, 292-4.

EXPERIMENTAL VALUES: (Continued)

[b] For binary systems the compiler computes the following:

   soly of $LiIO_3$ = 4.289 mol kg^{-1}

   soly of $Ti(IO_3)_4$ = 6.3 x 10^{-5} mol kg^{-1}

COMMENTS AND/OR ADDITIONAL DATA:

   The phase diagram is given below (based on mass % units).

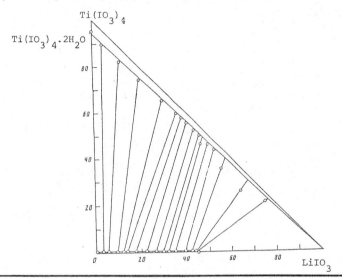

AUXILIARY INFORMATION

METHOD/APPARATUS/PROCEDURE:	SOURCE AND PURITY OF MATERIALS:
	ESTIMATED ERROR:
	REFERENCES:

COMPONENTS:	ORIGINAL MEASUREMENTS:
(1) Lithium iodate; LiIO$_3$; [13765-03-2]	Shklovskaya, R.M.; Arkhipov, S.M.; Kidyarov, B.I.; Poleva, G.V.; Troitskaya, N.I.
(2) Zirconium iodate; Zr(IO$_3$)$_4$; [22446-84-0]	
(3) Water; H$_2$O; [7732-18-5]	Zh. Neorg. Khim. 1982, 27, 257-8; Russ. J. Inorg. Chem. (Engl. Transl.) 1982, 27, 145-6.

VARIABLES:	PREPARED BY:
Composition at 298.2K	Hiroshi Miyamoto

EXPERIMENTAL VALUES: Composition of saturated solutions at 25.0°C

Lithium Iodate		Zirconium Iodate		Nature of
mass %	mol % (compiler)	mass %	mol % (compiler)	the solid phase[a]
-	-	0.0002[b]	0.05 x 10^{-4}	A
4.00	0.411	0.0026	0.61 x 10^{-4}	"
9.28	1.00	0.0060	1.5 x 10^{-4}	"
13.51	1.524	0.0087	2.3 x 10^{-4}	"
17.52	2.061	0.0092	2.5 x 10^{-4}	"
22.19	2.748	0.012	3.4 x 10^{-4}	"
28.06	3.721	0.017	5.2 x 10^{-4}	"
32.92	4.638	0.024	7.8 x 10^{-4}	"
38.16	5.766	0.058	2.0 x 10^{-3}	"
40.89	6.422	0.081	2.9 x 10^{-3}	"
43.54	7.108	0.092	3.5 x 10^{-3}	A+B
43.82[b]	7.173	-	-	B

[a] A = Zr(IO$_3$)$_4$;   B = LiIO$_3$

[b] For binary systems the compiler computes the following:

$$\text{soly of LiIO}_3 = 4.289 \text{ mol kg}^{-1}$$

$$\text{soly of Zr(IO}_3)_4 = 3 \times 10^{-6} \text{ mol kg}^{-1}$$

AUXILIARY INFORMATION

METHOD/APPARATUS/PROCEDURE:	SOURCE AND PURITY OF MATERIALS:
The system was studied by the isothermal method. Equilibrium was established after 15-20 days. Samples of the coexisting phases were analyzed for lithium by flame emission spectrometry. Zirconium concentrations >0.001 % in the samples of the liquid phase were determined gravimetrically by precipitation with mandelic acid; otherwise zirconium was determined photometrically with sodium hyposulfate. The solid phases were identified by the method of residues, and checked by X-ray diffraction.	"Special purity" grade lithium iodate was used. Zirconium iodate was prepared as follows: freshly precipitated zirconium hydroxide was treated with 60-70 % iodic acid at room temperature, and the reaction mass diluted to an iodic acid concentration of 2-4 %. The product was heated to 60-80°C and dried. The purity of the product was given as follows: Found, mass %: Zr 11.50; IO$_3$ 89.1. Molar ratio IO$_3$:Zr = 4.01:1. Calcd for Zr(IO$_3$)$_4$, mass %: Zr 11.53; IO$_3$ 88.47. The X-ray diffraction pattern of the salt obtained corresponded to that for anhydrous zirconium iodate.
	ESTIMATED ERROR: Soly: rel. error in Li analysis 1-3 %. Temp: nothing specified.

COMPONENTS:	ORIGINAL MEASUREMENTS:
(1) Lithium iodate; $LiIO_3$; [13765-03-2] (2) Hafnium iodate; $Hf(IO_3)_4$; [19630-06-9] (3) Water; H_2O; [7732-18-5]	Shklovskaya, R.M.; Arkhipov, S.M.; Kidyarov, B.I.; Tokareva, A.G. *Zh. Neorg. Khim.* 1981, 26, 1701-2; *Russ. J. Inorg. Chem. (Engl. Transl.)* 1981, 26, 919-20.
VARIABLES: Composition at 298.2 K	PREPARED BY: Hiroshi Miyamoto

EXPERIMENTAL VALUES: Composition of saturated solutions

$LiIO_3$		$Hf(IO_3)_4$		Nature of
mass %	mol % (compiler)	mass %	mol % (compiler)	the solid phase[a]
–	–	0.00037[b]	0.0000076	A
5.05	0.524	0.00060	0.000013	C
9.69	1.05	0.0011	0.000025	"
12.05	1.339	0.0025	0.000058	"
15.96	1.847	0.0065	0.00016	"
19.20	2.300	0.012	0.00030	"
22.70	2.828	0.014	0.00036	"
24.82	3.168	0.016	0.00042	"
27.62	3.644	0.019	0.00052	"
30.71	4.207	0.021	0.00060	"
32.39	4.533	0.025	0.00072	"
35.52	5.177	0.031	0.00094	"
39.34	6.040	0.035	0.0011	"
41.90	6.672	0.040	0.0013	"
43.33	7.049	0.063	0.0021	D
43.82[b]	7.173	–	–	B

[a] Solid phases are: A = $Hf(IO_3)_4$;   B = $LiIO_3$;

C = solid solution based on $Hf(IO_3)_4$ D = solid solution based on $LiIO_3$
(assumed by the compiler)

[b] For binary systems the compiler computes the following:

soly of $LiIO_3$ = 4.289 mol kg^{-1};   soly of $Tl(IO_3)_4$ = 4.2 x 10^{-6} mol kg^{-1}

AUXILIARY INFORMATION

METHOD/APPARATUS/PROCEDURE:	SOURCE AND PURITY OF MATERIALS:
The solubility in this system was studied by the isothermal method. Equilibrium was reached in 25-30 days. Samples of the co-existing phases were analyzed for lithium by emmission spectrometry, and the concentration of lithium was determined by comparing the solution being analyzed with solutions containing only lithium. The hafnium content was determined photometrically using Arsenazo III and reducing the iodate ion with hydroxylamine. The solid phases were identified by the method of residues, and the results confirmed by X-ray diffraction.	Lithium iodate was "highly pure" grade, and hafnium iodate obtained by the action of aqueous iodic acid solution on freshly precipitated hafnium hydroxide. The salt obtained contained 20.31 mass % Hf and 79.45 mass % IO_3; mol ratio IO_3/Hf = 3.99. The product agrees with the data (1) for anhydrous hafnium iodate.
	ESTIMATED ERROR: Soly: within 1-3 rel % (flame photometry of Li). Temp: precision $\pm$ 0.1 K.
	REFERENCES: 1. Deabridges, J.; Rohmer, R. *Bull. Soc. Chim. Fr.* 1968, 2, 521.

COMPONENTS:	ORIGINAL MEASUREMENTS:
(1) Lithium iodate; $LiIO_3$; [13765-03-2]	Shklovskaya, R.M.; Arkhipov, S.M.; Kidyarov, B.I.; Tokareva, A.G.
(2) Hafnium iodate; $Hf(IO_3)_4$; [19630-06-9]	*Zh. Neorg. Khim.* 1981, 26, 1701-2; *Russ. J. Inorg. Chem. (Engl. Transl.)* 1981, 26, 919-20.
(3) Water; H_2O; [7732-18-5]	

COMMENTS AND/OR ADDITIONAL DATA:

The phase diagram is given below (mass % units).

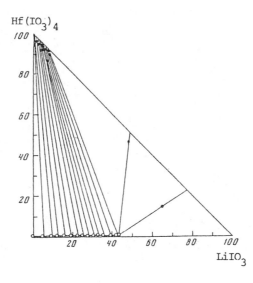

COMPONENTS:	ORIGINAL MEASUREMENTS:
(1) Lithium iodate; $LiIO_3$; [13765-03-2] (2) Lithium (I-4)-tetraoxorhenate(1-) Lithium perrhenate); $LiReO_4$; [13768-48-4] (3) Water; H_2O; [7732-18-5]	Shklovskaya, R.M.; Arkhipov, S.M.; Kidyarov, B.I. *Zh. Neorg. Khim.* <u>1979</u>, *24*, 2287-8; *Russ. J. Inorg. Chem. (Engl. Transl.)* <u>1979</u>, *24*, 1269-70.
VARIABLES: Composition at 298.2 K	PREPARED BY: Hiroshi Miyamoto

EXPERIMENTAL VALUES: Composition of saturated solutions at 25.0°C

$LiIO_3$		$LiReO_4$		Nature of
mass %	mol % (compiler)	mass %	mol % (compiler)	the solid phase[a]
43.82[b]	7.173	–	–	A
35.98	6.009	8.89	1.050	"
26.20	4.633	21.90	2.738	"
21.50	3.907	28.08	3.608	"
14.53	2.956	41.10	5.914	"
9.51	2.114	50.40	7.923	"
4.83	1.245	61.51	11.21	"
2.08	0.631	70.41	15.11	"
1.71	0.528	71.36	15.57	A+B
–	–	74.25	16.81	B

[a] $A = \alpha\text{-}LiIO_3$; $B = LiReO_4 \cdot H_2O$.

[b] For the binary system the compiler computes the following:

$$\text{soly of } LiIO_3 = 4.289 \text{ mol kg}^{-1}$$

AUXILIARY INFORMATION

METHOD/APPARATUS/PROCEDURE:
The ternary system $LiIO_3$-$LiReO_4$-H_2O was investigated by the isothermal method. Equilibrium was reached in 10-15 days.
Lithium in the liquid phases was determined
by ion exchange, and iodate content was determined by iodometric titration in the
presence of phthalate buffer at pH 5. The
perrhenate concentration was found by difference. Solid phase compositions determined by the method of residues and checked
by X-ray diffraction.

SOURCE AND PURITY OF MATERIALS:
Special purity grade $LiIO_3$ was used. Lithium perrhenate was made from lithium hydroxide and perrhenate obtained by ion exchange
from ammonium perrhenate.

ESTIMATED ERROR:
Soly: nothing specified.
Temp: precision ± 0.1 K.

COMMENTS AND/OR ADDITIONAL DATA:
The phase diagram is given below (based
on mass % units).

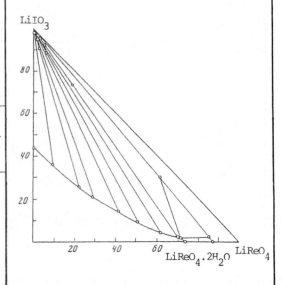

COMPONENTS:	ORIGINAL MEASUREMENTS:
(1) Lithium iodate; $LiIO_3$; [13765-03-2]	Vinogradov, E.E.; Tarasova, G.N.
(2) Neodymium iodate; $Nd(IO_3)_3$; [14732-16-2]	Zh. Neorg. Khim. 1982, 27, 269-70; Russ. J. Inorg. Chem. (Engl. Transl.) 1982, 27, 153-4.
(3) Water; H_2O; [7732-18-5]	

VARIABLES:	PREPARED BY:
Composition at 298.2 K	Hiroshi Miyamoto

EXPERIMENTAL VALUES: Composition of saturated solutions at 25.0°C

$Nd(IO_3)_3$ mass %	mol % (compiler)	$LiIO_3$ mass %	mol % (compiler)	Nature of the solid phase[b]
0.15[b]	4×10^{-3}	-	-	A
0.02	5×10^{-4}	1.02	0.102	"
0.01	3×10^{-4}	2.80	0.285	"
<0.01	$<3 \times 10^{-4}$	6.75	0.712	"
<0.01	"	14.97	1.714	"
<0.01	"	18.96	2.266	"
<0.01	"	23.11	2.892	"
<0.01	$<4 \times 10^{-4}$	29.13	3.913	A
<0.01	"	35.30	5.129	"
<0.01	"	34.99	5.063	A+B
<0.01	"	35.18	5.103	"
-	-	43.30[b]	7.034	B

[a] $A = Nd(IO_3)_3 \cdot 2H_2O$; $B = LiIO_3$

[b] For binary systems the compiler computes the following:

$$\text{soly of } LiIO_3 = 4.200 \text{ mol kg}^{-1}$$
$$\text{soly of } Nd(IO_3)_4 = 2.2 \times 10^{-3} \text{ mol kg}^{-1}$$

AUXILIARY INFORMATION

METHOD/APPARATUS/PROCEDURE:

The compiler assumes that the isothermal method was used. The experiments were carried out in a water thermostat, and equilibrium was established after 30-55 days.

The iodate concentration was determined by titration with sodium thiosulfate in the presence of sulfuric acid and potassium iodide, the neodymium ion by complexometric titration in the presence of hexamethylene-tetramine and Methylthymol blue, and lithium by flame photometry.

The composition of the solid phase was determined by Schreinemakers' method of residues, and thermogravimetry.

ESTIMATED ERROR:

Soly: nothing specified.

Temp: precision ± 0.1 K.

SOURCE AND PURITY OF MATERIALS:

Lithium iodate synthesized from iodic acid and lithium carbonate. Neodymium iodate was made from neodymium oxide and iodic acid. The purity of the product was checked chemically.

COMMENTS AND/OR ADDITIONAL DATA:

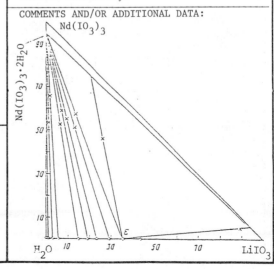

COMPONENTS:	ORIGINAL MEASUREMENTS:
(1) Lithium iodate; $LiIO_3$; [13765-03-2] (2) Samarium iodate; $Sm(IO_3)_3$; [134732-17-3] (3) Water; H_2O; [7732-18-5]	Shklovskaya, R.M.; Arkhipov, S.M.; Kidyarov, B.I.; Zherdienko, L.P. *Zh. Neorg. Khim.* 1977, *22*, 1139–41; *Russ. J. Inorg. Chem. (Engl. Transl.)* 1977, *22*, 624–5.

VARIABLES:	PREPARED BY:
Composition at 298 K	Hiroshi Miyamoto

EXPERIMENTAL VALUES: Composition of saturated solutions at 25°C

$Sm(IO_3)_3$		$LiIO_3$		Nature of
mass %	mol % (compiler)	mass %	mol % (compiler)	the solid
0.023[b]	0.00061	0.00	0.0000	A
0.020	0.00057	6.97	0.737	"
0.017	0.00051	13.10	1.472	"
0.015	0.00048	17.50	2.059	"
0.013	0.00044	23.42	2.941	"
0.012	0.00043	28.52	3.803	"
0.012	0.00046	33.10	4.673	"
0.013	0.00053	38.23	5.778	"
0.011	0.00047	41.56	6.583	A+B
0.011	0.00047	41.56	6.583	"
0.000	0.00000	43.82[b]	7.173	B

[a] $A = Sm(IO_3)_3 \cdot 2H_2O$; $B = LiIO_3$

[b] For binary systems the compiler computes the following:

soly of $LiIO_3$ = 4.289 mol kg^{-1}

soly of $Sm(IO_3)_3$ = 3.4 x 10^{-4} mol kg^{-1}.

AUXILIARY INFORMATION

METHOD/APPARATUS/PROCEDURE:	COMMENTS AND/OR ADDITIONAL DATA:
The ternary system, $LiIO_3$–$Sm(IO_3)_3$–H_2O, was investigated by the isothermal method. Equilibrium in the system was reached in 20–30 days. Samarium content in the liquid phase was determined by complexometric titration and lithium by flame photometry. The solid phases were identified by the method of "residues" and checked by X-ray diffraction.	The phase diagram is given below (based on mass % units). 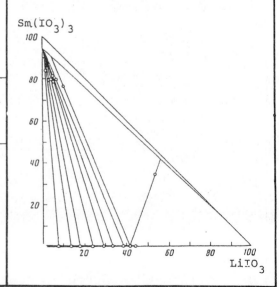

SOURCE AND PURITY OF MATERIALS:

$Sm(IO_3)_3 \cdot 2H_2O$ prepared from samarium carbonate and iodic acid. Special purity grade $LiIO_3$ was used.

ESTIMATED ERROR:

Nothing specified.

COMPONENTS:	ORIGINAL MEASUREMENTS:
(1) Lithium iodate; LiIO₃; [13765-03-2] (2) Lithium hydroxide; LiOH; [1310-65-2] (3) Water; H₂0; [7732-18-5]	Tarasova, G.N.; Vinogradov, E.E.; Lepeshkov, I.N. Zh. Neorg. Khim. 1976, 21, 874-5; Russ. J. Inorg. Chem. (Engl. Transl.) 1976, 21, 478-9.
VARIABLES: Composition at 298.2 K	PREPARED BY: Hiroshi Miyamoto

EXPERIMENTAL VALUES: Composition of saturated solutions at 25.0°C

	LiIO₃		LiOH		Nature of
mass %	mol % (compiler)	mass %	mol % (compiler)	the solid phase[a]
0.00	0.000	11.05	8.546	A
4.19	0.443	10.29	8.263	"
10.94	1.235	9.57	8.20	"
11.95	1.361	9.13	7.90	E₁
12.02	1.371	9.31	8.06	"
13.23	1.526	8.87	7.77	B
14.64	1.713	8.69	7.72	"
15.79	1.866	7.76	6.96	"
18.86	2.306	7.99	7.42	"
21.63	2.723	7.38	7.06	"
23.70	3.055	7.25	7.10	"
25.47	3.353	7.22	7.22	"
26.98	3.612	6.86	6.97	E₂
26.91	3.600	6.86	6.97	"
26.95	3.607	6.88	6.99	"
31.65	4.463	4.96	5.31	C
38.68	5.917	1.58	1.84	"
43.93[b]	7.203	0.00	0.00	"

[a] A = LiOH.H₂0;   B mLiIO₃.nLiOH;   C = LiIO₃;
The chemical formula of E₁ and E₂ was not given in the paper.

[b] For the binary system the compiler computes the following:
soly of LiIO₃ = 4.309 mol kg⁻¹

AUXILIARY INFORMATION

METHOD/APPARATUS/PROCEDURE:
Isothermal method used. Equilibrium was reached after continual stirring for 12-14 days. The hydroxide ion concentration was determined by titration with 0.1 mol dm⁻³ HCl in presence of Methyl Orange. Li⁺ was determined by flame photometry and by the periodate method. The IO₃⁻ ion was determined by titration with sodium thiosulfate solution in the presence of sulfuric acid and KI.

SOURCE AND PURITY OF MATERIALS:
Lithium iodate was prepared from lithium carbonate and iodic acid. Lithium hydroxide freed of Li₂CO₃ by recrystallization from aqueous solution in silver vessels in a stream of nitrogen.

ESTIMATED ERROR:
Soly: nothing specified.
Temp: precision ± 0.1 K.

COMMENTS AND/OR ADDITIONAL DATA:

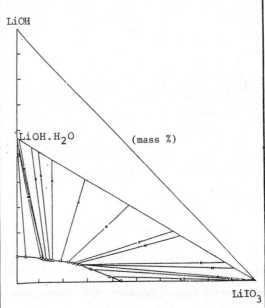

COMPONENTS:	ORIGINAL MEASUREMENTS:
(1) Lithium iodate; $LiIO_3$; [13765-03-2]	Ricci, J.E.; Amron, I.
(2) Iodic acid; HIO_3; [7782-68-5]	J. Am. Chem. Soc. 1951, 73, 3613-8.
(3) Water; H_2O; [7732-18-5]	

VARIABLES:	PREPARED BY:
Composition at 298.2 K	Hiroshi Miyamoto

EXPERIMENTAL VALUES: Composition of saturated solutions

LiIO₃ mass %	LiIO₃ mol % (compiler)	HIO₃ mass %	HIO₃ mol % (compiler)	density g cm⁻³	Nature of the solid phase[a]
43.86[b]	7.184	0.00	0.00	1.558	A
43.96	7.323	1.03	0.177	1.579	"
43.96	7.563	3.13	0.557	1.620	"
43.83	7.964	6.67	1.25	1.697	"
43.56	8.508	11.18	2.257	1.797	"
43.08	9.229	16.65	3.687	1.923	"
42.49	9.797	20.89	4.979	2.027	"
41.48	10.59	26.56	7.012		"
40.81	10.82	28.80	7.890	2.237	"
40.42	11.14	30.78	8.767	2.300	"
40.16	11.17	31.40	9.024	2.312	A+S
40.25	11.19	31.30	8.993	2.310	"
40.16	11.14	31.30	8.973	2.311	"
(av)40.19	11.16	31.33	8.995	2.311	"
39.75	11.23	32.46	9.484	2.334	S
39.57	11.19	32.65	9.542	2.340	"
38.84	11.33	34.58	10.43	2.385	"
38.53	11.40	35.42	10.83		"
37.21	11.47	38.25	12.19	2.475	"
37.13	11.50	38.52	12.34	2.476	"
36.18	11.47	40.29	13.21	2.525	"
36.28	11.57	40.38	13.31		"
35.35	11.58	42.23	14.30	2.567	"
34.70	11.59	43.54	15.04	2.602	"

continued.....

AUXILIARY INFORMATION

METHOD/APPARATUS/PROCEDURE:
Satd slns prepd by placing excess solid and
freshly boiled distilled water in glass-stoppered flasks, and rotating in a const temperature water bath.
After rotation and settling of the solid,
the liquid was sampled by means of 1 ml
specific gravity pipets. Solutions with at
least 20 % H_2O were withdrawn through filter
paper tips for separation of the solid.
Those with less water were sampled only
after sufficient settling. When the
crystals were not too fine, one or two days
of settling sufficed. For extremely fine
ppts, separation was accomplished by cen-
trifuging for one minute followed by replacing the bottle in the water-bath for
five minutes until sufficiently clear supernatant liquid was available for sampling.
Equilibrium was reached in 2 to 8 weeks depending on the composition. Equilibrium was
checked for a few representative solutions
in each series including the most viscous,
before the whole series was analyzed. The
solutions on the solubility curve of HIO_3
were all seeded with the solid before
stirring at 25°C.   continued.....

SOURCE AND PURITY OF MATERIALS:
Some of the lithium iodate used was made by
purification of two samples of commercial
c.p. material which assayed ∿97% $LiIO_3$. One
sample contained insoluble $Ba(IO_3)_2$ and gave
an acid reaction. Part of it was simply
recrystallized twice and part was neutralized with Kahlbaum LiOH before the second
crystallization. The other sample contained
insoluble Li_2CO_3 and gave an alkaline reaction; this was neutralized with iodic
acid and LiOH followed by two recrystallizations. The rest of the salt used was made
from Kahlbaum Li_2CO_3 and c.p. iodic acid
using LiOH for final neutralization.
The final product was obtained by slow
evaporation with stirring on a hot-plate.
After decantation the crystals were filtered
by suction and washed with water. Ground
and dried at 110-180°C, the product was
found to be 99.9 to 100.1 % pure by determination of lithium as Li_2SO_4 and iodate by
titration with $Na_2S_2O_3$ solution.

COMPONENTS:	ORIGINAL MEASUREMENTS:
(1) Lithium iodate; $LiIO_3$; [13765-03-2]	Ricci, J.E.; Amron, I.
(2) Iodic acid; HIO_3; [7782-68-5]	J. Am. Chem. Soc. 1951, 73, 3613-8.
(3) Water; H_2O; [7732-18-5]	

EXPERIMENTAL VALUES: (Continued)

Composition of saturated solutions

LiIO₃		HIO₃		density	Nature of
mass %	mol % (compiler)	mass %	mol % (compiler)	g cm^{-3}	the solid phase[a]
34.09	11.61	44.79	15.77	2.636	S
33.62	11.57	45.60	16.23	–	"
33.48	11.62	46.00	16.50	–	"
32.89	11.57	47.02	17.10	2.695	"
32.81	11.60	47.25	17.26	2.702	"
29.86	11.48	52.74	20.97	2.848	"
27.25	11.30	57.45	24.63	2.979	"
(av) 26.84	11.26	58.15	25.21	2.995	S+B
26.76	11.20	58.19	25.19	2.993	"
26.95	11.33	58.10	25.24	2.993	"
26.82	11.24	58.15	25.19	2.998	"
25.91	10.65	58.56	24.89	2.961	B
21.08	8.023	61.25	24.10	2.827	"
16.48	5.875	63.91	23.55	–	"
10.20	3.403	68.09	23.48	2.609	"
7.23	2.35	70.19	23.58	–	"
3.50	1.10	72.92	23.79	2.514	"
1.24	0.385	74.62	23.95	2.487	"
0.00	0.00	75.40[b]	23.89		"

[a] A = $LiIO_3$;   B = HIO_3;   S = solid solution

[b] For the binary systems, the compiler computes the following:

soly of $LiIO_3$ = 4.296 mol kg^{-1}

soly of HIO_3 = 17.42 mol kg^{-1}

AUXILIARY INFORMATION

METHOD/APPARATUS/PROCEDURE:

In most cases equilibrium was approached presumably from super-saturation, and some cases were obtained from undersaturation. For a few of the worst cases the rotation of the tubes was started at ∿45°C, and the temperature of the water-bath was slowly lowered to 25°C over a period of 30 hours, the slns being seeded with HIO_3 at ∿36°C. For the analysis of the saturated solutions the iodic acid content was determined by titration with standard NaOH solution, and the neutralized sample was then used for the determination of total iodate with standard $Na_2S_2O_3$ solution in the presence of H_2SO_4.

ESTIMATED ERROR:

Soly: precision was presumed within 0.1 %.
Temp: precision about ± 0.05 K (compiler).

COMMENTS AND/OR ADDITIONAL DATA:

The phase diagram is given below (based on mass % units).

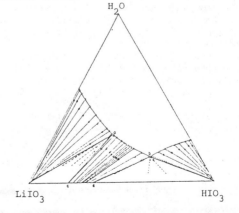

COMPONENTS:	ORIGINAL MEASUREMENTS:
(1) Lithium iodate; $LiIO_3$; [13765-03-2]	Azarova, L.A.; Vinogradov, E.E.
	Mikhailova, E.M.; Pakhomov, V.I.
(2) Iodic acid; HIO_3; [7782-68-5]	
	Zh. Neorg. Khim. 1973 *18*, 239-42;
(3) Water; H_2O; [7732-18-5]	*Russ. J. Inorg. Chem. (Engl. Transl.)*
	1973, *18*, 124-7

VARIABLES:	PREPARED BY:
Composition at 323.2 K	Hiroshi Miyamoto

EXPERIMENTAL VALUES: Composition of saturated solutions at 50.0°C

HIO₃		LiIO₃		Nature of
mass %	mol % (compiler)	mass %	mol % (compiler)	the solid phase[a]
0.00	0.00	43.28[b]	7.028	A
4.26	0.766	43.54	7.574	"
23.72	5.912	41.76	10.07	"
20.67	4.740	40.83	9.057	"
26.28	6.823	41.03	10.30	"
30.27	8.330	39.53	10.52	A+C
30.44	8.429	39.62	10.61	"
46.31	17.05	33.98	12.10	C
47.11	18.32	34.83	13.10	"
51.69	21.20	31.78	12.61	"
55.19	22.79	28.48	11.38	"
34.65	10.23	37.99	10.86	"
44.91	16.58	35.47	12.67	"
59.90	26.53	25.65	10.99	C+B
61.38	29.14	25.90	11.89	"
61.14	29.31	26.37	12.23	"
61.36	30.20	26.77	12.75	"
62.15	26.57	22.49	9.301	B
68.78	24.76	10.90	3.796	"
76.02	25.97	1.98	0.654	"
78.62	27.36	0.00	0.00	"

[a] A = $LiIO_3$;   B = HIO_3;   C = $mLiIO_3.nHIO_3$

[b] For binary systems the compiler computes the following:
soly of $LiIO_3$ = 4.20 mol kg^{-1} soly of HIO_3 = 20.90 mol kg^{-1}

METHOD/APPARATUS/PROCEDURE:	SOURCE AND PURITY OF MATERIALS:
Mixtures of $LiIO_3$, HIO_3 and H_2O were stirred in a thermostat for 7-14 days. Samples were allowed to stand in the thermostat, centrifuged over a period of 1 min, thermostated again, and only then the liquid phase was separated from the solid phase. HIO_3 was determined by titration with a standard NaOH solution, and then the total IO_3^- content in the sample was found by iodometric titration. Lithium was determined by flame photometry and the periodate method. The composition and nature of solid phases were determined by the Schreinemakers', X-ray diffraction, and thermographic methods. X-ray diffraction patterns of solid phases were recorded by the Debye-Scherrer method with an RKU-114 camera with nickel-filtered Cu-Kα radiation.	C.p. grade iodic acid used. $LiIO_3$ prepared from Li_2CO_3 and iodic acid. The product was analyzed chemically, and by X-ray diffraction, but the results were not given.
	ESTIMATED ERROR:
	Nothing specified.
	COMMENTS AND/OR ADDITIONAL DATA:
	The phase diagram is given below (based on mass % units).

COMPONENTS:	ORIGINAL MEASUREMENTS:
(1) Lithium iodate; $LiIO_3$; [13765-03-2]	Shklovskaya, R.M.; Arkhipov, S.M.; Kidyarov, B.I.; Mitnitskii, P.L.
(2) Iodic acid; HIO_3; [7782-68-5]	
	Izv. Sib. Otd. Akad. Nauk SSSR Ser. Khim.
(3) Water; H_2O; [7732-18-5]	*Nauk* 1976, (6), 89-91.

VARIABLES:	PREPARED BY:
Composition at 313 K.	Hiroshi Miyamoto

EXPERIMENTAL VALUES: Composition of saturated solutions at 40°C

$LiIO_3$		HIO_3		Nature of
mass %	mol % (compiler)	mass %	mol % (compiler)	the solid phase[a]
43.83[b]	7.176	0.000	0.000	A
42.872	7.758	7.382	1.381	"
41.614	7.900	11.503	2.258	"
39.830	9.805	26.596	6.768	"
39.177	10.464	30.763	8.494	"
39.701	10.802	30.994	8.717	A+C
35.771	10.255	37.004	10.966	C
34.748	10.708	40.715	12.970	"
33.884	10.258	40.942	12.813	"
31.447	10.450	46.632	16.019	"
26.672	10.922	57.684	24.417	"
25.563	10.670	59.309	25.591	"
24.468	9.872	59.493	24.812	"
23.417	9.551	60.845	25.654	"
23.689	10.124	61.805	27.303	B+C
20.073	8.4238	64.961	28.181	B
18.121	7.200	65.443	26.880	"
13.189	4.931	68.649	26.531	"
11.598	4.069	68.311	24.776	"
9.193	3.192	70.398	25.271	"
6.344	2.119	72.00	24.862	"
2.787	0.877	73.525	23.909	"
0.000	0.000	73.70[b]	22.299	" continued.....

AUXILIARY INFORMATION

METHOD/APPARATUS/PROCEDURE:	SOURCE AND PURITY OF MATERIALS:
For high concentrations of $LiIO_3$, saturated solutions were prepared isothermally. For high HIO_3 concentrations, saturated solutions were prepared isothermally from supersaturated solutions. Equilibrium was reached in 8 days for the former method, and in 30 days for the latter method.	"Chemically pure" grade $LiIO_3$ was used. The total amount of impurities did not exceed 0.001 %. Iodic acid was prepared as described in ref.(1).
The acid concentration in liquid and solid phases was determined by titration with standard NaOH solution, and the iodate concentration determined by titration with thiosulfate solution.	
The composition of the solid phase was determined by Schreinemakers' method of residues and checked by X-ray diffraction.	ESTIMATED ERROR: Nothing specified.

REFERENCES:
1. Vulikh, A.I.; Bogatyrev, V.L.; Kaz'minskaya, V.A.; Zherdienko, L.P. *Metody Polucheniya Khimcheskikh Reaktivov i Preparatov IREA, Vyp. 16,* M., s.5.

COMPONENTS:	ORIGINAL MEASUREMENTS:
(1) Lithium iodate; $LiIO_3$; [13765-03-2] (2) Iodic acid; HIO_3; [7782-68-5] (3) Water; H_2O; [7732-18-5]	Shklovskaya, R.M.; Arkhipov, S.M.; Kidyarov, B.I.; Mitnitskii, P.L.; *Izv. Sib. Otd. Akad. Nauk SSSR Ser. Khim.* *Nauk* <u>1976</u>, (6), 89-91.

EXPERIMENTAL VALUES: (Continued)

[a] $A = \alpha-LiIO_3$; $B = HIO_3$; $C =$ solid solution

[b] For binary systems the compiler computes the following:

   soly of $LiIO_3$ = 4.291 mol kg^{-1}

   soly of HIO_3 = 15.93 mol kg^{-1}

COMMENTS AND/OR ADDITIONAL DATA:

The phase diagram is given below
(based on mass % units).

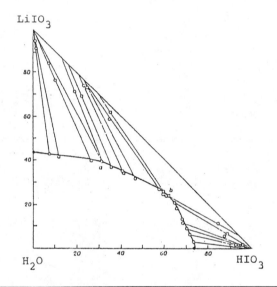

AUXILIARY INFORMATION	
METHOD/APPARATUS/PROCEDURE:	SOURCE AND PURITY OF MATERIALS:
	ESTIMATED ERROR:
	REFERENCES:

COMPONENTS:	ORIGINAL MEASUREMENTS:
(1) Lithium iodate; $LiIO_3$; [13765-03-2] (2) Iodic acid; HIO_3; [7782-68-5] (3) Water; H_2O; [7732-18-5]	Lukasiewicz, T.; Pietaszewska, J.; Amija, J. *Biul. Wojsk. Acad. Teck.* <u>1979</u>, *28*, 85-9.

VARIABLES:	PREPARED BY:
Temperature: 313 to 328 K pH: range of 1.9 to 3.5	A. Maczynski and H. Miyamoto

EXPERIMENTAL VALUES:

$t/°C$	pH	Composition of saturated solutions[a]	
		HIO_3/mol %	$LiIO_3$/mol %
40	1.9	0.68	7.06
50	"	0.60	6.22
55	"	0.40	6.58
40	2.1	0.38	6.44
50	"	0.34	6.64
55	"	0.20	6.56
40	2.3	0.30	6.66
50	"	0.17	6.63
55	"	0.16	6.60
40	2.5	0.18	6.65
50	"	0.16	6.63
55	"	0.12	6.58
40	3.0	0.06	6.62
50	"	0.05	6.50
55	"	0.05	6.48
40	3.5	0.03	6.44
50	"	0.02	6.62
55	"	0.03	6.48

[a]Composition of solid phases not specified.

AUXILIARY INFORMATION

METHOD/APPARATUS/PROCEDURE:	SOURCE AND PURITY OF MATERIALS:
The method employs the fact that the solubility of $LiIO_3$ in water decreases as the temperature is increased. A nearly saturated solution was prepared in a closed vessel at room temperature and placed in a thermostat. On heating to higher temperatures lithium iodate precipitated. The solution was kept at the higher experimental temperature for a few hours until the composition was constant. The analysis involved hydrogen ion concentration by NaOH titration, and iodate ion concentration determinations by iodometry as described in ref (1).	Nothing specified.
	ESTIMATED ERROR: Soly: precision $\pm$ 1 % or better. Temp: nothing specified.
	REFERENCES: 1. Ricci, J.; Amron, I. *J. Am. Chem. Soc.* <u>1953</u>, *73*, 3613

COMPONENTS:	ORIGINAL MEASUREMENTS:
(1) Lithium dihydrogen phosphate; LiH₂PO₄; [13453-80-8] (2) Lithium iodate; LiIO₃; [13765-03-2] (3) Phosphoric acid; H₃PO₄; [7664-38-2] (4) Water; H₂O; [7732-18-5]	Shklovskaya, R.M.; Arkhipov, S.M.; Kidyarov, B.I.; Vdovkina, T.E. Zh. Neorg. Khim. 1982, 27, 1597-8; Russ. J. Inorg. Chem. (Engl. Transl.) 1982, 27, 902.

VARIABLES:	PREPARED BY:
Composition at 298.2 K	Hiroshi Miyamoto

EXPERIMENTAL VALUES: Composition of saturated solutions at 25.0°C [a]

LiIO₃		LiH₂PO₄		Nature of
mass %	mol % (compiler)	mass %	mol % (compiler)	the solid phase[b]
43.82[c]	7.173	–	–	A
36.50	6.002	8.31	2.39	"
26.71	4.271	16.92	4.735	"
21.50	3.453	22.91	6.438	"
15.25	2.481	30.70	8.742	"
10.42	1.766	38.91	11.54	"
7.63	1.340	44.40	13.64	"
5.90	1.09	49.73	16.09	"
4.81	0.937	54.20	18.47	"
3.93	0.797	57.61	20.45	A+B
1.50	0.290	57.25	19.34	B
–	–	61.00	21.33	"

[a] Initial solvent was a 1 % phosphoric acid solution.

[b] A = LiIO₃;   B = LiH₂PO₄

[c] For the binary system the compiler computes the following:

$$\text{soly of LiIO}_3 = 4.289 \text{ mol kg}^{-1}$$

AUXILIARY INFORMATION

METHOD/APPARATUS/PROCEDURE:	SOURCE AND PURITY OF MATERIALS:
Isothermal method used. Equilibrium was reached in 15-20 days. The solubility determinations in the LiH₂PO₄-LiIO₃-H₂O system was carried out in 1 % phosphoric acid solution. Samples of the liquid phases were analyzed for iodate iodometrically, and for the dihydrogen phosphate by direct titration with sodium hydroxide using thymolphthalein as an indicator. The solid phases were identified by the method of residues and checked by X-ray diffraction.	"Special purity" grade lithium iodate and "pure" grade lithium dihydrogen phosphate were used.
	ESTIMATED ERROR: Soly: nothing specified. Temp: precision ± 0.1 K.
	REFERENCES:

COMPONENTS:	ORIGINAL MEASUREMENTS:
(1) Lithium iodate; $LiIO_3$; [13765-03-2] (2) Potassium iodate; KIO_3; [7758-05-6] (3) Iodic acid; HIO_3; [7782-68-5] (4) Water; H_2O; [7732-18-5]	Azarova, L.A.; Vinogradov, E.E.; Lepeshkov, I.M. *Zh. Neorg. Khim.* <u>1978</u>, *23*, 1952-7; *Russ. J. Inorg. Chem. (Engl. Transl.)* <u>1978</u>, *23*, 1072-5.

VARIABLES:	PREPARED BY:
Composition at 298 K	Hiroshi Miyamoto

EXPERIMENTAL VALUES: Composition of saturated solutions at 25°C

LiIO₃		KIO₃		HIO₃		Nature of
mass %	mol % (compiler)	mass %	mol % (compiler)	mass %	mol % (compiler)	the solid phase[a]
1.59	0.183	12.83	1.253	0.72	0.086	A+D
3.69	0.430	12.04	1.192	0.69	0.083	"
11.73	1.449	9.57	1.005	0.55	0.070	"
11.28	1.371	8.48	0.876	0.61	0.077	"
14.94	1.913	9.43	1.026	0.59	0.078	A+D+E
15.40	1.974	9.12	0.993	0.54	0.072	"
15.50	1.981	8.81	0.957	0.50	0.066	D+E
16.31	2.068	7.31	0.787	0.52	0.068	"
18.71	2.440	6.97	0.773	0.90	0.121	"
17.97	2.271	4.64	0.498	1.31	0.171	"
20.92	2.748	1.59	0.177	4.77	0.648	"
18.13	2.277	2.29	0.244	2.99	0.388	"
25.14	3.580	1.26	0.152	7.39	1.088	"
32.47	5.432	1.20	0.171	11.62	2.010	"
35.71	6.635	1.14	0.180	15.00	2.881	D+E+B
35.86	6.642	0.82	0.129	14.99	2.870	"
44.79	7.884	0.34	0.051	3.40	0.619	E+B
42.64	7.288	0.26	0.038	3.77	0.666	"
42.22	7.187	0.26	0.038	3.93	0.692	"
41.78	7.153	0.52	0.076	4.48	0.793	"
35.98	5.530	0.31	0.040	3.17	0.504	"
37.40	7.206	0.41	0.067	16.17	3.221	"

continued.....

AUXILIARY INFORMATION

METHOD/APPARATUS/PROCEDURE:	SOURCE AND PURITY OF MATERIALS:
The $LiIO_3$-KIO_3-HIO_3-H_2O system was studied by the isothermal method. Equilibrium in the system was established in 14 days except for viscous solutions when equilibrium was established after one month. In sampling, the solutions were centrifuged, then thermostated, and only after this procedure was liquid phase separated from the solid. The liquid and solid phases were analyzed for K^+ gravimetrically by precipitation with sodium tetraphenylborate in acetic acid, Li^+ by flame photometry, IO_3^- by iodometric titration, and H^+ by titration with standard alkali. The compositions of the solid phases were determined by Schreinemakers' method of residues.	"Chemically pure" grade KIO_3 and HIO_3 were used. Lithium iodate was made from HIO_3 and LI_2CO_3.
	ESTIMATED ERROR:
	Nothing specified.
	REFERENCES:

COMPONENTS:	ORIGINAL MEASUREMENTS:
(1) Lithium iodate; $LiIO_3$; [13765-03-2]	Azarova, L.A.; Vinogradov, E.E.; Lepeshkov, I.M.
(2) Potassium iodate; KIO_3; [7758-05-6]	
(3) Iodic acid; HIO_3; [7782-68-5]	$Zh.$ $Neorg.$ $Khim.$ 1978, 23, 1952-7; $Russ.$ $J.$ $Inorg.$ $Chem.$ $(Engl.$ $Transl.)$ 1978, 23, 1072-5.
(4) Water; H_2O; [7732-18-5]	

EXPERIMENTAL VALUES: (Continued)

Composition of saturated solutions at 25°C

$LiIO_3$		KIO_3		HIO_3		Nature of
mass %	mol % (compiler)	mass %	mol % (compiler)	mass %	mol % (compiler)	the solid phase[a]
40.91	11.856	1.09	0.268	31.15	9.332	F+B
38.41	9.504	0.99	0.208	27.24	6.968	B+D+F
41.97	11.981	0.93	0.226	29.67	8.755	"
1.61	0.182	2.02	0.194	10.31	1.207	D+F
2.98	0.342	1.83	0.178	10.38	1.231	"
4.24	0.490	1.65	0.162	9.88	1.179	"
5.00	0.584	1.54	0.153	10.34	1.249	"
7.59	0.910	1.48	0.151	10.19	1.262	"
10.50	1.330	1.50	0.161	12.21	1.599	"
10.48	1.285	1.39	0.145	9.48	1.202	"
18.83	2.693	1.33	0.162	13.98	2.067	"
6.43	2.355	0.41	0.128	74.40	28.17	C+F
9.22	3.085	0.01	0.003	69.16	23.92	"
11.17	4.133	0.07	0.022	70.30	26.89	"
15.34	6.195	0.06	0.021	68.62	28.65	"
18.19	8.317	0.59	0.229	68.41	32.33	"
21.78	8.444	trace	-	61.08	24.48	"
39.94	11.78	1.29	0.323	32.59	9.938	"
39.75	12.08	1.59	0.411	33.58	10.55	B+F+G
40.12	12.11	1.81	0.464	32.72	10.21	"
33.08	8.704	1.85	0.414	34.37	9.348	"
30.37	8.316	0.96	0.223	39.64	11.22	F+G
35.61	11.81	1.94	0.547	40.42	13.86	"
37.51	11.48	1.42	0.369	36.25	11.47	"
28.99	11.79	0.28	0.097	54.88	23.07	C+F+G
28.99	11.79	0.28	0.097	54.88	23.07	"

[a] A = KIO_3;   B = $LiIO_3$;   C = HIO_3;   D = $KIO_3 \cdot HIO_3$;   E = $KIO_3 \cdot 2LiIO_3$;

F = $KIO_3 \cdot 2HIO_3$;   G = $mLiIO_3 \cdot nHIO_3$

COMPONENTS:	ORIGINAL MEASUREMENTS:
(1) Lithium iodate; $LiIO_3$; [13765-03-2] (2) Aluminum iodate; $Al(IO_3)_3$; [15123-75-8] (3) Iodic acid; HIO_3; [7782-68-5] (4) Water; H_2O; [7732-18-5]	Shklovskaya, R.M.; Arkhipov, S.M.; Kidyarov, B.I.; Tsibulevskaya, K.A. *Zh. Neorg. Khim.* 1979, 24, 253-5; *Russ. J. Inorg. Chem. (Engl. Transl.)* 1979, 24, 141-2.
VARIABLES: Composition at 298.2 K	PREPARED BY: Hiroshi Miyamoto

EXPERIMENTAL VALUES: Compositions of saturated solutions at 25.0°C

HIO_3		$LiIO_3$		$Al(IO_3)_3$		Nature of
mass %	mol % (compiler)	mass %	mol % (compiler)	mass %	mol % (compiler)	the solid phase[a]
0.0	0.0	40.96	6.678	2.41	0.121	A+B
1.93	0.328	39.78	6.543	2.27	0.123	"
3.24	0.569	40.28	6.838	2.53	0.142	"
7.53	1.378	38.47	6.809	2.70	0.158	"
10.62	1.987	36.27	6.564	3.15	0.188	"
14.36	2.904	36.06	7.054	4.11	0.265	"
18.83	4.025	34.17	7.065	4.55	0.310	"
30.30	8.467	36.19	9.783	3.67	0.327	A+B+S
28.13	7.515	40.42	10.45	–	–	A+S
35.40	10.04	32.21	8.834	3.19	0.288	B+S
40.35	11.57	28.54	7.916	2.43	0.222	B+C+S
43.26	12.48	26.23	7.320	2.11	0.194	C+S
52.44	18.39	24.38	8.273	1.83	0.205	"
56.39	18.65	18.69	5.981	1.64	0.173	"
64.73	26.31	18.54	7.290	–	–	D+S
64.03	25.44	17.06	6.557	1.43	0.181	C+D+S
63.39	21.35	11.63	3.789	2.29	0.246	C+D
71.02	26.71	7.05	2.565	2.76	0.331	"
72.72	26.45	3.69	1.299	3.36	0.390	"
71.06	22.48	–	–	3.98	0.402	"
10.52	1.263	–	–	5.40	0.207	B+C
17.71	2.457	6.58	0.883	4.50	0.199	"

AUXILIARY INFORMATION   continued.....

METHOD/APPARATUS/PROCEDURE:	SOURCE AND PURITY OF MATERIALS:
The quaternary system $LiIO_3$-$Al(IO_3)_3$-HIO_3-H_2O was studied by the isothermal method. Equilibrium was established in 30-45 days. Aliquots of the liquid phases were analyzed for iodate by iodometric titration, for lithium by flame photometry, and for aluminum by complexometric titration. The iodic acid was determined by difference, but for the eutectic solution iodic acid was determined by titration with standard NaOH solution. The solid phases were identified by the method of residues, and by X-ray diffraction.	"Special purity" grade α-$LiIO_3$ and HIO_3 were recrystallized twice from aqueous solution. Aluminum iodate was prepared at 80-90°C by neutralization of a saturated solution of iodic acid with freshly precipitated aluminium hydroxide in equivalent amounts, cooling the solution to room temperature, and drying the salt. Found, mass %: Al 4.03; IO_3 78.7; H_2O 17.6. Calcd. for $Al(IO_3)_3 \cdot 6H_2O$, mass %: Al 4.09; IO_3 79.53; H_2O 16.38.

	ESTIMATED ERROR:
	
	  Soly: nothing specified.
	  Temp: precision $\pm$ 0.1 K.

	REFERENCES:

COMPONENTS:	ORIGINAL MEASUREMENTS:
(1) Lithium iodate; $LiIO_3$; [13765-03-2]	Shklovskaya, R.M.; Arkhipov, S.M.; Kidyarov, B.I.; Tsibulevskaya, K.A.
(2) Aluminum iodate; $Al(IO_3)_3$; [15123-75-8]	
(3) Iodic acid; HIO_3; [7782-68-5]	*Zh. Neorg. Khim.* 1979, 24, 253-5; *Russ. J. Inorg. Chem.* (*Engl. Transl.*) 1979, 24, 141-2.
(4) Water; H_2O; [7732-18-5]	

EXPERIMENTAL VALUES: (Continued)

[a] Solid phases are:

  A = α-$LiIO_3$;   B = $Al(IO_3)_3 \cdot 6H_2O$ C = $Al(IO_3)_3 \cdot 2HIO_3 \cdot 6H_2O$ D = HIO_3

  S = solid solution $(H,Li)IO_3$

COMMENTS AND/OR ADDITIONAL DATA:

The phase diagram is given below (based on mass % units).

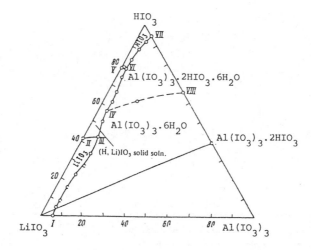

COMPONENTS:	ORIGINAL MEASUREMENTS:
(1) Lithium iodate; LiIO₃; [13765-03-2] (2) Lithium hydroxide; LiOH; [1310-65-2] (3) Potassium iodate; KIO₃; [7758-05-6] (4) Potassium hydroxide; KOH; [1310-58-3] (5) Water; H₂O; [7732-18-5]	Vinogradov, E.E.; Lepeshkov, I.N.; Tarasova, G.N. Zh. Neorg. Khim. 1978, 23, 3360-5; Russ. J. Inorg. Chem. (Engl. Transl.) 1978, 23, 1865-8.

VARIABLES:	PREPARED BY:
Composition at 298.2 K	Hiroshi Miyamoto

EXPERIMENTAL VALUES:

The phase diagram for the penternary system is shown below. Numerical solubility data are given on the following two pages.

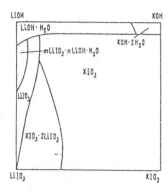

AUXILIARY INFORMATION

METHOD/APPARATUS/PROCEDURE:	SOURCE AND PURITY OF MATERIALS:
The LiIO₃-LiOH-KIO₃-KOH-H₂O system was studied by the isothermal method. The time required to reach equilibrium in the system with continuous stirring is different for different regions of the system (see below).	C.p. grade KIO₃ was used. LiIO₃ prepared from HIO₃ and Li₂CO₃. LiOH was freed of Li₂CO₃ impurity by recrystallization from saturated aqueous solution in silver vessels in a stream of purified nitrogen at 150°C. KOH was purified from alcoholic solution, the temperature being gradually raised to 250°C.

Region	Time
LiOH.H₂O + KIO₃	5 days
KIO₃ + KOH.2H₂O	5 days
LiIO₃ + KIO₃.2LiIO₃	10-14 days
LiIO₃ + KIO₃	10-14 days
LiIO₃ + solid soln	>3 months
LiIO₃ + LiOH.H₂O	>3 months

Specimens of the liquid phases were withdrawn and analyzed for Li⁺ by the periodate method, K⁺ gravimetrically using soldium tetraphenylborate, IO₃⁻ iodometrically, and OH⁻ by titration with 0.1 N HCl using Methyl Orange as an indicator.

The composition of the solid was determined by Schreinemakers' method of residues.

ESTIMATED ERROR:
Soly: nothing specified. Temp: precision ± 0.1° K.

REFERENCES:

COMPONENTS:	ORIGINAL MEASUREMENTS:
(1) Lithium iodate; $LiIO_3$; [13765-03-2]	Vinogradov, E.E.; Lepeshkov, I.N. Tarasova, G.N.
(2) Lithium hydroxide; LiOH; [1310-65-2]	*Zh. Neorg. Khim.* <u>1978</u> *23*, 3360-5;
(3) Potassium iodate; KIO_3; [7758-05-6]	*Russ. J. Inorg. Chem. (Engl. Transl.)*
(4) Potassium hydroxide; KOH; [1310-58-3]	<u>1978</u>, *23*, 1865-8.
(5) Water; H_2O; [7732-18-5]	

EXPERIMENTAL VALUES (Continued)

Composition of saturated solutions at 25°C

LiIO₃ mass %	LiIO₃ mol % (compiler)	LiOH mass %	LiOH mol % (compiler)	KIO₃ mass %	KIO₃ mol % (compiler)	KOH mass %	KOH mol % (compiler)	Nature of the solid phase
39.89	6.177	–	–	0.09	0.012	–	–	A+E
24.64	3.165	0.52	0.51	0.60	0.065	–	–	"
20.18	2.479	0.84	0.78	1.04	0.109	–	–	"
15.43	1.820	1.31	1.17	1.97	0.197	–	–	"
10.91	1.250	1.68	1.46	3.56	0.346	–	–	"
8.64	0.982	2.01	1.73	4.93	0.476	–	–	"
3.28	0.364	3.99	3.36	7.33	0.691	–	–	"
3.19	0.354	4.05	3.41	7.44	0.701	–	–	A+B+E
14.16	1.700	–	–	5.18	0.529	–	–	"
7.92	0.905	0.56	0.49	6.57	0.638	–	–	B+E
7.92	0.910	1.34	1.17	6.84	0.667	–	–	"
3.78	0.424	2.73	2.32	8.21	0.782	–	–	"
3.67	0.409	3.62	3.06	7.44	0.704	–	–	"
2.35	0.260	4.49	3.77	7.88	0.741	–	–	"
2.09	0.231	4.94	4.14	7.72	0.724	–	–	A+B
1.46	0.160	5.51	4.60	7.83	0.731	–	–	"
0.12	0.013	6.24	5.19	8.54	0.794	–	–	"
–	–	2.99	2.45	8.08	0.741	0.01	0.003	"
–	–	7.49	6.19	7.52	0.695	0.27	0.095	"
–	–	8.21	6.77	7.28	0.672	0.20	0.070	"
–	–	8.49	7.00	6.96	0.642	0.47	0.17	"
–	–	9.48	7.78	6.08	0.559	0.69	0.24	"
–	–	9.66	7.95	6.22	0.573	0.76	0.27	A+B+D
–	–	9.97	8.24	6.53	0.604	0.73	0.26	"
–	–	9.87	8.13	6.22	0.574	0.85	0.30	"
–	–	6.86	6.97	–	–	–	–	"
26.98	3.612	6.94	6.92	0.04	0.004	0	0	A+F
25.26	3.315	7.48	7.48	1.92	0.215	–	–	"
23.54	3.102	6.67	6.26	0.05	0.005	–	–	"
20.15	2.491	8.15	7.56	0.16	0.017	–	–	"
18.60	2.273	–	–	–	–	–	–	"
9.48	1.11	9.28	8.25	4.93	0.490	–	–	A+F
10.15	1.137	9.46	8.04	0.05	0.005	–	–	A+D+F
13.63	1.585	9.77	8.63	0.11	0.011	–	–	"

(continued)

COMPONENTS:	ORIGINAL MEASUREMENTS:
(1) Lithium iodate; LiIO3; [13765-03-2]	Vinogradov, E.E.; Lepeshkov, I.N.; Tarasova, G.N.
(2) Lithium hydroxide; LiOH; [1310-65-2]	
(3) Potassium iodate; KIO3; [7758-05-6]	*Zh. Neorg. Khim.* 1978, 23, 3360-5;
(4) Potassium hydroxide; KOH; [1310-58-3]	*Russ. J. Inorg. Chem. (Engl. Transl.)*
(5) Water; H2O; [7732-18-5]	1978, 23, 1865-8.

EXPERIMENTAL VALUES: (Continued)

Composition of saturated solutions at 25°C

LiIO3 mass %	LiIO3 mol % (compiler)	LiOH mass %	LiOH mol % (compiler)	KIO3 mass %	KIO3 mol % (compiler)	KOH mass %	KOH mol % (compiler)	Nature of the solid phase
11.95	1.361	9.13	7.90	–	–	–	–	D+F
7.28	0.803	8.21	6.87	1.71	0.160	–	–	=
1.39	0.151	10.01	8.242	5.36	0.494	–	–	=
2.08	0.226	8.66	7.15	5.25	0.485	–	–	=
1.05	0.114	9.77	8.09	6.29	0.583	–	–	=
–	–	10.87	9.007	6.86	0.636	0.36	0.13	=
–	–	9.56	7.71	2.78	0.251	2.64	0.909	B+D
–	–	8.14	6.62	4.01	0.365	2.71	0.941	=
–	–	8.21	6.68	2.50	0.230	4.79	1.66	=
–	–	7.56	6.24	1.70	0.157	8.02	2.83	=
–	–	6.35	5.36	1.11	0.105	12.22	4.403	=
–	–	5.93	5.02	1.05	0.0995	12.81	4.628	=
–	–	3.07	2.67	0.98	0.095	17.40	6.457	=
–	–	1.83	1.66	0.04	0.004	24.36	9.427	=
–	–	2.59	2.99	0.27	0.035	50.04	24.67	=
–	–	1.83	1.91	0.02	0.002	40.66	18.15	=
–	–	1.62	1.69	0.02	0.002	40.75	18.19	=
–	–	1.35	1.58	0.27	0.035	51.81	25.89	=
–	–	1.28	1.49	0.40	0.052	51.01	25.32	=
–	–	1.24	1.46	0.28	0.037	52.27	26.24	=
–	–	–	–	0.26	0.033	49.02	23.68	=
–	–	0.97	1.03	0.01	0.001	42.53	19.27	=
–	–	0.96	1.04	0.31	0.038	44.06	20.34	=
–	–	0.55	0.70	–	–	60.18	32.75	C+D
–	–	11.10	8.586	–	–	–	–	D
–	–	–	–	–	–	54.23	27.56	C
43.82	7.173	–	–	–	–	–	–	A
–	–	–	–	8.39	0.765	–	–	B

a A = LiIO3; B = KIO3; C = KOH.2H2O; D = LiOH.H2O; E = KIO3.2LiIO3; F = mLiIO3.nLiOH.H2O.

COMPONENTS:	ORIGINAL MEASUREMENTS:
(1) Lithium iodate; $LiIO_3$; [13765-03-2] (2) Ethanol; C_2H_6O; [64-17-5] (3) Water; H_2O; [7732-18-5]	Arkhipov, S.M.; Pruntsev, A.E.; Kidyarov, B.I. *Zh. Neorg. Khim.* <u>1977</u>, *22*, 3394-5; *Russ. J. Inorg. Chem. (Engl. Transl.)* <u>1977</u>, *22*, 1855.
VARIABLES:	PREPARED BY:
Concentration of ethanol at 298 K	Hiroshi Miyamoto

EXPERIMENTAL VALUES:

Numerical solubility data for the ternary $LiIO_3$-ethanol-H_2O system were not given in the original paper. The phase diagram shown here was the only data reported.

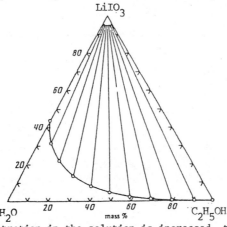

As the alcohol concentration in the solution is increased, there is a marked decrease in the solubility of lithium iodate. Thus, the solubility of lithium iodate is reduced to one-half in a solution with 15.5 mass % C_2H_5OH. A further increase in alcohol concentration leads to a less pronounced decrease in the solubility of $LiIO_3$. The solubility of lithium iodate in anhydrous ethanol is <0.01 mass %.

AUXILIARY INFORMATION

METHOD/APPARATUS/PROCEDURE:	SOURCE AND PURITY OF MATERIALS:
The solubility of lithium iodate-ethanol-water system was investigated by the isothermal method. Equilibrium was established in 5-6 days. The authors do report that saturated solutions were analyzed for $LiIO_3$ (i.e. presumably for both Li^+ and IO_3^-, compiler), but no details were given.	"Special purity" grade lithium iodate was used. Ethanol was purified and dried by a published method (ref 1).
	ESTIMATED ERROR: Soly: nothing specified. Temp: precision $\pm$ 0.1 K.

	REFERENCES:
	1. Plyushev, V.E.; Shakhno, I.V.;
	   Komissarova, L.N.; Nadexhdina, G. V.
	   Trudy Moskov. Inst. Tonk. Khim. Tekhol.
	   im Lomonosova, <u>1958</u>, 7, 53.

COMPONENTS:	ORIGINAL MEASUREMENTS:
(1) Lithium iodate; $LiIO_3$; [13765-03-2] (2) 2-Propanone (acetone); C_3H_6O; [67-64-1]	Miravitlles, Mille L. *Ann. Fís. Quím. (Madrid)* <u>1945</u>, *41*, 120-37.

VARIABLES:	PREPARED BY:
T/K = 288, 293 and 298	R. Herrera

EXPERIMENTAL VALUES:

Solubility[a]

t/°C	mass %	mol kg^{-1}
15	0.0333	0.00183
20	0.0327	0.00180
25	0.0319	0.00175

[a]Molalities calculated by the compiler.

AUXILIARY INFORMATION

METHOD/APPARATUS/PROCEDURE:	SOURCE AND PURITY OF MATERIALS:
Saturated solutions were prepared in an Erlenmeyer flask by mixing the dried acetone with an excess of halate for two hours. The solution was constantly stirred by bubbling dry air (air was dried by passing it through $CaCl_2$ while pumping it into the solution). Air going out from the flask after bubbling in the solution carried some acetone vapor during this operation. The solution temperature was kept constant by immersing the flask in a constant temperature water bath. After two hours, the air exit was closed. The resulting pressure forced the saturated solution from the Erlenmeyer through a tube filled with cotton (which acted as a filter), and was collected in a small flask. This flask was stoppered and weighed. The halate contained in the sample was weighed after complete evaporation of acetone.	Commercial redistilled acetone. This acetone was then dehydrated three times by leaving it in contact with calcium chloride for forty eight hours each time. Fresh $CaCl_2$ was used in each operation. Finally, the dehydrated acetone was distilled at 56.3°C. Source and purity of $LiIO_3$ not specified.
	ESTIMATED ERROR: Nothing specified.
	REFERENCES:

AMH—L

COMPONENTS:	EVALUATOR:
(1) Sodium iodate; NaIO₃; [7681-55-2] (2) Water; H₂0; [7732-18-5]	H. Miyamoto Niigata University Niigata, Japan and Mark Salomon US Army ET & DL Fort Monmouth, NJ, USA January 1985

CRITICAL EVALUATION:

THE BINARY SYSTEM

Data for the solubility of $NaIO_3$ in water have been reported in 19 publications (1-19).
Note that the compilation for reference (16) is given in the $LiIO_3$ chapter, and the
compilation for reference (18) is given in the previous volume on alkaline earth metal
halates (Vol. 14 of the *IUPAC SOLUBILITY DATA SERIES*, ref. (20)). Several authors appear
to report the same solubilities in two or more publications, and these values have been
treated as a single independent measurement (details are given in the text below).
Many of the studies deal with ternary systems, and the solubility in the binary system
was given as one point on a phase diagram. Depending upon temperature and composition,
three solid phases have been identified in the binary system:

$NaIO_3.5H_2O$ [17250-90-7]

$NaIO_3H_2O$ [22451-04-7]

$NaIO_3$ [7681-55-2]

The temperature dependence of the solubility of $NaIO_3$ in pure water has been studied
by Foote and Vance (2) over the range 273.2 - 363.5 K, and by Cornec and Spack (8) over
the range 273.2 - 373.2 K. Both studies reported the existence of the three solid
phases, and transition temperatures were determined graphically. Foote and Vance (2)
reported the pentahydrate → monohydrate transition temperature as 293.1 K, and the monohydrate → anhydrate transition temperature as 346.6 K. These two temperatures are in
good agreement with those of Cornec and Spack, but they differ from the evaluators'
recommended values evaluated graphically below (293.2 K and 347.4 K). The monohydrate →
pentahydrate transition temperature of 295 K reported by Hill and Donovan (6) appears
too high and should be rejected

The solubility data reported by Foote and Vance (2) and by Cornect and Spack (8) are
plottted as a function of temperature in Figure 1.

In evaluating the solubility data for the binary system, each polytherm in the phase
diagram was separately fitted to a smoothing equation. For mole fraction solubilities
the following smoothing equation was used,

$$Yx = A/(T/K) + Bln(T/K) + C + D(T/K) \qquad [1]$$

and for molalities, the following smoothing equation was used

$$Ym = A/(T/K) + Bln(T/K) + C \qquad [2]$$

The complex Y terms in eqs. [1] and [2] are defined in the PREFACE to this volume and
in the critical evaluations for $LiClO_3$ and $RbClO_3$. Solubility data were rejected when
the difference between the calculated and observed mole fraction solubilities exceeded
twice the standard error of estimate: i.e. when

$$abs \left[X_{obsd} - X_{calcd} \right] > 2\sigma_x \qquad [3]$$

Polytherm For NaIO₃.5H₂O As The Solid Phase

A summary of the experimental data is given in Table 1. The identical solubility at
273.2 K reported by Foote and Vance in three publications (2, 3, 10) was treated as one
independent measurement, and therefore 17 independent data points were used in the smoothing equations. No data points were rejected. Mole fraction solubilities were fitted to

$$Yx = -150902/(T/K) - 1053.291ln(T/K) + 1.95591(T/K)$$

$$\sigma_y = 0.011 \qquad \qquad \sigma_x = 2.9 \times 10^{-5}$$

COMPONENTS:	EVALUATOR:

(1) Sodium Iodate; $NaIO_3$; [7681-55-2]
 	
(2) Water; H_2O; [7732-18-5]
	H. Miyamoto
	Niigata University
	Niigata, Japan
	         and
	Mark Salomon
	US Army ET & DL
	Fort Monmouth, NJ, USA
	
	                                January, 1985

CRITICAL EVALUATION:

For mol/kg solubilities, the following smoothed relation was obtained:

$$Y_m = 4070.9/(T/K) + 31.360 ln(T/K) - 191.440$$

$$\sigma_y = 0.0061 \qquad \sigma_m = 0.0019$$

Solubilities calculated from these equations are designated as recommended, and recommended fraction solubilities are given in Table 4 at the end of this evaluation.

Table 1. Experimental Solubilities in the Binary $NaIO_3 \cdot 5H_2O$–H_2O System

T/K	mole fraction	mol/kg^{-1}	ref
272.8[a]	0.002215	0.1232	2
273.2	0.002253	0.1253	2,3,10
273.2	0.002281	0.1269	8
278.2	0.003097	0.1724	6
278.2	0.003078	0.1714	7
281.2	0.003681	0.2051	4
281.2	0.003671	0.2045	9
283.2	0.004172	0.2326 [c]	2
283.2	0.004212	0.2348	8
288.2	0.005665	0.3163	2
288.2	0.005655	0.3157	3
288.2	0.005625	0.3140	6
288.2	0.005706	0.3185	8
288.2	0.005625	0.3140	11
288.2	0.005645	0.3151	14
293.0[b]	0.007674	0.4293	2
293.2[b]	0.007653	0.4281	6

[a]Solid phase: ice + $NaIO_3 \cdot 5H_2O$

[b]Solid phase: $NaIO_3 \cdot 5H_2O$ + $NaIO_3 \cdot H_2O$ [c]Reference molality used in the smoothing eqn.

Polytherm For $NaIO_3 \cdot H_2O$ As The Solid Phase

Table 2 lists the solubilities at various temperatures given in 52 references, and which we have treated as representing 35 independent measurements. In fitting these 35 solubility values to the smoothing equations, five points were rejected as indicated in Table 2. The remaining 30 solubility values yielded the following:

$$Y_x = 3187.5/(T/K) + 45.366 \; ln \, (T/K) - 254.20 - 0.07121(T/K)$$

$$\sigma_y = 0.012 \qquad \qquad \sigma_x = 7.2 \times 10^{-5}$$

For mol/kg solubilities, the smoothed data are given by

$$Y_m = -1620.6/(T/K) + 1.394 \; ln \, (T/K) - 2.496$$

$$\sigma_y = 0.006 \qquad \qquad \sigma_m = 0.004$$

COMPONENTS:	EVALUATOR:
(1) Sodium iodate; NaIO$_3$; [7681-55-2] (2) Water; H$_2$O; [7732-18-5]	H. Miyamoto Niigata University Niigata, Japan and Mark Salomon US Army ET & DL Fort Monmouth, NJ, USA January, 1985

CRITICAL EVALUATION:

Solubilities calculated from these two smoothing equations are designated as *recommended* and recommended mole fraction solubilities are given in Table 4 at the end of this critical evaluation.

Table 2.   Solubilities in the Binary NaIO$_3$·H$_2$O–H$_2$O System

T/K	mole fraction	mol/kg	ref
278.2[m]	0.005249	0.2929	10
293.0[a]	0.007674	0.4293	2
293.2[a]	0.007653	0.4281	6
293.2	0.007695	0.4305	2
293.2	0.007661	0.4257	8
295.0	0.000797	0.4460	6
298.2	0.008557	0.4791	2,5,10,15,16,19
298.2	0.008568	0.4797	6,9
298.2	0.008461	0.4737	7,12
298.2	0.008460	0.4736	11
298.2[b]	0.008375	0.4688	13
298.2[c]	0.008472	0.4743[e]	18
302.7	0.009433	0.5286	5
303.2[b]	0.009313	0.5218	1
303.2	0.009608	0.5385	2
303.2	0.009608	0.5385	8
308.2	0.01065	0.5973	2–5
308.2	0.01066	0.5979	6,11
313.2	0.01193	0.6702	2,10
313.2	0.01192	0.6696	6,9,11
313.2	0.01185	0.6657	8
318.2	0.01322	0.7438	11
323.2	0.01468	0.8267	2
323.2	0.01468	0.8267	5,10
323.2	0.01454	0.8192	6
323.2[b]	0.01400	0.7880	7,11,12
323.2	0.01448	0.8158	8
323.2[b]	0.01403	0.7901	17
331.0	0.01694	0.9568	2
333.2	0.01786	1.0094	8
342.8	0.02095	1.1876	2
343.2	0.02123	1.2039	8
346.6[b,d]	0.02225	1.2633	2
352.2[m]	0.02479	1.4112	2
353.2[m]	0.02529	1.4403	8

COMPONENTS:	EVALUATOR:
(1) Sodium iodate; $NaIO_3$; [7681-55-2] (2) Water; H_2O; [7732-18-5]	H. Miyamoto Niigata University Niigata, Japan and Mark Salomon US Army ET & DL Fort Monmouth, NJ, USA January, 1985

CRITICAL EVALUATION:

[m] metastable

[a] Solid phase: $NaIO_3.5H_2O + NaIO_3.H_2O$

[b] Rejected data points

[c] Compilation of data in ref. 18 given in the earlier volume (20).

[d] Solid phase $NaIO_3.H_2O + NaIO_3$ [d] Reference molality used in the smoothing eqn.

Polytherm For Anhydrous $NaIO_3$ As The Solid Phase

The ten solubilities reported for anhydrous $NaIO_3$ (anhydrate) are given in Table 3, and all ten data points were successfully fitted to the smoothing equations. The resulting equations are:

$$Y_x = -75335.5/(T/K) - 421.305 \, ln \, (T/K) + 2464.25 + 0.60793(T/K)$$

$$\sigma_y = 0.012 \qquad\qquad \sigma_x = 1.5 \times 10^{-4}$$

for the mole fraction solubilities, and

$$Y_m = 1498.1/(T/K) + 7.780 \, ln \, (T/K) - 49.896$$

$$\sigma_y = 0.006 \qquad\qquad \sigma_m = 0.009$$

for mol/kg solubilities. *Recommended* mole fraction solubilities calculated from the smoothing equation are given in Table 4.

Table 3. Experimental Solubilities in the Binary $NaIO_3$-H_2O System

T/K	mole fraction	mol/kg	ref
340.2[m]	0.02096	1.188	2
343.2[m]	0.02166	1.229	2
346.6[a]	0.02225	1.263	2
349.0	0.02292	1.302	2
353.2	0.02398	1.364 [b]	8
353.8	0.02396	1.363	2
360.8	0.02535	1.444	2
363.2	0.02628	1.498	8
363.5	0.02652	1.512	2
373.2	0.02900	1.658	8

[m] metastable

[a] $NaIO_3.H_2O + NaIO_3$

[b] Reference molality used in the smoothing equation.

COMPONENTS:	EVALUATOR:
(1) Sodium iodate; $NaIO_3$; [7681-55-2] (2) Water; H_2O; [7732-18-5]	H. Miyamoto Niigata University Niigata, Japan and Mark Salomon US Army ET & DL Fort Monmouth, NJ, USA January, 1985

CRITICAL EVALUATION:

Recommended Solubilities In The Binary System

Table 4 lists recommended solubilities over the temperature range of 273 K to 373 K as calculated from the mole fraction smoothing equations. The transition points included in this table were evaluated graphically by the evaluators using the smoothed solubilities.

Table 4.   Recommended Mole Fraction Solubilities[a]

T/K	$NaIO_3 \cdot 5H_2O$	$NaIO_3 \cdot H_2O$	$NaIO_3$
273.2	0.002266		
278.2	0.003080	0.005265[m]	
283.2	0.004173	0.005974[m]	
288.2	0.005656	0.006755[m]	
293.2	0.007626	0.007613[m]	
293.22	0.00763[b]	0.00763[b]	
298.2		0.008551	
303.2		0.009574	
308.2		0.01069	
313.2		0.01189	
318.2		0.01319	
323.2		0.01459	
328.2		0.01610	
333.2		0.01770	
338.2		0.01942	0.0205[m]
343.2		0.02124	0.0216[m]
347.4		0.0220[b]	0.0220[b]
348.2		0.02318[m]	0.0227
353.2		0.02523[m]	0.0238
358.2		0.02738[m]	0.0250
363.2			0.0263
368.2			0.0276
373.2			0.0290

[a]Calculated from the smoothing equations

[b]Phase transitions determined graphically by evaluators.

[m]metastable

COMPONENTS:	EVALUATOR:
(1) Sodium iodate; $NaIO_3$; [7681-55-2] (2) Water; H_2O; [7732-18-5]	H. Miyamoto Niigata University Niigata, Japan June, 1984

CRITICAL EVALUATION:

TERNARY SYSTEMS

Many studies for solubilities in ternary aqueous systems with two saturating components have been reported. Summaries of these studies are given in Tables 5-8.

The System With Iodic Acid.

This system was studied by Meerburg (1) at 303 K and Shibuya and Watanabe (14) at 288 K. The compositions of the solid phase obtained are as follows:

$$NaIO_3.1.5H_2O; \qquad HIO_3; \qquad Na_2O.2I_2O_5; \qquad NaIO_3.2HIO_3 \qquad\qquad (1)$$

$$NaIO_3.5H_2O; \qquad 2NaIO_3.I_2O_5; \qquad NaIO_3.I_2O_5 \qquad\qquad\qquad (14)$$

Many solubility studies at 303 K reported the solid phase of $NaIO_3.H_2O$, but did not report $NaIO_3.1.5H_2O$. Solid HIO_3 was not reported in (14). The compositions of the double salts appear doubtful, and the evaluator is of the opinion that additional studies are required to confirm the compositions of the various solid phases.

System With Other Iodates

Solubilities in ternary aqueous systems containing sodium iodate with other iodates have been reported in 6 publications (7, 15-19) (see Table 5). No double salts were found in these systems as all are of the simple eutonic type. The sodium salt in the solid phase at 278 K is the pentahydrate, and that at 298 and 323 K is the monohydrate.

Table 5. Summary of solubility studies of ternary systems with sodium iodate and other iodates

Ternary System	T/K	Solid Phase	Reference
$NaIO_3$ + $LiIO_3$ + H_2O	298	$NaIO_3.H_2O$; $LiIO_3$	15
$NaIO_3$ + KIO_3 + H_2O	278	$NaIO_3.H_2O$; KIO_3	7
$NaIO_3$ + KIO_3 + H_2O	298, 323	$NaIO_3.H_2O$; KIO_3	7
$NaIO_3$ + $RbIO_3$ + H_2O	323	$NaIO_3.H_2O$; $RbIO_3$	17
$NaIO_3$ + $CsIO_3$ + H_2O	323	$NaIO_3.H_2O$; $CsIO_3$	17
$NaIO_3$ + $Mg(IO_3)_2$ + H_2O	278	$NaIO_3.5H_2O$; $Mg(IO_3)_2.10H_2O$	7
$NaIO_3$ + $Mg(IO_3)_2$ + H_2O	298, 323	$NaIO_3.H_2O$; $Mg(IO_3)_2.4H_2O$	7
$NaIO_3$ + $Ca(IO_3)_2$ + H_2O	298	$NaIO_3.H_2O$; $Ca(IO_3)_2.6H_2O$	18
$NaIO_3$ + $Al(IO_3)_3$ + H_2O	298	$NaIO_3.H_2O$; $Al(IO_3)_2.6H_2O$	16
$NaIO_3$ + $Hf(IO_3)_4$ + H_2O	298	$NaIO_3.H_2O$; $Hf(IO_3)_4$	19

System With Sodium Halides

Aqueous ternary systems containing sodium iodate with a sodium halide have been studied in 3 publications (3, 9, 11) (see Table 6). For $NaIO_3$ - $NaBr$ - H_2O and $NaIO_3$ - $NaCl$ - H_2O, double salts were found, but in the $NaIO_3$ - NaI - H_2O system the formation of double salts was not reported. The compositions of the double salts are given in Table 6.

COMPONENTS:	EVALUATOR:
(1) Sodium iodate; $NaIO_3$; [7681-55-2] (2) Water; H_2O; [7732-18-5]	H. Miyamoto Niigata University Niigata, Japan June, 1984

CRITICAL EVALUATION:

Table 6. Summary of solubility studies with sodium halides

Ternary System	T/K	Solid Phase	Reference
$NaIO_3$ + NaCl + H_2O	273, 288	$NaIO_3 \cdot 5H_2O$; NaCl; $2NaIO_3 \cdot 3NaCl \cdot 10H_2O$	3
$NaIO_3$ + NaCl + H_2O	298, 303	$NaIO_3 \cdot H_2O$; NaCl	3
$NaIO_3$ + NaBr + H_2O	278	$NaIO_3 \cdot 5H_2O$; $NaBr \cdot 2H_2O$; $2NaIO_3 \cdot 3NaBr \cdot 15H_2O$	11
$NaIO_3$ + NaBr + H_2O	288	$NaIO_3 \cdot 5H_2O$; $NaIO_3$; $NaBr \cdot 2H_2O$; $2NaIO_3 \cdot 3NaBr \cdot 15H_2O$	11
$NaIO_3$ + NaBr + H_2O	298	$NaIO_3 \cdot H_2O$; $NaBr \cdot 2H_2O$ $2NaIO_3 \cdot 3NaBr \cdot 15H_2O$	11
$NaIO_3$ + NaBr + H_2O	308	$NaIO_3 \cdot H_2O$; $NaIO_3$; $NaBr \cdot 2H_2O$; $2NaIO_3 \cdot 3NaBr \cdot 15H_2O$	11
$NaIO_3$ + NaBr + H_2O	313	$NaIO_3 \cdot H_2O$; $NaIO_3$; $NaBr \cdot 2H_2O$; $2NaIO_3 \cdot 2NaBr \cdot 15H_2O$; $2NaIO_3 \cdot 3NaBr \cdot 10H_2O$	11
$NaIO_3$ + NaBr + H_2O	318	$NaIO_3 \cdot H_2O$; $NaIO_3$; $NaBr \cdot 2H_2O$; $2NaIO_3 \cdot 3NaBr \cdot 10H_2O$	11
$NaIO_3$ + NaBr + H_2O	323	$NaIO_3 \cdot H_2O$; $NaIO_3$; $NaBr \cdot 2H_2O$; NaBr; $2NaIO_3 \cdot 3NaBr \cdot 10H_2O$	11
$NaIO_3$ + NaI + H_2O	281	$NaIO_3 \cdot 5H_2O$; $NaI \cdot 2H_2O$; Solid Solution	9
$NaIO_3$ + NaI + H_2O	298	$NaIO_3 \cdot H_2O$; $NaI \cdot 2H_2O$; Solid Solution	9
$NaIO_3$ + NaI + H_2O	313	$NaIO_3 \cdot H_2O$; $NaIO_3$; $NaI \cdot 2H_2O$	9

The System With Sodium Nitrate

Solubilities in this ternary system have been reported in 3 publications (4, 6, 8) (see
Table 7). Foote and Vance (4) and Hill and Donovan's (6) studies covered a wide range of
compositions for each temperature investigated. Below 281 K, the double salt $2NaIO_3 \cdot 3Na$
$NO_3 \cdot 15H_2O$ was reported in (4, 6, 8), and sodium iodate in the solid phase is the pentahydrate. Hill and Donovan stated that their results at 278 K confirmed those of Foote and
Vance, with a somewhat better agreement between calculated and experimental results for
the double salt $2NaIO_3 \cdot 3NaNO_3 \cdot 15H_2O$. Above 283 K, no double salts form.

COMPONENTS:	EVALUATOR:
(1) Sodium iodate; $NaIO_3$; [7681-55-2] (2) Water; H_2O; [7732-18-5]	H. Miyamoto Niigata University Niigata, Japan June, 1984

CRITICAL EVALUATION:

Table 7. Summary of solubilities in the ternary $NaIO_3$-$NaNO_3$-H_2O System

T/K	Solid Phase	Reference
273	$NaIO_3.5H_2O$; $NaNO_3$; $2NaIO_3.3NaNO_3.15H_2O$	4, 8
278	$NaIO_3.5H_2O$; $NaNO_3$; $2NaIO_3.3NaNO_3.15H_2O$	6
278	$NaIO_3.5H_2O$; $NaNO_3$;H_2O; $NaNO_3$; $2NaIO_3.3NaNO_3.15H_2O$	8
281	$NaIO_3.5H_2O$; $NaIO_3.H_2O$; $NaNO_3$; $2NaIO_3.3NaNO_3.15H_2O$	4
283	$NaIO_3.H_2O$; $NaNO_3$	8
293	$NaIO_3.H_2O$; $NaNO_3$	8
298	$NaIO_3.H_2O$; $NaNO_3$	4, 6
303	$NaIO_3.H_2O$; $NaNO_3$	8
308	$NaIO_3.H_2O$; $NaNO_3$	4, 8
313	$NaIO_3.H_2O$; $NaNO_3$	8
323	$NaIO_3$; $NaNO_3$	6, 8
333	$NaIO_3$; $NaNO_3$	8
343	$NaIO_3$; $NaNO_3$	8
353	$NaIO_3$; $NaNO_3$	8
363	$NaIO_3$; $NaNO_3$	8
373	$NaIO_3$; $NaNO_3$	8

Systems With Other Sodium Salts

Solubilities in the $NaIO_3$-Na_2SO_4-H_2O system at 273, 303, 308 and 323 K have been reported by Foote and Vance (5). Double salts were found only at 303, 308 and 323 K. The compositions of double salts are given in Table 8. Ternary systems $NaIO_3$-Na_2CO_3-H_2O, $NaIO_3$-$NaClO_3$-H_2O and $NaIO_3$-Na_2MoO_4-H_2O have been reported by Foote and Vance (10), Ricci (5) and Shklovskaya's group (13), respectively, and double salts were not found (see Table 8).

Table 8. Summary of the solubility of $NaIO_3$ in the
presence of several sodium salts

Ternary System	T/K	Solid Phase	Ref
$NaIO_3$ + Na_2SO_4 + H_2O	273	$NaIO_3.H_2O$; $Na_2SO_4.10H_2O$	5
$NaIO_3$ + Na_2SO_4 + H_2O	302.7	$NaIO_3.H_2O$; $Na_2SO_4.10H_2O$ $NaIO_3.3Na_2SO_4$	5
$NaIO_3$ + Na_2SO_4 +H_2O	308,323	$NaIO_3.H_2O$; $Na_2SO_4.10H_2O$ $NaIO_3.4Na_2SO_4$; $NaIO_3.3Na_2SO_4$	5
$NaIO_3$ + Na_2CO_3 + H_2O	273	$NaIO_3.5H_2O$; $Na_2CO_3.10H_2O$	10

continued.....

AMH—L*

COMPONENTS:	EVALUATOR:
(1) Sodium iodate; $NaIO_3$; [7681-55-2] (2) Water; H_2O; [7732-18-5]	H. Miyamoto Department of Chemistry Niigata University Niigata, Japan June, 1984

CRITICAL EVALUATION:

Table 8. (continued)

Ternary System	T/K	Solid Phase	Ref
$NaIO_3$ + Na_2CO_3 + H_2O	298	$NaIO_3.H_2O$; $Na_2CO_3.10H_2O$	10
$NaIO_3$ + Na_2CO_3 + H_2O	313,323	$NaIO_3.H_2O$; $Na_2CO_3.H_2O$	10
$NaIO_3$ + $NaClO_3$ + H_2O	298	$NaIO_3.H_2O$; $NaClO_3$	12
$NaIO_3$ + $NaClO_3$ + H_2O	323	$NaIO_3.H_2O$; $NaIO_3$; $NaClO_3$	12
$NaIO_3$ + Na_2MoO_4 + H_2O	298	$NaIO_3.H_2O$; $Na_2MoO_4.2H_2O$	13

REFERENCES:

1. Meerburg, P. A. Z. Anorg. Allg. Chem. 1905, 45, 324.

2. Foote, H. W.; Vance, J. E. Am. J. Sci. 1928, [5] 16, 68.

3. Foote, H. W.; Vance, J. E. Am. J. Sci. 1929, [5] 17, 425.

4. Foote, H. W.; Vance, J. E. Am. J. Sci. 1929, [5] 18, 375.

5. Foote, H. W.; Vance, J. E. Am. J. Sci. 1930, [5] 19, 203.

6. Hill, A. E.; Donovan, J. E. J. Am. Chem. Soc. 1931, 53, 934.

7. Hill, A. E.; Ricci, J. E. J. Am. Chem. Soc. 1931, 53, 4305.

8. Cornec, M. E.; Spack, A. Bull. Soc. Chim. Fr. 1931, 49, 582.

9. Hill, A. E.; Willson, H. S.; Bishop, J. A. J. Am. Chem. Soc. 1933, 55, 520.

10. Foote, H. W.; Vance, J. E. Am. J. Sci. 1933, 25, 499.

11. Ricci, J. E. J. Am. Chem. Soc. 1934, 56, 290.

12. Ricci, J. E. J. Am. Chem. Soc. 1938, 60, 2040.

13. Ricci, J. E.; Linke, W. F. J. Am. Chem. Soc. 1947, 69, 1080.

14. Shibuya, M.; Watanabe, T. Denki Kagaku 1967, 35, 550.

15. Shklovskaya, R. M.; Arkhipov. S. M.; Kidyarov, B. I.; Mitnitskii, P. L.
    Zh. Neorg. Khim. 1974, 19, 1975; Russ. J. Inorg. Chem. (Engl. Transl.)
    1974, 19, 1082.

16. Shkovskaya, R. M.; Arkhopov, S. M.; Kidyarov, B. I.; Tokareva, A. G. Zh. Neorg.
    Khim. 1980, 25, 1423; Russ. J. Inorg. Chem. (Engl. Transl.) 1980, 25, 791.

COMPONENTS:	EVALUATOR:
(1) Sodium Iodate; $NaIO_3$; [7681-55-2] (2) Water; H_2O; [7732-18-5]	H. Miyamoto Department of Chemistry Niigata University Niigata, Japan June, 1984

CRITICAL EVALUATION:

REFERENCES: (Continued)

17. Vinogradov, E. E.; Karataeva, I. M. *Zh. Neorg. Khim.* 1982, *27*, 2155; *Russ. J. Inorg. Chem. (Engl. Transl.)* 1982, *27*, 1218.

18. Hill, A. E.; Brown, S. F. *J. Am. Chem. Soc.* 1931, *53*, 4316.

19. Shklovskaya, R. M.; Arkhipov, S. M.; Kidyarov, B. I.; Poleva, G. V.; Timofeev, S. I. *Zh. Neorg. Khim.* 1983, *28*, 2435; *Russ. J. Inorg. Chem. (Engl. Transl.)* 1984 *28*, 1384.

20. Miyamoto, H.; Salomon, M.; Clever, H. L. *Alkaline Earth Metal Halates: Vol 14, IUPAC Solubility Data Series.* Pergamon Press, London, 1983.

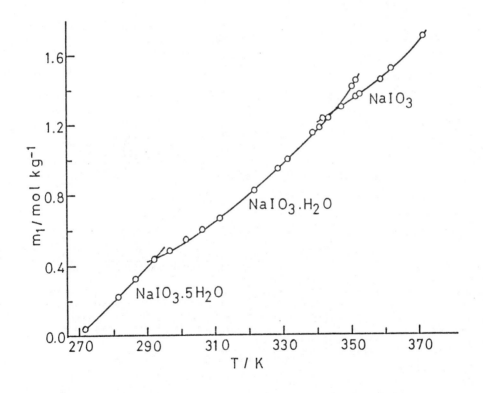

COMPONENTS:	ORIGINAL MEASUREMENTS:
(1) Sodium iodate; $NaIO_3$; [7681-55-2] (2) Water; H_2O; [7732-18-5]	Foote, H.W.; Vance, J.E. *Am. J. Sci.* 1928, *16*, 68-72.
VARIABLES: T/K = 272.8 to 363.5	PREPARED BY: Hiroshi Miyamoto

EXPERIMENTAL VALUES: Solubility as approached from:

t/°C	supersaturation mass %	undersaturation mass %	average mass %	mol kg^{-1} (compiler)	Nature of the solid phase[a]
- 0.35	-	-	2.38	0.123	Ice + A
0.0	2.39	2.46	2.42	0.125	A
+10.0	4.39	4.40	4.39	0.232	"
15.0	5.87	5.88	5.88	0.316	"
19.85	-	-	7.83	0.429	A+B
20.0	7.87	7.82	7.84	0.430	B
25.0	8.65	8.66	8.66	0.479	"
30.0	9.63	9.63	9.63	0.538	"
35.0	10.58	10.55	10.57	0.597	"
40.0	11.70	11.71	11.71	0.670	"
49.9	14.13	13.99	14.06	0.827	"
57.8	15.97	15.86	15.91	0.9560	"
69.6	19.05	19.00	19.03	1.188	"
73.4	-	-	20.00	1.263	B+C
79.0	21.91	21.74	21.82	1.41	B(m)
67.0	18.98	19.10	19.04	1.188	C(m)
70.6	19.55	19.57	19.56	1.229	"
75.8	20.48	20.49	20.49	1.302	C
80.6	21.22	21.26	21.24	1.363	"
87.6	22.12	22.32	22.22	1.444	"
90.3	23.02	23.02	23.03	1.512	"

[a] A = $NaIO_3 \cdot 5H_2O$;   B = $NaIO_3 \cdot H_2O$;   C = $NaIO_3$;   (m) = metastable

The authors reported the smoothing equation as follows:

log (soly/mass %) = 3.6344 - 802.8/(T/K)   (T/K = 293-322.9)

log (soly/mass %) = 7.7793 - 2019/(T/K)   (T/K = 273-288)

AUXILIARY INFORMATION

METHOD/APPARATUS/PROCEDURE:	SOURCE AND PURITY OF MATERIALS:
Binary mixts agitated in a thermostat for 4-6 hours. Equil was approached from both the supersatd and undersatd solutions, and analysis was determined in duplicate. Iodate was detd by adding excess KI, acidifying with H_2SO_4, and titrating with standard sodium thiosulfate sln.	Sodium iodate was a very pure commercial product having a composition closely approximating the monohydrate. The salt was recrystallized before use.
Solid phases analyzed as follows: Below 19.85 °C where the pentahydrate is stable, the solid was separated from sln in a cold room and quickly dried, and presumably analyzed for iodate. Over the temp range where the monohydrate is stable, numerous analyses were made of the solid phase, presumably by a method similar to that described above. For the region where the anhydr salt is stable, the solid was separated, washed quickly with alcohol, and dried between filter paper.	ESTIMATED ERROR: Nothing specified.

COMPONENTS:	ORIGINAL MEASUREMENTS:
(1) Sodium iodate; NaIO₃; [7681-55-2] (2) Water; H₂O; [7732-18-5]	Hill, A.E.; Donovan, J.E. J. Am. Chem. Soc. 1931, 53, 934-41.

VARIABLES:	PREPARED BY:
Temperature: 278.15 to 323.15	Hiroshi Miyamoto

EXPERIMENTAL VALUES: Solubility in the binary system

t/°C	mass %	Solubility mol % (compiler)	mol kg⁻¹ (compiler)	Density g cm⁻³	Nature of the solid phase[a]:
5	3.30	0.310	0.172	1.028	A
15	5.85	0.562	0.314	1.051	"
20	7.81	0.765	0.428	–	A+B
22	8.11	0.797	0.446	1.071	B
25	8.67	0.857	0.480	1.077	"
35	10.58	1.066	0.5979	1.093	"
40	11.70	1.192	0.6696	–	"
50	13.95	1.454	0.8192	–	"

[a] $A = NaIO_3 \cdot 5H_2O$; $B = NaIO_3 \cdot H_2O$

AUXILIARY INFORMATION

METHOD/APPARATUS/PROCEDURE:
Recrystallized and dehydrated excess sodium iodate was placed in 40 ml glass-stoppered Pyrex test-tubes with water and rotated in a water thermostat for about two weeks. Equilibrium was reached from super-saturation. After the tubes were allowed to settle, samples were withdrawn into a calibrated pipet fitted with a small cotton filter. One sample was weighed and evaporated in a platinum dish to constant weight at 110°C. From this the water content of the saturated solution was determined. To determine the NaIO₃ content, a second weighed sample was treated with KI and sulfuric acid and titrated with sodium thiosulfate. The densities of the solutions were also determined.

SOURCE AND PURITY OF MATERIALS:
"Good grade" sodium iodate was purified by recrystallization. No other information given.

ESTIMATED ERROR:
Soly: the error for the analysis of iodate by iodometry was within 0.2 %.
Temp: precision ± 0.05 K.

REFERENCES:

COMPONENTS:	ORIGINAL MEASUREMENTS:
(1) Sodium iodate; $NaIO_3$; [7681-55-2]	Cornec, M.E.; Spack, A.
(2) Water; H_2O; [7732-18-5]	*Bull. Soc. Chim. Fr.* <u>1931</u>, 49, 582-94.

VARIABLES:	PREPARED BY:
T/K = 273 to 373	Hiroshi Miyamoto

EXPERIMENTAL VALUES:

		Solubility of Sodium Iodate		Density	Nature of the
$t/°C$	mass %	g_1/100 g H_2O	$mol\ kg^{-1}$ (compiler)	$g\ cm^{-3}$	solid phase[a]
0	2.45	2.51	0.127	1.024	A
10	4.44	4.65	0.235	1.041	"
15	5.93	6.31	0.318	1.054	"
19.9[i]	7.7	8.35	0.422	1.070	A+B
20	7.77	8.43	0.426	1.071	B
30	9.63	10.65	0.538	1.085	"
40	11.64	13.17	0.6657	1.102	"
50	13.90	16.15	0.8158	1.119	"
60	16.65	20.0	1.010	1.142	"
70	19.24	23.8	1.204	1.164	"
80[m]	22.18	28.5	1.440	1.190	"
73.4[i]	20.2	25.3	1.28	1.172	B+C
80	21.25	27.0	1.364	1.180	C
90	22.87	29.65	1.498	1.192	"
100	24.70	32.8	1.658	1.204	"

[a] A = $NaIO_3 \cdot 5H_2O$;   B = $NaIO_3 \cdot H_2O$;   C = $NaIO_3$

[i] Interpolated data

[m] Metastable

AUXILIARY INFORMATION

METHOD/APPARATUS/PROCEDURE:	SOURCE AND PURITY OF MATERIALS:
The details of procedure were not given. The iodate content was determined by titration with thiosulfate solution.	Sodium iodate used was purchased as a "pure chemical". The salt was recrystallized four times. The product obtained was the monohydrate.
	ESTIMATED ERROR:
	Nothing specified.
	REFERENCES:

COMPONENTS:	ORIGINAL MEASUREMENTS:
(1) Sodium iodate; $NaIO_3$; [7681-55-2]	Ricci, J.E.; Linke, W.F.
(2) Di sodium (I-4)-tetraoxomolybdate (2-) (disodium molybdate); Na_2MoO_4; [7631-95-0]	J. Am. Chem. Soc. 1947, 69, 1080-3.
(3) Water; H_2O; [7732-18-5]	

VARIABLES:	PREPARED BY:
Composition at 298.2 K	Hiroshi Miyamoto

EXPERIMENTAL VALUES: Composition of saturated solutions

Na_2MoO_4		$NaIO_3$		Density	Nature of
mass %	mol % (compiler)	mass %	mol % (compiler)	g cm^{-3}	the solid phase[a]
39.38	5.378	0.00	0.000	1.432	A
39.16	5.375	0.58	0.083	1.437	"
38.63	5.354	1.79	0.258	1.450	"
38.46	5.349	2.20	0.318	1.453	A+B
38.51	5.358	2.18	0.316	1.452	"
38.43	5.343	2.21	0.320	1.451	"
38.43	5.343	2.21	0.320	1.454	"
38.47	5.350	2.18	0.315	1.455	"
38.46	5.349	2.20	0.318	1.453	"
37.23	5.090	2.24	0.319	1.436	B
31.49	3.995	2.54	0.335	1.368	"
24.24	2.825	3.08	0.373	1.277	"
17.89	1.943	3.42	0.386	1.204	"
11.41	1.163	4.16	0.441	1.143	"
5.57	0.543	5.67	0.575	1.099	"
0.00	0.000	8.49[b]	0.838	1.074	"

[a] A = $Na_2MoO_4 \cdot 2H_2O$;   B = $NaIO_3 \cdot H_2O$

[b] For the binary system the compiler computes the following:

soly of $NaIO_3$ = 0.469 mol kg^{-1}

AUXILIARY INFORMATION

METHOD/APPARATUS/PROCEDURE:	SOURCE AND PURITY OF MATERIALS:
Solubilities determined isothermally by stirring complexes of known compositions in Pyrex tubes, and sampling the equilibrated solutions by means of calibrated pipets fitted with filtering tips. Total solids were determined by evaporation of an aliquot of saturated solution and drying to constant weight at 125°C.	C.p. grade sodium molybdate dihydrate completely dehydrated by heating to 180° C, and stored at 150°C. The purity of this anhydrous salt was found to be 100.0 %. C.p. grade sodium iodate was found to be pure within 1/1000.

The iodate content in the saturated solution was determined iodometrically. A large excess of acid (HCl) was necessary to obtain the correct end-point within the short titration time in the presence of the molybdate.

ESTIMATED ERROR:

Soly: the accuracy of titrations was
      within 0.1 %.
Temp: precision ± 0.04 K.

COMMENTS AND/OR ADDITIONAL DATA:

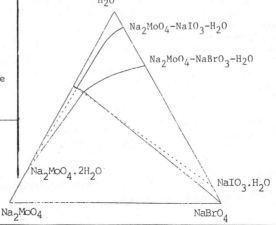

COMPONENTS:	ORIGINAL MEASUREMENTS:
(1) Sodium carbonate; Na_2CO_3; [4917-19-8] (2) Sodium iodate; $NaIO_3$; [7681-55-2] (3) Water; H_2O; [7732-18-5]	Foote, H.W.; Vance, J.E. *Am. J. Sci.* 1933, 25, 499-502.
VARIABLES: Composition T/K = 273 - 323	PREPARED BY: Hiroshi Miyamoto

EXPERIMENTAL VALUES: Composition of saturated solutions

t/°C	$NaIO_3$ mass %	$NaIO_3$ mol % (compiler)	Na_2CO_3 mass %	Na_2CO_3 mol % (compiler)	Nature of the solid phase[a]
0	–	–	6.42	1.15	A
	0.81	0.078	6.30	1.14	A+C
	0.83	0.080	6.27	1.13	"
	2.42[b]	0.225	–	–	C
25	–	–	22.60	4.728	A
	0.52	0.059	22.44	4.715	"
	2.16	0.247	22.22	4.745	A+D
	2.17	0.248	22.22	4.746	"
	2.54	0.282	18.82	3.898	D
	8.66[b]	0.856	–	–	"
40	–	–	32.83	7.670	B
	0.50	0.063	32.67	7.667	"
	1.79	0.227	32.09	7.603	B+D
	1.75	0.222	32.00	7.570	"
	2.00	0.248	29.87	6.918	D
	11.71[b]	1.193	–	–	"

continued.....

AUXILIARY INFORMATION

METHOD/APPARATUS/PROCEDURE:	SOURCE AND PURITY OF MATERIALS:
Sodium iodate, sodium carbonate and water were placed in glass stoppered bottles and rotated in a thermostat. Twenty-four hours were allowed for the attainment of equilibrium at which time samples of the solution were drawn off through glass wool filters for analysis. Sodium iodate content was determined by adding excess KI, acidifying with sulfuric acid, and titrating the liberated iodine with thiosulfate solution. Sodium carbonate was detd in a second sample by titration with HCl using methyl orange indicator. In these carbonate titrations, a constant light source was used and the end point was compared with a standard made by saturating water containing a few drops of methyl orange with carbon dioxide. The composition of the dry solid phase was determined by the method of Schreinemakers.	The authors only state that sodium iodate and carbonate were purified by customary methods.
	ESTIMATED ERROR: Nothing specified.
	REFERENCES:

COMPONENTS:	ORIGINAL MEASUREMENTS:
(1) Sodium carbonate; Na_2CO_3; [4917-19-8]	Foote, H.W.; Vance, J.E.
(2) Sodium iodate; $NaIO_3$; [7681-55-2]	*Am. J. Sci.* <u>1933</u>, *25*, 499-502.
(3) Water; H_2O; [7732-18-5]	

EXPERIMENTAL VALUES: (Continued)

Composition of saturated solutions

t/°C	NaIO₃		Na₂CO₃		Nature of the solid phase[a]
	mass %	mol % (compiler)	mass %	mol % (compiler)	
50	–	–	32.16	7.457	B
	1.30	0.163	31.52	7.374	"
	2.42	0.307	31.31	7.411	B+D
	2.54	0.322	31.23	7.396	"
	3.34	0.401	25.44	5.701	D
	14.06	1.468	–	–	"

[a] A = $Na_2CO_3.10H_2O$;   B = $Na_2CO_3.H_2O$;   C = $NaIO_3.5H_2O$;   D = $NaIO_3.H_2O$

[b] For the binary system the compiler computes the following:

soly of $NaIO_3$ = 0.125 mol kg^{-1} at 0°C

= 0.479 mol kg^{-1} at 25°C

= 0.6702 mol kg^{-1} at 40°C

= 0.8267 mol kg^{-1} at 50°C

COMMENTS AND/OR ADDITIONAL DATA:

The phase diagram is given below (based on mass % units).

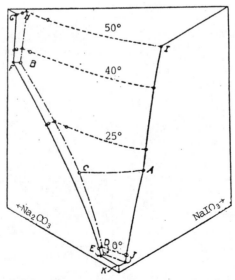

System NaIO₃–Na₂CO₃–H₂O, showing the solubility isotherms at four temperatures, and the stability areas of the five solid phases. No double salt exists.

COMPONENTS:	ORIGINAL MEASUREMENTS:
(1) Sodium nitrate; $NaNO_3$; [7631-99-4]	Foote, H.W.; Vance, J.E.
(2) Sodium iodate; $NaIO_3$; [7681-55-2]	Am. J. Sci. 1929, 18, 375-82.
(3) Water; H_2O; [7732-18-5]	

VARIABLES:	PREPARED BY:
Composition	
T/K = 273 - 308	Hiroshi Miyamoto

EXPERIMENTAL VALUES: Composition of saturated solutions

	$NaIO_3$		$NaNO_3$		Nature of the
t/°C	mass %	mol % (compiler)	mass %	mol % (compiler)	solid phase[a]
0	–	–	42.13	13.37	A
	0.82	0.113	41.76	13.34	A+D
	0.82	0.112	41.71	13.32	"
	0.86	0.117	41.15	13.06	D
	1.00	0.131	37.53	11.44	"
	1.16	0.147	34.61	10.24	"
	1.31	0.163	32.57	9.438	"
	1.38	0.171	32.19	9.298	B+D
	1.31	0.163	32.45	9.391	"
	1.26	0.151	29.18	8.153	B
	1.06	0.117	21.42	5.526	"
	2.42[b]	0.225	–	–	"
8	–	–	43.99	14.27	A
	1.67	0.236	43.28	14.25	A+D
	1.67	0.236	43.21	14.21	"
	1.88	0.259	40.80	13.08	D
	1.96	0.266	39.54	12.50	C+D
	2.02	0.274	39.36	12.42	C
	2.27	0.285	32.23	9.418	"

continued.....

AUXILIARY INFORMATION

METHOD/APPARATUS/PROCEDURE:	SOURCE AND PURITY OF MATERIALS:
Sodium iodate, sodium nitrate and water were placed in glass stoppered bottles and rotated in a thermostat. Samples of the solution were drawn off through glass wool filters. The iodate content was determined by adding KI to the solution, acidifying with sulfuric acid, and titrating the free iodine with sodium thiosulfate solution. The nitrate content was calculated from the iodate concentration and the total mass of salts in solution. Water was found by difference. The solid phases were analyzed as wet residues after largely freeing them from water by pressing between filter papers. The composition of the dry residue was determined by Schreinemakers' method.	Sodium iodate and nitrate used were c.p. products which were recrystallized once.
	ESTIMATED ERROR:
	Nothing specified.
	REFERENCES:

COMPONENTS:	ORIGINAL MEASUREMENTS:
(1) Sodium nitrate; $NaNO_3$; [7631-99-4]	Foote, H.W.; Vance, J.E.
(2) Sodium iodate; $NaIO_3$; [7681-55-2]	Am. J. Sci. 1929, 18, 375-82.
(3) Water; H_2O; [7732-18-5]	

EXPERIMENTAL VALUES: (Continued)

Composition of saturated solutions

t/°C	$NaIO_3$		$NaNO_3$		Nature of the solid phase[a]
	mass %	mol % (compiler)	mass %	mol % (compiler)	
8	2.25	0.279	31.03	8.948	B+C
	2.22	0.274	30.83	8.868	"
	3.90[b]	0.368	–	–	B
25	–	–	48.04	16.39	A
	1.09	0.161	47.39	16.29	"
	2.30	0.343	46.73	16.21	A+C
	2.25	0.335	46.65	16.16	"
	2.38	0.340	43.18	14.34	C
	2.55	0.350	39.88	12.76	"
	3.69	0.400	15.94	4.018	"
	8.66[b]	0.856	–	–	"
35	–	–	50.15	17.58	A
	1.58	0.241	49.25	17.47	"
	2.55	0.391	48.68	17.39	A+C
	2.55	0.391	48.68	17.39	"
	2.60	0.399	48.68	17.41	"
	2.66	0.400	46.99	16.45	C
	3.85	0.456	24.96	6.886	"
	10.57[b]	1.065	–	–	"

[a] A = $NaNO_3$;   B = $NaIO_3 \cdot 5H_2O$;   C = $NaIO_3 \cdot H_2O$;   D = $2NaIO_3 \cdot 3NaNO_3 \cdot 15H_2O$

[b] For the binary system the compiler computes the following:

soly of $NaIO_3$ = 0.125 mol kg^{-1} at 0°C; = 0.205 mol kg^{-1} at 8°C
= 0.479 mol kg^{-1} at 25°C; = 0.5973 mol kg^{-1} at 35°C

COMMENTS AND/OR ADDITIONAL DATA:

Isotherms based on mass % units are reproduced below on the following page.

COMPONENTS:	ORIGINAL MEASUREMENTS:
(1) Sodium nitrate; $NaNO_3$; [7631-99-4]	Foote, H.W.; Vance, J.E.
(2) Sodium iodate; $NaIO_3$; [7681-55-2]	$Am.\ J.\ Sci.$ 1929, 18, 375-82.
(3) Water; H_2O; [7732-18-5]	

COMMENTS AND/OR ADDITIONAL DATA: (Continued)

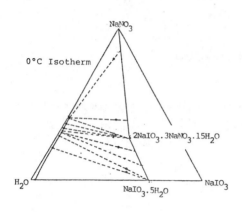

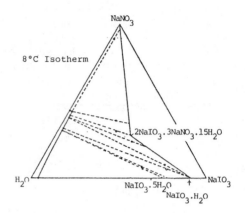

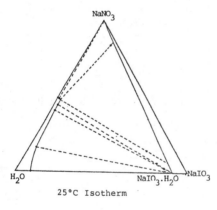

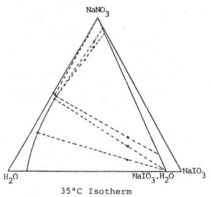

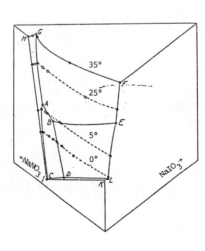

COMPONENTS:	ORIGINAL MEASUREMENTS:
(1) Sodium nitrate; $NaNO_3$; [7631-99-4]	Cornec, M.E.; Spack, A.
(2) Sodium iodate; $NaIO_3$ [7681-55-2]	*Bull. Soc. Chim. Fr.* 1931, 49, 582-94.
(3) Water; H_2O; [7732-18-5]	

VARIABLES:	PREPARED BY:
T/K = 273 to 373 Composition	Hiroshi Miyamoto

EXPERIMENTAL VALUES: Composition of saturated solutions

t/°C	Sodium iodate mass %	g_2/100 g_3	mol kg^{-1}	Sodium nitrate mass %	g_1/100 g_3	mol kg^{-1}	Density g cm^{-3}	Nature of the solid phase[a]
0	0.53	0.92	0.046	42.0	72.9	8.58	1.357	A
	0.82	1.43	0.0723	41.9	73.0	8.59	1.360	A+S
	0.85	1.46	0.0738	40.8	69.9	8.22	1.351	S
	1.25	1.92	0.0970	33.5	51.3	6.04	1.286	"
	1.34	2.02	0.102	32.5	49.1	5.78	1.278	S+B
	1.25	1.81	0.0915	29.6	42.8	5.04	1.251	B
5	1.28	2.28	0.115	42.8	76.4	8.99	1.373	A+S
	2.02	3.15	0.159	34.1	53.4	6.28	1.300	B+C+S
10^m	2.04						1.388	A+S
9.7^i	1.97	3.57	0.180	43.5	79.6	9.37	1.387	A+C+S
10	1.98	3.64	0.184	43.6	80.1	9.42	1.388	A+C
15^i	2.95	3.63	0.183	15.8	19.5	2.29	1.144	B+C
20	2.14	4.11	0.208	45.7	87.6	10.3	1.405	A+C
30	2.43	4.88	0.247	47.8	95.8	11.3	1.423	"
35	2.58						1.432	"
40	2.77	5.83	0.295	49.7	104.7	12.32	1.442	"
50	3.24						1.462	"

continued.....

AUXILIARY INFORMATION

METHOD/APPARATUS/PROCEDURE:	SOURCE AND PURITY OF MATERIALS:
The details of procedure were not given. The iodate content was determined by titration with thiosulfate solution. The total solids were determined by evaporation of the solution at about 140°C. The compiler assumes that the concentration of the nitrate was determined by difference.	Sodium iodate used was purchased as a "pure chemical". The salt ws recrystallized four times. The product obtained was the monohydrate.
	ESTIMATED ERROR: Nothing specified.
	REFERENCES:

COMPONENTS:	ORIGINAL MEASUREMENTS:
(1) Sodium nitrate; $NaNO_3$; [7631-99-4]	Cornec, M.E.; Spack, A.
(2) Sodium iodate; $NaIO_3$; [7681-55-2]	*Bull. Soc. Chim. Fr.* 1931, 49, 582-94.
(3) Water; H_2O; [7732-18-5]	

EXPERIMENTAL VALUES: (Continued)

Composition of saturated solutions

t/°C		Sodium iodate			Sodium nitrate		Density	Nature of the
	mass %	g_2/100 g_3	mol kg^{-1}	mass %	g_1/100 g_3	mol kg^{-1}	g cm^{-3}	solid phase
41.5[i]	2.85	6.0	0.30	50.0	106.0	12.47	1.445	A+C+D
35[m]	2.85						1.435	A+D
50	2.90	6.39	0.323	51.7	114.1	13.42	1.458	"
60	2.97	6.87	0.347	53.8	124.4	14.64	1.474	"
70	3.16	7.68	0.388	55.7	135.3	15.92	1.491	"
80	3.35	8.60	0.435	57.7	148.2	17.44	1.509	"
90	3.60	9.79	0.495	59.6	162.3	19.10	1.528	"
100	3.94	11.40	0.5761	61.5	177.9	20.93	1.549	"
15	3.05	3.49	0.176	9.4	10.8	1.27	1.096	B
15	2.97	3.50	0.177	12.2	14.4	1.69	1.116	"
15	2.95	3.60	0.182	15.0	18.3	2.15	1.138	"
15[m]	3.04	3.89	0.197	18.8	24.1	2.84	1.170	"
15	2.85	3.62	0.183	18.4	23.3	2.74	1.164	C
15[m]	3.03	3.66	0.185	14.3	17.2	2.02	1.133	"
15[m]	3.48	3.97	0.201	8.8	10.1	1.19	1.096	"

[a] A = $NaNO_3$;   B = $NaIO_3.5H_2O$;   C = $NaIO_3.H_2O$;   D = $NaIO_3$

S = double salt: $2NaIO_3.3NaNO_3.15H_2O$.

[m] Metastable

[i] Interpolated.

COMPONENTS:	ORIGINAL MEASUREMENTS:
(1) Sodium nitrate; $NaNO_3$; [7631-99-4] (2) Sodium iodate; $NaIO_3$; [7681-55-2] (3) Water; H_2O; [7732-18-5]	Hill, A.E.; Donovan, J.E. J. Am. Chem. Soc. 1931, 53, 934-41.

VARIABLES:	PREPARED BY:
Composition T/K = 278.15 - 313.15	Hiroshi Miyamoto

EXPERIMENTAL VALUES: Composition of saturated solutions

t/°C	$NaIO_3$ mass %	mol % (compiler)	$NaNO_3$ mass %	mol % (compiler)	Density g cm^{-3}	Nature of the solid phase[a]
5	3.3[b]	0.310	0.00	0.00	1.028	A
	1.58	0.151	4.50	1.00	1.042	"
	1.55	0.175	22.59	5.927	1.182	"
	1.65	0.193	26.38	7.195	1.214	"
	1.75	0.210	28.44	7.932	1.230	"
	1.87	0.230	30.70	8.781	1.255	"
	1.78	0.220	31.38	9.030	1.262	"
	2.10	0.261	31.54	9.128	1.263	"
	2.07	0.260	32.48	9.493	1.275	"
	1.86	0.233	32.37	9.424	1.269	"
	1.94	0.248	34.34	10.23	1.291	A+E
	1.76	0.231	36.69	11.19	1.308	E
	1.60	0.213	38.38	11.91	1.324	"
	1.52	0.204	39.17	12.25	1.331	"
	1.41	0.192	40.40	12.80	1.336	"
	1.28	0.179	42.74	13.90	1.359	E+D
	1.01	0.141	42.94	13.95	1.359	D
	0.00	0.000	43.42	13.99	1.368	"

continued....

AUXILIARY INFORMATION

METHOD/APPARATUS/PROCEDURE:	SOURCE AND PURITY OF MATERIALS:
For sodium iodate-sodium nitrate-water system, weighed quantities of these salts were treated with weighed amounts of water in Pyrex test-tubes. The tubes were slowly rotated in a water-thermostat at the desired temperature for about two weeks. After the slns were allowed to settle, samples were withdrawn into a calibrated pipet fitted with a small cotton filter. One sample was weighed and evaporated in a platinum dish to constant weight at 110°C. From this the water content of the saturated solution was determined. To determine the $NaIO_3$ content, a second weighed sample was treated with KI and H_2SO_4, and titrated with $Na_2S_2O_3$. The sodium nitrate content was obtained by difference.	"Good grade" sodium iodate and sodium nitrate were purified by recrystallization.
	ESTIMATED ERROR: Soly: the error for the analysis of iodate by iodometry was within 0.2 %. Temp: precision ± 0.05 K.
	REFERENCES:

COMPONENTS:	ORIGINAL MEASUREMENTS:
(1) Sodium nitrate; $NaNO_3$; [7631-99-4]	Hill, A.E.; Donovan, J.E.
(2) Sodium iodate; $NaIO_3$; [7681-55-2]	J. Am. Chem. Soc. 1931, 53, 934-41.
(3) Water; H_2O; [7732-18-5]	

EXPERIMENTAL VALUES: (Continued)

Composition of saturated solutions

t/°C	$NaIO_3$ mass %	$NaIO_3$ mol % (compiler)	$NaNO_3$ mass %	$NaNO_3$ mol % (compiler)	Density g cm^{-3}	Nature of the solid phase[a]
25	8.67[b]	0.857	0.00	0.00	1.077	B
	6.38	0.634	3.26	0.754	1.078	"
	5.99	0.596	3.91	0.906	1.078	"
	4.80	0.486	7.32	1.726	1.092	"
	4.30	0.444	10.10	2.429	1.109	"
	3.68	0.399	16.08	4.058	1.149	"
	3.41	0.381	19.47	5.060	1.171	"
	3.06	0.367	27.16	7.593	1.232	"
	2.84	0.361	32.67	9.661	1.276	"
	2.60	0.350	38.19	11.98	1.328	"
	2.31	0.337	45.12	15.34	1.392	"
	2.23	0.332	46.81	16.24	1.408	B+D
	1.09	0.161	47.44	16.32	1.396	D
	0.00	0.00	47.98	16.35	1.388	"
50	13.95[b]	1.454	0.00	0.00	–	B
	9.63	1.011	5.74	1.403	–	"
	6.22	0.700	17.03	4.460	–	"
	4.82	0.598	28.18	8.136	–	"
	3.92	0.544	39.15	12.65	–	"
	4.00	0.561	39.94	13.05	–	"
	3.84	0.548	41.53	13.80	–	B+C
	3.77	0.542	42.26	14.16	–	C
	3.64	0.531	43.46	14.75	–	"
	3.46	0.515	45.23	15.66	–	"
	3.09	0.481	49.11	17.80	–	"
	2.91	0.469	51.86	19.46	–	C+D
	1.58	0.252	52.55	19.49	–	D
	0.00	0.000	53.50	19.61	–	"

[a] A = $NaIO_3 \cdot 5H_2O$;   B = $NaIO_3 \cdot H_2O$;   C = $NaIO_3$;   D = $NaNO_3$;

E = $2NaIO_3 \cdot 3NaNO_3 \cdot 15H_2O$.

[b] For the binary system the compiler computes the following:

soly of $NaIO_3$ = 0.172 mol kg^{-1} at 5°C

= 0.480 mol kg^{-1} at 25°C

= 0.8192 mol kg^{-1} at 50°C

continued....

COMPONENTS:	ORIGINAL MEASUREMENTS:
(1) Sodium nitrate; $NaNO_3$; [7631-99-4]	Hill, A.E.; Donovan, J.E.
(2) Sodium iodate; $NaIO_3$; [7681-55-2]	J. Am. Chem. Soc. 1931, 53, 934-41.
(3) Water; H_2O; [7732-18-5]	

COMMENTS AND/OR ADDITIONAL DATA:

Isotherms based on mass % units are reproduced below.

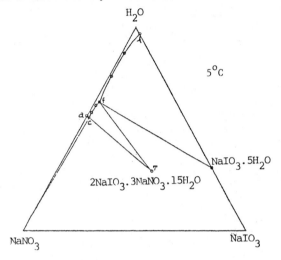

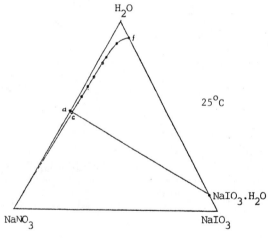

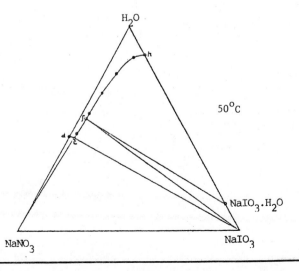

COMPONENTS:	ORIGINAL MEASUREMENTS:
(1) Sodium sulfate; Na_2SO_4; [7757-82-6] (2) Sodium iodate; $NaIO_3$; [7681-55-2] (3) Water; H_2O; [7732-18-5]	Foote, H.W.; Vance, J.E. Am. J. Sci. 1930, 19, 203-13.

VARIABLES:	PREPARED BY:
Composition T/K = 298 - 323	Hiroshi Miyamoto

EXPERIMENTAL VALUES: Composition of saturated solutions

t/°C	$NaIO_3$ mass %	$NaIO_3$ mol % (compiler)	Na_2SO_4 mass %	Na_2SO_4 mol % (compiler)	Nature of the solid phase[b]
25	–	–	21.75	3.405	A
	2.20	0.252	21.30	3.402	"
	2.80	0.323	21.18	3.402	A+C
	2.78	0.320	21.18	3.401	"
	2.80	0.323	21.19	3.404	"
	3.17	0.353	17.52	2.716	C
	3.76	0.400	12.64	1.874	"
	8.66[c]	0.856	–	–	"
29.5	–	–	28.12	4.727	A
	1.62	0.198	27.72	4.730	A+E
	1.53	0.187	27.72	4.725	"
	1.52	0.184	26.85	4.530	E
	1.90	0.228	25.68	4.294	"
	2.70	0.325	25.17	4.225	C+E
	2.73	0.328	24.86	4.159	"
	9.47[c]	0.943	–	–	C
35	–	–	33.10	5.905	B
(A)	0.15	0.019	32.86	5.856	B+D
	0.15	0.019	32.91	5.868	"

continued.....

AUXILIARY INFORMATION

METHOD/APPARATUS/PROCEDURE:	SOURCE AND PURITY OF MATERIALS:
Sodium iodate, sodium sulfate and water were placed in glass stoppered bottles and the bottles rotated in a thermostat. Two weeks were allowed for the attainment of equilibrium except in the case of the solubility isotherm at 25°C and 50°C where a minimum of 48 hours were allowed. Samples of the solution were drawn off through glass wool filters for analysis. The composition of the dry solid phases were determined by the method of Schreinemakers'. Sodium iodate in the liquid and solid phases was determined by adding potassium iodide to the samples, acidifying with sulfuric acid, and titrating the liberated iodine with thiosulfate solution. The sulfate content was calculated from the iodate concentration and the mass of total salts. Water was found by difference.	The authors only stated that sodium iodate and sodium sulfate were purified by customary methods.
	ESTIMATED ERROR: Nothing specified.
	REFERENCES:

COMPONENTS:	ORIGINAL MEASUREMENTS:
(1) Sodium sulfate; Na_2SO_4; [7757-82-6] (2) Sodium iodate; $NaIO_3$; [7681-55-2] (3) Water; H_2O; [7732-18-5]	Foote, H.W.; Vance, J.E. Am. J. Sci. 1930, 19, 203-13.

EXPERIMENTAL VALUES;   (Continued)

Composition of saturated solutions

t/°C	NaIO₃		Na₂SO₄		Nature of the
	mass %	mol % (compiler)	mass %	mol % (compiler)	solid phase[b]
35 (A)	0.11	0.014	32.83	5.845	D
	0.16	0.020	32.18	5.688	"
	0.29	0.036	30.92	5.392	"
	0.57	0.070	29.63	5.105	"
	0.62	0.076	28.98	4.958	D+E
	0.80	0.098	28.77	4.921	"
	1.30	0.156	26.06	4.345	E
	2.34	0.276	23.62	3.878	"
	3.33	0.392	22.57	3.705	C+E
	3.59	0.415	20.70	3.338	C
	3.73	0.426	19.33	3.075	"
	4.47	0.493	15.30	2.350	"
	10.57[c]	1.065	–	–	"
35 (B)	–	–	33.10[a]	5.905	B
	0.15[a]	0.019	32.86	5.856	B+D
	0.15[a]	0.019	32.91	5.868	"
	0.10	0.013	32.82	5.842	D
	0.28	0.035	31.06	5.424	"
	0.30	0.037	30.35	5.257	"
	0.83	0.102	28.84	4.939	D+E
	0.90	0.110	28.50	4.864	"
	1.29	0.154	25.97	4.325	E
	2.39	0.282	23.79	3.916	"
	3.33[a]	0.392	22.57	3.705	C+E
	3.59[a]	0.415	20.70	3.338	C
	3.73[a]	0.426	19.33	3.075	"
	4.47[a]	0.493	15.30	2.350	"
	10.57[a,c]	1.065	–	–	"

[a] Data taken from 35(A) isotherm

t/°C	NaIO₃		Na₂SO₄		Nature of the
50	–	–	31.76	5.574	B
	0.17	0.21	31.60	5.547	B+D
	0.13	0.016	31.70	5.569	"
	0.15	0.019	31.67	5.563	D
	0.28	0.035	30.02	5.178	"
	0.63	0.077	28.53	4.856	"
	0.98	0.119	27.56	4.658	D+E
	1.06	0.128	27.28	4.600	"
	1.25	0.149	25.93	4.315	E
	1.87	0.220	23.96	3.927	"
	2.75	0.323	22.85	3.737	"
	4.01	0.469	21.21	3.456	"
	5.29	0.619	19.97	3.258	C+E
	5.32	0.623	19.97	3.259	"
	14.06[c]	1.468	–	–	C

continued.....

COMPONENTS:	ORIGINAL MEASUREMENTS:
(1) Sodium sulfate; Na_2SO_4; [7757-82-6]	Foote, H.W.; Vance, J.E.
(2) Sodium iodate; $NaIO_3$; [7681-55-2]	*Am. J. Sci.* <u>1930</u>, *19*, 203-13.
(3) Water; H_2O; [7732-18-5]	

EXPERIMENTAL VALUES: (Continued)

[b] A = $Na_2SO_4 \cdot 10H_2O$;   B = Na_2SO_4;   C = $NaIO_3 \cdot H_2O$;   D = $NaIO_3 \cdot 4Na_2SO_4$;

E = $NaIO_3 \cdot 3Na_2SO_4$

[c] For the binary system the compiler computes the following:

soly of $NaIO_3$ = 0.479 mol kg^{-1} at 25°C ;   0.529 mol kg^{-1} at 29.5°C

0.5973 mol kg^{-1} at 35°C ;   0.8267 mol kg^{-1} at 50°C

COMMENTS AND/OR ADDITIONAL DATA:

The solubility isotherms are reported below (based on mass % units)

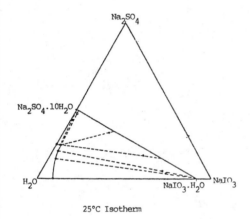

25°C Isotherm

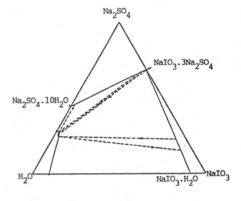

29.5°C Isotherm

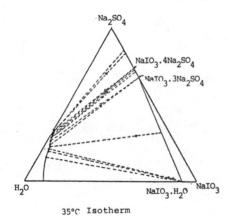

35°C Isotherm

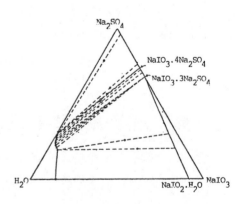

50°C Isotherm

COMPONENTS:	ORIGINAL MEASUREMENTS:
(1) Sodium chloride; NaCl; [7647-14-5] (2) Sodium iodate; NaIO$_3$; [7681-55-2] (3) Water; H$_2$O; [7732-18-5]	Foote, H.W.; Vance, J.E. *Am. J. Sci.* <u>1929</u>, *17*, 425-30.
VARIABLES: Composition at 273, 288, 298 and 308 K	PREPARED BY: Hiroshi Miyamoto

EXPERIMENTAL VALUES: Composition of saturated solutions

t/°C	NaIO$_3$ mass %	NaIO$_3$ mol % (compiler)	NaCl mass %	NaCl mol % (compiler)	Nature of the solid phase[a]
0	–	–	26.34	9.928	A
	0.29	0.032	26.36	9.970	A+D
	0.37	0.041	26.28	9.942	"
	0.38	0.042	26.30	9.952	"
	0.54	0.059	24.16	8.995	D
	0.66	0.072	23.08	8.527	"
	0.73	0.079	22.62	8.332	"
	1.03	0.111	20.85	7.593	D+B
	1.03	0.111	20.88	7.606	"
	0.83	0.086	16.30	5.712	B
	2.42[b]	0.225	–	–	"
15	–	–	26.38	9.947	A
	0.97	0.109	26.14	9.943	A+D
	0.97	0.109	26.12	9.934	"
	0.97	0.109	26.30	10.02	"
	1.29	0.144	24.64	9.287	D
	1.68	0.185	23.14	8.650	C+D
	1.71	0.189	23.14	8.653	"
	1.75	0.190	20.73	7.601	C
	1.87	0.196	16.32	5.782	"

continued.....

AUXILIARY INFORMATION

METHOD/APPARATUS/PROCEDURE:
Sodium iodate, sodium chloride and water were placed in glass stoppered bottles, and the bottles rotated in a thermostat for 24 hours. Samples of the solution were drawn off through glass wool filters. The iodate content was determined by addking KI to the solution, acidifying with sulfuric acid, and titrating the free iodine with sodium thiosulfate solution. The chloride content was calculated from the IO$_3$ concentration and the total weight of salt in solution.
Water was found by difference.
The solid phases were analyzed as wet residues after largely freeing them from water by pressing between filter papers.
The composition of the dry residue was then determined by Schreinemakers' method.

SOURCE AND PURITY OF MATERIALS:
The source of NaCl and NaIO$_3$ was not given in the original paper. The authors state that the salts were purified by usual methods, however, the details of purification were not reported.

ESTIMATED ERROR:
Nothing specified.

REFERENCES:

COMPONENTS:	ORIGINAL MEASUREMENTS:
(1) Sodium chloride; NaCl; [7647-14-5]	Foote, H.W.; Vance, J.E.
(2) Sodium iodate; NaIO$_3$; [7681-55-2]	Am. J. Sci. 1929, 17, 425-30.
(3) Water; H$_2$O; [7732-18-5]	

EXPERIMENTAL VALUES: (Continued)

Composition of saturated solutions

t/°C	NaIO$_3$ mass %	mol % (compiler)	NaCl mass %	mol % (compiler)	solid phase[a]
15	2.34	0.233	9.46	3.193	B+C
	2.35	0.234	9.47	3.197	"
	2.53	0.247	6.60	2.184	B
	5.88[b]	0.566	-	-	"
25	-	-	26.50	10.00	A
	1.96	0.225	26.08	10.03	A+C
	1.99	0.226	25.93	9.960	"
	1.98	0.225	26.08	10.03	"
	8.66[b]	0.856	-	-	C
35	-	-	26.66	10.08	A
	1.70	0.193	26.20	10.05	"
	2.39	0.273	26.04	10.06	A+C
	2.40	0.274	26.00	10.04	"
	2.39	0.273	26.02	10.05	"
	2.41	0.275	26.07	10.07	"
	2.47	0.275	23.15	8.730	C
	2.57	0.282	21.24	7.891	"
	4.51	0.454	7.87	2.68	"
	10.57[b]	1.065	--	--	"

[a] A = NaCl;   B = NaIO$_3$.5H$_2$O;   C = NaIO$_3$.H$_2$O;   D = 2NaIO$_3$.3NaCl.10H$_2$O

[b] For the binary system the compiler computes the following:

soly of NaIO$_3$ = 0.125 mol kg^{-1} at 0°C

= 0.316 mol kg^{-1} at 15°C

= 0.479 mol kg^{-1} at 25°C

= 0.5973 mol kg^{-1} at 35°C

COMMENTS AND/OR ADDITIONAL DATA:

The solubility isotherms are reproduced below (based on mass % units).

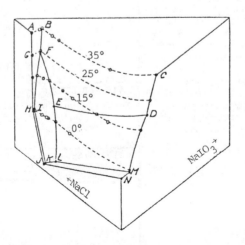

COMPONENTS:	ORIGINAL MEASUREMENTS:
(1) Sodium bromide; NaBr; [7647-15-6] (2) Sodium iodate; $NaIO_3$; [7681-55-2] (3) Water; H_2O; [7732-18-5]	Ricci, J.E. J. Am. Chem. Soc. 1934, 56, 290-5.

VARIABLES:	PREPARED BY:
Composition T/K = 278 - 323	Hiroshi Miyamoto

EXPERIMENTAL VALUES: Composition of saturated solutions

t/°C	NaBr mass %	mol %	$NaIO_3$ mass %	mol %	Density g cm^{-3}	Nature of the solid phase[a]
5	45.08	12.57	0.00	0.00	1.489	A
	45.04	12.56	0.076	0.011	1.491	A+S5
	45.00	12.54	0.075	0.011	1.492	"
	45.07	12.57	0.073	0.011	1.492	"
	45.04	12.56	0.075	0.011	1.492	"
	43.99	12.10	0.084	0.012	1.473	S5
	39.98	10.46	0.124	0.0169	1.415	"
	35.04	8.655	0.231	0.0297	1.352	"
	29.50	6.874	0.584	0.0708	1.287	"
	24.56	5.461	1.09	0.126	1.238	"
	23.53	5.186	1.27	0.146	1.229	S5+B
	23.49	5.176	1.28	0.147	1.230	"
	23.51	5.181	1.28	0.147	1.230	"
	22.95	5.027	1.25	0.142	1.225	B
	18.42	3.850	1.13	0.123	1.175	"
	14.13	2.832	1.09	0.114	1.132	"
	10.37	2.008	1.13	0.114	1.097	"
	4.97	0.920	1.44	0.139	1.052	"
	0.00	0.000	3.297	0.3094	1.027	"
	23.30	5.127	1.34	0.153	1.227	S5(m)
	21.71	4.723	1.78	0.201	1.215	S5+C(m)
	20.78	4.476	1.74	0.195	1.204	C(m)
	16.39	3.384	1.85	0.199	1.162	"

continued.....

AUXILIARY INFORMATION

METHOD/APPARATUS/PROCEDURE:	SOURCE AND PURITY OF MATERIALS:
Ternary complexes were stirred for 1-2 weeks at temperatures below 40°C, and for 2-4 days at higher temperatures (40-50°C). This length of time allowed for the attainment of equilibrium as determined in several cases by successive analysis of the solutions. Care had to be taken to seed each complex with the expected stable solid phase whenever possible, and to break up the caked hydrates which sometimes formed on mixing the salt with water in the preparation of the complexes. In one sample of the saturated solution, the iodate was determined by titration with standard thiosulfate solution. In another sample, the total solid was determined by evaporation of the solution at 100°C followed by one to two hours at 350°C. The concentration of the bromide was then determined by difference.	The salts used were prepared by recrystallization of the best available c.p. material which, in the case of the bromide, usually contained from 0.5 to 1.0 % chloride. The purified salts were dried to the anhydrous state and stored at 100°C.
	ESTIMATED ERROR: Nothing specified.
	REFERENCES:

COMPONENTS:	ORIGINAL MEASUREMENTS:
(1) Sodium bromide; NaBr; [7647-15-6]	Ricci, J.E.
(2) Sodium iodate; NaIO₃; [7681-55-2]	*J. Am. Chem. Soc.* 1934, 56, 290-5.
(3) Water; H₂0; [7732-18-5]	

EXPERIMENTAL VALUES: (Continued)

Composition of saturated solutions

t/°C	NaBr mass %	NaBr mol % (compiler)	NaIO₃ mass %	NaIO₃ mol % (compiler)	Density g cm⁻³	Nature of the solid phase[a]
5	12.26	2.438	2.03	0.210	1.123	C(m)
	6.97	1.327	2.52	0.249	1.079	"
	0.00	0.000	5.479[b]	0.525	1.050	"
15	46.54	13.23	0.00	0.000	–	A
	27.15	6.266	1.91	0.229	1.278	S5+C
	22.08	4.837	2.05	0.234	1.219	C
	18.54	3.925	2.20	0.242	1.184	"
	13.41	2.710	2.53	0.266	1.136	"
	9.16	1.78	2.69	0.272	1.101	B
	4.27	0.802	3.53	0.345	1.064	"
	0.00	0.000	5.85[b]	0.562	1.051	"
25	48.41	14.11	0.00	0.000	1.530	A
	48.23	14.11	0.42	0.064	1.538	A+S5
	48.17	14.08	0.42	0.064	1.536	"
	48.21	14.10	0.42	0.064	1.541	"
	48.22	14.11	0.42	0.064	1.534	"
	48.21	14.10	0.42	0.064	1.537	"
	47.73	13.87	0.42	0.063	–	S5
	47.35	13.69	0.42	0.063	1.522	"
	46.73	13.40	0.45	0.067	1.509	"
	43.58	12.01	0.55	0.079	1.472	"
	39.55	10.40	0.86	0.12	1.417	"
	38.83	10.13	0.95	0.13	1.406	"
	36.61	9.343	1.31	0.174	1.380	"
	35.23	8.867	1.51	0.198	1.367	"
	34.62	8.668	1.66	0.216	1.360	"
	34.34	8.575	1.71	0.222	1.359	"
	32.79	8.084	2.13	0.273	1.344	S5+C
	32.72	8.063	2.15	0.275	1.343	"
	32.56	8.011	2.18	0.279	1.343	"
	32.63	8.034	2.17	0.278	1.343	"
	32.68	8.050	2.16	0.277	1.343	"
	32.44	7.970	2.17	0.277	1.338	C
	26.39	6.072	2.35	0.281	1.266	"
	16.49	3.451	3.00	0.326	1.172	"
	7.78	1.52	4.46	0.454	1.104	"
	0.00	0.00	8.569[b]	0.8460	1.075	"
35	50.48	15.14	0.00	0.000		A
	50.04	15.16	1.01	0.159		A+S5
	50.00	15.14	1.02	0.161		"
	50.02	15.15	1.02	0.161		"
	49.46	14.86	1.03	0.161		S5
	47.82	14.05	1.08	0.165		"
	46.46	13.42	1.18	0.177		"
	44.96	12.75	1.30	0.192		"
	42.59	11.76	1.60	0.230		"
	40.55	10.96	1.98	0.278		"
	38.11	10.08	2.63	0.362		"
	38.1	10.08	2.6	0.358		S5+D(m?)

COMPONENTS:	ORIGINAL MEASUREMENTS:
(1) Sodium bromide; NaBr; [7647-15-6]	Ricci, J.E.
(2) Sodium iodate; NaIO₃ [7681-55-2]	J. Am. Chem. Soc. 1934, 56, 290-5.
(3) Water; H₂O; [7732-18-5]	

EXPERIMENTAL VALUES: (Continued)

Composition of saturated solutions

t/°C	NaBr mass %	NaBr mol %	NaIO₃ mass %	NaIO₃ mol %	Density g cm⁻³	Nature of the solid phase[a]
35	38.30	10.17	2.70	0.373		D(m?)
	36.52	9.498	2.81	0.380		"
	34.51	8.776	2.95	0.390		"
	0.00	0.00	10.58[b]	1.066		C
40	51.5	15.7	0.00	0.00		A
	50.84	15.67	1.37[b]	0.220		A+SO
	50.84	15.67	1.37	0.220		"
	50.84	15.67	1.37	0.220		"
	50.37	15.42	1.39	0.221		SO
	50.43	15.46	1.42	0.226		"
	49.38	14.94	1.53	0.241		"
	48.41	14.49	1.71	0.266		"
	47.95	14.27	1.77	0.274		SO+S5
	47.96	14.28	1.80	0.279		S5
	46.82	13.74	1.89	0.288		"
	45.93	13.34	2.00	0.302		"
	44.86	12.86	2.13	0.318		"
	43.37	12.25	2.48	0.364		"
	42.38	11.84	2.59	0.376		"
	42.10	11.73	2.65	0.384		S5+D
	41.82	11.62	2.72	0.393		"
	41.96	11.67	2.69	0.389		"
	42.44	11.85	2.52	0.366		D(m)
	41.36	11.40	2.59	0.371		D
	40.05	10.86	2.63	0.371		"
	39.5	10.64	2.65	0.371		D+C
	37.73	9.952	2.75	0.377		D(m)
	37.29	9.789	2.80	0.382		"
	39.06	10.47	2.68	0.373		C
	36.79	9.575	2.62	0.355		"
	32.08	7.909	2.77	0.355		"
	27.16	6.355	3.04	0.370		"
	20.65	4.539	3.65	0.417		"
	0.00	0.000	11.70[b]	1.192		"
45	52.55	16.24	0.00	0.000		A
	51.79	16.30	1.81	0.296		A+SO
	49.91	15.35	2.08	0.333		SO
	49.5	15.23	2.5	0.40		SO+D
	50.29	15.67	2.56	0.415		D(m)
	48.64	14.79	2.51	0.397		D
	45.95	13.45	2.50	0.380		"
	42.63	11.94	2.58	0.376		"
	-	-	-	-		D+C
	0.00	0.00	12.83[b]	1.322		C

continued.....

COMPONENTS:	ORIGINAL MEASUREMENTS
(1) Sodium bromide; NaBr; [7647-15-6]	Ricci, J.E.
(2) Sodium iodate; NaIO$_3$; [7681-55-2]	J. Am. Chem. Soc. 1934, 56, 290-5.
(3) Water; H$_2$O; [7732-18-5]	

EXPERIMENTAL VALUES: (Continued)

Composition of saturated solutions

t/°C	NaBr mass %	NaBr mol % (compiler)	NaIO$_3$ mass %	NaIO$_3$ mol % (compiler)	Density g cm^{-3}	Nature of the solid phase[a]
50	53.63	16.84	0.00	0.000		A
	53.0(+)	16.92	1.57(-)	0.261		A+E
	52.97	16.90	1.57	0.260		E
	52.57	16.89	2.37	0.396		E+SO
	52.39	16.80	2.40	0.400		SO
	52.12	16.65	2.40	0.399		"
	51.40	16.26	2.49	0.410		"
	50.97	16.06	2.62	0.429		SO+D
	50.90	16.02	2.63	0.430		"
	50.91	16.03	2.64	0.432		"
	50.93	16.04	2.63	0.431		"
	50.30	15.69	2.60	0.422		D
	47.72	14.32	2.54	0.396		"
	44.74	12.89	2.56	0.384		"
	41.56	11.50	2.67	0.384		"
	38.56	10.29	2.86	0.397		"
	32.41	8.087	3.41	0.442		"
	26.02	6.105	4.30	0.525		"
	23.54	5.403	4.73	0.564		"
	21.46	4.841	5.15	0.604		D+C
	19.75	4.365	4.93	0.566		C
	19.57	4.313	4.86	0.557		"
	13.28	2.790	6.28	0.686		"
	6.63	1.34	9.03	0.952		"
	0.00	0.00	13.49[b]	1.400		"

[a] A = NaBr.2H$_2$O;   B = NaIO$_3$.5H$_2$O;   C = NaIO$_3$.H$_2$O;   D = NaIO$_3$;   E = NaBr;

S5 = 2NaIO$_3$.3NaBr.15H$_2$O;   SO = 2NaIO$_3$.3NaBr.10H$_2$O;   m = metastable

[b] For the binary system the compiler computes the following:

soly of NaIO$_3$ = 0.2929 mol kg^{-1} at 5°C

= 0.3163 mol kg^{-1} at 15°C

= 0.4736 mol kg^{-1} at 25°C

= 0.5979 mol kg^{-1} at 35°C

= 0.6696 mol kg^{-1} at 40°C

= 0.7438 mol kg^{-1} at 45°C

= 0.7880 mol kg^{-1} at 50°C

COMMENTS AND/OR ADDITIONAL DATA:

Isotherms based on mass % units are reproduced below.

continued.....

COMPONENTS:	ORIGINAL MEASUREMENTS:
(1) Sodium bromide; NaBr; [7647-15-6]	Ricci, J.E.
(2) Sodium iodate; $NaIO_3$; [7681-55-2]	J. Am. Chem. Soc. 1934, 56, 290-5.
(3) Water; H_2O; [7732-18-5]	

COMMENTS AND/OR ADDITIONAL DATA: (Continued)

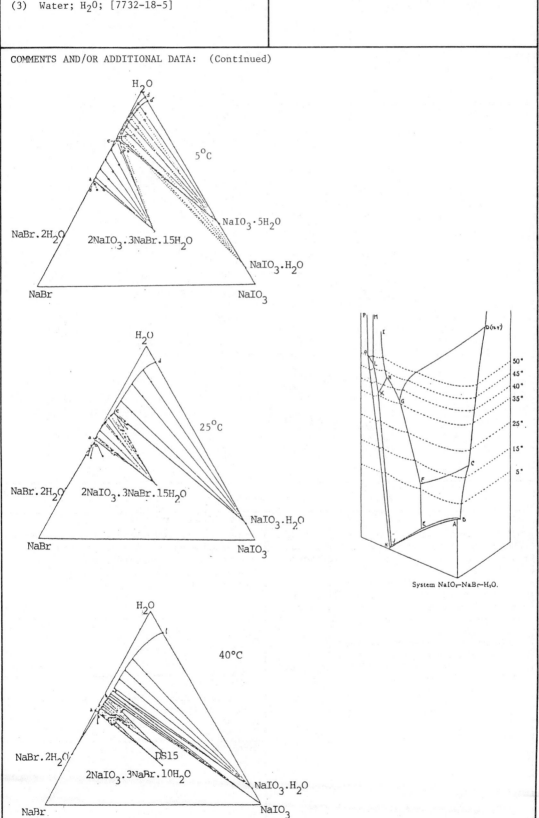

System $NaIO_3-NaBr-H_2O$.

COMPONENTS:	ORIGINAL MEASUREMENTS:
(1) Sodium iodide; NaI; [7681-82-5]	Hill, A.E.; Willson, H.S.; Bishop, J.A.
(2) Sodium iodate; NaIO$_3$; [7681-55-2]	J. Am. Chem. Soc. 1933, 55, 520-6.
(3) Water; H$_2$0; [7732-18-5]	

VARIABLES:	PREPARED BY:
Composition T/K = 281 - 313	Hiroshi Miyamoto

EXPERIMENTAL VALUES: Composition of saturated solutions

t/°C	NaI mass %	NaI mol % (compiler)	NaIO$_3$ mass %	NaIO$_3$ mol % (compiler)	Density g cm^{-3}	Nature of the solid phase[a]
8	0.00	0.00	3.89[b]	0.367	1.035	A
	6.05	0.783	1.99	0.195	1.069	"
	17.18	2.471	1.44	0.157	1.169	"
	19.47	2.882	1.84	0.206	1.196	A+S
	25.20	3.924	0.72	0.085	1.249	S
	40.70	7.629	0.08	0.011	1.445	"
	57.87	14.17	0.02	0.004	–	"
	62.44	16.66	0.02	0.004	1.861	S+D
	62.49	16.68	0.00	0.000	–	D
25	0.00	0.00	8.67[b]	0.857	1.077	B
	11.57	1.617	4.23	0.448	1.107	"
	24.54	3.882	2.68	0.321	1.253	"
	28.70	4.758	2.48	0.311	1.290	"
	31.72	5.454	2.41	0.314	1.340	"
	31.74	5.455	2.36	0.307	1.340	"
	31.99	5.515	2.36	0.308	1.341	B+S
	33.04	5.749	2.04	0.269	1.352	S
	36.64	6.607	1.22	0.167	1.377	"
	46.08	9.359	0.31	0.048	–	"
	56.83	13.70	0.17	0.031	1.722	"

continued.....

AUXILIARY INFORMATION

METHOD/APPARATUS/PROCEDURE:	SOURCE AND PURITY OF MATERIALS:
The salts were weighed into stoppered Pyrex tubes with weighed amounts of water and stirred by mechanical inversion in a thermostat for 4-7 days. Small samples of the saturated solution were withdrawn by suction through a filter into a pipet. One sample was dried to constant weight in the oven, while a second was analyzed for iodate. Iodate was determined by iodometry, thiosulfate solution being used in the titration. The water was determined gravimetrically (after evaporation).	Sodium iodate was recrystallized from water and dehydrated in an electric oven at 100°C. Sodium iodide (c.p. grade) was purified by recrystallization and dried in an electric oven at 100°C.
	ESTIMATED ERROR: Soly: precision of the analyses about 0.3 %. Temp: not given.
	REFERENCES:

COMPONENTS:	ORIGINAL MEASUREMENTS:
(1) Sodium iodide; NaI; [7681-82-5]	Hill, A.E.; Willson, H.S.; Bishop, J.A.
(2) Sodium iodate; NaIO$_3$; [7681-55-2]	J. Am. Chem. Soc. 1933, 55, 520-6.
(3) Water; H$_2$O; [7732-18-5]	

EXPERIMENTAL VALUES: (Continued)

Composition of saturated solutions

t/°C	NaI mass %	NaI mol % (compiler)	NaIO$_3$ mass %	NaIO$_3$ mol % (compiler)	Density g cm^{-3}	Nature of the solid phase[a]
25	64.67	18.06	0.08	0.017	–	S+D
	64.72	18.06	0.00	0.000	–	D
40	0.00	0.000	11.70[b]	1.192		B
	15.33	2.254	5.24	0.584		"
	18.16	2.738	4.74	0.541		"
	19.64	3.001	4.47	0.517		"
	22.06	3.441	3.90	0.461		"
	26.85	4.408	3.49	0.434		C
	33.65	5.970	2.92	0.392		"
	40.52	7.877	2.78	0.409		"
	41.16	8.076	2.79	0.415		C+S
	41.30	8.121	2.80	0.417		S
	42.55	8.510	2.72	0.412		"
	51.97	11.62	0.58	0.098		"
	64.40	18.04	0.47	0.100		"
	66.15	19.15	0.32	0.070		"
	67.58	20.16	0.28	0.063		S+D
	67.35	19.87	0.00	0.000		D

[a] A = NaIO$_3$.5H$_2$O;   B = NaIO$_3$.H$_2$O;   C = NaIO$_3$;   D = NaI.2H$_2$O;   S = solid solution

[b] For the binary system the compiler computes the following:

soly of NaIO$_3$ = 0.205 mol kg^{-1} at 8°C

= 0.480 mol kg^{-1} at 25°C

= 0.6696 mol kg^{-1} at 40°C

COMMENTS AND/OR ADDITIONAL DATA:

The solubility isotherms are reproduced below (based on mass % units).

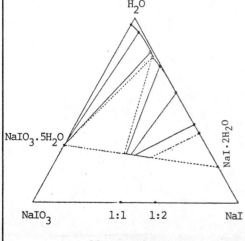

8°C Isotherm

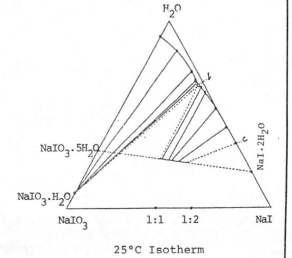

25°C Isotherm

COMPONENTS:	ORIGINAL MEASUREMENTS:
(1) Sodium iodate; $NaIO_3$; [7681-55-2] (2) Potassium iodate; KIO_3; [7758-05-6] (3) Water; H_2O; [7732-18-5]	Hill, A.E.; Ricci, J.E. *J. Am. Chem. Soc.* 1931, 53, 4305-15.
VARIABLES: Composition at 278.2, 298.2, 323.2 K	PREPARED BY: Hiroshi Miyamoto

EXPERIMENTAL VALUES: Composition of saturated solutions

t/°C	$NaIO_3$ mass %	$NaIO_3$ mol % (compiler)	KIO_3 mass %	KIO_3 mol % (compiler)	Density g cm^{-3}	Nature of the solid phase[a]
5	0.00	0.000	5.16	0.456	1.043	A
	1.41	0.136	4.71	0.420	1.051	"
	2.17	0.211	4.72	0.424	1.060	A+B
	2.48	0.238	3.19	0.283	1.046	B
	3.28[b]	0.308	0.00	0.000	1.028	"
25	0.00	0.000	8.45	0.771	1.071	A
	4.26	0.433	7.09	0.666	1.098	"
	7.13	0.743	6.73	0.649	1.126	A+C
	7.79	0.793	3.79	0.357	1.103	C
	8.57[b]	0.846	0.00	0.000	1.074	"
50	0.00	0.000	13.21	1.265	–	A
	3.92	0.417	11.92	1.173	–	"
	7.70	0.847	11.14	1.133	–	"
	10.92	1.237	10.61	1.112	–	A+C
	11.41	1.261	7.93	0.810	–	C
	12.55	1.349	4.24	0.421	–	"
	13.49[b]	1.400	0.00	0.000	–	"

[a] A = KIO_3;   B = $NaIO_3 \cdot 5H_2O$;   C = $NaIO_3 \cdot H_2O$

[b] For the binary system the compiler computes the following:
soly of $NaIO_3$ = 0.171 mol kg^{-1} at 5°C; = 0.474 mol kg^{-1} at 25°C;
= 0.7880 mol kg^{-1} at 50°C.

AUXILIARY INFORMATION

METHOD/APPARATUS/PROCEDURE:	COMMENTS AND/OR ADDITIONAL DATA:
The complexes used for the ternary system were made up from weighed amounts of water, dried $NaIO_3$ and KIO_3. For the 5°C isotherm, the solids were first dissolved by heating, and the solutions were seeded after cooling. The solutions were agitated in a thermostat at the desired temperature for about thirteen days. For the analysis, samples of filtered solution were evaporated to dryness at 110°C, and other samples were titrated for iodate by iodometry.	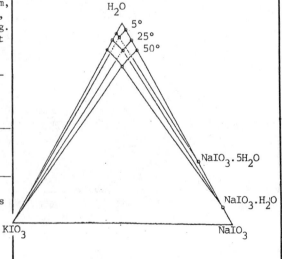

SOURCE AND PURITY OF MATERIALS:

Nothing specified.

ESTIMATED ERROR:

Nothing specified, but the compiler assumes that the agreement between duplicate analyses was around ± 0.5 %.

5°C, 25°C, 50°C Isotherm
(mass % units)

COMPONENTS:	ORIGINAL MEASUREMENTS:
(1) Sodium iodate; $NaIO_3$; [7681-55-2] (2) Rubidium iodate; $RbIO_3$; [13446-76-9] (3) Water; H_2O; [7732-18-5]	Vinogradov, E.E.; Karataeva, I.M. *Zh. Neorg. Khim.* <u>1982</u>, *27*, 2155-7; *Russ. J. Inorg. Chem. (Engl. Transl.)* <u>1982</u>, *27*, 1218-9.

VARIABLES:	PREPARED BY:
Composition at 323.2 K	Hiroshi Miyamoto

EXPERIMENTAL VALUES: Composition of saturated solutions

t/°C	$NaIO_3$ mass %	$NaIO_3$ mol % (compiler)	$RbIO_3$ mass %	$RbIO_3$ mol % (compiler)	Nature of the solid phase[a]
50	13.52[b]	1.403	–	–	A
	13.40	1.419	1.96	0.158	"
	13.77	1.465	2.03	0.164	A+B
	13.74	1.461	2.05	0.166	"
	12.20	1.283	2.50	0.200	B
	7.25	0.723	2.30	0.174	"
	3.96	0.383	2.32	0.170	"
	2.92	0.282	3.29	0.241	"
	–	–	4.39[b]	0.317	"

[a] A = $NaIO_3 \cdot H_2O$;   B = $RbIO_3$

[b] For the binary systems the compiler computes the following:

soly of $NaIO_3$ = 0.7900 mol kg^{-1}

soly of $RbIO_3$ = 0.176 mol kg^{-1}

AUXILIARY INFORMATION

METHOD/APPARATUS/PROCEDURE: Probably the isothermal method was used. Equilibrium was established after 4-5 days. Rubidium and iodate ions in the liquid and solid phases were analyzed. The sodium content was determined by difference. The composition of the solid phase was determined by X-ray analysis.	COMMENTS AND/OR ADDITIONAL DATA: The phase diagram is given below (based on mass % units).
SOURCE AND PURITY OF MATERIALS: No information given.	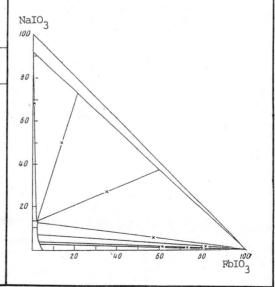
ESTIMATED ERROR: Nothing specified.	

COMPONENTS:	ORIGINAL MEASUREMENTS:
(1) Sodium iodate; $NaIO_3$; [7681-55-2] (2) Cesium iodate; $CsIO_3$; [13454-81-4] (3) Water; H_2O; [7732-18-5]	Vinogradov, E.E.; Karataeva, I.M. *Zh. Neorg. Khim.* 1982, *27*, 2155-7; *Russ. J. Inorg. Chem. (Engl. Transl.)* 1982, *27*, 1218-9.

VARIABLES:	PREPARED BY:
Composition at 323.2 K	Hiroshi Miyamoto

EXPERIMENTAL VALUES: Composition of saturated solutions

t/°C	NaIO₃ mass %	NaIO₃ mol % (compiler)	CsIO₃ mass %	CsIO₃ mol % (compiler)	Nature of the solid phase[b]
50	13.52[b]	1.403	–	–	A
	12.15	1.262	1.41	0.0942	"
	12.46	1.309	2.16	0.146	A+B
	12.46	1.306	1.96	0.132	"
	12.49	1.314	2.23	0.151	"
	12.61	1.325	2.02	0.136	"
	11.77	1.231	2.40	0.161	B
	10.27	1.054	2.12	0.140	"
	6.93	0.688	2.18	0.139	"
	3.34	0.324	3.44	0.215	"
	–	–	5.07[b]	0.312	"

[a] A = $NaIO_3 \cdot H_2O$;   B = $CsIO_3$

[b] For binary systems the compiler computes the following:

   soly of $NaIO_3$ = 0.7900 mol kg^{-1}

   soly of $CsIO_3$ = 0.174 mol kg^{-1}

AUXILIARY INFORMATION

METHOD/APPARATUS/PROCEDURE:

Probably the isothermal method was used.
Equilibrium was established after 4-5 days.
Cesium and iodate ions in the liquid and
solid phases were analyzed. The sodium
content was determined by difference.
The composition of the solid phase was determined by X-ray analysis.

SOURCE AND PURITY OF MATERIALS:

No information given.

ESTIMATED ERROR:

Nothing specified.

COMMENTS AND/OR ADDITIONAL DATA:

The phase diagram is given below (based
on mass % units).

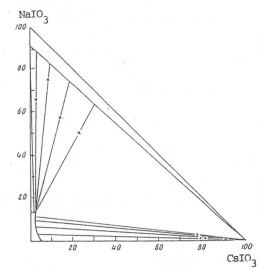

COMPONENTS:	ORIGINAL MEASUREMENTS:
(1) Sodium iodate; $NaIO_3$; [7681-55-2] (2) Aluminum iodate; $Al(IO_3)_3$; [15123-75-3] (3) Water; H_2O; [7732-18-5]	Shklovskaya, R.M.; Arkhipov, S.M.; Kidyarov, B.I.; Tokareva, A.G. *Zh. Neorg. Khim.* 1980, 25, 1423-4; *Russ. J. Inorg. Chem. (Engl. Transl.)* 1980, 25, 791.
VARIABLES: Composition at 298.2 K	PREPARED BY: Hiroshi Miyamoto

EXPERIMENTAL VALUES: Composition of saturated solutions at 25°C

$NaIO_3$		$Al(IO_3)_3$		Nature of the
mass %	mol % (compiler)	mass %	mol % (compiler)	solid phase[a]
8.66[b]	0.856	–	–	A
8.30	0.821	0.38	0.013	"
8.15	0.819	2.02	0.0728	A+B
6.83	0.678	2.21	0.0787	B
6.44	0.638	2.38	0.0846	"
6.30	0.624	2.45	0.0871	"
5.14	0.504	2.66	0.0936	"
3.60	0.350	3.18	0.111	"
1.52	0.146	4.15	0.143	"
0.9	0.09	4.48	0.154	"
–	–	5.70[b]	0.197	"

[a] $A = NaIO_3 \cdot H_2O$ $B = Al(IO_3)_3 \cdot 6H_2O$

[b] For binary systems the compiler computes the following:

soly of $NaIO_3 = 0.479$ mol kg^{-1}

soly of $Al(IO_3)_3 = 0.110$ mol kg^{-1}

AUXILIARY INFORMATION

METHOD/APPARATUS/PROCEDURE:	SOURCE AND PURITY OF MATERIALS:
The isothermal method was used. Equilibrium was reached within 15-20 days. The aluminum content in the co-existing phases was determined by complexometric titration. Sodium was determined by the flame photometry. The photometry was carried out on solutions in which the sodium concentration did not exceed 10 μg dm^{-3}.	Aluminum iodate hexahydrate was synthesized from iodic acid and aluminum hydroxide. Chemically pure grade sodium iodate monohydrate was recrystallized twice from aqueous solution.
	ESTIMATED ERROR: Soly: 1-3 rel %. Temp: precision $\pm$ 0.1 K.
	REFERENCES:

COMPONENTS:	ORIGINAL MEASUREMENTS:
(1) Sodium iodate; $NaIO_3$; [7681-55-2]	Shklovskaya, R.M.; Arkhipov, S.M.; Kidyarov, B.I.; Poleva, G.V.; Timofeev, S.I.
(2) Hafnium iodate; $Hf(IO_3)_4$; [19630-06-9]	*Zh. Neorg. Khim.* 1983, *28*, 2435-6; *Russ. J. Inorg. Chem. (Engl. Transl.)* 1983, *28*, 1384-5.
(3) Water; H_2O; [7732-18-5]	

VARIABLES:	PREPARED BY:
Composition at 298.2 K	Hiroshi Miyamoto

EXPERIMENTAL VALUES: Composition of saturated solutions at 25°C

$NaIO_3$		$Hf(IO_3)_4$		Nature of the solid phase[a]
mass %	mol % (compiler)	mass %	mol % (compiler)	
–	–	0.00037	7.6×10^{-6}	A
0.87	0.080	0.00036	7.4×10^{-6}	"
1.45	0.134	0.00034	7.1×10^{-6}	"
2.38	0.221	0.00027	5.7×10^{-6}	"
3.29	0.309	0.00025	5.3×10^{-6}	"
4.03	0.381	0.00023	4.9×10^{-6}	"
4.89	0.466	0.00020	4.3×10^{-6}	"
5.68	0.545	0.00018	3.9×10^{-6}	"
6.38	0.617	0.00020	4.4×10^{-6}	"
7.18	0.699	0.00032	7.0×10^{-6}	"
8.23	0.810	0.00034	7.5×10^{-6}	"
8.62	0.851	0.00035	7.8×10^{-6}	A+B
8.66[b]	0.856	–	–	B

[a] $A = Hf(IO_3)_4$ $B = NaIO_3 \cdot H_2O$

[b] For binary systems the compiler computes the following:

soly of $NaIO_3$ = 0.479 mol kg^{-1}

soly of $Hf(IO_3)_4$ = 4.2×10^{-6} mol kg^{-1}

AUXILIARY INFORMATION

METHOD/APPARATUS/PROCEDURE:	SOURCE AND PURITY OF MATERIALS:
The isothermal method was used. Equilibrium was reached in 25-30 days. Samples of the coexisting phases were analyzed for sodium by emission spectrometry. The hafnium content was determined potentiometrically using Arsenazo III after reducing the iodate ion with hydroxylamine. The composition in the solid phase was determined by the method of residues and the result was checked by X-ray analysis.	C.p. grade $NaIO_3 \cdot H_2O$ was recrystallized from distilled water. Hafnium iodate was prepared by the action of aqueous iodic acid solution on freshly precipitated hafnium hydroxide (ref 1.)
	ESTIMATED ERROR: Soly: within 1-3 % rel. % (emission spectrometry for Na). Temp: precision ± 0.1 K.
	REFERENCES: 1. Deabriges, J.; Rohmer, R. *Bull. Soc. Chim. Fr.* 1968, *2*, 521.

COMPONENTS:	ORIGINAL MEASUREMENTS:
(1) Sodium iodate; $NaIO_3$; [7681-55-2] (2) Iodic acid; HIO_3; [7782-68-5] (3) Water; H_2O; [7732-18-5]	Meerburg, P.A. Z. Anorg. Allg. Chem. <u>1905</u>, 45, 324-44.
VARIABLES: T/K = 303 Composition	PREPARED BY: Hiroshi Miyamoto

EXPERIMENTAL VALUES: Composition of saturated solutions at 30°C

Iodic Acid		Sodium Iodate		Nature of
mass %	mol % (compiler)	mass %	mol % (compiler)	the solid phase
0	0	9.36[b]	0.931	A
1.98	0.526	9.52	0.968	"
4.86	0.576	10.22	1.077	"
5.86	0.708	11.04	1.187	"
7.40	0.915	11.60	1.275	A(m)
9.73	1.280	14.73	1.722	"
6.76	0.826	11.18	1.215	A+C
6.75	0.824	11.10	1.205	"
6.66	0.814	11.28	1.226	"
7.80	0.955	10.30	1.121	C
9.15	1.120	9.00	0.980	"
9.93	1.222	8.71	0.953	"
11.20	1.280	7.54	0.826	"
11.89	1.471	7.21	0.793	C+D
11.75	1.451	7.18	0.788	"
14.62	1.822	5.65	0.629	D
23.23	3.139	3.69	0.443	"
32.68	4.920	2.91	0.389	"
40.91	6.882	2.64	0.395	"
46.62	8.567	2.67	0.436	"
55.48	11.77	2.12	0.400	"
65.47	16.94	1.83	0.420	"
76.19	25.73	1.42	0.426	D+B
76.70[b]	25.21	0	0	B

[a] A = $NaIO_3.1.5H_2O$;   B = HIO_3;   C = $Na_2O.2I_2O_5$;   D = $NaIO_3.2HIO_3$.

[b] For binary systems the compiler computes the following:

$\quad\quad\quad$ soly of HIO_3 = 18.71 mol kg^{-1}

$\quad\quad\quad$ soly of $NaIO_3$ = 0.522 mol kg^{-1}

METHOD/APPARATUS/PROCEDURE:	SOURCE AND PURITY OF MATERIALS
A mixture of $NaIO_3$, HIO_3 and water was placed in a bottle and the bottle agitated in a thermostat for a week or more at a desired temperature. Equilibrium was established from supersaturation.	Nothing specified.
	ESTIMATED ERROR:
	Nothing specified.
The iodic acid and sodium iodate contents were detd as follows: an excess of KI was added to an aliquot of satd sln, and the HIO_3 content detd by titration of the iodine liberated with standard sodium thiosulfate. Dil sulfuric acid was then added to the solution and the iodine liberated was again titrated with sodium thiosulfate to obtain the total iodate concentration.	COMMENTS AND/OR ADDITIONAL DATA: (mass % units) 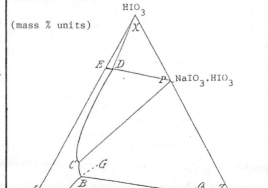
The sodium iodate concentration was calculated from the difference between the second and the first titration.	
Composition of solid phases determined by the method of residues.	

COMPONENTS:	ORIGINAL MEASUREMENTS:
(1) Sodium iodate; $NaIO_3$; [7681-55-2] (2) Iodic acid; HIO_3; [7782-68-5] (3) Water; H_2O; [7732-18-5]	Shibuya, M.; Watanobe, T. *Denki Kagaku* 1967, 35, 550-8

VARIABLES:	PREPARED BY:
Composition T/K = 288.2	Hiroshi Miyamoto and Mark Salomon

EXPERIMENTAL VALUES: Composition of saturated solutions at 15.0°C

Iodic acid		Sodium Iodate		density	Nature of the
mass %	mol % (compiler)	mass %	mol % (compiler)	$g\ cm^{-3}$	solid phase[a]
0.00	0.000	5.87[b]	0.564	1.051	A
0.62	0.067	5.87	0.568	1.057	"
1.79	0.197	5.87	0.574	1.070	"
2.59	0.287	5.92	0.584	1.077	"
3.81	0.428	5.90	0.589	1.089	"
5.12	0.583	5.99	0.606	1.103	"
5.66	0.649	6.14	0.626	1.109	"
7.63	0.895	6.38	0.665	1.132	A+B
8.06	0.947	6.18	0.646	1.135	B
9.11	1.08	5.64	0.592	1.143	B+C
10.57	1.256	4.74	0.501	1.145	C
12.21	1.459	3.68	0.391	1.152	"
19.16	2.435	2.43	0.274	1.215	"
33.45	5.018	1.88	0.251	-	"
43.26	7.444	1.82	0.278	-	"

[a] $A = NaIO_3 \cdot 5H_2O$; $B = 2NaIO_3 \cdot I_2O_5$; $C = NaIO_3 \cdot I_2O_5$

[b] For the binary system the compiler computes the following:

Soly of $NaIO_3$ = 0.315 mol kg^{-1}

AUXILLARY INFORMATION

METHOD/APPARATUS/PROCEDURE:	SOURCE AND PURITY OF MATERIALS:
Isothermal method by three techniques depending upon mole fraction, x, of HIO_3. (1) For x = 0 to 0.574. Excess acid added to $NaIO_3$ sln and stirred for 5 h. (2) For x = 0.574 to 0.646. Aq acid sln for x = 0.638 was cooled to obtain HIO_3 crystals. The crystals were added to an unsatd sln of $NaIO_3$, and the mixture was stirred for a long time. (3) x = 0.646 to 1.0. Method essentially identical to (2) except that the acid crystals pptd from a sln where x = 0.883, and stirring time was stated as 48 h.	Sodium iodate was recryst three times from an aqueous sln prepd by electrolytic oxidation of iodine in alkaline sln. Iodic acid was prepd by ion exchange as follows: aq $NaIO_3$ sln was passed through a column of H^+-resin (Amberlite IR 120), and the eluate was concentrated to about 30 % acid content by evaporation. The acid content was detd by acid-base titration.
After equil was established and the slns allowed to settle, aliquots of satd sln were withdrawn with a pipet and weighed. The densities of the satd slns were detd. The total iodate concn was detd iodometrically, and the HIO_3 detd by acid-base titrn. Sodium was detd by difference.	**ESTIMATED ERROR:** Soly: rel error probably ± 0.2 % (compilers). Temp: precision ± 0.05 K.
The composition of the solid phase was detd as follows: chem analyses were used to detn the acid and $NaIO_3$ contents, and thermogravimetry and NMR were used to detn the water content.	**REFERENCES:**

COMPONENTS:	ORIGINAL MEASUREMENTS:
(1) Sodium iodate; $NaIO_3$; [6781-55-2]	Kolthoff, I.M.; Chantooni, Jr., M.K.
(2) 6,7,10,17,18,20,21-Octahydrodibenzo [b,k] [1,4,7,10,13,16] hexaoxacyclooctadecin (dibenzo-18-crown-6); $C_{20}H_{24}O_6$; [14187-32-7]	*Anal. Chem.* 1980, *52*, 1039-44.
(3) Methanol; CH_4O; [67-56-1]	

VARIABLES:	PREPARED BY:
T/K = 298	Hiroshi Miyamoto

EXPERIMENTAL VALUES:

The solubility product of $NaIO_3$ in methanol at 25°C is

$$1.5 \times 10^{-7} \ mol^2 \ dm^{-6}$$

COMMENTS AND/OR ADDITIONAL DATA:

In solutions saturated with respect to $NaIO_3$ and dibenzo-18-crown-6 (DB-18), the authors studied the equilibrium

$$Na^+ \ + \ L \ = \ LNa^+ \quad ; \quad K_f(LNa^+) \ = [L][Na^+]/[LNa^+]$$

where L = (ligand) concentration of dibenzo-18-crown-6. Details of experimental method presumed to be similar to those for KIO_3-DB-18-MeOH system (compiled elsewhere in this volume). Authors only report log $[K_f(LNa^+)/mol^{-1} \ dm^3]$ = 4.4.

AUXILIARY INFORMATION

METHOD/APPARATUS/PROCEDURE:	SOURCE AND PURITY OF MATERIALS:
A Markson No. 1001 Na^+ specific ion electrode used to determine a_{Na}^+. The electrode was calibrated and found to respond in a Nernstian manner.	Fisher "c.p." grade $NaIO_3$ was recrystallized 3 times from distilled water, and dried at 70°C. (Fisher "Spectro purity" grade) was distilled from magnesium turnings.
	ESTIMATED ERROR:
	Nothing specified.
	REFERENCES:

COMPONENTS:	EVALUATOR:
(1) Potassium iodate; KIO_3; [7758-05-6] (2) Water; H_2O; [7732-18-5]	H. Miyamoto Niigata University Niigata, Japan and Mark Salomon US Army ET & DL Fort Monmouth, NJ, USA <div align="right">June, 1986</div>

CRITICAL EVALUATION:

THE BINARY SYSTEM

Solubility data for KIO_3 in pure water have been reported in 30 publications (1-28, 35, 36). A summary of the solubility data over the temperature range 273-373 K is given in Table 1. Note that the data from references (15, 16, 20 and 31) have been compiled in the $LiIO_3$ chapter, and the data in reference (3) have been compiled in the $NaIO_3$ chapter. With the exception of the high temperature study of Benrath et al. (7) which employed the synthetic method, all other studies are based on the isothermal method. From Table 1 it is seen that a number of authors reported identical solubilities in more than one publication, and the evaluators have therefore treated these results as one independent measurement. Thus, at 298.2 K, the data in (3, 8, 10) have been treated as one independent solubility value as have the data reported in (17, 19, 23), and at 323.2 K in (18, 21, 26). Most authors reported the solid phase over the temperature range of 278-323 K as the anhydrous salt, and although Breusov et al. (11) and Benrath et al. (7) did not report the nature of the solid phases over the temperature range of 273-573 K, it probably is the anhydrous salt in all cases.

In fitting the solubility data to the two smoothing equations, a number of data points were rejected as the differences in smoothed (calculated) solubilities differed from the experimental values by more than 2σ (where σ is the standard error of estimate). The rejected data are referenced in Table 1, and it should be noted that the results of Wright (36) are so divergent from all other data that one must carefully question whether his results for ethanol-water mixtures are of any practical value.

Thirty-two data points over the temperature range 273-373 K (see Table 1) were fitted to the smoothing equations with the following results:

$$Y_x = -20587.2/(T/K) - 92.129 \ln (T/K) + 549.07 + 0.12250 (T/K)$$

$$\sigma_y = 0.013 \qquad\qquad \sigma_x = 4.9 \times 10^{-5}$$

and

$$Y_m = -4037.0/(T/K) - 6.671 \ln (T/K) + 51.547$$

$$\sigma_y = 0.0079 \qquad\qquad \sigma_m = 0.0047$$

Table 2 lists the solubilities calculated from these two smoothing equations, and the smoothed solubilities are designated as *recommended* solubilities.

Benrath's data (7) over the temperature range of 390-573 K were treated separately, and the following smoothing equations were obtained:

$$Y_x = -9532/(T/K) - 30.702 \ln (T/K) + 188.09 + 0.03523(T/K)$$

$$\sigma_y = 0.024 \qquad\qquad \sigma_x = 8.3 \times 10^{-4}$$

and

$$Y_m = 298.8/(T/K) + 2.888 \ln (T/K) - 17.21$$

$$\sigma_y = 0.015 \qquad\qquad \sigma_m = 0.090$$

Smoothed solubilities based on Benrath's data are given in Table 3, and are designated as *tentative* solubilities.

COMPONENTS:	EVALUATOR:
(1) Potassium iodate; KIO_3; [7758-05-6] (2) Water; H_2O; [7732-18-5]	H. Miyamoto Niigata University Niigata, Japan and Mark Salomon US Army ET & DL Fort Monmouth, NJ, USA June, 1986

CRITICAL EVALUATION:

Table 1.   Summary of solubilities in the KIO_3-H_2O system[a]

T/K	mol kg^{-1}	mole fraction	reference
273.2	0.221	0.00396	9
273.2	0.224	0.00402	11
278.2	0.259	0.00464	4
278.2	0.254	0.00456	3
278.2	0.2556	0.00458	5
283.2	0.300	0.00538	11
288.2	0.335	0.005994	2
293.2	0.378	0.006758	2
293.2	0.386	0.00690	35
293.2[a]	0.59	-----	36
298.2[a]	0.438	0.00783	11
298.2	0.431	0.00771	3,8,10
298.2	0.431	0.00771	13
298.2	0.432	0.00772	4
298.2	0.432	0.00772	12
298.2	0.4314	0.00771	5
298.2	0.4312	0.007709	6
298.2	0.431	0.00771	27
298.2	0.431	0.00771	35
298.2[a]	0.425	0.00760	2
298.2	0.428	0.00765	24
298.2	0.429	0.00766	17,19,23
298.2	0.4325	0.007732	28
298.2[a]	0.45	0.0080	14
303.2[a]	0.491	0.00877	1
303.2	0.482	0.00861	11
303.2	0.487	0.00869	35
313.2	0.585	0.01043	11
323.2[a]	0.5989	0.01067	22
323.2[a]	0.6228	0.01109	18,21,26
323.2[a]	0.703	0.01250	11
323.2	0.7112	0.01265	3
323.2	0.7112	0.01265	9
323.2	0.7106	0.01264	5
323.6	0.7206	0.01280	4
333.2	0.844	0.01498	11
343.2	0.985	0.01744	11
353.2	1.136	0.02005	11
363.2	1.306	0.02300	11
373.2	1.475	0.02588	11
373.2[a]	4.60	-----	36

[a]rejected data points

COMPONENTS:	EVALUATOR:
(1) Potassium iodate; KIO$_3$; [7758-05-6] (2) Water; H$_2$0; [7732-18-5]	H. Miyamoto Niigata University Niigata, Japan and Mark Salomon US Army ET & DL Fort Monmouth, NJ, USA June, 1986

CRITICAL EVALUATION:

Table 2.   Smoothed solubilities from 273-373 K[a]

T/K	mol kg^{-1}[b]	mole fraction
273.2	0.223	0.00398
278.2	0.258	0.00461
283.2	0.296	0.00531
288.2	0.337	0.00605
293.2	0.382	0.00685
298.2	0.430	0.00770
303.2	0.481	0.00860
313.2	0.592	0.01039
323.2	0.715	0.01266
333.2	0.849	0.01497
343.2	0.992	0.01744
353.2	1.143	0.02009
363.2	1.300	0.00229
373.2	1.446	0.02593

[a]All data in this table are designated as *recommended*.

[b]Reference molality used in the smoothing equation is 0.431 mol/kg.

COMPONENTS:	EVALUATOR:
(1) Potassium iodate; KIO$_3$; [7758-05-6] (2) Water; H$_2$O; [7732-18-5]	H. Miyamoto Niigata University Niigata, Japan and Mark Salomon US Army ET & DL Fort Monmouth, NJ, USA June, 1986

CRITICAL EVALUATION:

Table 3. Smoothed solubilities from 393–573 K
based on data from (7)[a]

T/K	mol kg^{-1}[b]	mole fraction
393.2	1.69	0.0294
403.2	1.85	0.0323
413.2	2.02	0.0353
423.2	2.20	0.0384
433.2	2.40	0.0417
443.2	2.60	0.0450
453.2	2.82	0.0485
463.2	3.04	0.0521
473.2	3.28	0.0558
483.2	3.53	0.0597
493.2	3.79	0.0638
503.2	4.07	0.0680
513.2	4.35	0.0724
523.2	4.66	0.0770
533.2	4.97	0.0818
543.2	5.30	0.0869
553.2	5.64	0.0922
563.2	6.00	0.0978
573.2	6.37	0.1038

[a]All data in this table are designated as *tentative*.

[b]Reference molality used in the smoothing equation is 3.69 mol/kg.

COMPONENTS:	EVALUATOR:
(1) Potassium iodate; KIO_3; [7758-05-6] (2) Water; H_2O; [7732-18-5]	Hiroshi Miyamoto Department of Chemistry Niigata University Niigata, Japan June, 1984

CRITICAL EVALUATION:

TERNARY SYSTEMS

1. One Saturating Component

Ricci and Nesse (28) measured solubilities of potassium iodate in mixtures of water and 1,4-dioxane at 298 K over the complete range of concentration from 0 to 100 % dioxane at intervals of 10 % by mass. The measurements cover a dielectric constant range from 2.10 to 78.50. The solubility of potassium iodate decreases with   increasing dioxane concentration, that is, the solubility decreases with decreasing dielectric constant of the solvent mixture.

Bronsted (29) measured solubilities of potassium iodate in aqueous potassium hydroxide solutions at 298 K over the potassium hydroxide concentration range from 4.71 to 15.02 mol dm^{-3}. The temperature dependence of the solubility showed a minimum near the hydroxide concentration of 12 mol dm^{-3}.

2. Two Saturating Components

Summaries of solubilities in aqueous ternary systems with 2 saturating components are given in Tables 4-6.

The System With Iodic Acid. This system was studied by Meerburg (1) at 303 K and by Smith (9) at 278, 298 and 303 K. Two double salts $KIO_3 \cdot HIO_3$ and $KIO_3 \cdot 2HIO_3$ are found in this system.

Systems With The Other Alkali Metal Iodates. Solubility studies of ternary systems containing potassium iodate and other alkali iodates have been reported in 5 publications (2, 12, 13, 15, 16) (see Table 4). The two ternary systems $KIO_3-NaIO_3-H_2O$ are of the simple eutonic type, and no double salts form. The sodium salt in the solid phase at 278 K is the pentahydrate, and that at 298 and 323 K is the monohydrate. The solubility for the ternary $KIO_3-LiIO_3-H_2O$ system has been reported in 2 publications (15, 16). The double salt $KIO_3 \cdot 2LiIO_3$ was formed.

Table 4. Summary of solubility studies on ternary systems
with potassium iodate and other alkali metal iodates

Ternary system	T/K	Solid phase
$KIO_3 - LiIO_3 - H_2O$	298	KIO_3, $LiIO_3$, $KIO_3 \cdot 2LiIO_3$
$KIO_3 - LiIO_3 - H_2O$	323	KIO_3, $LiIO_3$, $KIO_3 \cdot 2LiIO_3$
$KIO_3 - NaIO_3 - H_2O$	278	KIO_3, $NaIO_3 \cdot 5H_2O$
$KIO_3 - NaIO_3 - H_2O$	298, 323	KIO_3, $NaIO_3 \cdot H_2O$
$KIO_3 - RbIO_3 - H_2O$	298	Not given
$KIO_3 - CsIO_3 - H_2O$	298	KIO_3, $CsIO_3$

Systems With Alkaline Earth Metal Iodates. The ternary $KIO_3-Mg(IO_3)_2-H_2O$ system at 323 K has been studied by Vinogradov and Karataeva (25). The dominant feature in this system is the existence of the double salt $2KIO_3 \cdot Mg(IO_3)_2 \cdot 4H_2O$. The ternary $KIO_3-Ba(IO_3)_2-H_2O$ system was studied by Azarova and Vinogradov (24) and is of the simple eutonic type: no double salts were formed. Ternary systems with calcium and strontium iodate have not been studied.

COMPONENTS:	EVALUATOR:
(1) Potassium iodate; KIO_3; [7758-05-6] (2) Water; H_2O; [7732-18-5]	Hiroshi Miyamoto Department of Chemistry Niigata University Niigata, Japan June, 1984

CRITICAL EVALUATION:

Systems With Transition Metal and Rare Earth Iodates. Solubilities in ternary systems with transition metal iodates have been reported in 4 publications (18, 21, 22, 26), and are summarized in Table 5.

Table 5. Summary of solubility studies on ternary systems with KIO_3 and transition metal iodates

Ternary system	T/K	Solid phase	Reference
KIO_3 - $Mn(IO_3)_2$ - H_2O	323	KIO_3, $Mn(IO_3)_2 \cdot K_2[Mn(IO_3)_4(H_2O)_2]$	21
KIO_3 - $Co(IO_3)_2$ - H_2O	323	KIO_3, $Co(IO_3)_2 \cdot 2H_2O$ $2KIO_3 \cdot Co(IO_3)_2 \cdot 2H_2O$	18
KIO_3 - $Ni(IO_3)_2$ - H_2O	323	$KIO_3 \cdot Ni(IO_3)_2 \cdot 2H_2O$ $K_2[Ni(IO_3)_2 \cdot (H_2O)_2]$	22
KIO_3 - $Cu(IO_3)_2$ - H_2O	323	KIO_3, $Cu(IO_3)_2 \cdot 2H_2O$ $2KIO_3 \cdot Cu(IO_3)_2 \cdot 2H_2O$	26
KIO_3 - $Zn(IO_3)_2$ - H_2O	323	KIO_3, $Zn(IO_3)_2 \cdot 2H_2O$, $2KIO \cdot Zn(IO_3)_2 \cdot 2H_2O$	18

Double salts $2KIO_3 \cdot Co(IO_3)_2 \cdot 2H_2O$ and $KIO_3 \cdot Zn(IO_3)_2 \cdot 2H_2O$ were reported in (18) by Lepeshkov, Vinogradov and Karataeva. Further investigation of the double compounds by various physicochemical methods (21) suggest that the compounds are complexes with Co^{2+} or Zn^{2+} central ions in an octahedral environment of four IO_3^- ions and two water molecules. Vinogradov, Karataeva and Lepeshkov (26) reported that like the double compounds of potassium iodate with cobalt or zinc iodate, the double salt of potassium iodate and copper iodate is probably a complex with the following formula: $K_2[Cu(IO_3)_4(H_2O)_2]$. Therefore, the dominant feature in these systems is the existence of complex compounds with the general formula $K_2[M(IO_3)_4 \cdot (H_2O)_2]$ (M = Mn, Co, Ni, Cu and Zn), and the transition metal iodate in the solid phase is $M(IO_3)_2 \cdot 2H_2O$ (M = Co, Ni, Cu and Zn). Manganese iodate is the anhydrate.

The ternary KIO_3-$Nd(IO_3)_3$-H_2O system was studied by Tarasova, Vinogradov and Kudinov (23). The system is of the simple eutonic type, and no double salts form.

Although aluminum cannot be classified as a transition metal, this system is reviewed in this section. The ternary KIO_3-$Al(IO_3)_3$-H_2O system studied by Vinogradov and Tarasova (19) is eutonic, and no double compounds form.

System With Potassium Halides. Solubility studies for these ternary systems are summarized in Table 6. These systems are eutonic, and no double salts form.

Table 6. Summary of the ternary systems KIO_3 - Potassium halide - H_2O

Ternary system	T/K	Solid phase	Reference
KIO_3 - KCl - H_2O	278, 298, 323	KIO_3, KCl	3
KIO_3 - KBr - H_2O	278, 298, 322	KIO_3, KBr	5
KIO_3 - KI - H_2O	298	KIO_3, KI	6

COMPONENTS:	EVALUATOR:
(1) Potassium iodate; KIO_3; [7758-05-6] (2) Water; H_2O; [7732-18-5]	Hiroshi Miyamoto Department of Chemistry Niigata University Niigata, Japan June, 1984

CRITICAL EVALUATION:

<u>System With The Other Potassium Salts</u>. The ternary system $KIO_3-K_2SO_4-H_2O$ was studied by Hill and Ricci (3). At 278, 298 and 323 K, compound formation such as those which form in the corresponding sodium system (32) does not occur. The potassium system is thus of the simple eutonic type.

The ternary system $KIO_3-KNO_3-H_2O$ was studied by Hill and Brown (4). At 278, 298 and 323 K, no compound formation occurs which differs from the corresponding sodium salt systems (33, 34) where compound formation was observed.

The ternary system $KIO_3-KClO_3-H_2O$ was studied by Ricci (8). Neither compound formation nor solid solution were observed at the temperature studied.

The ternary system $KIO_3-K_2MoO_4-H_2O$ was studied by Ricci and Loprest (10). The results showed that the only solids are the pure anhydrous salts, and no double compounds form at 298 K.

<u>The System With Potassium Hydroxide</u>. Solubilities in the ternary $KIO_3-KOH-H_2O$ system at 298 K have been studied by Leposhkov, Vinogradov and Tarasova (17). The phase diagram is of the simple eutonic type, and no double salts form.

OTHER MULTICOMPONENT SYSTEMS

The quaternary $KIO_3-KI-KOH-H_2O$ system was studied by Malyshev, Kuz'menko, Novikov and Traul'ko (30) at 293, 333 and 353 K. The pH was adjusted to 13.8 with potassium hydroxide, and the authors stated that the solubility of potassium iodate in the alkaline medium is less than in the KIO_3-KI-H_2O ternary system. The compositions of the solid phase in equilibrium with the saturated solutions are KIO_3 and KI.

The quaternary system $KIO_3-LiIO_3-HIO_3-H_2O$ has been studied by Azarova and Vinogradov (31) at 323 K. Three double salts, $KIO_3.2LiIO_3$, $KIO_3.HIO_3$ and $KIO_3.2HIO_3$ were formed, and lithium iodate and iodic acid formed a restricted range of solid solutions.

The $KIO_3-LiIO_3-LiOH-KOH-H_2O$ system at 298 K was studied by Vinogradov, Lepeshkov and Tarasova (20). Solubilities in the quaternary systems $KIO_3-LiIO_3-LiOH-H_2O$ and $KIO_3-LiOH-KOH-H_2O$ have been reported, but data for the five component system were not given. Solid phases found in this study are the pure components $LiIO_3$, KIO_3, $LiOH.2H_2O$ and $KOH.2H_2O$. Solid solutions and the double salt $KIO_3.2LiIO_3$ were also found.

REFERENCES:

1. Meerburg, P. A. Z. Anorg. Allg. Chem. <u>1905</u>, 45, 324.

2. Flottmann, F. Z. Anal. Chem. <u>1928</u>, 73, 1.

3. Hill, A. E.; Ricci, J. E. J. Am. Chem. Soc. <u>1931</u>, 53, 4305.

4. Hill, A. E.; Brown, S. F. J. Am. Chem. Soc. <u>1931</u>, 53, 4316.

5. Ricci, J. E. J. Am. Chem. Soc. <u>1934</u>, 56, 290.

6. Ricci, J. E. J. Am. Chem. Soc. <u>1937</u>, 59, 866.

7. Benerath, A.; Gjedebo, F.; Schiffer, B.; Wunderlich, H. Z. Anorg. Allg. Chem. <u>1937</u>, 231, 285.

8. Ricci, J. E. J. Am. Chem. Soc. <u>1938</u>, 60, 2040.

9. Smith, S. B. J. Am. Chem. Soc. <u>1947</u>, 69, 2285.

COMPONENTS:	EVALUATOR:
(1) Potassium iodate; KIO_3; [7758-05-6] (2) Water; H_2O; [7732-18-5]	Hiroshi Miyamoto Department of Chemistry Niigata University Niigata, Japan June, 1984

CRITICAL EVALUATION:

REFERENCES: (Continued)

10. Ricci, J. E.; Loprest, F. J. J. Am. Chem. Soc. 1953, 75, 1224.

11. Breusov, O. N.; Kashina, N. I.; Revzina, T. V.; Sobolevskaya, N. G. Zh.
    Neorg. Khim. 1967, 12, 2240; Russ. J. Inorg. Chem. (Engl. Transl.) 1967, 12, 1179.

12. Kirgintsev, A. N.; Yakobi, N. Y. Zh. Neorg. Khim. 1968, 13, 2851; Russ. J. Inorg.
    Chem. (Engl. Transl.) 1968, 13, 1467.

13. Kirgintsev, A. N.; Shklovskaya, R. M.; Arkhipov, S. M. Izv. Akad. Nauk SSSR, Ser.
    Khim. 1971, 2631; Bull. Acad. Sci. USSR, Div. Chem. (Engl. Transl.) 1971, 2501.

14. Kolthoff, I. M.; Chantooni, Jr., M. K. J. Phys. Chem. 1973, 77, 523.

15. Azarova, L. A.; Vinogradov, E. E.; Mikhailova, E. M.; Pakomov, V. I. Zh. Neorg.
    Khim. 1973, 18, 2559; Russ. J. Inorg. Chem. (Engl. Transl.) 1973, 18, 1357.

16. Shklovskaya, R. M.; Kashina, N. I.; Arkhipov, S. M.; Kuzina, V. A.; Kidyarov, B. I.
    Zh. Neorg. Khim. 1975, 20, 783; Russ. J. Inorg. Chem. (Engl. Transl.) 1975, 20, 441.

17. Lepeshkov, I. N.; Vinogradov, E. E.; Tarasova, G. N. Zh. Neorg. Khim. 1976, 21,
    1353; Russ. J. Inorg. Chem. (Engl. Transl.) 1976, 21, 739.

18. Lepeshkov, I. N.; Vinogradov, E. E.; Karataeva, I. M. Zh. Neorg. Khim. 1977, 22,
    2277; Russ. J. Inorg. Chem. (Engl. Transl.) 1977, 22, 1232.

19. Vinogradov, E. E.; Tarasova, G. N. Zh. Neorg. Khim. 1978, 23, 3161; Russ. J. Inorg.
    Chem. (Engl. Transl.) 1978, 23, 1754.

20. Vinogradov, E. E.; Lepeshkov, I. N.; Tarasova, G. N. Zh. Neorg. Khim. 1978, 23,
    3360; Russ. J. Inorg. Chem. (Engl. Transl.) 1978, 23, 1865.

21. Vinogradov, E. E.; Karataeva, I. M.; Lepeshkov, I. N. Zh. Neorg. Khim. 1979, 24,
    223; Russ. J. Inorg. Chem. (Engl. Transl.) 1979, 24, 124.

22. Lepeshkov, I. N.; Vinogradov, E. El; Karataeva, I. M. Zh. Neorg. Khim. 1980, 25,
    832; Russ. J. Inorg. Chem. (Engl. Transl.) 1980, 25, 463.

23. Tarasova, G. N.; Vinogradov, E. E.; Kudinov, I. B. Zh. Neorg. Khim. 1981, 26,
    2841; Russ. J. Inorg. Chem. (Engl. Transl.) 1981, 26, 1520.

24. Azarova, L. A.; Vinogradov, E. E. Zh. Neorg. Khim. 1982, 27, 2967; Russ. J.
    Inorg. Chem. (Engl. Transl.)1982, 27,1681.

25. Vinogradov, E. E.; Karataeva, I. M. Zh. Neorg. Khim. 1976, 21, 1666; Russ. J.
    Inorg. Chem. (Engl. Transl.) 1976, 21, 912.

26. Vinogradov, E. E.; Karataeva, I. M.; Lepeshkov, I. N. Zh. Neorg. Khim. 1979, 24,
    1375; Russ. J. Inorg. Chem. (Engl. Transl.) 1979, 24, 762.

27. Chang, T. L.; Hsieh, Y. Y. J. Chinese Chem. Soc. 1949, 16, 10.

28. Ricci, J. E.; Nesse, G. J. J. Am. Chem. Soc. 1942, 64, 2305.

29. Bronsted, J. N. J. Am. Chem. Soc. 1920, 40, 1448.

30. Malyshev, A. A.; Kuz'menko, A. L.; Novikov, G. I.; Traul'ko, I. V. Zh. Neorg.
    Khim. 1981, 26, 832; Russ. J. Inorg. Chem. (Engl. Transl.) 1981, 26, 448.

31. Azarova, L. A.; Vinogradov, E. E.; Lepeshkov, I. M. Zh. Neorg. Khim. 1978, 23,
    1952; Russ. J. Inorg. Chem. (Engl. Transl.) 1978, 23, 1072.

COMPONENTS:	EVALUATOR:
(1) Potassium iodate; KIO_3; [7758-05-6] (2) Water; H_2O; [7732-18-5]	Hiroshi Miyamoto Department of Chemistry Niigata University Niigata, Japan June, 1984

CRITICAL EVALUATION:

REFERENCES: (Continued)

32. Foote, H. W.; Vance, J. E. *Am. J. Sci.* 1930, *19*, 203.

33. Foote, H. W.; Vance, J. E. *Am. J. Sci.* 1929, *18*, 375.

34. Hill, A. E.; Donovan, J. E. *J. Am. Chem. Soc.* 1931, *53*, 934.

35. Miyamoto, H.; Hasegawa, T.; Sano, H. *J. Solution Chem.* in press.

36. Wright, R. *J. Chem. Soc.* 1927, 1334.

COMPONENTS:	ORIGINAL MEASUREMENTS:
(1) Potassium iodate; KIO_3; [7758-05-6] (2) Water; H_2O; [7732-18-5]	Flottmann, F. Z. Anal. Chem. 1928, 73, 1-39.

VARIABLES:	PREPARED BY:
T/K = 288, 293 and 298	Hiroshi Miyamoto

EXPERIMENTAL VALUES:

t/°C	Solubility of potassium iodate		Density	Refractive index
	mass %	mol kg^{-1}[a]	g cm^{-3}	n_D
15	6.6894 6.6802 6.6827 (Av)6.684 (σ=0.005)	0.335	1.0584	1.33831
20	7.4765 7.4825 7.4755 (Av)7.478 (σ=0.004)	0.378	1.0648	1.33873
25	8.3386 8.3445 8.3452 (Av)8.343 (σ=0.004)	0.425	1.0708	1.33911

[a]Molalities calculated by the compiler using 1977 IUPAC recommended atomic masses.

AUXILIARY INFORMATION

METHOD/APPARATUS/PROCEDURE:	SOURCE AND PURITY OF MATERIALS:
An excess potassium iodate was added to distilled water, and the mixture was shaken in a thermostat for about 10 hours. Equilibrium was established from both under-saturation and supersaturation. The sample of the saturated solution was filtered off, and the solution was evaporated to dryness.	The purest commercial potassium iodate (Kahlbaum) was dissolved in distilled water, the solution was decanted three times to remove the impurity. The re-crystallized potassium iodate was used for the solubility determination.

	ESTIMATED ERROR:
	Soly: standard deviation is given in the table described above (compiler calculated) Temp: ± 0.02°C (author)

	REFERENCES:

COMPONENTS:	ORIGINAL MEASUREMENTS:
(1) Potassium iodate; KIO_3; [7758-05-6] (2) Water; H_2O; [7732-18-5]	Benrath, A.; Gjedebo, F.; Schiffer, B.; Wunderlich, H. Z. Anorg. Allgem. Chem. 1937, 231, 285-97.
VARIABLES: T/K = 390 to 573	PREPARED BY: Hiroshi Miyamoto and Mark Salomon

EXPERIMENTAL VALUES:

Solubility of KIO_3[a,b]

t/°C	mass %	mol kg^{-1}	mole %
117	26.1	1.65	2.89
126	27.4	1.76	3.08
147	31.4	2.14	3.71
160	34.1	2.42	4.17
177	37.4	2.79	4.79
201	41.6	3.33	5.66
206	42.6	3.47	5.88
220	44.1	3.69	6.23
231	46.8	4.11	6.89
243	48.6	4.42	7.37
253	50.5	4.77	7.91
265	51.6	4.98	8.24
269	53.1	5.29	8.70
291	56.5	6.07	9.86
300	58.0	6.45	10.41

[a] Molalities and mole % calculated by the compilers.

[b] Nature of the solid phases not specified.

AUXILIARY INFORMATION

METHOD/APPARATUS/PROCEDURE:	SOURCE AND PURITY OF MATERIALS:
Synthetic method used with visual observation of temperature of crystallization and solubilization (ref 1). The weighed salt and water were placed in a small tube. The tubes were set in an oven equipped with a mica window. A thermometer was immersed in the oven.	No information is given.
	ESTIMATED ERROR: Nothing specified.
	REFERENCES: 1. Jaenecke, E. Z. Physik. Chem. 1936, A177, 7.

COMPONENTS:	ORIGINAL MEASUREMENTS:
(1) Potassium iodate; KIO_3; [7758-05-6] (2) Water; H_2O; [7732-18-5]	Bresusov, O.N.; Kashina, N.I.; Revzina, T.V.; Sobolevskaya, N.G. *Zh. Neorg. Khim.* 1967, *12*, 2240-3; *Russ. J. Inorg. Chem. (Engl. Transl.)* 1967, *12*, 1179-81.

VARIABLES:	PREPARED BY:
Temperature: 273.2 to 373.2 K	Hiroshi Miyamoto

EXPERIMENTAL VALUES:

Solubility of KIO_3[a]

t/°C	mass %	mol %	mol kg^{-1} (compiler)
0	4.57	0.402	0.224
10	6.04	0.538	0.300
20	7.68	0.695	0.389
25	8.57	0.783	0.438
30	9.35	0.861	0.482
40	11.13	1.043	0.585
50	13.07	1.250	0.703
60	15.30	1.498	0.844
70	17.41	1.744	0.985
80	19.55	2.005	1.136
90	21.85	2.300	1.306
100	23.99	2.588	1.475

[a] The nature of the solid phase was not specified.

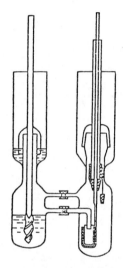

High temp. apparatus

AUXILIARY INFORMATION

METHOD/APPARATUS/PROCEDURE:	SOURCE AND PURITY OF MATERIALS:
Isothermal method. Equilibrium reached in 4-5 h. From 90-100°C, soly detd in apparatus shown in figure. At equilibrium, the apparatus was tilted to allow satd sln to filter through connecting tube into weighed test tubes. The test tube was closed with a stopper, withdrawn, and weighed. Condensation on the walls of the apparatus and loss of water by evaporation was thus prevented. At the lower temperatures, ordinary soly vessels were used, and pipets with glass filters were used for sampling (no other details given). Above 50°C, the pipets were preheated in the thermostat. The iodate content was determined iodometrically.	Results of analysis of KIO_3: KIO_3 content; 99.5 % Impurities, %, Rb 0.01; Cs 0.01; Na 0.005; SO_4 <0.01; Fe 0.005.
	ESTIMATED ERROR: Soly: nothing specified. Temp: precision $\pm$ 0.1 K.
	REFERENCES:

COMPONENTS:	ORIGINAL MEASUREMENTS:
(1) Potassium iodate; KIO_3; [7758-05-6] (2) Water; H_2O; [7732-18-5]	Kolthoff, I.M.; Chantooni, M.K. J. Phys. Chem. 1973, 77, 523-6.
VARIABLES: T/K = 298	**PREPARED BY:** Hiroshi Miyamoto

EXPERIMENTAL VALUES:

(1) Volumetric determination:

The solubility of KIO_3 in water at 25°C was found to be

$$0.44 \text{ mol dm}^{-3}.$$

(2) Potentiometric determination:

The solubility product of KIO_3 in water is given:

$$pK_{s0} = 1.6 \text{ (authors)}$$
$$K_{s0} = 2.5 \times 10^{-2} \text{ mol}^2 \text{ dm}^{-6} \text{ (compiler)}$$

The solubility product of KIO_3 was calculated from EMF data using the following equation:

$$E_{II} - E_I = 0.0591 [pK_{s0}(AgCl) - pK_{s0}(AgIO_3)$$
$$+ pK_{s0}(KIO_3) + 2 \log[c(KCl) y_{\pm}(KCl)]]$$

where E_I and E_{II} are Emfs of Cell I and II, respectively. With a particular cation glass electrode $E_I = +0.051V$ and $E_{II} = +0.213V$, which combined with the accepted values in water $pK_{s0}(AGCl) = 9.7$, $pK_{s0}(AgIO_3) = 7.5$, and the mean activity coefficient $y_{\pm}(KCl)$ in 0.44 mol dm^{-3} KCl solution of 0.65 (ref 1) yield the value for $p\overline{K}_{s0}(KIO_3)$.

AUXILIARY INFORMATION

METHOD/APPARATUS/PROCEDURE:	SOURCE AND PURITY OF MATERIALS:
(1) The solubility product of KIO_3 in water was estimated from the difference in emf of Cell I and II without liquid junction Ag,AgIO$_3$/salt(c$_1$) /K(gl) I Ag,AgCl/salt(c$_2$) /K(gl) II where c_1 is the concentration of IO_3 in saturated solution, and c_2 is the concentration of Cl^- saturated in 0.44 mol dm^{-3} KIO_3 solution. (2) The details of the isothermal method are not given. The iodate content was determined iodometrically.	KIO_3 was dried in vacuo at 70°C for 3 hours. Electrodes were prepared electrolytically (ref 2).

SOURCE AND PURITY continues:

	ESTIMATED ERROR: The uncertainty in pK_{s0} is $\pm$ 0.05. Temp: not given.

REFERENCES:
1. Bates, R.G.; Staples, B.G.; Robinson, R.A. Anal. Chem. 1970, 42, 867.
2. Ives, D.J.; Janz, G.J. Reference Electrodes. Academic Press. N.Y. 1961, p179; Kolthoff, I.M.; Chantooni, M.K. J. Am. Chem. Soc. 1965, 87, 4428.

COMPONENTS:	ORIGINAL MEASUREMENTS:
(1) Potassium iodate; KIO_3; [7758-05-6] (2) Water-d_2; D_2O; [7789-20-0] (3) Water; H_2O; [7732-18-5]	Chang, T.L.; Hsieh, Y.Y. J. *Chinese Chem. Soc. Peking*, 1949, *16*, 10-2.

VARIABLES:	PREPARED BY:
T/K = 298	G. Jancso

EXPERIMENTAL VALUES:

Water-d_2 mol %	Potassium Iodate of solubilities mole/55.51 moles of solvent
0	0.431
99.3	0.3586 0.3597
	(Av)0.359
100[a]	0.358

[a] Solubility in 100 mole % D_2O calculated by the compiler using linear
extrapolation.

AUXILIARY INFORMATION

METHOD/APPARATUS/PROCEDURE:	SOURCE AND PURITY OF MATERIALS:
Saturated solutions of potassium iodate were prepared by the method of supersaturation. The saturated solutions were made by agitating the excess salt with water for one hour at 70°C and then for several hours in a 25°C bath. A sample of the clear solution was delivered in a weighing bottle, then the solvent evaporated and the residual pure salt was dried in vacuum at 100°C and weighed. Two duplicate determinations were made on the same sample of prepared solution.	Baker's analyzed "chemically pure" reagent grade KIO_3 was used. Heavy water was obtained from Norsk Hydro-Electrisk Kvalato-faktieselskab in Oslo, and had a deuterium concentration of 99.7 mol %. The D_2O content of the water mixture was determined by pycnometer both before and after each measurement. The mole percentage was calculated from the specific gravity at 25°C (ref 1).

ESTIMATED ERROR:
Soly: precision better than 1 %.
Temp: nothing specified.

REFERENCES:
1. Swift, E. Jr. *J. Am. Chem. Soc.*
      1939, *61*, 198.

COMPONENTS:	ORIGINAL MEASUREMENTS:
(1) Potassium nitrate; KNO_3; [7757-79-1] (2) Potassium iodate; KIO_3; [7758-05-6] (3) Water; H_2O; [7732-18-5]	Hill, A.E.; Brown, S.F. *J. Am. Chem. Soc.* <u>1931</u>, *53*, 4316-20.

| VARIABLES:
T/K = 278, 298 and 323.6

Composition | PREPARED BY:
Hiroshi Miyamoto |

EXPERIMENTAL VALUES: Composition of saturated solutions

t/°C	KIO_3 mass %	mol % (compiler)	KNO_3 mass %	mol % (compiler)	Density g cm^{-3}	Nature of the solid phase[a]
5	5.25[b]	0.464	0.00	0.000	1.043	A
	3.29	0.299	5.36	1.032	1.060	"
	2.93	0.278	10.53	2.116	1.090	"
	2.89	0.282	13.53	2.796	1.110	"
	2.87	0.282	14.14	2.938	1.120	A+B
	1.08	0.104	14.26	2.911	1.100	B
	0.00	0.000	14.43	2.917	1.097	"
25	8.46[b]	0.772	0.00	0.000	1.072	A
	5.92	0.553	5.48	1.084	1.084	"
	5.25	0.501	8.51	1.719	1.110	"
	4.57	0.469	16.77	3.643	1.156	"
	4.48	0.466	18.19	4.004	1.160	"
	4.21	0.478	26.84	6.455	1.232	A+B
	2.65	0.297	27.25	6.459	1.215	B
	2.06	0.229	27.30	6.428	1.210	"
	0.00	0.000	27.79	6.417	1.192	"
50.4	13.35	1.280	0.00	0.000	1.110	A
	7.32	0.780	17.42	3.930	1.167	"
	5.79	0.708	31.43	8.132	1.259	"
	5.30	0.697	37.83	10.52	1.31	"
	5.17	0.690	39.17	11.07	1.33	"
	4.91	0.682	42.42	12.46	1.35	"
	4.75	0.674	44.23	13.29	1.37	A+B
	2.44	0.340	45.44	13.40	1.34	B
	0.00	0.000	46.57	13.44	1.326	"

[a] A = KIO_3;   B = KNO_3

[b] For the binary system the compiler computes the following:

soly of KIO_3 = 0.259 mol kg^{-1} at 5°C

= 0.432 mol kg^{-1} at 25°C

= 0.7206 mol kg^{-1} at 50.4°C

continued.....

COMPONENTS:	ORIGINAL MEASUREMENTS:
(1) Potassium nitrate; KNO_3; [7757-79-1]	Hill, A.E.; Brown, S.F.
(2) Potassium iodate; KIO_3; [7758-05-6]	J. Am. Chem. Soc. 1931, 53, 4316-20.
(3) Water; H_2O; [7732-18-5]	

COMMENTS AND/OR ADDITIONAL DATA: (Continued)

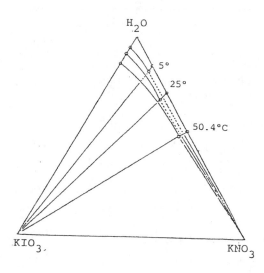

5°C, 25°C, 50.4°C Isotherms (mass % Units)

AUXILIARY INFORMATION

METHOD/APPARATUS/PROCEDURE:	SOURCE AND PURITY OF MATERIALS:
The ternary mixtures were prepared by weight and were rotated in closed tubes for periods from one to two weeks (which time was shown by analysis of solutions to be sufficient for attainment of equilibrium). After the solid settled, samples were withdrawn with a pipet. The iodate content was determined iodometrically, and water was determined by evaporation and heating to constant weight. Potassium nitrate was determined by difference. The identity of the solid phases was established by the method of extrapolation of the tie-line passing through the points for the composition of the solutions and those of the original complexes.	The salts used were commercial products of good quality, and purified by recrystalization from water.
	ESTIMATED ERROR:
	Soly: the compiler assumes that the precision of the solubility is within 0.3 %. Temp: not given.
	REFERENCES:

COMPONENTS:	ORIGINAL MEASUREMENTS:
(1) Potassium sulfate; K_2SO_4; [7778-80-5] (2) Potassium iodate; KIO_3; [7758-05-6] (3) Water; H_2O; [7732-18-5]	Hill, A.E.; Ricci, J.E. *J. Am. Chem. Soc.* <u>1931</u>, 53, 4305-15.
VARIABLES: Composition T/K = 278, 298, 323	PREPARED BY: Hiroshi Miyamoto

EXPERIMENTAL VALUES: Composition of saturated solutions

t/°C	KIO$_3$ mass %	mol % (compiler)	K$_2$SO$_4$ mass %	mol % (compiler)	Density g cm^{-3}	Nature of the solid phase[a]
5	5.16[b]	0.456	0.00	0.000	1.043	A
	3.14	0.283	4.07	0.450	1.060	"
	2.57	0.237	7.08	0.802	1.081	A+B
	1.80	0.165	7.25	0.816	1.077	B
	0.00	0.000	7.64	0.848	1.062	"
25	8.45[b]	0.771	0.00	0.000	1.071	A
	5.66	0.526	4.75	0.542	1.085	"
	4.72	0.448	7.74	0.902	1.103	"
	4.30	0.414	9.65	1.14	1.117	A+B
	2.44	0.232	10.10	1.177	1.103	B
	0.00	0.000	10.76	1.231	1.083	"
50	13.21[b]	1.265	0.00	0.000	–	A
	8.68	0.860	7.90	0.961	–	"
	7.39	0.750	11.43	1.424	–	"
	7.06	0.718	12.06	1.507	–	A+B
	3.78	0.375	13.00	1.583	–	B
	0.00	0.000	14.18	1.680	–	"

[a] A = KIO_3;   B = K_2SO_4

[b] In the binary system, the solubilities of KIO_3 are (compiler):
    0.254 mol kg^{-1} at 5°C; 0.431 mol kg^{-1} at 25°C; 0.7112 mol kg^{-1} at 50°C.

AUXILIARY INFORMATION

METHOD/APPARATUS/PROCEDURE:	SOURCE AND PURITY OF MATERIALS:
Ternary mixtures prepared by weight. For the 5°C isotherm, the solids were dissolved by heating, and the solutions were inoculated after cooling. The solutions were agitated in a thermostat at the desired temperature. Stirring times were about two weeks at 5°C, and about five days at 50°C. Iodate detd by iodometric titration, and total solids by evaporation to dryness and heating at 220°C for 2 h.	Nothing specified.

ESTIMATED ERROR:

Nothing specified, but the compiler assumes that the precision in analyses was 0.5 %.

COMMENTS AND/OR ADDITIONAL DATA:

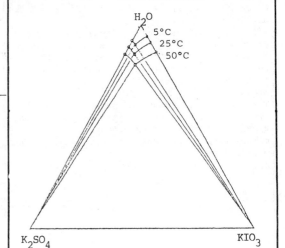

COMPONENTS:	ORIGINAL MEASUREMENTS:
(1) Potassium chloride; KCl; [7447-40-7]	Hill, A.E.; Ricci, J.E.
(2) Potassium iodate; KIO$_3$; [7758-05-6]	J. Am. Chem. Soc. 1931, 53, 4305-15.
(3) Water; H$_2$O; [7732-18-5]	

VARIABLES:	PREPARED BY:
Composition T/K = 278, 298, 323	Hiroshi Miyamoto

EXPERIMENTAL VALUES: Composition of saturated solutions

t/°C	KIO$_3$ mass %	KIO$_3$ mol % (compiler)	KCl mass %	KCl mol % (compiler)	Density g cm^3	Nature of the solid phase[a]
5	5.16[b]	0.456	0.00	0.000	1.043	A
	2.91	0.258	3.03	0.770	1.044	"
	1.79	0.166	10.13	2.700	1.084	"
	1.50	0.151	19.56	5.641	1.147	"
	1.44	0.149	22.64	6.712	1.170	A+B
	0.00	0.000	22.84	6.675	1.155	B
25	8.45[b]	0.771	0.00	0.000	1.071	A
	5.83	0.530	2.78	0.726	1.066	"
	4.29	0.394	5.93	1.56	1.073	"
	3.01	0.288	12.37	3.402	1.109	"
	2.40	0.244	19.64	5.724	1.153	"
	2.10	0.225	25.82	7.949	1.197	A+B
	0.00	0.000	26.36	7.961	1.179	B
50	13.21[b]	1.265	0.00	0.000	–	A
	7.42	0.710	6.83	1.87	–	"
	4.71	0.473	15.64	4.509	–	"
	3.74	0.399	23.17	7.087	–	"
	3.07	0.344	29.08	9.353	–	A+B
	1.77	0.196	29.46	9.362	–	B
	0.00	0.000	30.03	9.397	–	"

[a] A = KIO$_3$;   B = KCl

[b] Solubilities of KIO$_3$ in the binary systems are (compiler):
0.254 mol kg^{-1} at 5°C; 0.431 mol kg^{-1} at 25°C; 0.7112 mol kg^{-1} at 50°C.

METHOD/APPARATUS/PROCEDURE:	SOURCE AND PURITY OF MATERIALS:
Ternary mixtures prepared by weight. For the 5°C isotherm, the solids were first dissolved by heating and the solutions were inoculated after cooling. Solutions were rotated in a thermostat at the desired temperature for periods of two or three days up to two weeks. Iodate detd by iodometric titration, and total solids by evaporation to dryness.	Nothing specified.

ESTIMATED ERROR:

Nothing specified, but the compiler assumes that the precision in analyses was 0.5 %.

COMMENTS AND/OR ADDITIONAL DATA:

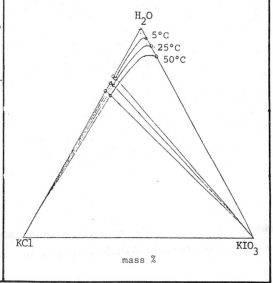

COMPONENTS:	ORIGINAL MEASUREMENTS:
(1) Potassium bromide; KBr; [7758-02-3] (2) Potassium iodate; KIO_3; [7758-05-6] (3) Water; H_2O; [7732-18-5]	Ricci, J.E. J. Am. Chem. Soc. 1934, 56, 290-5.
VARIABLES: T/K = 278, 298, 323 Composition	PREPARED BY: Hiroshi Miyamoto

EXPERIMENTAL VALUES: Composition of saturated solutions[a]

t/°C	KBr mass %	mol %	KIO_3 mass %	mol %	Density g cm^{-3}	Nature of the solid phase[b]
5	36.26	7.929	0.00	0.00	1.333	A
	35.71	7.944	1.80	0.223	1.351	A+B
	35.72	7.948	1.80	0.223	1.352	"
	35.71	7.943	1.79	0.221	1.353	"
	35.71	7.944	1.80	0.223	1.352	"
	30.50	6.369	1.77	0.206	1.290	B
	22.59	4.319	1.80	0.191	1.208	"
	14.80	2.616	1.95	0.192	1.136	"
	7.58	1.257	2.45	0.226	1.080	"
	0.00	0.00	5.186[c]	0.458	1.043	"
25	40.62	9.384	0.00	0.00	1.381	A
	40.28	9.393	0.98	0.13	1.396	"
	39.75	9.387	2.36	0.310	1.407	A+B
	39.75	9.387	2.36	0.310	1.408	"
	39.76	9.389	2.35	0.309	1.407	"
	39.75	9.387	2.36	0.310	1.407	"
	34.38	7.591	2.47	0.303	1.341	B
	25.91	5.194	2.73	0.304	1.249	"
	17.40	3.203	3.26	0.334	1.168	"
	8.35	1.424	4.53	0.430	1.106	"
	0.00	0.00	8.452[c]	0.771	1.071	"

continued....

AUXILIARY INFORMATION

METHOD/APPARATUS/PROCEDURE:	SOURCE AND PURITY OF MATERIALS:
At 5°C and 25°C ternary complexes were stirred for one to two weeks, and for two to four days at 50°C. This length of time allowed for the attainment of equilibrium as determined in several cases by successive analyses. In one sample of the saturated solution, the iodate was determined by titration with standard thiosulfate solution. In another sample, the total solid was determined by evaporation of the solution at 100°C followed by one to two hours at 250°C. The concentration of the bromide was then determined by difference. For the determination of the composition of these solid phases, the method of algebraic extrapolation of tie-lines was used.	KIO_3 and KBr were prepared by recrystalization from the best available c.p. grade materials, which, in the case of the bromide, usually contained from 0.5 to 1.0 % chloride. The purified salts were dried to the anhydrous state.
	ESTIMATED ERROR: Soly: average deviation in accuracy of analysis is 0.19 % (maximum 0.52 %). Temp: not given.
	REFERENCES:

COMPONENTS:	ORIGINAL MEASUREMENTS:
(1) Potassium bromide; KBr; [7758-02-3]	Ricci, J.E.
(2) Potassium iodate; KIO_3; [7758-05-6]	*J. Am. Chem. Soc.* 1934, *56*, 290-5.
(3) Water; H_2O; [7732-18-5]	

EXPERIMENTAL VALUES: (Continued)

Composition of saturated solutions[a]

	KBr		KIO_3		Density	Nature of the
t/°C	mass %	mol %	mass %	mol %	g cm^{-3}	solid phase[b]
50	44.78	10.93	0.00	0.00	–	A
	43.88	10.94	2.22	0.308	–	"
	43.50	10.93	3.09	0.432	–	A+B
	43.50	10.92	3.06	0.427	–	"
	43.50	10.93	3.07	0.429	–	"
	38.03	8.900	3.32	0.432	–	B
	30.57	6.563	3.87	0.462	–	"
	23.19	4.616	4.66	0.516	–	"
	15.65	2.913	5.89	0.610	–	"
	8.18	1.45	8.16	0.803	–	"
	0.00	0.00	13.20[c]	1.264	–	"

[a] Mole percent data calculated by the compiler.

[b] A = KBr;   B = KIO_3

[c] For the binary system the compiler computes the following:

soly of KIO_3 = 0.2556 mol kg^{-1} at 5°C

= 0.4314 mol kg^{-1} at 25°C

= 0.7106 mol kg^{-1} at 50°C

COMMENTS AND/OR ADDITIONAL DATA:

The phase diagram is given below (based on mass % units).

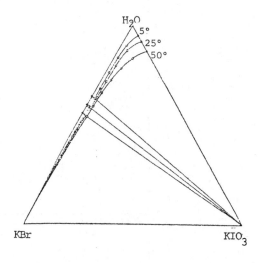

COMPONENTS:	ORIGINAL MEASUREMENTS:
(1) Potassium iodide; KI; [7681-11-0]	Ricci, J.E.
(2) Potassium iodate; KIO$_3$; [7758-05-6]	*J. Am. Chem. Soc.* 1937, 59, 866-7.
(3) Water; H$_2$O; [7732-18-5]	

VARIABLES:	PREPARED BY:
Composition at 298.15 K	Hiroshi Miyamoto

EXPERIMENTAL VALUES: Composition of saturated solutions at 25.00°C

| KIO$_3$ | | KI | | Density | Nature of the |
mass %	mol % (compiler)	mass %	mol % (compiler)	g cm^{-3}	solid phase[a]
8.449[b]	0.7709	0.00	0.00	1.071	A
7.15	0.659	2.40	0.285	1.053	"
4.33	0.427	12.04	1.532	–	"
3.27	0.357	22.38	3.152	1.227	"
2.54	0.350	41.10	7.308	1.451	"
2.35	0.421	57.02	13.16	1.722	"
2.36	0.435	58.54	13.92	1.749	A+B
2.36	0.435	58.54	13.92	1.754	"
2.35	0.433	58.47	13.88	1.749	"
2.35	0.433	58.51	13.90	1.751	"
2.25	0.415	58.62	13.93	–	B
1.10	0.200	59.14	13.87	1.731	"
0.00	0.000	59.76	13.88	1.718	"

[a] A = KIO$_3$;   B = KI

[b] For the binary system the compiler computes the following:

soly of KIO$_3$ = 0.4312 mol kg^{-1}

AUXILIARY INFORMATION

METHOD/APPARATUS/PROCEDURE:	SOURCE AND PURITY OF MATERIALS:
Mixtures of KIO$_3$, KI and H$_2$O were stirred in a bath thermostatically controlled at 25°C for at least 2 days. KIO$_3$ was determined by adding excess KI, acidifying, and titrating the liberated iodine with standard sodium thiosulfate solution. The total solid was determined by evaporation at 100°C followed by 250°C. KI was then calculated by difference. The densities were calculated from the weight delivered by a volumetric pipet calibrated for delivery.	No information given.
	ESTIMATED ERROR: Soly: nothing specified. Temp: precision ± 0.02 K.
	REFERENCES:

COMPONENTS:	ORIGINAL MEASUREMENTS:
(1) Potassium iodate; KIO_3; [7758-05-6] (2) Rubidium iodate; $RbIO_3$; [13446-76-9] (3) Water; H_2O; [7732-18-5]	Kirgintsev, A.N.; Shklovskaya, R.M.; Arkhipov, S.M. *Izv. Akad. Nauk SSSR, Ser. Khim.* 1971, 2631-4; *Bull. Acad. Sci. USSR, Div.* *Chem. Sci.* 1971, 2501-4.

VARIABLES:	PREPARED BY:
Composition at 298.2 K	Hiroshi Miyamoto

EXPERIMENTAL VALUES: Composition of saturated solutions[c]

KIO₃		RbIO₃	
mass %	mol %[a]	mass %	mol %[a]
8.45[b]	0.771	0.0	0.0
7.51	0.682	0.40	0.030
6.51	0.587	0.69	0.051
4.57	0.405	0.97	0.071
3.79	0.334	1.05	0.076
2.77	0.242	1.23	0.088
2.11	0.184	1.51	0.108
1.40	0.121	1.69	0.120
0.89	0.077	1.88	0.134
0.41	0.035	2.05	0.145
0.0	0.0	2.31[b]	0.163

[a] Compiler's calculations using 1977 IUPAC recommended atomic masses.

[b] For binary systems the compiler computes the following:

$$\text{soly of } KIO_3 = 0.431 \text{ mol kg}^{-1}$$
$$\text{soly of } RbIO_3 = 0.0908 \text{ mol kg}^{-1}$$

[c] The compiler assumes that the heading $RbIO_3-CsIO_3-H_2O$ in Table 2 in the original paper is a typographical error, and should be read as $KIO_3-RbIO_3-H_2O$.

AUXILIARY INFORMATION	

METHOD/APPARATUS/PROCEDURE:	SOURCE AND PURITY OF MATERIALS:
The solubility was studied by the method of isothermal removal of supersaturation with mixing for 24 hours. The iodate content was determined by iodometric titration. The contents of alkali metals were determined in the same sample by the method of flame photometry from three parallel analyses. The composition of the solid phase was determined by Schreinemakers' method of residues.	C.p. grade KIO_3 and $CsIO_3$ were recrystallized from double distilled water.
	ESTIMATED ERROR: Soly: precision within ± 3.5 % (compiler). Temp: precision ± 0.1 K.
	REFERENCES:

COMPONENTS:	ORIGINAL MEASUREMENTS:
(1) Potassium iodate; KIO_3; [7758-05-6]	Kirgintsev, A.I.; Yakobi, N.Y.
(2) Cesium iodate; $CsIO_3$; [13454-81-4]	$Zh.\ Neorg.\ Khim.$ <u>1968</u>, 13, 2851-3; $Russ.\ J.\ Inorg.\ Chem.\ (Engl.\ Transl.)$ <u>1968</u>, 13, 1467-8.
(3) Water; H_2O; [7732-18-5]	

VARIABLES:	PREPARED BY:
Composition at 298.2 K	Hiroshi Miyamoto

EXPERIMENTAL VALUES: Composition of saturated solutions at 25.0°C

KIO_3 mass %	mol % (compiler)	$CsIO_3$ mass %	mol % (compiler)	Nature of the solid phase[a]
–	–	2.53[b]	0.152	A
0.38	0.033	2.29	0.137	"
1.01	0.0874	1.90	0.114	"
1.33	0.115	1.84	0.111	"
2.50	0.219	1.56	0.0949	"
3.52	0.310	1.24	0.0759	"
4.43	0.394	1.32	0.0816	"
6.53	0.590	0.94	0.059	"
8.29	0.764	1.12	0.0718	A+B
8.34	0.766	0.79	0.050	B
8.46	0.772	–	–	"

[a] $A = CsIO_3$; $B = KIO_3$

[b] For binary systems the compiler computes the following:

soly of KIO_3 = 0.432 mol kg^{-1}

soly of $CsIO_3$ = 0.0843 mol kg^{-1}

AUXILIARY INFORMATION

METHOD/APPARATUS/PROCEDURE:	SOURCE AND PURITY OF MATERIALS:
The isothermal relief of supersaturation method was employed. The supersaturated slns were stirred for 7-8 hours. To det the comp of the coexisting phases, the authors used a method of indirect analysis. If p is the total mass of iodates, and b the total moles of iodates in it, then $p = bN_1M_1 + bN_2M_2$ (1) where N_1 and N_2 are the mole fractions of the first and second components, M_1 and M_2 their molecular masses. Solving eq (1) yields	Analytical reagent grade KIO_3 and $CsIO_3$ were recrystallized from twice-distilled water.

| $N_1 = (p/b) - M_2/M_1 - M_2$ (2) Eq. (2) was used to calculate the composition of the liquid. After settling, samples of the liquid phase were removed for analysis. The parameters, p and b, were determined by an ion-exchange method. | ESTIMATED ERROR: Soly: accuracy $\leq$ 5 %. Temp: precision $\pm$ 0.1 K. |
| | REFERENCES: |

COMPONENTS:	ORIGINAL MEASUREMENTS:
(1) Potassium iodate; KIO_3; [7758-05-6] (2) Barium iodate; $Ba(IO_3)_2$; [10567-69-8] (3) Water; H_2O; [7732-18-5]	Azarova, L.A.; Vinogradov, E.E. *Zh. Neorg. Khim.* 1982, *27*, 2967-70; *Russ. J. Inorg. Chem. (Engl. Transl.)* 1982, *27*, 1681-3.
VARIABLES: Composition at 298 K	PREPARED BY: Hiroshi Miyamoto

EXPERIMENTAL VALUES: Composition of saturated solutions at 25°C

$Ba(IO_3)_2$		KIO_3		Nature of the
mass %	mol % (compiler)	mass %	mol % (compiler)	solid phase[a]
0.058[b]	0.0021	–	–	A
0.668	0.027	7.90	0.722	"
1.58	0.0615	3.73	0.330	"
0.28	0.011	7.96	0.725	"
0.032	0.0013	7.94	0.721	"
1.113	0.0447	7.44	0.680	A+B
–	–	8.39	0.765[b]	B

[a] $A = Ba(IO_3)_2$; $B = KIO_3$

[b] For binary systems the compiler computes the following:

soly of KIO_3 = 0.428 mol kg^{-1}

soly of $Ba(IO_3)_2$ = 1.2 x 10^{-3} mol kg^{-1}

AUXILIARY INFORMATION

METHOD/APPARATUS/PROCEDURE:
Probably the isothermal method was used.
Equilibrium was reached in 10-12 days.
The potassium content was determined gravi-
metrically with sodium tetraphenylborate.
The iodate concentration was determined by
iodometric titration with sodium thiosulfate.
The barium content was determined by preci-
pitation as $BaSO_4$ using H_2SO_4.
The composition of the solid phases were de-
termined by Schreinemakers' method of
residues, and by X-ray diffraction.

SOURCE AND PURITY OF MATERIALS:
Analytical grade barium iodate and chemically pure grade potassium iodate were used.

ESTIMATED ERROR:

Nothing specified.

COMMENTS AND/OR ADDITIONAL DATA:
The phase diagram is given below (based on
mass % units).

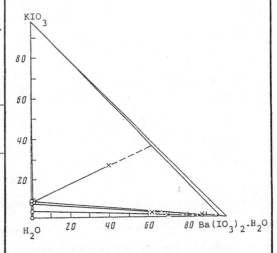

COMPONENTS:	ORIGINAL MEASUREMENTS:
(1) Potassium iodate; KIO_3; [7758-05-6] (2) Aluminum iodate; $Al(IO_3)_3$; [15123-75-8] (3) Water; H_2O; [7732-18-5]	Vinogradov, E.E.; Tarasova, G.N. *Zh. Neorg. Khim.* 1978, *23*, 3161-4; *Russ. J. Inorg. Chem. (Engl. Transl.)* 1978, *23*, 1754-6.

VARIABLES:	PREPARED BY:
Composition at 298.2 K	Hiroshi Miyamoto

EXPERIMENTAL VALUES: Composition of saturated solutions at 25.0°C

$Al(IO_3)_3$		KIO_3		Nature of the solid phase[a]
mass %	mol % (compiler)	mass %	mol % (compiler)	
5.71[b]	0.197	–	–	A
5.05	0.174	0.31	0.028	"
4.09	0.142	2.15	0.192	"
3.54	0.123	2.71	0.242	"
3.09	0.108	3.69	0.332	"
2.40	0.0872	8.47	0.793	A+B
2.41	0.0876	8.55	0.801	"
2.40	0.0872	8.48	0.794	"
2.38	0.0864	8.51	0.797	"
–	–	8.40[b]	0.766	B

[a] A = $Al(IO_3)_3 \cdot 6H_2O$;   B = KIO_3

[b] For binary systems the compiler computes the following:

soly of KIO_3 = 0.429 mol kg^{-1}

soly of $Al(IO_3)_3$ = 0.110 mol kg^{-1}

AUXILIARY INFORMATION

METHOD/APPARATUS/PROCEDURE:	COMMENTS AND/OR ADDITIONAL DATA:
Ternary mixtures stirred and thermostated for 12-14 d. Liquid and solid phases analyzed for all ions. IO_3^- detd by iodometric titrn, K detd gravimetrically as the tetraphenylborate, and Al by EDTA titrn with Xylenol Orange indicator. Solid phase compositions detd by Schreinemakers' method of residues.	The phase diagram based on mass % units is given below.

SOURCE AND PURITY OF MATERIALS:
C.p. grade KIO_3 used. Aluminum iodate prepd at 80-90°C by neut of satd HIO_3 solution with freshly prepared $Al(OH)_3$ in stoichiometric quantities. The salt was dried and analyzed: found, mass % Al 4.03; IO_3 78.7; H_2O 17.6. Calcd, mass %: Al 4.09; IO_3 79.53; H_2O 16.38 (by difference)

ESTIMATED ERROR:
Soly: nothing specified.
Temp: precision ± 0.1 K.

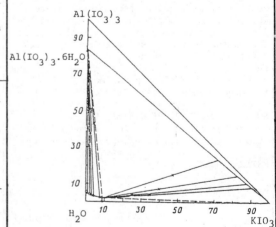

COMPONENTS:	ORIGINAL MEASUREMENTS:
(1) Potassium iodate; KIO$_3$; [13446-17-8] (2) Manganese iodate; Mn(IO$_3$)$_2$; [25659-29-4] (3) Water; H$_2$O; [7732-18-5]	Vinogradov, E.E.; Karataeva, I.M.; Lepeshkov, I.N. *Zh. Neorg. Khim.* <u>1979</u>, *24*, 223-7; *Russ. J. Inorg. Chem. (Engl. Transl.)* <u>1979</u>, *24*, 124-7.
VARIABLES: Composition at 323 K	PREPARED BY: Hiroshi Miyamoto

EXPERIMENTAL VALUES: Composition of saturated solutions at 50°C

KIO$_3$ mass %	KIO$_3$ mol % (compiler)	Mn(IO$_3$)$_2$ mass %	Mn(IO$_3$)$_2$ mol % (compiler)	Nature of the solid phase[a]
11.76[b]	1.109	–	–	A
11.76	1.109	trace	–	A+C
11.24	1.055	trace	–	"
11.35	1.066	trace	–	C
9.80	0.906	0.0021	0.00010	"
7.78	0.705	0.0080	0.00038	"
5.20	0.460	0.0065	0.00030	"
3.80	0.331	0.0072	0.00033	"
3.80	0.331	0.0063	0.00029	"
3.83	0.334	0.0086	0.00040	B
2.01	0.172	0.028	0.00127	"
0.83	0.071	0.187	0.00840	"
–	–	0.266[b]	0.0119	"

[a] A = KIO$_3$;   B = Mn(IO$_3$)$_2$;   C = K$_2$[Mn(IO$_3$)$_4$(H$_2$O)$_2$]

[b] For binary systems the compiler computes the following:

   soly of KIO$_3$ = 0.6228 mol kg^{-1}

   soly of Mn(IO$_3$)$_2$ = 0.00659 mol kg^{-1}

AUXILIARY INFORMATION

METHOD/APPARATUS/PROCEDURE:	COMMENTS AND/OR ADDITIONAL DATA:
Equilibrium in KIO$_3$-Mn(IO$_3$)$_2$-H$_2$O system was reached after about a month. The iodate content was determined iodometrically, potassium by flame photometry. Manganese was determined by titration with ammonium chloride at pH 9.5-10 using Methyl Thymol Blue as an indicator. The solid phases were investigated by thermogravimetric, X-ray diffraction, and IR spectroscopic methods.	The phase diagram is given below (based on mol % units). 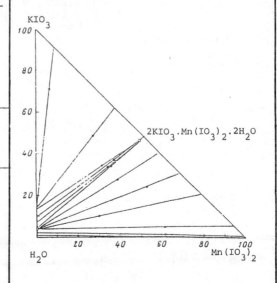
SOURCE AND PURITY OF MATERIALS: C.p. grade KIO$_3$ was used. Manganese iodate was made from manganese sulfate and iodic acid.	
ESTIMATED ERROR: Nothing specified.	

COMPONENTS:	ORIGINAL MEASUREMENTS:
(1) Potassium iodate; KIO₃; [7758-05-6]	Lepeshkov, I.N.; Vinogradov, E.E.; Karataeva, I.M.
(2) Cobalt iodate; Co(IO₃)₂; [13455-28-2]	Zh. Neorg. Khim. 1977, 22, 2277-8. Russ. J. Inorg. Chem. (Engl. Transl.) 1977, 22, 1232-5.
(3) Water; H₂O; [7732-18-5]	

VARIABLES:	PREPARED BY:
Composition at 323.2 K	Hiroshi Miyamoto

EXPERIMENTAL VALUES: Composition of saturated solutions at 50°C

KIO₃ mass %	KIO₃ mol % (compiler)	Co(IO₃)₂ mass %	Co(IO₃)₂ mol % (compiler)	Nature of the solid phase[a]
–	–	0.78[b]	0.035	A
1.23	0.105	0.49	0.022	"
2.41	0.208	0.27	0.012	A+C
2.45	0.212	0.36	0.016	"
2.27	0.196	0.31	0.014	C
2.56	0.221	0.24	0.011	"
5.46	0.485	0.29	0.013	"
8.18	0.746	0.17	0.0081	"
10.02	0.930	0.09	0.0044	"
11.20	1.051	traces	–	B+C
11.76[b]	1.109	–	–	B

[a] A = Co(IO₃)₂·2H₂O;   B = KIO₃;   C = 2KIO₃·Co(IO₃)₂·2H₂O

[b] For binary systems the compiler computes the following:

soly of KIO₃ = 0.6228 mol kg⁻¹

soly of Co(IO₃)₂ = 0.019 mol kg⁻¹

AUXILIARY INFORMATION

METHOD/APPARATUS/PROCEDURE:	COMMENTS AND/OR ADDITIONAL DATA:
Approximately three weeks were needed to reach equilibrium. Potassium content was determined gravimetrically by precipitation with sodium tetraphenylborate, cobalt by titration with EDTA, and iodate by titration with thiosulfate solution. The composition of the double compound was proved by X-ray diffraction, derivatograms, and IR spectra.	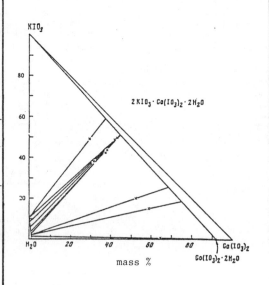

SOURCE AND PURITY OF MATERIALS:
C.p. grade KIO₃ used. Cobalt iodate prepared by pptn from cobalt nitrate solution. No other details given.

ESTIMATED ERROR:
Nothing specified.

COMPONENTS:	ORIGINAL MEASUREMENTS:
(1) Potassium iodate; KIO_3; [7758-05-6] (2) Nickel iodate; $Ni(IO_3)_2$; [13477-98-0] (3) Water; H_2O; [7732-18-5]	Lepeshkov, I.N.; Vinogradov, E.E.; Karataeva, I.M. *Zh. Neorg. Khim.* 1980, 25, 823-4; *Russ. J. Inorg. Chem. (Engl. Transl.)* 1980, 25, 463-4.
VARIABLES: Composition at 323 K	PREPARED BY: Hiroshi Miyamoto

EXPERIMENTAL VALUES: Composition of saturated solutions at 50°C

KIO₃		Ni(IO₃)₂		Nature of the
mass %	mol % (compiler)	mass %	mol % (compiler)	solid phase[a]
11.36[b]	1.067	–	–	A
10.41	0.9687	–	–	A+B
10.38	0.9656	–	–	B
9.68	0.894	–	–	"
7.16	0.645	0.018	8.5×10^{-4}	"
7.05	0.635	0.028	1.3×10^{-3}	"
3.66	0.319	0.139	6.35×10^{-3}	B+C
3.26	0.283	0.141	6.42×10^{-3}	"
3.55	0.309	0.160	7.30×10^{-3}	"
3.50	0.305	0.131	5.98×10^{-3}	"
2.49	0.215	0.394	1.78×10^{-2}	C
–	–	0.83[b]	3.7×10^{-2}	"

[a] $A = KIO_3$; $B = K_2[Ni(IO_3)_4(H_2O)_2]$; $C = Ni(IO_3)_2.2H_2O$

[b] For binary systems the compiler computes the following:

soly of KIO_3 = 0.5989 mol kg^{-1}

soly of $Ni(IO_3)_2$ = 0.020 mol kg^{-1}

AUXILIARY INFORMATION

METHOD/APPARATUS/PROCEDURE:	COMMENTS AND/OR ADDITIONAL DATA:
The compiler assumes that the isothermal method was used. The system reached equilibrium in about a month. The liquid and solid phases were analyzed for all ions present. The nickel content was determined by titration with EDTA in the presence of Murexide as an indicator in strongly alkaline solution. Analysis for other ions were not described in the paper. The solid phases were investigated by thermal, thermogravimetric, X-ray diffraction, and infrared spectroscopic methods.	The phase diagram is given below (based on mass % units). 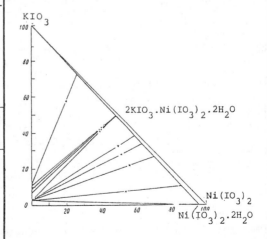

ESTIMATED ERROR:

Nothing specified.

SOURCE AND PURITY OF MATERIALS:
"Chemically pure" grade KIO_3 was used.
Nickel iodate was made from iodic acid and nickel nitrate.

COMPONENTS:	ORIGINAL MEASUREMENTS:
(1) Potassium iodate; KIO₃; [7758-05-6] (2) Zinc iodate; Zn(IO₃)₂; [7790-37-6] (3) Water; H₂O; [7732-18-5]	Lepeshkov, I.N.; Vinogradov, E.E.; Karataeva, I.M. *Zh. Neorg. Khim.* <u>1977</u>, *22*, 2277-81; *Russ. J. Inorg. Chem. (Engl. Transl.)* <u>1977</u>, *22*, 1232-5.
VARIABLES: Composition at 323 K	PREPARED BY: Hiroshi Miyamoto

EXPERIMENTAL VALUES: Composition of saturated solutions at 50°C

KIO₃		Zn(IO₃)₂		Nature of the
mass %	mol % (compiler)	mass %	mol % (compiler)	solid phase[a]
–	–	0.68[b]	0.030	A
1.20	0.102	0.54	0.045	"
1.70	0.146	0.18	0.0079	A+C
1.68	0.144	0.20	0.0088	"
1.83	0.157	0.10	0.0044	C
5.30	0.469	0.12	0.0055	"
8.40	0.767	0.11	0.0052	"
10.60	0.990	0.12	0.0058	"
13.02	1.245	0.06	0.0030	B+C
12.91	1.233	0.05	0.0025	"
11.76[b]	1.109	–	–	B

[a] A = Zn(IO₃)₂·2H₂O;   B = KIO₃;   C = 2KIO₃·Zn(IO₃)₂·2H₂O

[b] For binary systems the compiler computes the following:

soly of KIO₃ = 0.6228 mol kg⁻¹

soly of Zn(IO₃)₂ = 0.016 mol kg⁻¹

AUXILIARY INFORMATION

METHOD/APPARATUS/PROCEDURE:	COMMENTS AND/OR ADDITIONAL DATA:
Approximately three weeks were needed to reach equilibrium. Potassium content was determined gravimetrically by precipitation with sodium tetraphenylborate, zinc by titration with EDTA, and iodate by titration with thiosulfate solution. The composition of the double compound was proved by X-ray diffraction, derivatograms, and IR spectra. SOURCE AND PURITY OF MATERIALS: C.p. grade KIO₃ was used. Zinc iodate was made by pptn from zinc nitrate: no other details given. ESTIMATED ERROR: Nothing specified.	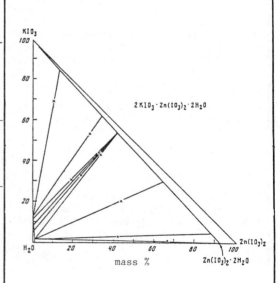

COMPONENTS:	ORIGINAL MEASUREMENTS:
(1) Potassium iodate; KIO_3; [7758-05-6] (2) Dipotassium (I-4)-tetraoxomolybdate (2-) (potassium molybdate); K_2MoO_4; [13446-49-6] (3) Water; H_2O; [7732-18-5]	Ricci, J.E.; Loprest, F.J. *J. Am. Chem. Soc.* 1953, 75, 1224-6.
VARIABLES: Composition at 298 K	PREPARED BY: Hiroshi Miyamoto

EXPERIMENTAL VALUES: Composition of saturated solutions at 25°C

K_2MoO_4		KIO_3		Density $g\ cm^{-3}$	Nature of the solid phase[a]
mass %	mol % (compiler)	mass %	mol % (compiler)		
64.57	12.12	0.00	0.00	1.800	A
63.94	12.14	1.15	0.243	1.822	A+B
63.97	12.15	1.15	0.243	1.823	"
63.93	12.15	1.20	0.254	1.818	"
63.95	12.15	1.17	0.247	1.821	"
60.52	10.64	1.13	0.221	–	B
56.57	9.169	1.13	0.204	1.683	"
51.08	7.490	1.30	0.212	1.600	"
43.86	5.724	1.63	0.237	1.484	"
38.10	4.573	1.92	0.256	1.406	"
30.97	3.384	2.33	0.283	1.331	"
21.75	2.131	2.92	0.318	1.225	"
13.93	1.258	3.69	0.371	1.151	"
10.00	0.872	4.37	0.424	1.115	"
0.00	0.000	8.45[b]	0.771	1.071	"

[a] $A = K_2MoO_4$; $B = KIO_3$

[b] For the binary system the compiler computes the following:

soly of KIO_3 = 0.431 mol kg^{-1}

AUXILIARY INFORMATION	
METHOD/APPARATUS/PROCEDURE: Isothermal method used. Ternary complexes of known composition were stirred in Pyrex tubes. At equilibrium aliquots of saturated solution were drawn by means of calibrated pipets supplied with filtering tips. The analysis involved determination of total solid by evaporation and iodometric determination of iodate by treatment with iodide and acid, and titration with thiosulfate solution.	**SOURCE AND PURITY OF MATERIALS:** K_2MoO_4 used was about 99.9 % pure on the basis of volumetric determination of molybdate with standard $AgNO_3$ and standard KCNS in an adaptation of Volhard's method. The salt was recrystallized before use. The source of KIO_3 was not given.
	ESTIMATED ERROR: Nothing specified.
	REFERENCES:

COMPONENTS:	ORIGINAL MEASUREMENTS:
(1) Potassium iodate; KIO_3; [7758-03-6] (2) Neodymium iodate; $Nd(IO_3)_3$; [14732-16-2] (3) Water; H_2O; [7732-18-5]	Tarasova, G.N.; Vinogradov, E.E.; Kudinov, I.B. *Zh. Neorg. Khim.* **1981**, *26*, 2841-7; *Russ. J. Inorg. Chem. (Engl. Transl.)* **1981**, *26*, 1520-3.
VARIABLES: Composition at 298.2 K	PREPARED BY: Hiroshi Miyamoto

EXPERIMENTAL VALUES: Composition of saturated solutions

$Nd(IO_3)_3$		KIO_3		
mass %	mol % (compiler)	mass %	mol % (compiler)	Nature of the solid phase[a]
0.15[b]	0.0040	--	--	A
0.12	0.0032	0.06	0.005	A+C
0.10	0.0027	0.06	0.005	"
0.10	0.0027	0.06	0.005	"
0.01	0.00027	2.10	0.180	C
0.01	0.00028	2.99	0.259	"
0.01	0.00028	2.87	0.248	"
0.01	0.00028	2.58	0.222	"
0.01	0.00028	5.21	0.461	"
0.01	0.00029	8.02	0.729	C+B
0.01	0.00029	8.00	0.727	"
0.01	0.00029	8.04	0.731	"
0.01	0.00029	8.40[b]	0.766	B

[a] $A = Nd(IO_3)_3 \cdot 2H_2O$; $B = KIO_3$; $C - 3Nd(IO_3)_3 \cdot 2KIO_3 \cdot 2H_2O$

[b] For the binary system the compiler computes the following:

soly of $Nd(IO_3)_3 = 2.2 \times 10^{-3}$ mol kg^{-1}

AUXILIARY INFORMATION

METHOD/APPARATUS/PROCEDURE:	COMMENTS AND/OR ADDITIONAL DATA:
Isothermal method. Equilibrium reached in 30-35 d. Liq and solid phases analyzed for IO_3 by iodometric titrn and for Nd by complexometric titrn in the presence of hexamethylenetetramine with Methyl Thymol Blue indicator. Solid phase compositions detd by Schreinemakers' method of residues.	The phase diagram based on mass % units is reproduced below. 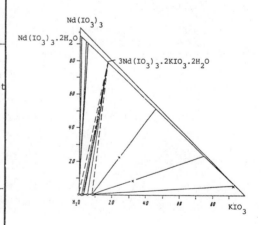
SOURCE AND PURITY OF MATERIALS: NdIO₃ prepd by stoichiometric mixing of HIO₃ and neodymium oxide and stirring of the aqueous mixture for 20 h at 80-90°C. The ppt was filtered, washed repeatedly with hot water, and dried at 110-120°C. The authors state that the purity of the resulting neodymium iodate was checked by chemical analysis, but the results were not reported in the source publication. "Chemically pure" grade KIO_3 was used.	
ESTIMATED ERROR: Soly: nothing specified. Temp: precision $\pm$ 0.1 K.	

COMPONENTS:	ORIGINAL MEASUREMENTS:
(1) Potassium iodate; KIO_3; [7758-05-6] (2) Potassium hydroxide; KOH; [1310-58-3] (3) Water; H_2O; [7732-18-5]	Bronsted, J.N. *J. Am. Chem. Soc.* <u>1920</u>, *40*, 1448-54.

VARIABLES:	PREPARED BY:
Concentration of KOH at 293 K	Hiroshi Miyamoto

EXPERIMENTAL VALUES:

Concn of KOH $mol\ dm^{-3}$	Soly of KIO_3 $mol\ dm^{-3}$
4.71	0.0390
5.06	0.0362
6.35	0.0256
7.95	0.0179
9.41	0.0144
10.95	0.0130
11.10	0.0128
12.19	0.0131
12.92	0.0135
14.02	0.0154
14.85	0.0194

AUXILIARY INFORMATION

METHOD/APPARATUS/PROCEDURE:	SOURCE AND PURITY OF MATERIALS:
The details of the establishment of equilibrium and the analytical method were not given in the original paper.	Nothing specified.
	ESTIMATED ERROR: Nothing specified.
	REFERENCES:

COMPONENTS:	ORIGINAL MEASUREMENTS:
(1) Potassium iodate; KIO_3; [7758-05-6] (2) Potassium hydroxide; KOH; [1310-58-3] (3) Water; H_2O; [7732-18-5]	Lepeshkov, I.N.; Vinogradov, E.E.; Tarasova, G.N. *Zh. Neorg. Khim.* 1976, *21*, 1353-6; *Russ. J. Inorg. Chem. (Engl. Transl.)* 1976, *21*, 739-41.
VARIABLES: Composition at 298.2 K	PREPARED BY: Hiroshi Miyamoto

EXPERIMENTAL VALUES: Composition of saturated solutions

KIO_3		KOH		Nature of the
mass %	mol % (compiler)	mass %	mol % (compiler)	solid phase[a]
8.40[b]	0.766	0.00	0.00	A
4.35	0.393	4.10	1.41	"
3.13	0.284	6.45	2.23	"
2.04	0.189	10.37	3.656	"
1.42	0.135	14.70	5.321	"
0.98	0.097	20.16	7.579	"
0.76	0.078	25.03	9.765	"
0.33	0.036	34.30	14.41	"
0.31	0.036	39.18	17.21	"
0.30	0.036	44.15	20.33	"
0.26	0.033	49.02	23.68	A+B
0.26	0.033	49.01	23.67	"
0.24	0.030	49.02	23.67	"
-	-	54.23	27.56	B

[a] A = KIO_3;   B = $KOH \cdot 2H_2O$

[b] For the binary system the compiler computes the following:

   soly of KIO_3 = 0.429 mol kg^{-1}

AUXILIARY INFORMATION

METHOD/APPARATUS/PROCEDURE:	SOURCE AND PURITY OF MATERIALS:
The solubility in the KIO_3-KOH-H_2O system was studied by the isothermal method. Mixtures were stirred in a water thermostat, and equilibrium was reached in 1-2 days. The concentration of hydroxide ion was found by titration with 0.1 mol dm^{-3} HCl in the presence of Methyl Orange. The IO_3 ion was determined by titration with sodium thiosulfate solution in the presence of sulfuric acid and KI. The concentration of K$^+$ ion was determined gravimetrically by precipitation with sodium tetraphenylborate. The composition of the solid phases was found by Schreinemakers' method of residues.	"Chemically pure" grade KIO_3 was used. Commercial KOH contains considerable amounts of K_2CO_3 impurity which cannot be removed by recrystallization from water. The materials were purified by recrystallization in silver vessels in a stream of purified nitrogen as the temperature was gradually increased to 250°C.
	ESTIMATED ERROR: Soly: nothing specified. Temp: precision $\pm$ 0.1 K.
	REFERENCES:

COMPONENTS:	ORIGINAL MEASUREMENTS:
(1) Potassium iodate; KIO_3; [7758-05-6] (2) Iodic acid; HIO_3; [7782-68-5] (3) Water; H_2O; [7732-18-5]	Meerburg, P.A. Z. Anorg. Allg. Chem. <u>1905</u>, 45, 324-44

VARIABLES. T/K = 303 Composition	PREPARED BY: Hiroshi Miyamoto

EXPERIMENTAL VALUES:

Composition of saturated solutions at 30°C

Iodic Acid		Potassium Iodate		Nature of
mass %	mol % (compiler)	mass %	mol % (compiler)	the solid phase[a]
0	0	9.51[b]	0.877	A
0.64	0.072	9.48	0.879	A+C
0.66	0.075	9.52	0.884	"
0.65	0.073	9.46	0.878	"
0.65	0.073	8.90	0.821	C
0.67	0.074	6.6	0.60	"
1.14	0.123	4.57	0.406	"
1.69	0.182	3.63	0.321	"
2.02	0.217	3.10	0.274	"
3.34	0.360	2.14	0.190	"
5.00	0.543	1.32	0.118	"
7.09	0.783	1.0	0.091	"
8.04	0.895	0.85	0.078	C+D
3.47	0.380	3.57	0.321	D(m)
4.80	0.528	2.90	0.262	"
6.45	0.710	1.35	0.122	"
9.35	1.05	0.64	0.059	D
12.04	1.389	0.44	0.042	"
17.50	2.133	0.30	0.030	"
31.20	4.468	0.52	0.061	"
53.64	10.72	0.68	0.11	"
62.52	14.81	0.72	0.14	"
76.40	25.49	0.80	0.22	D+B
76.70[b]	25.21	0	0	B

[a] A = KIO_3;   B = HIO_3;   C = $KIO_3.HIO_3$;   D = $KIO_3.2HIO_3$;   m = metastable.

[b] For binary systems the compiler computes the following

$$\text{soly of } HIO_3 = 18.71 \text{ mol kg}^{-1}$$
$$\text{soly of } KIO_3 = 0.491 \text{ mol kg}^{-1}$$

METHOD/APPARATUS/PROCEDURE:	SOURCE AND PURITY OF MATERIAL:
A mixture of KIO_3, HIO_3 and water was placed in a bottle and the bottle agitated in a thermostat for a week or more at a desired temperature. Equilibrium was established from supersaturation. The iodic acid and potassium iodate contents were detd as follows: an excess of KI was added to an aliquot of satd sln, and the HIO_3 content detd by titration of the iodine liberated with standard sodium thiosulfate. Dil sulfuric acid was then added to the solution and the iodine liberated was again titrated with sodium thiosulfate to obtain the total iodate concentration. The potassium iodate concentration was calculated from the difference between the second and the first titration. Composition of solid phases was determined by the method of residues.	Nothing specified.
	ESTIMATED ERROR:
	Nothing specified.
	COMMENTS AND/OR ADDITIONAL DATA:
	(mass % units) 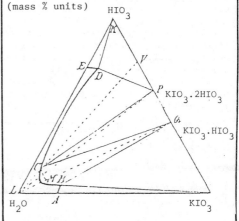

COMPONENTS:	ORIGINAL MEASUREMENTS:
(1) Potassium iodate; KIO_3; [7790-32-1]	Smith, S.B.
(2) Iodic acid; HIO_3; [7782-68-5]	J. Am. Chem. Soc. 1947, 69, 2285-6.
(3) Water; H_2O; [7732-18-5]	

VARIABLES:	PREPARED BY:
Composition and temperature T/K = 273-323	Hiroshi Miyamoto and Mark Salomon

EXPERIMENTAL VALUES: Composition of saturated solutions

t/°C	HIO$_3$ mass %	HIO$_3$ mol % (compiler)	KIO$_3$ mass %	KIO$_3$ mol % (compiler)	Nature of the solid phase[a]
0	–	–	4.51[b]	0.396	A
	0.35	0.0375	4.47	0.394	A+C
	1.30	0.137	1.68	0.145	C
	5.76	0.623	0.13	0.012	C+D
	73.69	22.53	0.39	0.098	B+D
	73.56	22.17	0.00	0.00	B
25	–	–	8.39[b]	0.765	A
	0.61	0.068	8.46	0.777	A+C
	0.77	0.083	4.90	0.435	C
	7.68	0.850	0.61	0.055	C+D
	75.51	24.29	0.42	0.111	B+D
	75.56	24.05	–	–	B
50	–	–	13.21[b]	].265	A
	1.34	0.159	13.58	1.324	A+C
	3.74	0.415	4.64	0.423	C
	11.02	1.276	1.85	0.176	C+D
	78.72	28.52	1.17	0.348	B+D
	78.78	27.55	–	–	B

[a] A = KIO_3;   B = HIO_3;   C = $KIO_3 \cdot HIO_3$;   D = $KIO_3 \cdot 2HIO_3$

[b] soly of KIO_3 = 0.221 mol kg^{-1} at 0°C; = 0.428 mol kg^{-1} at 25°C
= 0.7112 mol kg^{-1} at 50°C

AUXILIARY INFORMATION

METHOD/APPARATUS/PROCEDURE:	COMMENTS AND/OR ADDITIONAL DATA:
Isothermal method used. Ternary mixtures were prepd using the same compounds which constitute the solid phases in the equilibrated systems. This reduced the time required to reach equil to 48 hours. A sample of satd sln (or residue) was first titrd with standard alkali to det acid content. Total iodate in the same sample then detd by titrn with std thiosulfate in the presence of excess KI and HCl. Not all the soly data were reported. Only those results for invariant points and points of congruent solubilities were given. The values at the invariant points are the results of 2 or more closely agreeing results (this may indicate that some data were rejected: compilers).	The 50°C isotherm is reproduced below. Concentration units are mass %. 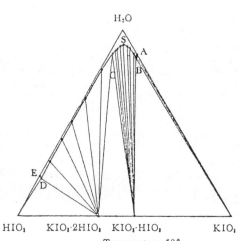 Temperature 50°

SOURCE AND PURITY OF MATERIALS:
Nothing specified.

ESTIMATED ERROR:
Soly: precision probably much better than ± 1 % (compilers). Temp: nothing specified.

COMPONENTS:	ORIGINAL MEASUREMENTS:
(1) Potassium iodide; KI; [7681-11-0] (2) Potassium iodate; KIO$_3$; [7758-05-6] (3) Potassium hydroxide; KOH; [1310-58-3] (4) Water; H$_2$0; [7732-18-5]	Malyshev, A.A.; Kuz'menko, A.L.; Novikov, G.I.; Traul'ko, I.V. *Zh. Neorg. Khim.* 1981, 26, 832-4; *Russ. J. Inorg. Chem. (Engl. Transl.)* 1981, 26, 448-9.

VARIABLES:	PREPARED BY:
T/K = 293, 333 and 353 Composition	Hiroshi Miyamoto

EXPERIMENTAL VALUES:

Common solubility of potassium iodate and iodide in alkaline(KOH) solution (pH 13.8)

t/°C	KI/mass %	KIO$_3$/mass %	Nature of the solid phase
20	54.06	1.93	KIO$_3$ + KI
	39.70	1.97	KIO$_3$
	34.36	2.03	"
	28.73	2.17	"
	26.44	2.22	"
	14.81	2.46	"
	0	3.45	"
60	58.39	2.50	KIO$_3$ + KI
	54.14	2.54	KIO$_3$
	47.04	2.88	"
	41.99	3.05	"
	35.98	3.53	"
	30.61	3.91	"
	9.73	6.54	"
	6.02	7.32	"
	3.70	7.95	"
	2.42	8.16	"
	0	8.60	"

continued.....

AUXILIARY INFORMATION

METHOD/APPARATUS/PROCEDURE:	SOURCE AND PURITY OF MATERIALS:
The investigation was carried out by isothermal saturation in a thermostat. Potassium iodide and iodate were dissolved in alkaline medium of pH 13.8. After equilibrium was reached, the liquid phase was analyzed for iodate iodometrically, and for iodide argentometrically. The composition of the solid phase was identified by the immersion method described in ref 1.	"Chemically pure" grade potassium iodate and iodide were recrystallized from twice-distilled water. Chemically pure grade potassium hydroxide used was freed from carbonate.

ESTIMATED ERROR:

Nothing specified.

REFERENCES:
1. Melankholin, N.M. *Izmerenie Pokazatelei Prelomleniya pod Mikroskopom Immersionnyn Methodom (Measurement of Refractive Indices under a Microscope by the Immersion Method)* Iz. Acad. Nauk SSSR. Moscow-Lenningrad. 1949.

COMPONENTS:	ORIGINAL MEASUREMENTS:
(1) Potassium iodide; KI; [7681-11-0]	Malyshev, A.A.; Kuz'menko, A.L.;
(2) Potassium iodate; KIO$_3$; [7758-05-6]	Novikov, G.I.; Traul'ko, I.V.
(3) Potassium hydroxide; KOH; [1310-58-3]	Zh. Neorg. Khim. 1981, 26, 832-4; Russ. J. Inorg. Chem. (Engl. Transl.) 1981, 26, 448-9.
(4) Water; H$_2$0; [7732-18-5]	

EXPERIMENTAL VALUES: (Continued)

t/°C	KI/mass %	KIO$_3$/mass %	Nature of the solid phase
80	61.51	2.80	KIO$_3$ + KI
	38.61	4.36	KIO$_3$
	13.80	7.10	"
	9.10	9.60	"
	0	12.27	"

To obtain the fitting equation for solubility of potassium iodate in alkaline medium at pH 13.8 the following equation was used:

$$c_2 = a_o + a_1 c_1 + a_2 c_1^2$$

where the concentration (c) based on mass % was used.

The calculated coefficients are given below:

t/°C	a_0	a_1	a_2
20	0.0340	−0.0642	0.0636
60	0.0857	−0.2047	0.1739
80	0.1219	−0.2984	0.2381

The mean relative error of the experimental and calculated results does not exceed 1.89 % at 20°C, 1.51 % at 60°C and 1.56 % at 80°C.

COMPONENTS:	ORIGINAL MEASUREMENTS:
(1) Potassium chlorate; $KClO_3$; [3811-04-9] (2) Ethanol; C_2H_6O; [64-17-5] (3) Water; H_2O; [7732-18-5]	Wright, R. J. Chem. Soc. 1927, 1334-6.
VARIABLES: Concentration of ethanol T/K = 293, 373	PREPARED BY: Hiroshi Miyamoto

EXPERIMENTAL VALUES:

t/°C	Concn of ethanol mass %	mol % (compiler)	soly of $KClO_3$ g/100g solvent	mol kg^{-1} (compiler)
20	0	0	7.2 7.2 (Av) 7.2	0.59
	50	28	1.1 1.1 (Av) 1.1	0.090
100	0	0	56.8 56.0 (Av) 56.4	4.60
	50	28	14.0 14.2 (Av) 14.1	1.15

AUXILIARY INFORMATION

METHOD/APPARATUS/PROCEDURE:	SOURCE AND PURITY OF MATERIALS:
At 20°C, $KClO_3$ and water were placed in stoppered tubes and thermostated. Weighed aliquots were taken and the amount of solute estimated either by titrn or by evapn to dryness. At 100°C care had to be taken to guard against alteration in the composition of the mixed solvent by evaporation. The method employed was as follows: To a test tube of 10 mm diameter and 5 cm^3 capacity was sealed a 20 cm length of tubing about 4 mm in diameter. A sufficient quantity of dry salt and about 3 cm^3 of solvent were placed in the tube. The tube was then bent round until it lay parallel with the test-tube and sealed off. The sealed U-tube was rocked in a bath at 100°C for 4 hours. After saturation, the excess solid was brought out into the narrow limb, the wide limb being left about three quarters full of the clear saturated sln. The tubes were removed from the bath and cooled to room temperature. The wide limb was cut off above the level of the sln. The solubility of salt was found by evaporation to dryness and weighing.	Nothing specified.
	ESTIMATED ERROR: Soly: rel error about 1 % (compiler). Temp: nothing specified.
	REFERENCES:

COMPONENTS:	ORIGINAL MEASUREMENTS:
(1) Potassium iodate; KIO_3; [7758-05-6]	Ricci, J.E.; Nesse, G.J.
(2) 1,4-Dioxane; $C_4H_8O_2$; [123-91-1]	J. Am. Chem. Soc. 1942, 64, 2305-11.
(3) Water; H_2O; [7732-18-5]	

VARIABLES:	PREPARED BY:
T/K = 298	Hiroshi Miyamoto
Concentration of 1,4-dioxane	

EXPERIMENTAL VALUES:

Concn of Dioxane		Soly of KIO_3[a]	
mass %	mol % (compiler)	mass %	mol dm^{-3}
0	0	8.472[b]	0.4238
10	2.2	5.300	0.2598
20	4.9	3.172	0.1531
30	8.1	1.815	0.08770
40	12	0.8855	0.04273
50	17	0.4712	0.02277
60	23	0.1350	0.00653
70	32	0.0384	0.00186
80	45	0.0060	0.00029
90	65	0.0012	0.000059
100	100	0.0000	0.00000

[a] Each value is the average of at least one determination from supersaturation and undersaturation.
In the iodometric determinations the agreement between such values was about 2/1000.

[b] For the binary aqueous system at 25°C, the compiler computes the following:

$$soly\ of\ KIO_3 = 0.4325\ mol\ kg^{-1}$$

$$= 0.7732\ mole\ \%$$

AUXILIARY INFORMATION

METHOD/APPARATUS/PROCEDURE:	SOURCE AND PURITY OF MATERIALS:
Mixtures of dioxane and water of known proportions were stirred with excess KIO_3 in glass stoppered bottles for 2 to 7 days. Equilibrium was established from both under- and super-saturation. Iodate was analyzed by reduction to iodide by sodium thiosulfate, the excess reagent being removed by acidification and boiling with dil H_2SO_4. Except in high dioxane solvents, the resulting iodine solution was analyzed volumetrically at a pH of 9-10 by titrn with standard $AgNO_3$ using eosin indicator. An appropriate blank was calcd from a series of standardizations using pure KIO_3 similarly treated. For very low solubilities the iodide was detd gravimetrically as AgI. The densities of the satd slns were detd by weighing filtered samples of slns delivered from calibrated pipets. The values were not reported in the original paper.	"C.p. grade" potassium iodate was used. The dioxane was purified and its purity verified as described in ref 1.
	ESTIMATED ERROR:
	Soly: described above. Temp: nothing specified.
	REFERENCES:
	1. Davis, T.W.; Ricci, J.E.; Sauter, C.G. J. Am. Chem. Soc. 1939, 61, 3274.

COMPONENTS:	ORIGINAL MEASUREMENTS:
(1) Potassium chlorate; $KClO_3$; [3811-04-9] (2) Glycine; $C_2H_5NO_2$; [56-40-6] (3) Water; H_2O; [7732-18-5]	Schnellbach, W.; Rosin, J. J. Am. Pharm. Assoc. 1931, 20, 227-33.
VARIABLES:	PREPARED BY:
T/K = 298	Hiroshi Miyamoto and Mark Salomon

EXPERIMENTAL VALUES:

Solubility at 25°C/g in 100 g solvent[a]

Equilibration time/days[b]

	5	12	13	14	21	42
I Undersaturation	1.04	--	--	--	--	--
II Undersaturation	--	--	--	1.06	1.04	--
	--	--	--	--	1.02	--
III Supersaturation	--	--	1.07	00	00	1.07
	--	--	1.06	--	--	1.07
IV Supersaturation	--	1.10	--	--	--	1.06
	--	1.11	--	--	--	--
	--	1.12	--	--	--	1.06

[a] According to the USP XXI (1985), two glycine solutions are defined.

(1) Glycine: not less than 98.5 % and not more than 101.5 % of $C_2H_5NO_2$ calculated on the dried basis.

(2) Glycine Irrigation: not less than 95.0 % and not more than 105.0 % of $C_2H_5NO_2$.

[b] The average of these results, excluding those of Experiment IV after 12 days which indicates supersaturation, is 1.055 g/100 g solvent.

The authors conclude the following:

(1) 1.055 g of potassium chlorate is soluble in 100 g of U.S.P. glycine sln at 25°C.

(2) One gram of potassium chlorate is soluble in 75.26 cm^3 (= 93.79 g) of U.S.P. glycine solution at 25°C.

AUXILIARY INFORMATION

METHOD/APPARATUS/PROCEDURE:	SOURCE AND PURITY OF MATERIALS:
The saturated solutions were prepared by the undersaturation and the supersaturation methods. Chlorate was reduced to chloride, and the resulting chloride determined volumetrically by Volhard's method or gravimetrically by precipitation as silver chloride. The reduction was effected by diluting a weighed quantity of the solution with water in an Erlenmeyer flask provided with a Bunsen valve. A moderate excess of acidic ferrous sulfate solution was added and the mixture heated. After cooling, the solution was treated with nitric acid, and excess 0.1 mol dm^{-3} silver nitrate added: the excess was titrated with 0.1 mol dm^{-3} sulfocyanate solution. For gravimetric determinations, the reduced solution, after treating with silver nitrate, etc. and the resulting silver chloride weighed.	U.S. Pharmacopeia quality potassium chlorate and glycine were used.
	ESTIMATED ERROR: Soly: σ = 0.005 (compilers). Temp: nothing specified.
	REFERENCES:

COMPONENTS:	ORIGINAL MEASUREMENTS:
(1) Potassium iodate; KIO_3; [7758-05-6]	Miyamoto, H.; Hasegawa, T.; Sano, H.
(2) N,N-Dimethylformamide; C_3H_7NO; [68-12-2]	J. *Solution Chem.* in press.
(3) Water: H_2O; [7732-18-5]	

VARIABLES:	PREPARED BY:
Solvent composition	M. Salomon
Temperature	

EXPERIMENTAL VALUES:

Solubilities in mol dm^{-3} reported for 20°C, 25°C, and 30°C

t/°C = 20		t/°C = 25	
mass % DMF	KIO_3/mol dm^{-3}	mass % DMF	KIO_3/mol dm^{-3}
0	0.379	0	0.423
4.79	0.284	5.12	0.314
10.05	0.210	10.00	0.234
15.67	0.147	15.43	0.169
19.75	0.114	20.43	0.123
24.57	0.082	24.78	0.0923
30.22	0.055	29.71	0.0658
35.07	0.039	34.78	0.0454
41.99	0.023	40.02	0.0299

AUXILIARY INFORMATION

METHOD/APPARATUS/PROCEDURE:	SOURCE AND PURITY OF MATERIALS:
Same as in reference (1).	Guaranteed grade KIO_3 (Wako Pure Chemicals) was recrystallized two times from doubly distilled water.
	Guaranteed grade dimethylformamide (Wako) was stored over BaO for two days, and then distilled three times under reduced pressure.
	Doubly distilled water had an electrolytic conductance of 9.8×10^{-7} S cm^{-1}.

ESTIMATED ERROR:
Soly: standard deviations for measurements in pure water are 0.0001 at 20°C, and 0.001 at 25 & 30°C. For mixed solvents $\sigma = 0.0002$ to 0.0001.
Temp: not stated.

REFERENCES:
1. Miyamoto, H.; Shimura, N.; Sasaki, K. J. *Solution Chem.* <u>1985</u>, *14*, 485.

2. Ricci, J. E. J. *Am. Chem. Soc.* <u>1934</u>, *56*, 290.

COMPONENTS:	ORIGINAL MEASUREMENTS:
(1) Potassium iodate; KIO_3; [7758-05-6]	Miyamoto, H.; Hasegawa, T.; Sano, H.
(2) N,N-Dimethylformamide; C_3H_7NO; [68-12-2]	J. *Solution Chem.* in press.
(3) Water; H_2O; [7732-18-5]	

EXPERIMENTAL VALUES: (Continued)

$t/°C = 30$

mass % DMF	$KIO_3/mol\ dm^{-3}$
0	0.475
5.53	0.345
9.81	0.268
14.71	0.201
20.10	0.144
25.03	0.103
29.79	0.074
35.02	0.051
40.33	0.033

For the binary KIO_3-H_2O system, measured densities of saturated solutions permits conversion from mol dm^{-3} to mol kg^{-1} and mole fraction units.

$t/°C$	density/g cm^{-3}	$c/mol\ dm^{-3}$	$m/mol\ kg^{-1}$[b]	χ[b]
20	1.064	0.379	0.386	0.00690
25	1.071[a]	0.423	0.431	0.00771
30	1.078	0.475	0.487	0.00869

[a]Ref. (2)

[b]Calculated by the compiler.

COMPONENTS:	ORIGINAL MEASUREMENTS:
(1) Potassium iodate; KIO_3; [7758-05-6] (2) Dimethylsulfoxide ; C_2H_6OS; [67-88-5] (3) Water; H_2O; [7732-18-5]	Miyamoto, H.; Hasegawa, T.; Sano, H. J. *Solution Chem.* in press.

VARIABLES:	PREPARED BY:
Solvent composition Temperature	M. Salomon

EXPERIMENTAL VALUES:

Solubility of KIO_3 in water-dimethylsulfoxide mixtures

Solubility in mol dm^{-3}

mass % dimethylsulfoxide	t/°C	20	25	30
0		0.379	0.423	0.475
5.03		0.295	0.326	0.367
10.02		0.226	0.250	0.281
20.09		0.126	0.142	0.155
30.01		0.0649	0.0746	0.0832
40.03		0.0311	0.0354	0.0401

AUXILIARY INFORMATION

METHOD/APPARATUS/PROCEDURE:	SOURCE AND PURITY OF MATERIALS:
Experimental details given in reference (1)	Guaranteed grade KIO_3 (Wako Pure Chemicals) was recrystallized two times from doubly distilled water. Guaranteed grade dimethylsulfoxide (Wako) was distilled three times under reduced pressure. Doubly distilled water had an electrolytic conductance of 9.8×10^{-7} S cm^{-1}.
	ESTIMATED ERROR: Soly: stnd deviation between 0.0002 and 0.001. Temp: not stated.
	REFERENCES: 1. Miyamoto, H.; Shimura, H.; Sasaki, K. J. *Solution Chem.* 1985, *14*, 485.

COMPONENTS:	ORIGINAL MEASUREMENTS:
(1) Potassium iodate; KIO_3; [7758-05-6] (2) Methanol; CH_4O; [67-56-1]	Kolthoff, I.M.; Chantooni, M.K. J. Phys. Chem. 1973, 77, 523-6.

VARIABLES:	PREPARED BY:
T/K = 298	Hiroshi Miyamoto

EXPERIMENTAL VALUES:

(1) Volumetric determination:

The authors reported the solubility of KIO_3 in methanol at 25°C (found iodometrically) to be 2.70×10^{-4} mol dm^{-3}.

The solubility product of KIO_3 in methanol was calculated from the solubility data assuming complete dissociation.

The solubility product is given as follows:

$$pK_{s0} = 7.2 \qquad \text{(authors)}$$

$$K_{s0} = 6.3 \times 10^{-8} \text{ mol}^2 \text{ dm}^{-6} \qquad \text{(compiler)}$$

(2) Conductometric determination:

The authors reported $\log K_{s0} = -7.3_5$ from a measured specific conductivity $2.7_3 \times 10^{-5}$ S cm^{-1}. In these calculations the authors used $\Lambda^{\infty} = 99.2$ S cm^2 mol^{-1}, the Debye-Huckel equation, and probably the limiting law.

AUXILIARY INFORMATION

METHOD/APPARATUS/PROCEDURE:	SOURCE AND PURITY OF MATERIALS:
(1) The iodate content was determined iodometrically. No other information given. (2) Details of the conductivity cell and method are given in ref. 1.	KIO_3 was dried in *vacuo* at 70°C for 3 hours. Matheson Spectroquality grade methanol was distilled once over magnesium turnings. The water content was 0.01 % by Karl Fischer Titration.

ESTIMATED ERROR:
The uncertainty in pK_{s0} is $\pm$ 0.1 log units.
Temp: not given.

REFERENCES:
1. Kolthoff, I.M.; Bruckenstein, S.; Chantooni, M.K. J. Am. Chem. Soc. 1961, 83, 3927.

COMPONENTS:	ORIGINAL MEASUREMENTS:
(1) Potassium iodate; KIO$_3$; [7758-05-6]	Kolthoff, I.M.; Chantooni, M.K.
(2) 6,7,9,10,17,18,20,21-Octahydrodibenzo [b,k] [1,4,7,10,13,16] hexaoxacyclo-octadecin(dibenzo-18-crown-6); C$_{20}$H$_{24}$O$_6$; [14187-32-7]	Anal. Chem. 1980, 52, 1039-44.
(3) Methanol; CH$_4$O; [67-56-1]	

VARIABLES:	PREPARED BY:
Composition	Hiroshi Miyamoto and Mark Salomon
T/K = 298	

EXPERIMENTAL VALUES:

Solubilities of KIO$_3$ and dibenzo-18-crown-6 (DB-18) in methanol at 25°C

KIO$_3$	DB-18	electrolytic conductance of satd sln
mol dm^{-3}	mol dm^{-3}	$10^5\kappa$/S cm^{-1}
$2.6_1 \times 10^{-4}$	0	2.30
4.66×10^{-3}	6.51×10^{-3}	24.6
0	1.38×10^{-3a}	---

[a] Result quoted from reference 1.

COMMENTS AND/OR ADDITIONAL DATA:

Dissociation constants K_d(KIO$_3$) and K_d(LKIO$_3$) were determined conductometrically and found to equal 2.5×10^{-3} mol^{-1} dm^3 and 6×10^{-3} mol^{-1} dm^3, respectively (L = DB-18).

From the total solubility of the ether in the absence of salt, C_L, the solubility of the ether in the presence of the salt, C_L(salt), the solubility of the salt and the activity of the salt, the formation constant of the ether-K$^+$ complex, K_f(LK$^+$), was found to be 1.12×10^5 mol^{-1} dm^3. The selectivity for the K$^+$/Na$^+$ coordination with DB-18, K_f(LK$^+$)/K_f(LNa$^+$), was reported to equal 3.

From activity measurements, authors report in pure methanol [K$^+$] = 2.40×10^{-4} mol dm^{-3}, $y_{\pm}^2 = 0.88$, and thus $K_{s0}^{\circ} = [M^+]^2 y_{\pm}^2 = 5 \times 10^{-8}$ mol^2 dm^{-6}.

AUXILIARY INFORMATION

METHOD/APPARATUS/PROCEDURE:	SOURCE AND PURITY OF MATERIALS:
Solubilities of KIO$_3$ and DB-18 detd isothermally in pure methanol and methanol satd with DB-18. Satd slns prepd by adding 0.5 mmol of each compound to about 2 drops of alcohol, stirring, and decanting the liquid. To the residue about 10 ml of methanol was added, and the mixt magnetically stirred for 2 days. Two additional days of stirring did not affect the soly. The ether, DB-18, was detd spectrophotometrically in dichloroethane. Quantitative transfer of the ether to dichloroethane was found to be complete after 4 extractions, and Beer's law was followed to at least 1.3×10^{-4} mol dm^{-3}. KIO$_3$ was detd by 3 aq extractions from the satd sln and dichloroethane. The combined aq extracts were titrated idometrically. Complete extraction of the iodate was confirmed experimentally.	"C.p." grade (Fisher) KIO$_3$ recrystallized three times from water and dried at atm pressure at 70°C. Aldrich dibenzo-18-crown-6 (DB-18) was recrystallized 4 times from water and dried at atm pressure at 50°C; m.p. = 163°C, lit, 164°C (2). Methanol (Fisher "spectroquality" grade) was distilled once from Mg turnings.
	ESTIMATED ERROR:
	Soly: nothing specified. Temp: nothing specified.
A Markson No. 1002 K$^+$ specific ion electrode was used to detn the K$^+$ activity, and Nernstian behavior was observed. Conductivities were measured with an Industrial Instrument Model RC 16B1 bridge with Jones type cells.	REFERENCES: 1. Pederson, C.J. J. Am. Chem. Soc. 1970, 92, 388. 2. Pederson, C.J.; Frensdorff, H.K. Angew. Chem., Int. Ed. Engl. 1972, 11, 16.

COMPONENTS:	ORIGINAL MEASUREMENTS:
(1) Potassium iodate; KIO_3; [7758-05-6] (2) Dimethylsulfoxide(sulfinyl bis-methane); C_2H_6OS; [67-68-5]	Kolthoff, I.M.; Chantooni, M.K. J. Phys. Chem. 1973, 77, 523-6.

VARIABLES:	PREPARED BY:
T/K = 298	Hiroshi Miyamoto

EXPERIMENTAL VALUES:

(1) Potentiometric determination:

The authors reported $- \log K_{s0}^{\circ} = 7.7$ from $-\log a(K^+) = 3.85$, which was determined potentiometrically assuming $a(IO_3^-) = a(K^+)$.

The compiler computes $K_{s0}^{\circ} = 2.0 \times 10^{-8}$ mol^2 dm^{-6} from this value.

(2) Conductance determination:

The authors reported $\log K_{s0} = - 7.4$ from a measured specific conductivity of 0.85×10^{-5} S cm^{-1}. Λ^{∞} was given as 37.4 S cm^2 mol^{-1} (obtained from the literature), and presumably the limiting law was used to calculate the solubility. Activity coefficients were calculated from the "partially extended" Debye-Huckel equation.

The compiler calculates $K_{s0} = 4.0 \times 10^{-8}$ mol^2 dm^{-6} from this value.

AUXILIARY INFORMATION

METHOD/APPARATUS/PROCEDURE:	SOURCE AND PURITY OF MATERIALS:
(1) The solubility product of KIO_3 in dimethylsulfoxide was determined potentiometrically from emf measurements made on the following cell: $K(gl)/KIO_3(c_1)//AgNO_3(c_2)/Ag$ where $KIO_c(c_1)$ and $AgNO_3(c_2)$ are concentrations of the saturated KIO_3 solution and 0.01 mol dm^{-3} $AgNO_3$ solution, and // is a salt bridge containing 0.01 mol dm^{-3} tetraethylammonium perchlorate. The liquid junction potential calculated by the Henderson equation (ref 1) is -5 mV, and was neglected.	KIO_3 was dried in vacuo at 70°C for 3 hours. Dimethylsulfoxide, Fisher certified reagent grade product, was purified as described in ref 3. The water content of the purified solvent as found by Karl Fischer titration was less than 0.0005 %. Electrodes were prepared electrolytically (ref 4).

ESTIMATED ERROR:	REFERENCES:
The uncertainty of pK_{s0} is $\pm$ 0.1. Temperature not given.	1. Ives, D.J.G.; Janz, G.J. Reference Electrodes. Academic Press. N.Y. 1961, p. 54. 2. Kolthoff, I.M.; Bruckenstein, S.; Chantooni, Jr., M.K. J. Am. Chem. Soc. 1961, 83, 3927. 3. Kolthoff, I.M.; Reddy, T.B. Inorg. Chem. 1962, 1, 189. 4. Ives, D.J.G.; Janz, G.J. Reference Electrodes. Academic Press. N.Y. 1961, p. 179: Kolthoff, I.M.; Chantooni, M.K. J. Am. Chem. Soc. 1965, 87, 4428.

COMPONENTS:	ORIGINAL MEASUREMENTS:
(1) Potassium iodate; KIO3; [7758-05-6] (2) Ammonia; NH3; [7664-41-7]	Hunt, H.; Boncyk, L. J. Am. Chem. Soc. 1933, 55, 3528-30.

VARIABLES:	PREPARED BY:
T/K = 298	Hiroshi Miyamoto and Mark Salomon

EXPERIMENTAL VALUES:

The solubility of KIO3 in liquid ammonia at 25°C was reported as

$$0.000 \text{ g/100 g NH}_3$$

Compilers' note: In a subsequent paper (2) Hunt gives the solubility of KIO3 in liquid ammonia as 3.044×10^{-5} mol kg^{-1} at 25°C. See the compilation of ref (2).

AUXILIARY INFORMATION

METHOD/APPARATUS/PROCEDURE
Two methods were used as described in (1).

Method I. 25 ml test tubes with a constriction at the middle were employed. About 10-25 g NH3 were condensed in the bottom, and the dry salt contained in a small tube tightly covered with cotton cloth was added to the test tube; this small tube remained in the upper part of the test tube as it could not pass the constriction in the middle of the test tube. The top of the test tube was drawn to a tip and sealed, and the tube inverted and placed in a thermostat at 25°C. Equilibrium between NH3 and the excess salt in the small covered tube required 1-3 weeks with periodic shaking. The test tube was then inverted and only the satd sln drained into the lower end (excess solid remained in the small tube covered with the cotton cloth). The sln was frozen and sealed at the constriction, and weighed. The seal was then broken and the NH3 boiled off, and the residue weighed.

Method II.   Excess NH3 was condensed on a weighed amount of salt in a tube fitted with a stopcock. After thermostating at 25°C, NH3 was slowly permitted to escape through the stopcock until a crystal of solid appeared and remained undissolved upon prolonged shaking.

Authors state that the error due to the condensation of gaseous NH3 was not significant since the dead space was kept to a minimum of about 30 cm^3. However this amount of dead space was stated to limit the precision of the method to 0.5 %.

SOURCE AND PURITY OF MATERIALS:
Reagent grade KIO3 was recrystallized three times from water and then from "a suitable" anhydrous solvent. The salt was dried to constant weight in a vacuum oven.

Purification of NH3 not specified, but probably similar to that described in (1). In (1) commercial anhyd ammonia was stored over metallic sodium for several weeks before use.

ESTIMATED ERROR:

Soly: accuracy probably around ± 1-2 %.
      (compilers).
Temp: 25.00 + 0.025°C accuracy established
      by NBS calibration (see ref 1).

REFERENCES:

1. Hunt, H.; J. Am. Chem. Soc. 1932, 54, 3509.
2. Anhorn, V.J.; Hunt, H. J. Phys. Chem. 1941, 45, 351.

COMPONENTS:	ORIGINAL MEASUREMENTS:
(1) Potassium iodate; KIO_3; [7758-05-6] (2) Potassium chloride; KCl; [7447-40-7] (3) Liquid ammonia; NH_3; [7664-41-7]	Anhorn, V.J.; Hunt, H. J. Phys. Chem. 1941, 45, 351-62.

VARIABLES:	PREPARED BY:
Concentration of KCl at 298.15 K	Hiroshi Miyamoto and Mark Salomon

EXPERIMENTAL VALUES:

concentration of KCl		solubility of KIO_3
$mol\ dm^{-3}$	$mol\ kg^{-1}$	$10^5\ mol\ kg^{-1}$
0	0	3.0441
0	0	3.0438
0	0	3.0412
0	0	3.0461
0	0	3.0448
0	0	3.044 (average value)
0.001907	0.003162	0.7439
0.002658	0.004407	0.8428
0.007624[a]	0.01264	1.4238

[a] Saturated

continued.......

AUXILIARY INFORMATION

METHOD/APPARATUS/PROCEDURE:

The U-shaped Pyrex soly apparatus is shown in the figure on the next page. A glass float in which an iron nail was sealed was used to stir the slns by engaging solenoid E. B is a cotton plug in which a small glass tube was embedded to permit passage of NH_3 vapors. Both salts were placed in tube G which was then sealed at A, evacuated at F, and the entire apparatus heated in a flame. The apparatus was evacuated for 10 h and then flushed with dry air followed by flushing with NH_3. Ammonia was distd from a reservoir and condensed in tube G, and the apparatus sealed at F. Equilibrium was established by thermostating at 25°C, and was approached from above and below with stirring every 2 h. The satd sln was decanted into tube H, and the distance of the sln from etch mark C measured with a cathatometer. The ammonia was then dist back into tube G, cooled in a bath of solid CO_2-$CHCl_3$-CCl_4, and the tube H removed by breaking about 2 cm above the etch mark C. The KIO_3 content (residue) in H was detd by the method described in (1), and the non-saturating salt was detd gravimetrically by the method given in (2). The volume of

SOURCE AND PURITY OF MATERIALS:

Ammonia was stored over sodium and distd through glass wool into a reservoir prior to distn into the soly tube.
AR grade KIO_3 recrystallized three times from conductivity water, dried at 180°C, ground to a powder and analyzed as in (3).
KCl was crystallized three times from conductivity water and dried at 110°C.

ESTIMATED ERROR:

Soly: for the binary system, standard dev is .0018 (compilers). No information available for ternary systems.
Temp: accuracy ± 0.005 K (authors).

REFERENCES:
1. Anhorn, V.J.; Hunt, H. Ind. Eng. Chem. Anal. Ed. 1937, 9, 591.
2. Willard; Furman, N.H. Elementary Quantitative Analysis. D. Van Nostrand. New York. 1935, p 295.
3. Scott; Standard Methods of Chemical Analysis. D. Van Nostrand. New York. 1939. p 372.

COMPONENTS:	ORIGINAL MEASUREMENTS:
(1) Potassium iodate; KIO_3; [7758-05-6]	Anhorn, V.J.; Hunt, H.
(2) Potassium chloride; KCl; [7447-40-7]	J. Phys. Chem. <u>1941</u>, 45, 351-62.
(3) Liquid ammonia; NH_3; [7664-41-7]	

COMMENTS AND/OR ADDITIONAL DATA:

The solubility apparatus used is shown below.

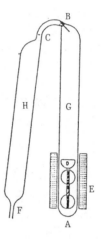

METHOD/APPARATUS/PROCEDURE: (Continued)

tube H was then calibrated. For very dilute
solutions, standard (aq) solutions of the
nonsaturating salt were prepared by weight
and placed in a cup D, and the water evaporated slowly at 50°C. The cup D was then placed
on the float, the KIO_3 added and the tube G
sealed at A. The soly was then detd as described above. Densities of saturated solutions prepared by weight were measured pyno-
metrically at 25°C, and the densities of
saturated solutions detd above were obtained
by graphical interpolation.
The soly in the binary system was detd five
times, and an unspecified number of times in
ternary systems. The nature of the solid
phase was not discussed.

COMPONENTS:	ORIGINAL MEASUREMENTS:
(1) Potassium iodate; KIO_3; [7758-05-6] (2) Sodium chloride; NaCl; [7647-14-5] (3) Liquid ammonia; NH_3; [7664-41-7]	Anhorn, V.J.; Hunt, H. *J. Phys. Chem.* <u>1941</u>, 45, 351-62.
VARIABLES:	PREPARED BY:
Concentration of NaCl at 298.15 K	Hiroshi Miyamoto

EXPERIMENTAL VALUES:

Concentration of NaCl		Solubility of KIO_3
mol dm^{-3}	mol kg^{-1}	10^5 mol kg^{-1}
0	0	3.044 (av)
0.0003685	0.0006109	4.495
0.001655	0.002744	5.489
0.003543	0.005874	5.831
0.007738	0.01283	6.188
0.03063	0.05078	7.098
0.04352	0.07214	7.377
0.09376	0.1554	8.024
0.1154	0.1913	8.233
0.1554	0.2575	8.611
0.2101	0.3482	8.913
0.2746	0.4551	9.145

AUXILIARY INFORMATION

METHOD/APPARATUS/PROCEDURE:	SOURCE AND PURITY OF MATERIALS:
See the compilation of the KIO_3-KCl-NH_3 system reported by Anhorn and Hunt.	Amonia was stored over sodium and distd through glass wool into a reservoir prior to distn into the soly tube. Analytical reagent grade KIO_3 was recrystallized three times from conductivity water and dried at 180°C. The purity was tested by the method described in ref 1. Purified HCl gas was passed into a saturated solution of analytical reagent grade NaCl. The precipitated NaCl was then crystallized from conductivity water and fused in a platinum crucible. The fused salt was ground to a fine powder in an agate mortar.
	ESTIMATED ERROR: See the KIO_3-KCl-NH_3 compilation of the source paper.
	REFERENCES: Scott; *Standard Methods of Chemical Analysis* D. Van Nostrand. New York. <u>1939</u>, p 372.

COMPONENTS:	ORIGINAL MEASUREMENTS:
(1) Potassium iodate; KIO_3; [7758-05-6] (2) Ammonium chloride; NH_4Cl; [12125-02-9] (3) Liquid ammonia; NH_3; [7664-41-7]	Anhorn, V.J.; Hunt, H. J. Phys. Chem. 1941, 45, 351-62.

VARIABLES:	PREPARED BY:
Concentration of NH_4Cl at 298.15 K	Hiroshi Miyamoto

EXPERIMENTAL VALUES:

Concentration of NH_4Cl		Solubility of KIO_3
mol dm^{-3}	mol kg^{-1}	10^5 mol kg^{-1}
0	0	3.044 (average value)
0.009398	0.001558	5.7782
0.002220	0.003670	7.1681
0.004161	0.006898	8.7650
0.007543	0.01251	10.662
0.01035	0.01715	11.820
0.01822	0.03020	14.025
0.02243	0.03718	14.920
0.03251	0.05389	16.580
0.04568	0.07572	18.045
0.07695	0.1275	20.688
0.1060	0.1758	22.189
0.1463	0.2424	23.975
0.1635	0.2710	24.590
0.2004	0.3320	25.745
0.2657	0.4402	27.480
0.3800	0.6299	30.018
0.4330	0.7180	30.980
0.5728	0.9528	33.160

AUXILIARY INFORMATION

METHOD/APPARATUS/PROCEDURE:	SOURCE AND PURITY OF MATERIALS:
See the compilation of the KIO_3-KCl-NH_3 system reported by Anhorn and Hunt.	Ammonia was stored over sodium and distd through glass wool into a reservoir prior to distn into the soly tube. Analytical reagent grade KIO_3 was recrystallized three times from conductivity water and dried at 180°C. The source of NH_4Cl is not given.
	ESTIMATED ERROR: See the KIO_3-KCl-NH_3 compilation of the source paper.
	REFERENCES:

COMPONENTS:	ORIGINAL MEASUREMENTS:
(1) Potassium iodate; KIO_3; [7758-05-6] (2) Potassium bromide; KBr; [7758-02-3] (3) Liquid ammonia; NH_3; [7664-41-7]	Anhorn, V.J.; Hunt, H. J. *Phys. Chem.* <u>1941</u>, 45, 351-62.

VARIABLES:	PREPARED BY:
Concentration of KBr at 298.15 K	Hiroshi Miyamoto

EXPERIMENTAL VALUES:

Concentration of KBr		Solubility of KIO_3
mol dm^{-3}	mol kg^{-1}	10^5 mol kg^{-1}
0	0	3.044 (average value)
0.002492	0.004131	0.7392
0.004468	0.007407	0.8180
0.005864	0.009721	0.8595
0.007025	0.01165	0.8876
0.01170	0.01940	0.9062
0.02591	0.04295	1.0684
0.04867	0.08070	1.1350
0.07078	0.1173	1.1678
0.1195	0.1981	1.2090
0.1549	0.2568	1.2330

AUXILIARY INFORMATION

METHOD/APPARATUS/PROCEDURE:	SOURCE AND PURITY OF MATERIALS:
See the compilation of the KIO_3-KCl-NH_3 system reported by Anhorn and Hunt.	Ammonia was stored over sodium and distd through glass wool into a reservoir prior to distillation into the solubility tube. Analytical reagent grade KIO_3 was recrystallized three times from conductivity water and dried at 180°C. KBr was recrystallized three times from conductivity water and dried at 110°C. The product was ground to a fine powder before final drying.
	ESTIMATED ERROR: See the KIO_3-KCl-NH_3 compilation of the source paper.
	REFERENCES:

COMPONENTS:	ORIGINAL MEASUREMENTS:
(1) Potassium iodate; KIO_3; [7758-05-6] (2) Potassium iodide; KI; [7681-11-0] (3) Liquid ammonia; NH_3; [7664-41-7]	Anhorn, V.J.; Hunt, H. J. Phys. Chem. 1941, 45, 351-62.

VARIABLES:	PREPARED BY:
Concentration of KI at 298.15 K	Hiroshi Miyamoto

EXPERIMENTAL VALUES:

Concentration of KI		Solubility of KIO_3
mol dm^{-3}	mol kg^{-1}	10^5 mol kg^{-1}
0	0	3.044 (average value)
0.001709	0.002834	0.4238
0.004258	0.007058	0.4893
0.005154	0.008534	0.5165
0.01098	0.01819	0.5682
0.01801	0.02985	0.6188
0.03735	0.06191	0.7211
0.03774	0.06254	0.7370
0.05310	0.08801	0.7849
0.08446	0.1400	0.8868
0.1031	0.1709	0.9362
0.1234	0.2046	0.9886
0.2224	0.3693	1.1530
0.3018	0.5021	1.2140

AUXILIARY INFORMATION

METHOD/APPARATUS/PROCEDURE:	SOURCE AND PURITY OF MATERIALS:
See the compilation of the KIO_3-KCl-NH_3 system reported by Anhorn and Hunt.	Ammonia was stored over sodium and distd through glass wool into a reservoir prior to distillation into the soly tube. Analytical reagent grade KIO_3 was recrystallized three times from conductivity water and dried at 180°C. KI was recrystallized three times from conductivity water and dried at 110°C. The product was ground to a fine powder before final drying.
	ESTIMATED ERROR:
	See the KIO_3-KCl-NH_3 compilation of the source paper.
	REFERENCES:

COMPONENTS:	ORIGINAL MEASUREMENTS:
(1) Potassium iodate; KIO_3; [7758-05-6] (2) Hydrazine; N_2H_4; [302-01-2]	Welsh, T.W.B.; Broderson, H.J. J. Am. Chem. Soc. 1915, 37, 816-24.
VARIABLES: Room temperature (compiler's assumption)	PREPARED BY: Mark Salomon and Hiroshi Miyamoto

EXPERIMENTAL VALUES:

The solubility of KIO_3 in hydrazine at room temperature was given as

$$0.01 \ g/1cm^3 \ N_2H_4$$

The authors stated that the chief object of this research was to obtain qualitative

and approximate quantitative data.

AUXILIARY INFORMATION

METHOD/APPARATUS/PROCEDURE:

The solubility vessel was a glass tube to
which a U-shaped capillary tube was attached
to the bottom. A stopcock at the end of the
capillary permitted the adjustment of the
rate of flow of dry nitrogen. About 1 cc of
anhydrous hydrazine was placed in the tube,
and small amounts of $NaClO_3$ added from a
weighing bottle.
After each addition of $NaClO_3$, a loosely fitting cork was placed in the top of the
solubility tube. Nitrogen was bubbled
through solution until the salt dissolved.
The process was repeated until no more salt
would dissolve. Temperature was not kept
constant.
The accuracy of this method is very poor.
In addition, the authors stated that it was
difficult to prevent the oxidation of
hydrazine.

SOURCE AND PURITY OF MATERIALS:

Anhydrous hydrazine was prepared by first
partially dehydrating commercial hydrazine
with sodium hydroxide according to the
method of Raschig (1).   Further removal
of water was distilled over barium oxide
after the method of de Bruyn (2).
The form of distillation apparatus employed
and the procedure followed in the respective
distillations were those described by Welsh
(3). The product was found on analysis to
contain 99.7 % hydrazine. The hydrazine was
stored in 50 cm^3 sealed tubes. Sodium
chlorate was the ordinary pure chemicals of
standard manufacture.

ESTIMATED ERROR:
Soly: accuracy $\pm$ 50 % at best (compilers).

REFERENCES:
1. Raschig, F. Ber. Dtsch. Chem. Ges. 1927,
   43, 1927.: Hale, C.F.; Shetterly, F.F.
   J. Am. Chem. Soc. 1911, 33, 1071.
2. de Bruyn, L. Rec. Trav. Chim. Pays-Bas.
   1895, 14, 458.
3. Welsh, T.W. J. Am. Chem. Soc. 1915, 37,
   497.

COMPONENTS:	EVALUATOR:
(1) Rubidium iodate; $RbIO_3$; [13446-76-9]	H. Miyamoto Niigata University Niigata, Japan
(2) Water; H_2O; [7732-18-5]	and Mark Salomon US Army ET & DL Fort Monmouth, NJ, USA
	June, 1986

CRITICAL EVALUATION:

THE BINARY SYSTEM

Data for the solubility of $RbIO_3$ in water have been reported in 15 publications (1-15). Publications (4-15) are studies of ternary systems using the isothermal method, and although some investigators (1-4, 14, 15) did not analyze the solid phase, the evaluators assume it to be the anhydrous salt by analogy to the solid phase found in studies on ternary systems in (5-13). The compilations for references (6,7) are given in the $LiIO_3$ chapter, for (12) in the $NaIO_3$ chapter, for (4) in the KIO_3 chapter, and for ref. (13) dealing with the $RbIO_3$-$Mg(IO_3)_2$-H_2O system in the first volume on Alkaline Earth Metal Halates (16).

Analyses of saturated solutions varied: iodometric titration was used in (3-12), the Carius method in (1), gravimetry with sodium tetraphenylborate in (5-10) or with sulfate in (2), and flame photometry in (4).

A summary of the solubilities reported in (1-15) is given in Table 1. The results for the approximate temperature of 296 K in (1,2) were rejected as was the obviously high solubility at 323 K reported in (5). The low solubility of 0.00163 mole fraction at 298 K reported in (4) was rejected on the usual basis of its poor fit to the smoothing equations. The remaining data were fitted to the two smoothing equations based on mole fraction and mol/kg solubilities, and the respective results are:

$$Y_x = -27922/(T/K) - 131.152 \; \ell n \; (T/K) + 776.99 + 0.18259(T/K)$$

$$\sigma_y = 0.016 \qquad\qquad \sigma_x = 1.4 \times 10^{-5}$$

and

$$Y_m = -4381/(T/K) - 6.236 \; \ell n \; (T/K) + 50.222$$

$$\sigma_y = 0.012 \qquad\qquad \sigma_m = 0.0017$$

All solubilities calculated from the smoothing equations are designated as *recommended* values.

TERNARY SYSTEMS

1. One saturating component.

Solubilities of $RbIO_3$ in aqueous KNO_3 solutions and in dilute HNO_3 solutions at 298.2 K have been reported by Larson and Renier (14). The solubility of $RbIO_3$ increases with increasing concentration of KNO_3, but in HNO_3, the solubility first increases and reaches a maximum of around 0.16 mol dm^{-3} at an acid concentration of around 0.3 mol dm^{-3}.

2. Two saturating components.

Solubilities in aqueous solutions containing a second saturating component in addition to $RbIO_3$ have been reported in 10 publications (4-13), and all studies used the isothermal method. A summary of the various ternary systems studied is given in Table 3.

The ternary system $RbIO_3$-$RbOH$-H_2O is of the simple eutonic type (8), but the dominant features in the ternary systems with HIO_3 (5) and $LiIO_3$ (6) are the formation of the double salts $RbIO_3 \cdot 2HIO_3$ and $RbIO_3 \cdot 2LiIO_3$. No double salts were found in the remaining ternary systems.

COMPONENTS:	EVALUATOR:
(1) Rubidium iodate; RbIO₃; [13446-76-9]	H. Miyamoto Niigata University Niigata, Japan
(2) Water; H₂O; [7732-18-5]	and Mark Salomon US Army ET & DL Fort Monmouth, NJ, USA
	June, 1986

CRITICAL EVALUATION:

Table 1.   Summary of solubilities in the $RbIO_3$-H_2O system[a]

T/K	mol kg^{-1}	mole fraction	ref
273.2	0.0411	0.000741	3
283.2	0.0609	0.001053	3
293.2	0.0828	0.00149	3
293.2	0.0811	0.00146	15
296[a]	0.0806	---	1
296.2[a]	0.081	---	2
298.2[a]	0.0908	0.00163	4
298.2	0.0926[b]	0.00167	14
298.2	0.0928	0.00167	7
298.2	0.0940	0.00169	8
298.2	0.0940	0.00169	9
298.2	0.0943	0.00170	15
298.2	0.0944	0.00170	11
298.2	0.0948	0.00171	3
303.2	0.107	0.00193	3
303.2	0.109	0.00196	15
313.2	0.139	0.00250	3
323.2	0.176	0.00315	3
323.2	0.176	0.00317	6,10
323.2	0.176	0.00317	12,13
323.2[a]	0.223	0.00400	5
333.2	0.220	0.00394	3
343.2	0.266	0.00477	3
353.2	0.320	0.00574	3
363.2	0.380	0.00680	3
373.2	0.449	0.00802	3

[a]Rejected data points.

[b]Calculated by the evaluators using a density of 1.016 g/cm^3.

COMPONENTS:	EVALUATOR:
(1) Rubidium iodate; $RbIO_3$; [13446-76-9] (2) Water; H_2O; [7732-18-5]	H. Miyamoto Niigata University Niigata, Japan and Mark Salomon US Army ET & DL Fort Monmouth, NJ, USA June, 1986

CRITICAL EVALUATION:

Table 2. Smoothed solubilities from 273-373 K[a]

T/K	mol/kg[b]	mole fraction
273.2	0.0422	0.00074
278.2	0.0503	0.00089
283.2	0.0595	0.00106
288.2	0.0697	0.00125
293.2	0.0812	0.00146
298.2	0.0938	0.00169
303.2	0.108	0.00194
313.2	0.140	0.00251
323.2	0.177	0.00317
333.2	0.220	0.00392
343.2	0.268	0.00477
353.2	0.322	0.00573
363.2	0.380	0.00681
373.2	0.444	0.00802

[a]All data in this table are designated as *recommended*.

[b]Reference molality used in the smoothing equation is 0.094 mol/kg.

Table 3. Summary of solubility studies in ternary systems

Ternary system	T/K	Solid phase	Reference
$RbIO_3$ - KIO_3 - H_2O	298	Not given	4
$RbIO_3$ - $CsIO_3$ - H_2O	298	Not given	4
$RbIO_3$ - HIO_3 - H_2O	323	$RbIO_3$; HIO_3; $RbIO_3.2HIO_3$	5
$RbIO_3$ - $LiIO_3$ - H_2O	323	$RbIO_3$; $LiIO_3$; $2LiIO_3.RbIO_3$	6
$RbIO_3$ - $LiIO_3$ - H_2O	298	$RbIO_3$; $LiIO_3$; $RbIO_3.2LiIO_3$	7
$RbIO_3$ - $RbOH$ - H_2O	298	$RbIO_3$; $RbOH.2H_2O$	8
$RbIO_3$ - $Al(IO_3)_3$ - H_2O	298	$RbIO_3$; $Al(IO_3)_3.6H_2O$	9
$RbIO_3$ - $Zn(IO_3)_2$ - H_2O	323	$RbIO_3$; $Zn(IO_3)_3.2H_2O$	10
$RbIO_3$ - $Nd(IO_3)_2$ - H_2O	298	$RbIO_3$; $Nd(IO_3)_3.2H_2O$	11
$RbIO_3$ - $NaIO_3$ - H_2O	323	$RbIO_3$; $NaIO_3.H_2O$	12
$RbIO_3$ - $Mg(IO_3)_2$ - H_2O	323	$RbIO_3$; $Mg(IO_3)_2.4H_2O$	13

COMPONENTS:	EVALUATOR:
(1) Rubidium iodate; $RbIO_3$; [13446-76-9] (2) Water; H_2O; [7732-18-5]	H. Miyamoto Niigata University Niigata, Japan and Mark Salomon US Army ET & DL Fort Monmouth, NJ, USA June, 1986

CRITICAL EVALUATION:

REFERENCES:

1. Wheeler, H. L. *Am. J. Sci.* 1892, (3) *44*, 123.

2. Barker, T. V. *J. Chem. Soc.* 1908, *93*, 15.

3. Breusov, O. N.; Kashina, N. I.; Revzina, T. V.; Sobolevskaya, N. G. *Zh. Neorg. Khim.* 1967, *12*, 2240; *Russ. J. Inorg. Chem. (Engl. Transl.)* 1967, *12*, 1179.

4. Kirgintsev, A. N.; Shklovskaya, R. M.; Arkhipov, S. M. *Izv. Akad. Nauk SSSR Ser. Khim.* 1971, 2631; *Bull. Acad. Sci. USSR, Div. Chem. Sci. (Engl. Transl.)* 1971, 2501.

5. Tatarinov, V. A. *Uch. Zap. Yarosl. Gos. Pedagog. Inst.* 1972, No. 103, 83.

6. Karataeva, I. M.; Vinogradov, E. E.; *Zh. Neorg. Khim.* 1974, *19*, 3156; *Russ. J. Inorg. Chem. (Engl. Transl.)* 1974, *19*, 1726.

7. Shklovskaya, R. M.; Kashina, N. I.; Arkhipov, V. A.; Kuzina, V. A.; Kidyarov, B. I. *Zh. Neorg. Khim.* 1975, *20*, 783; *Russ. J. Inorg. Chem. (Engl. Transl.)* 1975, *20*, 411.

8. Lepeshkov, I. N.; Vinogradov, E. E.; Tarasova, G. N. *Zh. Neorg. Khim.* 1976, *21*, 1353; *Russ. J. Inorg. Chem. (Engl. Transl.)* 1976, *21*, 739.

9. Vinogradov, E. E.; Tarasova, G. N. *Zh. Neorg. Khim.* 1978, *23*, 3161; *Russ. J. Inorg. Chem. (Engl. Transl.)* 1978, *23*, 1754.

10. Vinogradov, E. E. ; Karataeva, I. M. *Zh. Neorg. Khim* 1979, *24*, 2529; *Russ. J. Inorg. Chem. (Engl. Transl.)* 1979, *24*, 1406.

11. Tarasova, G. N.; Vinogradov, E. E.; Kudinov, I. B. *Zh. Neorg. Khim.* 1981, *26*, 2841; *Russ. J. Inorg. Chem. (Engl. Transl.)* 1981, *26*, 1520.

12. Vinogradov, E. E.; Karataeva, I. M. *Zh. Neorg. Khim.* 1982, *27*, 2155; *Russ. J. Inorg. Chem. (Engl. Transl.)* 1982, *27*, 1681.

13. Vinogradov, E. E.; Karataeva, I. M. *Zh. Neorg. Khim.* 1976, *21*, 1666; *Russ. J. Inorg. Chem. (Engl. Transl.)* 1976, *21*, 912.

14. Larson, W. D.; Renier, J. J. *J. Am. Chem. Soc.* 1952, *74*, 3184.

15. Miyamoto, H.; Hasegawa, T.; Sano, H. *J. Solution Chem.* in press.

16. Miyamoto, H.; Salomon, M.; Clever, H. L. *Solubility Data Series Volume 14: Alkaline Earth Metal Halates.* Pergamon Press, London. 1983.

COMPONENTS:	ORIGINAL MEASUREMENTS:
(1) Rubidium iodate; $RbIO_3$; [13446-76-9] (2) Water; H_2O; [7732-18-5]	Wheeler, H.L. *Am. J. Sci.* 1892, *44*, 123-33.

VARIABLES:	PREPARED BY:
T/K = 296	Hiroshi Miyamoto

EXPERIMENTAL VALUES:

The solubility of $RbIO_3$ in water was given as

100 parts of water dissolve 2.1 parts of $RbIO_3$.

The compiler's conversions to mass % and mol kg^{-1} are:

2.05 mass %

0.0806 mol kg^{-1}

AUXILIARY INFORMATION

METHOD/APPARATUS/PROCEDURE:	SOURCE AND PURITY OF MATERIALS:
No information was given.	Rubidium iodate was prepared by stoichiometric mixing of iodine pentoxide, in either strong or dilute aqueous solution, with a solution of rubidium carbonate. The precipitate, after vacuum filtering, was washed with a little water and dried on paper. Found: Rb 32.17; I 48.50; O 20.59. Calcd for $RbIO_3$; Rb 32.83, I 48.72, O 18.43.
	ESTIMATED ERROR:
	Nothing specified.
	REFERENCES:

COMPONENTS:	ORIGINAL MEASUREMENTS:
(1) Rubidium iodate; $RbIO_3$; [13446-76-9] (2) Water; H_2O; [7732-18-5]	Barker, T.V. J. *Chem. Soc.* 1908, 93, 15-6.

VARIABLES:	PREPARED BY:
T/K = 296	Hiroshi Miyamoto

EXPERIMENTAL VALUES:

The solubility of $RbIO_3$ in water at 23°C is given as follows:

100 parts of water dissolve 2.1 parts of $RbIO_3$.

This is equivalent to 0.081 mol kg^{-1} (compiler).

The specific gravity of the saturated solution at 14°C was reported as 4.559.

The compiler assumes that precipitation occurred upon cooling the saturated

solution at 23°C to 14°C.

AUXILIARY INFORMATION

METHOD/APPARATUS/PROCEDURE:	SOURCE AND PURITY OF MATERIALS:
The iodine content was estimated by the Carius method (the reference was not given in the original paper), but the compiler assumes that the total solubility was determined by evaporation and heating to constant mass. The heating was carried out in two operations lasting four hours: the first to 150°C, and the second to 250°C. The rubidium content was determined by the usual sulfate method. No other information was given in the original paper.	Rubidium iodate was prepared by adding aqueous HIO_3 solution to aqueous rubidium carbonate solution. Another method was also used to prepare rubidium iodate: a good yield was obtained by passing chlorine into a hot concentrated solution of a mixture of rubidium iodide and hydroxide. No other information given.
	ESTIMATED ERROR: Nothing specified.
	REFERENCES:

COMPONENTS:	ORIGINAL MEASUREMENTS:
(1) Rubidium iodate; RbIO$_3$; [13446-76-9] (2) Water; H$_2$O; [7732-18-5]	Breusov, O.N.; Kashina, N.I.; Revzina, T.V.; Sobolevskaya, N.G. *Zh. Neorg. Khim.* 1967, *12*, 2240-3; *Russ. J. Inorg. Chem.* (*Engl. Transl.*) 1967, *12*, 1179-81.
VARIABLES: Temperature: 273.2 to 323.2 K	PREPARED BY: Hiroshi Miyamoto

EXPERIMENTAL VALUES:

Solubility of RbIO$_3$

t/°C	mass %	mol %	mol kg^{-1} (compiler)
0	1.06	0.0741	0.0411
10	1.56	0.1053	0.0609
20	2.11	0.149	0.0828
25	2.41	0.171	0.0948
30	2.71	0.193	0.107
40	3.49	0.250	0.139
50	4.37	0.315	0.176
60	5.41	0.394	0.220
70	6.48	0.477	0.266
80	7.70	0.574	0.320
90	9.00	0.680	0.380
100	10.46	0.802	0.449

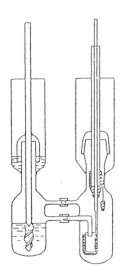

High temp. apparatus

AUXILIARY INFORMATION

METHOD/APPARATUS/PROCEDURE:	SOURCE AND PURITY OF MATERIALS:
Isothermal method. Equilibrium reached in 4-5 h. From 90-100°C, soly detd in apparatus shown in figure. At equilibrium, the apparatus was tilted to allow satd sln to filter through connecting tube into weighed test tubes. The test tube was closed with a stopper, withdrawn, and weighed. Condensation on the walls of the apparatus and loss of water by evaporation was thus prevented. At the lower temperatures, ordinary soly vessels were used, and pipets with glass filters were used for sampling (no other details given). Above 50°C, the pipets were preheated in the thermostat. The iodate content was determined iodometrically.	Results of analysis of RbIO$_3$: RbIO$_3$ content; 99.5 % Impurities, %, K 0.06; Cs 0.13; Na 0.016; SO$_4$ <0.05; Fe 0.005.
	ESTIMATED ERROR: Soly: nothing specified. Temp: precision ± 0.1 K.
	REFERENCES:

COMPONENTS:	ORIGINAL MEASUREMENTS:
(1) Rubidium iodate; $RbIO_3$; [13446-76-9]	Larson, W.D.; Renier, J.J.
(2) Nitric acid; HNO_3; [7697-37-2]	J. Am. Chem. Soc. 1952, 74, 3184-5.
(3) Water; H_2O; [7732-18-5]	

VARIABLES:	PREPARED BY:
Concentration of HNO_3 at 298.15 K	Hiroshi Miyamoto and Mark Salomon

EXPERIMENTAL VALUES:

Concn of HNO_3		Soly of $RbIO_3$	Density
$mol\ kg^{-1}$	$mol\ dm^{-3}$	$mol\ dm^{-3}$	$g\ cm^{-3}$
0	0	0.0919	1.0160
0.05020	0.04977	0.1051	1.0216
0.1006	0.09941	0.1175	1.0250
0.1511	0.1490	0.1297	1.0290
0.2016	0.1985	0.1415	1.0337
0.2516	0.2473	0.1533	1.0385
0.2933	0.2878	0.1626	1.0414
0.4056	0.3984	0.1546	1.0472
0.5006	0.4908	0.1495	1.0506

AUXILIARY INFORMATION

METHOD/APPARATUS/PROCEDURE:	SOURCE AND PURITY OF MATERIALS:
Isothermal method. KNO_3 and excess $RbIO_3$ were placed in bottles either coated with paraffin wax or uncoated and rotated in a thermostat for at least 12 h. Results in coated or uncoated bottles were identical. At least two independent experiments were carried out for each KNO_3 concn, and two or more samples for analysis were taken from each saturated solution. Conventional volumetric analysis was used, and recrystallized KIO_3 was used as the primary standard. Duplicate analyses agreed to within about 1 part per 800 or 900. Nature of the solid phase(s) not specified.	$RbIO_3$ prepared by addition of excess HIO_3 to sln of Rb_2CO_3. The salt was washed by decantation 3 times with cold water, filtered and washed again. It was air-dried and stored over anhydrous $CaCl_2$. Analysis for IO_3 gave +99.9 % of theoretical. HIO_3 prepared from "AR" grade I_2O_5 and water. "C.p." grade HNO_3 was used.

	ESTIMATED ERROR: Soly: precision in iodate analyses about $\pm$ 0.1 % (compilers). Temp: accuracy $\pm$ 0.05 K (authors).
	REFERENCES:

COMPONENTS:	ORIGINAL MEASUREMENTS:
(1) Rubidium iodate; $RbIO_3$; [13446-76-9] (2) Potassium nitrate; KNO_3; [7757-79-1] (3) Water; H_2O; [7732-18-5]	Larson, W.D.; Renier, J.J. J. Am. Chem. Soc. 1952, 74, 3184-5.

VARIABLES:	PREPARED BY:
Concentration of KNO_3 at 298.15 K	Hiroshi Miyamoto and Mark Salomon

EXPERIMENTAL VALUES:

Concn of KNO_3	Solubility of $RbIO_3$		Density
$mol\ kg^{-1}$	$mol\ dm^{-3}$	10^3/mole fraction	$g\ cm^{-3}$
0	0.0919	1.665	1.0160
0.05001	0.0972	1.747	1.0196
0.1004	0.1018	1.851	1.0236
0.1511	0.1054	1.917	1.0285
0.2021	0.1070	1.949	1.0318
0.2547	0.1118	2.037	1.0370
0.2940	0.1143	2.084	1.0396
0.4069	0.1192	2.179	1.0475
0.5025	0.1220	2.235	1.0538

To determine the solubility at zero ionic strength, the authors used the Debye-Huckel
expression for the difference in solubilities between two saturated solutions of
differing ionic strength,

$$I_1^{\frac{1}{2}}I_2^{\frac{1}{2}}A^2 + (I_1^{\frac{1}{2}} + I_2^{\frac{1}{2}})A + 1 - B(I_1^{\frac{1}{2}} - I_2^{\frac{1}{2}})/\log(L_2/L_1)^{\frac{1}{2}} = 0$$

where I is the ionic strength, L is the solubility, and B = 0.509 at 25°C. Values for A
were calculated using 14 pairs of L_2/L_1 values, and the authors report A = 0.674 $\pm$
0.026. Using this value of A, the solubility at zero ionic strenth was given as

$$c_0 = 0.06834\ mol\ dm^{-3},\ or\ x_0 = (1.237 \pm 0.004) \times 10^{-3}\ mole\ fraction.$$

$(w_0 = 1.759\ mass\ \%,\quad m_0 = 0.06875\ mol\ kg^{-1} : compilers)$

AUXILIARY INFORMATION

METHOD/APPARATUS/PROCEDURE:	SOURCE AND PURITY OF MATERIALS:
Isothermal method. KNO_3 and excess $RbIO_3$ were placed in bottles either coated with paraffin wax or uncoated and rotated in a thermostat for at least 12 h. Results in coated or uncoated bottles were identical. At least two independent experiments were carried out for each KNO_3 concentration, and two or more samples for analysis were taken from each saturated solution. Conventional volumetric analysis was used, and recrystallized KIO_3 was used as the primary standard. Duplicate analyses agreed to within about 1 part per 800 or 900. Nature of the solid phase(s) not specified.	$RbIO_3$ prepared by addition of excess HIO_3 to sln of Rb_2CO_3. The salt was washed by decantation 3 times with cold water, filtered and washed again. It was air-dried and stored over anhydrous $CaCl_2$. Analysis for IO_3 gave +99.9 % of theoretical HIO_3 prepd from "AR" grade I_2O_5 and water. "C.p." grade KNO_3 dried at 150°C for several hours.
	ESTIMATED ERROR: Soly: precision in iodate analyses about $\pm$ 0.1 % (compilers). Temp: accuracy $\pm$ 0.05 K (authors).
	REFERENCES:

COMPONENTS:	ORIGINAL MEASUREMENTS:
(1) Rubidium iodate; $RbIO_3$; [13446-76-9] (2) Rubidium hydroxide; RbOH; [1310-82-3] (3) Water; H_2O; [7732-18-5]	Lepeshkov, I.N.; Vinogradov, E.E.; Tarasova, G.N. *Zh. Neorg. Khim.* <u>1976</u>, *21*, 1353-6; *Russ. J. Inorg. Chem. (Engl. Transl.)* <u>1976</u>, *21*, 739-41.
VARIABLES: Composition at 298.2 K	PREPARED BY: Hiroshi Miyamoto

EXPERIMENTAL VALUES: Composition of saturated solutions

RbIO₃		RbOH		Nature of the
mass %	mol % (compiler)	mass %	mol % (compiler)	solid phase[a]
2.39[b]	0.169	–	–	A
0.77	0.055	3.59	0.655	"
0.32	0.024	10.03	1.929	"
0.11	0.0092	20.93	4.452	"
0.06	0.0061	38.99	10.11	"
0.06	0.0072	51.55	15.77	"
0.09	0.014	66.34	25.78	A+B
0.05	0.0076	66.30	25.72	"
0.05	0.0076	66.21	25.65	"
–	–	66.78	26.11	B

[a] A = $RbIO_3$;   B = $RbOH \cdot 2H_2O$

[b] For the binary system the compiler computes the following:

soly of $RbIO_3$ = 0.0940 mol kg^{-1}

AUXILIARY INFORMATION

METHOD/APPARATUS/PROCEDURE:	SOURCE AND PURITY OF MATERIALS:
The solubility in the $RbIO_3$-RbOH-H_2O system was studied by the isothermal method. Mixtures were stirred in a water thermostat. Equilibrium was reached in 3-4 days. The concentration of hydroxide ion was found by titration with 0.1 mol dm^{-3} HCl in the presence of Methyl Orange. The IO_3 content was detd by titration with sodium thiosulfate solution in the presence of sulfuric acid and KI. Rubidium was determined gravimetrically as the tetraphenylborate. The composition of the solid phases was found by Schreinemakers' method of residues.	"C.p." grade $RbIO_3$ was used. Commercial RbOH contains considerable amounts of Rb_2CO_3 impurity which cannot be removed by recryst from water. The hydroxide was purified by recryst in silver vessels in a stream of purified nitrogen as the temp was slowly increased to 250°C.

ESTIMATED ERROR:	COMMENTS AND/OR ADDITIONAL DATA:
Soly: nothing specified. Temp: precision ± 0.1 K.	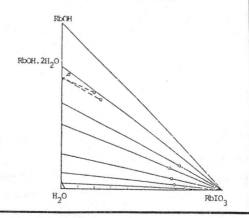

COMPONENTS:	ORIGINAL MEASUREMENTS:
(1) Rubidium iodate; $RbIO_3$; [13446-76-9] (2) Cesium iodate; $CsIO_3$; [13454-81-4] (3) Water; H_2O; [7732-18-5]	Kirgintsev, A.N.; Shklovskaya, R.M.; Arkhipov, S.M. *Izv. Akad. Nauk SSSR, Ser. Khim.* 1971, 2631-4; *Bull. Acad. Sci. USSR, Div. Chem. Sci.* 1971, 2501-4.
VARIABLES: Composition at 298.2 K	PREPARED BY: Hiroshi Miyamoto

EXPERIMENTAL VALUES:

Composition of saturated solutions

RbIO$_3$		CsIO$_3$	
mass %	mol %[a]	mass %	mol %[a]
2.31[b]	0.163	0.0	0.0
2.06	0.146	0.44	0.026
1.91	0.136	0.74	0.044
1.63	0.116	1.19	0.0715
1.41	0.100	1.44	0.0866
1.37	0.0974	1.51	0.0908
1.13	0.0803	1.72	0.103
0.87	0.062	1.78	0.107
0.58	0.041	2.10	0.126
0.28	0.020	2.26	0.136
0.0	0.0	2.50[b]	0.150

[a] Calculated by the compiler using IUPAC recommended atomic masses.

[b] For binary systems the compiler computes the following:

$$\text{soly of } RbIO_3 = 0.0908 \text{ mol kg}^{-1}$$

$$\text{soly of } CsIO_3 = 0.0833 \text{ mol kg}^{-1}$$

AUXILIARY INFORMATION

METHOD/APPARATUS/PROCEDURE:	SOURCE AND PURITY OF MATERIALS:
Isothermal relief of supersaturation method. Super saturated solutions were prepared, and the solid and liquid phases separated. The mother liquor was equilibrated at 25°C for 24 hours. The number of moles of the anion was determined by iodometric titration. Alkali metal contents were determined in the same sample by the method of flame photometry from three parallel analyses. The composition of the solid phases was established by the Schreinemakers' method of residues. The authors did not give a phase diagram.	"C.p." grade $RbIO_3$ and $CsIO_3$ were recrystallized from double distilled water.
	ESTIMATED ERROR: Soly: accuracy within $\pm$ 3.5 % (authors). Temp: precision $\pm$ 0.1 K.
	REFERENCES:

COMPONENTS:	ORIGINAL MEASUREMENTS:
(1) Rubidium iodate; $RbIO_3$; [13446-76-9] (2) Zinc iodate; $Zn(IO_3)_2$; [7790-37-6] (3) Water; H_2O; [7732-18-5]	Vinogradov, E.E.; Karataeva, I.M. *Zh. Neorg. Khim.* <u>1979</u>, *24*, 2529-32; *Russ. J. Inorg. Chem. (Engl. Transl.)* <u>1979</u>, *24*, 1406-8.
VARIABLES: Composition at 323 K	PREPARED BY: Hiroshi Miyamoto

EXPERIMENTAL VALUES: Composition of saturated solutions at 50°C

RbIO₃		Zn(IO₃)₂		
mass %	mol % (compiler)	mass %	mol % (compiler)	Nature of the solid phase[a]
4.39[b]	0.317	–	–	A
4.42	0.319	0.07	0.003	A+B
4.43	0.320	0.09	0.004	"
4.42	0.319	0.09	0.004	"
4.48	0.324	0.09	0.004	"
4.51	0.326	0.09	0.004	"
4.30	0.310	0.11	0.0050	"
4.56	0.330	0.09	0.004	"
4.52	0.327	0.11	0.0050	"
4.42	0.319	0.06	0.003	"
2.42	0.172	0.13	0.0058	B
0.11	0.0077	0.59	0.026	"
–	–	0.68[b]	0.030	"

[a] $A = RbIO_3$; $B = Zn(IO_3)_2 \cdot 2H_2O$

[b] For binary systems the compiler computes the following:

soly of $RbIO_3 = 0.176$ mol kg^{-1}

soly of $Zn(IO_3)_2 = 0.016$ mol kg^{-1}

AUXILIARY INFORMATION

METHOD/APPARATUS/PROCEDURE:	COMMENTS AND/OR ADDITIONAL DATA:
Equilibrium in the system was reached after about a month. Both liquid and solid phases were analyzed for all the ions by the methods described in refs 1 and 2. The solid phases were identified by X-ray diffraction and thermographically.	The phase diagram is given below (based on mass % units). 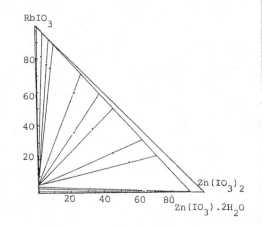

SOURCE AND PURITY OF MATERIALS:
"Chemically pure" grade rubidium iodate was used. Zinc iodate was prepared from zinc oxide and iodic acid.

ESTIMATED ERROR:

Nothing specified.

REFERENCES:
1. Lepeshkov, I.N.; Vinogradov, E.E.;
   Karataeva, I.M. *Zh. Neorg. Khim.*
   <u>1977</u>, *22*, 2277.

2. Karataeva, I.M.; Vinogradov, E.E.
   Zh. Neorg. Khim. <u>1974</u>, *19*, 3156.

COMPONENTS:	ORIGINAL MEASUREMENTS:
(1) Rubidium iodate; RbIO₃; [13446-76-9] (2) Aluminum iodate; Al(IO₃)₃; [15123-75-8] (3) Water; H₂O; [7732-18-5]	Vinogradov, E.E.; Tarasova, G.N. *Zh. Neorg. Khim.* 1978, *23*, 3161-4; *Russ. J. Inorg. Chem. (Engl. Transl.)* 1978, *23*, 1754-6.

VARIABLES:	PREPARED BY:
Composition at 298.2 K	Hiroshi Miyamoto

EXPERIMENTAL VALUES: Composition of saturated solutions at 25.0°C

Al(IO₃)₃		RbIO₃		Nature of the
mass %	mol % (compiler)	mass %	mol % (compiler)	solid phase[a]
5.71[b]	0.197	–	–	A
4.49	0.155	1.03	0.0753	A+B
4.50	0.155	0.96	0.070	"
4.51	0.155	0.98	0.072	"
4.65	0.161	1.02	0.0746	"
4.63	0.160	0.98	0.072	"
3.97	0.136	1.13	0.0822	B
2.15	0.0729	1.68	0.121	"
–	–	2.39[b]	0.169	"

[a] A = Al(IO₃)₃·6H₂O;   B = RbIO₃

[b] For binary systems the compiler computes the following:

$$\text{soly of } RbIO_3 = 0.0940 \text{ mol kg}^{-1}$$

$$\text{soly of } Al(IO_3)_3 = 0.110 \text{ mol kg}^{-1}$$

AUXILIARY INFORMATION

METHOD/APPARATUS/PROCEDURE:

Mixtures of Al(IO₃)₃, RbIO₃ and H₂O were
stirred in a thermostat for 18-21 days.
The liquid and solid phases were analyzed
for IO₃⁻, Rb⁺ and Al³⁺. The iodate content
was determined by titrating with sodium thiosulfate solution in the presence of KI and
H₂SO₄. Rubidium was determined gravimetrically as the tetraphenylborate, and aluminum
determined by titrating with EDTA using
Xylenol Orange as an indicator.
The composition of the solid phases were determined by Schreinemakers' method of
residues.

ESTIMATED ERROR:

Soly: nothing specified.
Temp: precision ± 0.1 K.

SOURCE AND PURITY OF MATERIALS:

"C.p." grade RbIO₃ used. Al(IO₃)₃ prepared
at 80-90°C by stoichiometrically neutraliz-
ing a saturated solution of HIO₃ with freshly pptd Al(OH)₃. Found, mass %: Al 4.03;
IO₃ 78.7; H₂O 17.6. Calculated for
Al(IO₃)₃·6H₂O, mass %: Al 4.09; IO₃ 79.53;
H₂O 16.38 (by difference).

COMMENTS AND/OR ADDITIONAL DATA:

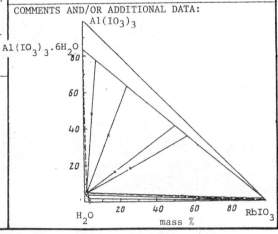

COMPONENTS:	ORIGINAL MEASUREMENTS:
(1) Rubidium iodate; $RbIO_3$; [13446-76-9] (2) Hafnium iodate; $Hf(IO_3)_4$; [19630-06-9] (3) Water; H_2O; [7732-18-5]	Shklovskaya, R.M.; Arkhipov, S.M. Kidyarov, B.I.; Poleva, G.V.; Vdovkina, T.E. *Zh. Neorg. Khim.* 1984, *29*, 1346-8; *Russ. J. Inorg. Chem. (Engl. Transl.)* 1984, *29*, 773-4.
VARIABLES: T/K = 298.2 Composition	PREPARED BY: Mark Salomon

EXPERIMENTAL VALUES: The $RbIO_3$ - $Hf(IO_3)_4$ - H_2O system at 25.0°C

Composition of saturated solutions[a]

$RbIO_3$		$Hf(IO_3)_4$		
mass %	mole %	mass %	mole %	Nature of the solid phase
----	----	0.00037	7.59×10^{-6}	$Hf(IO_3)_4$
0.27	0.0187	0.000074	1.52×10^{-6}	solid solution based on $Hf(IO_3)_4$
0.52	0.0362	0.000073	1.50×10^{-6}	"
0.74	0.0516	0.000073	1.51×10^{-6}	"
1.04	0.0727	0.000072	1.49×10^{-6}	"
1.28	0.0896	0.000072	1.49×10^{-6}	"
1.44	0.1099	0.000072	1.50×10^{-6}	"
1.64	0.1152	0.000072	1.50×10^{-6}	"
1.76	0.1238	0.000071	1.48×10^{-6}	"
1.99	0.1403	0.000071	1.48×10^{-6}	"
2.22	0.1568	0.000071	1.49×10^{-6}	"
2.31[b]	0.1633	0.000070	1.47×10^{-6}	solid solution + $RbIO_3$
2.31[b]	0.1633	0.000070	1.47×10^{-6}	"
2.36	0.1670	----	----	$RbIO_3$

[a] Mole % values calculated by the compiler.

[b] Eutonic solution.

For binary systems, the compiler computes the following:

solubility of $RbIO_3$ = 0.0928 mol kg^{-1}

solubility of $Hf(IO_3)_4$ = 4.21×10^{-6} mol kg^{-1}

AUXILIARY INFORMATION

METHOD/APPARATUS/PROCEDURE:	SOURCE AND PURITY OF MATERIALS:
Isothermal method used. Equilibrium required 25-30 days. Solid and liquid phases analyzed for Rb by emission spectrometry using solutions of Rb concentration between 0.1 - 100 μg cm^{-3} in the presence of 2 % NaCl solution (added to suppress the ionization of Rb atoms). Preliminary experiments established that Hf does not influence the intensity of the emission of Rb. The concentration of Rb was therefore determined by comparing samples of saturated solution previously buffered with 2 % NaCl solution with standard Rb solutions also buffered with 2 % NaCl solution. For liquid phase samples, Hf was determined photometrically using Arsenazo III after reduction of IO_3 with hydroxylamine. For solid phase samples, Rb was analyzed as described above and iodate by iodometric titration. The Hf content was determined by difference. Solid phase samples were identified by the method of residues and by X-ray diffraction. The maximum concentration of $RbIO_3$ in the solid solution is 2.6 %.	"Highly pure" $RbIO_3$ was used. $Hf(IO_3)_4$ was prepared from aqueous HIO_3 and freshly precipitated hydrated hafnium oxide under conditions described previously (1). No other information given.
	ESTIMATED ERROR: Soly: uncertainty in analyses did not exceed 3-8 rel %. Temp: precision given as $\pm$ 0.1 K.

	REFERENCES:
	1. Deabriges, J.; Rohmer, R.
	    Bull. Soc. Chim. France
	    1968, 521.

COMPONENTS:	ORIGINAL MEASUREMENTS:
(1) Rubidium iodate; $RbIO_3$; [13446-76-9] (2) Neodymium iodate; $Nd(IO_3)_3$; [14732-16-2] (3) Water; H_2O; [7732-18-5]	Tarasova, G.N.; Vinogradov, E.E.; Kudinov, I.B. *Zh. Neorg. Khim.* 1981, 26, 2841-7; *Russ. J. Inorg. Chem.* (Engl. Transl.) 1981, 26, 1520-3.

VARIABLES:	PREPARED BY:
Composition at 298.2 K	Hiroshi Miyamoto

EXPERIMENTAL VALUES: Composition of saturated solutions at 25.0°C

Neodymium Iodate		Rubidium Iodate		Nature of the
mass %	mol % (compiler)	mass %	mol % (compiler)	solid phase[a]
0.15[b]	0.0040	--	--	A
<0.01	0.0003	1.11	0.0776	A+B
<0.01	<0.0003	1.10	0.0769	"
<0.01	<0.0003	2.19	0.155	"
<0.01	<0.0003	2.45	0.173	"
<0.01	<0.0003	2.56	0.181	"
<0.01	<0.0003	2.48	0.176	"
<0.01	<0.0003	2.46	0.174	"
<0.01	<0.0003	2.18	0.154	"
--	--	2.40[b]	0.170	B

[a] $A = Nd(IO_3)_3 \cdot 2H_2O$; $B = RbIO_3$

[b] For binary systems the compiler computes the following:

soly of $RbIO_3$ = 0.0944 mol kg^{-1}

soly of $Nd(IO_3)_3$ = 2.2 x 10^{-3} mol kg^{-1}

AUXILIARY INFORMATION

METHOD/APPARATUS/PROCEDURE:	SOURCE AND PURITY OF MATERIALS:
Mixtures of $Nd(IO_3)_3$, $RbIO_3$ and water were stirred in a water thermostat. Equilibrium was reached in 30-35 days. The liquid and solid phases were analyzed for IO_3^- and Nd^{3+} ions. The iodate ion concentration was determined by titration with sodium thiosulfate in the presence of sulfuric acid and KI. The neodymium content was determined by complexometric titration in the presence of hexamethylenetetramine with Methyl thymol blue indicator. The composition of the solid phases was found by Schreinemakers' method of residues.	Neodymium iodate was prep by reacting neodymium oxide and HIO_3 in stoichiometric proportions. The aqueous sln and precipitates were stirred continuously for 20 h at 80-90° C. Then the precipitate was transferred to a filter, washed repeatedly with hot water, and dried at 110-120°C. The authors state that the purity of the resulting neodymium iodate was checked by chemical analysis, but the result was not given in the original paper. Chemically pure grade $RbIO_3$ was used.
	ESTIMATED ERROR: Soly: nothing specified. Temp: precision ± 0.1 K.
	REFERENCES:

COMPONENTS:	ORIGINAL MEASUREMENTS:
(1) Rubidium iodate; $RbIO_3$; [13446-76-9] (2) Iodic acid; HIO_3; [7782-68-5] (3) Water; H_2O; [7732-18-5]	Tatarinov, V.A. *Uch. Zap. Yarosl. Gos. Pedagog. Inst.* 1972, No. 103, 83-5.

VARIABLES:	PREPARED BY:
Composition at 323 K	Hiroshi Miyamoto

EXPERIMENTAL VALUES: Composition of saturated solutions

RbIO₃		HIO₃		Nature of the
mass %	mol % (compiler)	mass %	mol % (compiler)	solid phase[a]:
5.48[b]	0.400	–	–	A
5.05	0.371	1.36	0.148	"
3.84	0.289	4.95	0.551	A+C
3.82	0.287	4.98	0.555	"
2.55	0.191	5.80	0.643	C
0.76	0.062	16.80	2.043	"
0.71	0.072	35.00	5.277	"
0.61	0.086	56.00	11.66	"
0.82	0.18	74.48	23.55	"
0.81	0.17	74.50	23.57	C+B
–	–	76.53[b]	25.03	B

[a] A = $RbIO_3$;   B = HIO_3;   C = $RbIO_3 \cdot 2HIO_3$

[b] For binary systems the compiler computes the following:

soly of $RbIO_3$ = 0.223 mol kg^{-1}

soly of HIO_3 = 18.54 mol kg^{-1}

AUXILIARY INFORMATION

METHOD/APPARATUS/PROCEDURE:	COMMENTS AND/OR ADDITIONAL DATA:
The isothermal method was used. Equilibrium between the liquid and solid phases was established in 24 hours. The rubidium iodate content in the samples was determined iodometrically, and HIO_3 determined by titration with base.	The phase diagram is given below (based on mass % units).

SOURCE AND PURITY OF MATERIALS:
Rubidium iodate was prepared from iodic acid and rubidium sulfate, and the product was recrystallized.
"C.p." grade HIO_3 was recrystallized.

ESTIMATED ERROR:

Nothing specified.

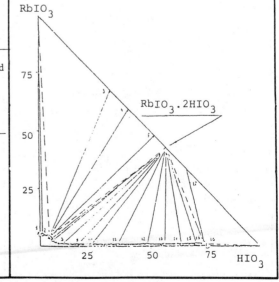

COMPONENTS:	ORIGINAL MEASUREMENTS:
(1) Rubidium iodate; $RbIO_3$; [13446-76-9]	Miyamoto, H.; Hasegawa, T.; Sano, H.
(2) N,N-Dimethylformamide; C_3H_7NO; [68-12-2]	J. *Solution Chem*. in press.
(3) Water; H_2O; [7732-18-5]	

VARIABLES:	PREPARED BY:
Solvent composition	M. Salomon
Temperature	

EXPERIMENTAL VALUES:

Solubilities in the $RbIO_3-H_2O$ system at 20°C, 25°C, 30°C

	t/°C = 20		t/°C = 25	
mass % dimethylformamide	$RbIO_3$/mol dm^{-3}		mass % dimethylformamide	$RbIO_3$/mol dm^{-3}
0	0.0805		0	0.0937
4.79	0.0609		5.12	0.0703
10.05	0.0465		10.00	0.0554
19.75	0.0262		20.43	0.0311
30.22	0.0139		29.71	0.0172
41.99	0.0057		40.02	0.0079

AUXILIARY INFORMATION

METHOD/APPARATUS/PROCEDURE:	SOURCE AND PURITY OF MATERIALS:
Same as in reference (1).	Extra pure grade Rb_2CO_3 and guaranteed grade HIO_3 used as received. $RbIO_3$ pptd by addn of excess HIO_3 sln to aq Rb_2CO_3 sln while heating. After stirring for 5 h, the sln was allowed to settle for 1 day, and the ppt washed with cold water until the dried salt produced a constant soly. The salt was stored in the dark.

Guaranteed grade dimethylformamide (Wako) was stored over BaO for two days, and then distilled three times under reduced pressure.

Doubly distilled water had an electrolytic conductance of 9.8×10^{-7} S cm^{-1}.

ESTIMATED ERROR:
Soly: stnd deviation between 0.0002 and 0.001
Temp: not stated

REFERENCES:

1. Miyamoto, H.; Shimura, H.; Sasaki, K.
   J. *Solution Chem*. 1985, *14*, 485.

COMPONENTS:	ORIGINAL MEASUREMENTS:
(1) Rubidium iodate; RbIO$_3$; [13446-76-9]	Miyamoto, H.; Hasegawa, T.; Sano, H.
(2) N,N-Dimethylformamide; C$_3$H$_7$NO; [68-12-2]	J. *Solution Chem.* in press.
(3) Water; H$_2$O; [7732-18-5]	

EXPERIMENTAL VALUES: (Continued)

t/°C = 30

mass % dimethylformamide	RbIO$_3$/mol dm^{-3}
0	0.108
5.53	0.0817
9.81	0.0652
20.10	0.0356
29.79	0.0197
40.33	0.0093

For the binary RbIO$_3$-H$_2$O system, measured densities of saturated solutions permits conversion from mol dm^{-3} to mol kg^{-1} and mole fraction units.

t/°C	density/g cm^{-3}	c/mol dm^{-3}	m/mol kg^{-1}[a]	χ[a]
20	1.014	0.0805	0.0811	0.00146
25	1.018	0.0937	0.0943	0.00170
30	1.020	0.108	0.109	0.00196

[a]Calculated by the compiler.

COMPONENTS:	ORIGINAL MEASUREMENTS:
(1) Rubidium iodate; $RbIO_3$; [13446-76-9]	Miyamoto, H.; Hasegawa, T.; Sano, H.
(2) Dimethylsulfoxide ; C_2H_6OS: [67-88-5]	J. *Solution Chem.* in press.
(3) Water; H_2O; [7732-18-5]	

VARIABLES:	PREPARED BY:
Solvent composition Temperature	M. Salomon

EXPERIMENTAL VALUES:

$RbIO_3$ soly/mol dm^{-3}

mass % dimethylsulfoxide	t/°C =	20	25	30
0		0.0805	0.0937	0.108
5.03		0.0639	0.0751	0.0864
10.02		0.0505	0.0588	0.0688
20.09		0.0298	0.0355	0.0402
30.01		0.0163	0.0196	0.0225
40.03		0.0081	0.0095	0.0109

AUXILIARY INFORMATION

METHOD/APPARATUS/PROCEDURE:	SOURCE AND PURITY OF MATERIALS:
Same as in reference (1).	Extra pure grade Rb_2CO_3 and guaranteed grade HIO_3 used as received. $RbIO_3$ pptd by addn of excess HIO_3 sln to aq Rb_2CO_3 sln while heating. After stirring for 5 h, the sln was allowed to settle for 1 day, and the ppt washed with cold water until the dried salt produced a constant soly. The salt was stored in the dark.
	Guaranteed grade dimethyl sulfoxide (Wako) was distilled three times under reduced pressure.
	Doubly distilled water had an electrolytic conductance of 9.8×10^{-7} S cm^{-1}

ESTIMATED ERROR:
Soly: stnd deviation between 0.0002 and
        0.001.
Temp: not stated

REFERENCES:

1. Miyamoto, H.; Shimura, H.; Sasaki, K.
       J. *Solution Chem.* 1985, 14, 485.

COMPONENTS:	ORIGINAL MEASUREMENTS:
(1) Rubidium iodate; RbIO$_3$; [13446-76-9] (2) 6,7,10,17,18,20,21-Octahydrodibenzo [b,k] [1,4,7,10,13,16] hexaoxacyclooctadecin (dibenzo-18-crown-6); C$_{20}$H$_{24}$O$_6$; [14187-32-7] (3) Methanol; CH4O; [67-56-1]	Kolthoff, I.M.; Chantooni, M.K. *Anal. Chem.* 1980, *52*, 1039-49.

VARIABLES:	PREPARED BY:
T/K = 298	Hiroshi Miyamoto and Mark Salomon

EXPERIMENTAL VALUES:

The solubility product of RBIO$_3$ in methanol at 25°C is given as

$$2.7 \times 10^{-9} \text{ mol}^2 \text{ dm}^{-6}$$

COMMENTS AND/OR ADDITIONAL DATA:

The formation constant for RbL$^+$ (L = crown ether) was also determined.

The authors reported

$$\log\{K_f(RbL^+)/mol^{-1} \text{ dm}^3\} = 4.23$$

AUXILIARY INFORMATION

METHOD/APPARATUS/PROCEDURE:	SOURCE AND PURITY OF MATERIALS:
A Markson No. 1002 K$^+$ specific ion electrode was used to measure Rb$^+$ activity after conditioning the electrode by soaking in 0.01 mol dm^{-3} RbClO$_4$ solution for 3-4 days. The electrode response to $a_{Rb}{}^+$ was "practically" Nernstian.	Methanol (Fisher "Spectroquality" grade) distilled from Mg turnings. RbOH prepared by passing RbBr through a column of Dowex IX-8 resin in the hydroxide form. RbIO$_3$ prepared by neutralizing RbOH with HIO$_3$, recrystallized three times from water, and dried at 70°C.
	ESTIMATED ERROR: Nothing specified.
	REFERENCES:

COMPONENTS:	EVALUATOR:
(1) Cesium iodate; $CsIO_3$; [13434-81-4]	H. Miyamoto Niigata University Niigata, Japan and Mark Salomon US Army ET & DL Fort Monmouth, NJ, USA
(2) Water; H_2O; [7732-18-5]	
	June, 1986

CRITICAL EVALUATION:

THE BINARY SYSTEM

Data for the solubility of $CsIO_3$ in water have been reported in 12 publications (1-12). A number of compilations containing solubilities in the binary system can be found elsewhere in this volume: ref. (4) has been compiled in the KIO_3 chapter, ref. (5) is in the $RbIO_3$ chapter, refs. (7, 8) are in the $LiIO_3$ chapter, and ref. (11) is in the $NaIO_3$ chapter.

Although some investigators (1-3, 5, 12) did not report the nature of the solid phase in the binary system, the evaluators assume it to be the anhydrous salt by analogy to the anhydrous salt found in studies on ternary systems.

A summary of the experimental data converted to mole fraction and mol/kg units by the evaluators and compilers is given in Table 1. In fitting these data to the smoothing equations, it was assumed that there are 24 independent data points to be considered as indicated in the table although it appears that Barker (2) may have used the earlier value reported in (1). In any case, these two data points were rejected on the bases of the uncertainty in experimental temperature (about 297 K), and the fact that their results are close to the experimental values reported by other investigators for 298.2 K. All other data (22 independent values) were used in the smoothing equations, and for mole fraction solubilities the smoothing equation is:

$$Y_x = -26800/(T/K) - 118.503 \ln (T/K) + 706.355 + 0.15820(T/K)$$

$$\sigma_y = 0.032 \qquad\qquad \sigma_x = 3.2 \times 10^{-5}$$

For solubilities in mol/kg units, the smoothing equation is:

$$Y_m = -5309.8/(T/K) - 8.4748 \ln (T/K) + 66.085$$

$$\sigma_y = 0.019 \qquad\qquad \sigma_m = 0.0026$$

The smoothed solubilities at rounded temperatures calculated from these two equations are given in Table 2, and these values are designated as *recommended* solubilities.

TERNARY SYSTEMS

Data for the solubilities in ternary systems with two saturating components have been reported in 8 publications (4-11). A summary of these studies is given below.

Summary of solubility studies in aqueous ternary systems

Ternary system	T/K	Solid Phase	Reference
$CsIO_3 - KIO_3 - H_2O$	298	$CsIO_3$; KIO_3	4
$CsIO_3 - RbIO_3 - H_2O$	298	Not given	5
$CsIO_3 - HIO_3 - H_2O$	298	$CsIO_3$; HIO_3; $CsIO_3.HIO_3$	6
$CsIO_3 - LiIO_3 - H_2O$	298	$CsIO_3$; $LiIO_3$	7
$CsIO_3 - LiIO_3 - H_2O$	323	$CsIO_3$; $LiIO_3$	8
$CsIO_3 - CsNO_3 - H_2O$	323	$CsIO_3$; $CsNO_3$	9
$CsIO_3 - Al(IO_3)_3 - H_2O$	298	$CsIO_3$; $Al(IO_3)_3.6H_2O$	10
$CsIO_3 - NaIO_3 - H_2O$	323	$CsIO_3$; $NaIO_3.H_2O$	11

COMPONENTS:	EVALUATOR:
(1) Cesium iodate; $CsIO_3$; [13434-81-4] (2) Water; H_2O; [7732-18-5]	H. Miyamoto Niigata University Niigata, Japan and Mark Salomon US Army ET & DL Fort Monmouth, NJ, USA　　　June, 1986

CRITICAL EVALUATION:

The System With Iodic Acid. This ternary system was studied by Tatarinov (6), and the dominant feature is the formation of the double salt, $CsIO_3.HIO_3$.

Systems With Other Iodates. Other ternary systems were reported in 7 publications (4, 5, 7-11). Kirgintsev, Shklovskaya and Arkhipov (5) measured solubilities in the ternary $CsIO_3-RbIO_3-H_2O$ system at 298 K, but did not report the composition of the solid phases. In other publications (4, 7-11), no double salts were reported, and all systems studied were of the simple eutonic type.

The $CsIO_3-CsNO_3-H_2O$ system is similar to the systems described in this section, that is, no double salts were formed and this system is of the simple eutonic type.

Table 1. Experimental solubilities in the $CsIO_3-H_2O$ system

T/K	mole fraction	mol/kg	reference
273.2	0.00633	0.0351	3
283.2	0.00933	0.0518	3
293.2	0.00137	0.0761	3
293.2	0.00135	0.0750	12
297[a]	0.00152	0.084	1
297[a]	0.00156	0.0844	2
298.2	0.00157	0.0874	3
298.2	0.00150	0.0833	5
298.2	0.00152	0.0843	4
298.2	0.00154	0.0857	12
298.2	0.00157	0.0871[b]	7,10
298.2	0.00160	0.0891	6
303.2	0.00180	0.100	3
303.2	0.00180	0.0998	12
313.2	0.00243	0.135	3
323.2	0.00310	0.173	3
323.2	0.00312	0.174	8
323.2	0.00312	0.174	9
323.2	0.00312	0.174	11
333.2	0.00385	0.215	3
343.2	0.00481	0.268	3
353.2	0.00581	0.324	3
363.2	0.00707	0.395	3
373.2	0.00835	0.468	3

[a]Rejected data points.

[b]Reference molality used in the smoothing equation.

COMPONENTS:	EVALUATOR:
(1) Cesium iodate; $CsIO_3$; [13434-81-4] (2) Water; H_2O; [7732-18-5]	H. Miyamoto Niigata University Niigata, Japan and M. Salomon US Army ET & DL Fort Monmouth, NJ, USA June, 1986

CRITICAL EVALUATION:

Table 2.   Recommended solubilities in the binary
$CsIO_3$-H_2O system calculated from the
smoothing equations

T/K	mole fraction	mol/kg
273.2	0.000633	0.0355
283.2	0.000938	0.0521
293.2	0.00133	0.0735
298.2	0.00156	0.0863
303.2	0.00182	0.101
313.2	0.00240	0.134
323.2	0.00310	0.173
333.2	0.00390	0.218
343.2	0.00482	0.271
353.2	0.00586	0.329
363.2	0.00703	0.393
373.2	0.00835	0.462

REFERENCES

1. Wheeler, H. L. *Am. J. Sci.* 1892, [3] *44*, 123.

2. Barker, T. V. *J. Chem. Soc.* 1908, *93*, 15.

3. Breusov, O. N.; Kashina, N. I.; Revzina, T. V.; Sobolevskaya, N. G. *Zh. Neorg. Khim.* 1967, *12*, 2240; *Russ. J. Inorg. Chem. (Engl. Transl.)* 1967, *12*, 1179.

4. Kirgintsev, A. I.; Yakobi, N. Y. *Zh. Neorg. Khim.* 1968, *13*, 2851; *Russ. J. Inorg. Chem. (Engl. Transl.)* 1968, *13*, 1467.

5. Kirgintsev, A. N.; Shklovskaya, R. M.; Arkhipov, S. M. *Izv. Akad, Nauk SSSR, Ser. Khim.* 1971, 2631; *Bull. Acad. Sci. USSR, Div. Chem. Sci. (Engl. Transl.)* 1971, 2501.

6. Tatarinov, V. A. *Uch. Zap. Yarosl. Pedagog. Inst.* 1973, No. 120, 71.

7. Shklovskaya, R. M.; Arkhipov, S. M. Kidyarov, E. I.; Mitnitskii, P. L. *Zh. Neorg. Khim.* 1974, *19*, 1975; *Russ. J. Inorg. Chem. (Engl. Transl.)* 1979, *19*, 1082.

8. Karataeva, I. M.; Vinogradov, E. E. *Zh. Neorg. Khim.* 1974, *19*, 3156; *Russ. J. Inorg. Chem. (Engl. Transl.)* 1974, *19*, 1726.

9. Vinogradov, E. E.; Karataeva, I. M. *Zh. Neorg. Khim.* 1976, *21*, 1664; *Russ. J. Inorg. Chem. (Engl. Transl.)* 1976, *21*, 910.

10. Shklovskaya, R. M.; Arkhipov, S. M.; Kidyarov, B. I.; Poleva, G. B. *Zh. Neorg. Khim.* 1982, *27*, 1610; *Russ. J. Inorg. Chem. (Engl. Transl.)* 1982; *27*, 910.

11. Vinogradov, E. E.; Karataeva, I. M. *Zh. Neorg. Khim.* 1982, *27*, 2155; *Russ. J. Inorg. Chem. (Engl. Transl.)* 1982, *27*, 1681.

12. Miyamoto, H.; Hasegawan, T.; Sano, H. *J. Solution Chem.* in press.

COMPONENTS:	ORIGINAL MEASUREMENTS:
(1) Cesium iodate; $CsIO_3$; [13454-81-4] (2) Water; H_2O; [7732-18-5]	Wheeler, H.L. *Am. J. Sci.* 1892, [3] *44*, 123-33.
VARIABLES: T/K = 297	PREPARED BY: Hiroshi Miyamoto

EXPERIMENTAL VALUES:

The solubility of $CsIO_3$ in water is given as

100 parts water dissolve 2.6 parts of $CsIO_3$.

The compiler's conversions to mass % and mol kg^{-1} are

2.53 mass %

0.0844 mol kg^{-1}

AUXILIARY INFORMATION

METHOD/APPARATUS/PROCEDURE:	SOURCE AND PURITY OF MATERIALS:
No information was given.	Cesium iodate was prepared by stoichiometric mixing of iodic acid and cesium carbonate. The solution was boiled, and upon cooling small cubic crystals were separated. The product was filtered, washed with cold water, pressed on papers, and then dried at 100°C. Found: Cs 43.08; I 40.84; O 15.74. Calcd for $CsIO_3$: Cs 43.18; I 41.23; O 15.59.
	ESTIMATED ERROR: Nothing specified.
	REFERENCES:

COMPONENTS:	ORIGINAL MEASUREMENTS:
(1) Cesium iodate; $CsIO_3$; [13454-81-4]	Barker, T.V.
(2) Water; H_2O; [7732-18-5]	*J. Chem. Soc.* <u>1908</u>, *93*, 15-6.

VARIABLES:	PREPARED BY:
T/K = 297	Hiroshi Miyamoto

EXPERIMENTAL VALUES:

The solubility of $CsIO_3$ in water at 24°C was given as

100 parts of water dissolves 2.6 parts of salt.

This is equivalent to 0.084 mol kg^{-1} (compiler).

The specific gravity of the saturated solution at 16°C
was reported as 4.559. The compiler assumes that pptn occurred upon
cooling a satd sln at 24°C to 16°C.

AUXILIARY INFORMATION

METHOD/APPARATUS/PROCEDURE:	SOURCE AND PURITY OF MATERIALS:
The iodine content was estimated by the Carius method (the reference was not given in the original paper), but the compiler assumes that the total solubility was determined by evaporation and heating to constant mass. The heating was carried out in two operations lasting four hours: the first to 150°C, and the second to 250°C. The cesium content was determined by the usual sulfate method. No other information was given in the original paper.	Cesium iodate was prepared by adding aqueous HIO_3 solution to aqueous cesium carbonate solution. Another method was also used to prepare cesium iodate: a good yield was obtained by passing chlorine into a hot concentrated solution of a mixture of cesium iodide and hydroxide. No other information given.
	ESTIMATED ERROR:
	Nothing specified.
	REFERENCES:

COMPONENTS:	ORIGINAL MEASUREMENTS:
(1) Cesium iodate; $CsIO_3$; [13454-81-4] (2) Water; H_2O; [7732-18-5]	Breusov, O.N.; Kashina, N.I.; Revzina, T.V.; Sobolevskaya, N.G. *Zh. Neorg. Khim.* 1967, *12*, 2240-3; *Russ. J. Inorg. Chem. (Engl. Transl.)* 1967, *12*, 1179-81.

VARIABLES:	PREPARED BY:
Temperature: 273.2 to 373.2 K	Hiroshi Miyamoto

EXPERIMENTAL VALUES:

	Solubility of $CsIO_3$		
$t/°C$	mass %	mol %	mol kg^{-1} (compiler)
0	1.07	0.0633	0.0351
10	1.57	0.0933	0.0518
20	2.29	0.137	0.0761
25	2.62	0.157	0.0874
30	2.99	0.180	0.100
40	4.00	0.243	0.135
50	4.05	0.310	0.173
60	6.20	0.385	0.215
70	7.62	0.481	0.268
80	9.08	0.581	0.324
90	10.85	0.707	0.395
100	12.58	0.835	0.468

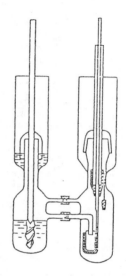

High temperature aparatus

AUXILIARY INFORMATION

METHOD/APPARATUS/PROCEDURE:

Isothermal method. Equilibrium reached in 4-5 h. From 90-100°C, soly detd in apparatus shown in figure. At equilibrium, the apparatus was tilted to allow satd sln to filter through connecting tube into weighed test tubes. The test tube was closed with a stopper, withdrawn, and weighed. Condensation on the walls of the apparatus and loss of water by evaporation was thus prevented. At the lower temperatures, ordinary soly vessels were used, and pipets with glass filters were used for sampling (no other details given). Above 50°C, the pipets were preheated in the thermostat.
Iodate was determined iodometrically.

SOURCE AND PURITY OF MATERIALS:

Results of analysis of $CsIO_3$:

$CsIO_3$ content; 99.5 %
Impurities, %; K 0.005; Rb 0.20; Na 0.02; SO_4 <0.05; Fe 0.005.

ESTIMATED ERROR:
Soly: nothing specified.
Temp: precision ± 0.1 K.

REFERENCES:

COMPONENTS:	ORIGINAL MEASUREMENTS:
(1) Cesium nitrate; $CsNO_3$; [7789-18-6] (2) Cesium iodate; $CsIO_3$; [13454-81-4] (3) Water; H_2O; [7732-18-5]	Vinogradov, E.E.; Karataeva, I.M. *Zh. Neorg. Khim.* 1976, *21*, 1664-6; *Russ. J. Inorg. Chem. (Engl. Transl.)* 1976, *21*, 910-1.
VARIABLES: Composition at 323 K	PREPARED BY: Hiroshi Miyamoto

EXPERIMENTAL VALUES: Composition of saturated solutions

| CsIO$_3$ | | CsNO$_3$ | | Nature of the |
mass %	mol % (compiler)	mass %	mol % (compiler)	solid phase[a]
5.07[b]	0.312	–	–	A
1.95	0.136	15.56	1.711	"
1.74	0.124	17.97	2.024	"
1.38	0.110	28.09	3.547	"
31.42	14.18	63.30	45.11	
1.02	0.0927	38.19	5.483	A+B
1.02	0.0929	38.34	5.516	"
1.02	0.0937	38.89	5.639	"
1.04	0.0948	38.33	5.515	"
1.02	0.0934	38.70	5.596	"
0.98	0.0906	39.39	5.749	"
0.99	0.0900	38.22	5.487	
0.23	0.0209	39.01	5.601	B
–	–	39.49	5.689	"

[a] A = $CsIO_3$;   B = $CsNO_3$.

[b] For the binary system the compiler computes the following:

soly of $CsIO_3$ = 0.174 mol kg^{-1}

AUXILIARY INFORMATION

METHOD/APPARATUS/PROCEDURE:	SOURCE AND PURITY OF MATERIALS:
The compiler assumes that the isothermal method was used. The ternary $CsIO_3$-$CsNO_3$-H_2O system was studied by the method described in ref 1. COMMENTS AND/OR ADDITIONAL DATA: The phase diagram is given below (based on mass % units).	The compiler assumes that chemically pure grade cesium iodate was used as in ref (1).
	ESTIMATED ERROR: Nothing specified.
	REFERENCES: 1. Karataeva, I.M.; Vinogradov, E.E. *Zh. Neorg. Khim.* 1974, *19*, 3156.

COMPONENTS:	ORIGINAL MEASUREMENTS:
(1) Cesium iodate; $CsIO_3$; [13454-81-4] (2) Aluminum iodate; $Al(IO_3)_3$; [15123-75-8] (3) Water; H_2O; [7732-18-5]	Shklovskaya, R.M.; Arkhipov, S.M.; Kidyarov, B.I.; Poleva, G.B. Zh. Neorg. Khim. 1982, 27, 1610-1; Russ. J. Inorg. Chem. (Engl. Transl.) 1982, 27, 910-1.

VARIABLES:	PREPARED BY:
Composition at 298 K	Hiroshi Miyamoto

EXPERIMENTAL VALUES: Composition of saturated solutions at 25°C

CsIO_3		Al(IO_3)_3		Nature of the
mass %	mol % (compiler)	mass %	mol % (compiler)	solid phase[a]
2.61[b]	0.157	–	–	A
2.24	0.135	0.82	0.028	"
2.19	0.133	1.68	0.0570	"
2.06	0.126	2.33	0.0794	"
1.99	0.123	3.31	0.114	"
1.85	0.115	4.44	0.154	"
1.57	0.0977	4.58	0.159	A+B
1.29	0.0803	4.90	0.170	B
0.85	0.053	5.07	0.176	"
0.40	0.025	5.61	0.194	"
–	–	5.7 [b]	0.197	"

[a] A = $CsIO_3$;   B = $Al(IO_3)_3 \cdot 6H_2O$

[b] For binary systems the compiler computes the following:

soly of $CsIO_3$ = 0.0871 mol kg^{-1}

soly of $Al(IO_3)_3$ = 0.11 mol kg^{-1}

AUXILIARY INFORMATION

METHOD/APPARATUS/PROCEDURE:	SOURCE AND PURITY OF MATERIALS:
The isothermal method was used. Equilibrium was reached in 15-20 days. The iodate ion concentration in the liquid phase was determined by iodometric titration. The aluminum was determined complexometrically with Trilon(disodium salt of EDTA) and spectrographically. The cesium content was found by difference. The solid phases were identified by the method of residues and checked by X-ray diffraction.	Aluminum iodate hexahydrate was synthesized from iodic acid and aluminum hydroxide. "Special purity" grade cesium iodate was used.
	ESTIMATED ERROR: Nothing specified.
	REFERENCES:

COMPONENTS:	ORIGINAL MEASUREMENTS:
(1) Cesium iodate; $CsIO_3$; [13434-81-4] (2) Hafnium iodate; $Hf(IO_3)_4$; [19630-06-9] (3) Water; H_2O; [7732-18-5]	Shklovskaya, R.M.; Arkhipov, S.M.; Kidyarov, B.I.; Poleva, G.V.; Vdovkina, T.E. *Zh. Neorg. Khim.* 1984, 29, 1346-8; *Russ. J. Inorg. Chem.* (*Engl. Transl.*) 1984, 29, 773-4.

VARIABLES:	PREPARED BY:
T/K = 298.2 Composition	Mark Salomon

EXPERIMENTAL VALUES: The $CsIO_3$ - $Hf(IO_3)_4$ - H_2O system at 25.0°C

Composition of saturated solutions[a]

CsIO₃		Hf(IO₃)₄		
mass %	mol %	mass %	mol %	Nature of the solid phase
-	-	0.00037	7.59×10^{-6}	$Hf(IO_3)_4$
0.32	0.0188	0.000059	1.21×10^{-6}	solid solution based on $Hf(IO_3)_4$
0.60	0.0353	0.000057	1.18×10^{-6}	"
0.72	0.0424	0.000064	1.34×10^{-6}	"
1.04	0.0615	0.000076	1.57×10^{-6}	"
1.23	0.0728	0.000081	1.68×10^{-6}	"
1.58	0.0939	0.000087	1.81×10^{-6}	"
1.65	0.0981	0.000094	1.96×10^{-6}	"
1.95	0.1163	0.000099	2.07×10^{-6}	"
2.27	0.1358	0.00011	2.31×10^{-6}	"
2.43	0.1456	0.00013	2.73×10^{-6}	"
2.53[b]	0.1517	0.00015	3.15×10^{-5}	solid solution + $CsIO_3$
2.53[b]	0.1517	0.00015	3.15×10^{-6}	"
2.61	0.1566	-	-	$CsIO_3$

[a] Mol % values calculated by the compiler.

[b] Eutonic solution.

For binary systems, the compiler computes the following:
solubility of $CsIO_3$ = 0.0871 mol kg^{-1}
solubility of $Hf(IO_3)_4$ = 4.21×10^{-6} mol kg^{-1}

AUXILIARY INFORMATION

METHOD/APPARATUS/PROCEDURE:	SOURCE AND PURITY OF MATERIALS:
Isothermal method used. Equilibrium required 25-30 days. Solid and liquid phases analyzed for Cs by emission spectrometry using solutions of Cs concentration between 0.1 - 100 $\mu g\,cm^{-3}$ in the presence of 2 % NaCl solution (added to suppress the ionization of Cs atoms). Preliminary experiments established that Hf does not influence the intensity of the emission of Cs. The concentration of Cs was therefore determined by comparing samples of saturated solution previously buffered with 2 % NaCl solution with standard Cs solutions also buffered with 2 % NaCl solution. For liquid phase samples, Hf was determined photometrically using Arsenazo III after reduction of IO_3 with hydroxylamine. For solid phase samples, Cs was analyzed as described above and iodate by iodometric titrn. The Hf content was determined by difference. Solid phase samples were identified by the method of residues and by X-ray diffraction. The maximum concentration of $CsIO_3$ in the solid solution is 5.8 %.	"Highly pure" $CsIO_3$ was used. $Hf(IO_3)_4$ was prepared from aqueous HIO_3 and freshly precipitated hydrated hafnium oxide under conditions described previously (1). No other information given.
	ESTIMATED ERROR: Soly: uncertainty in analyses did not exceed 3-8 rel %. Temp: precision given as $\pm$ 0.1 K.
	REFERENCES: 1. Deabriges, J.; Rohmer, R. *Bull. Soc. Chim. France* 1968, 521.

COMPONENTS:	ORIGINAL MEASUREMENTS:
(1) Cesium iodate; $CsIO_3$; [13454-81-4] (2) Iodic acid; HIO_3; [7782-68-5] (3) Water; H_2O; [7732-18-5]	Tatarinov, V.A. *Uch. Zap. Yarosl. Pedagog. Inst.* 1973, *No. 120*, 71-3.

VARIABLES:	PREPARED BY:
Composition at 298 K	Hiroshi Miyamoto

EXPERIMENTAL VALUES: Composition of saturated solutions

$CsIO_3$ mass %	mol % (compiler)	HIO_3 mass %	mol % (compiler)	Nature of the solid phase[a]
2.67[b]	0.160	–	–	A
2.65	0.160	0.61	0.064	A+C
1.12	0.0683	3.31	0.353	C
0.62	0.048	25.91	3.484	"
0.25	0.036	66.33	16.89	"
0.17	0.030	74.75	23.38	C+B
–	–	75.25[b]	23.74	B

[a] A = $CsIO_3$;   B = HIO_3;   C = $CsIO_3 \cdot HIO_3$

[b] For binary systems the compiler computes the following:

soly of $CsIO_3$ = 0.0891 mol kg^{-1}

soly of HIO_3 = 17.28 mol kg^{-1}

AUXILIARY INFORMATION

METHOD/APPARATUS/PROCEDURE:	COMMENTS AND/OR ADDITIONAL DATA
The isothermal method was used. Equilibrium between the liquid and solid phases was established in 24 hours. The cesium iodate content in the samples was determined iodometrically, and HIO_3 determined by titration with base.	The phase diagram is given below (based on mass % units). 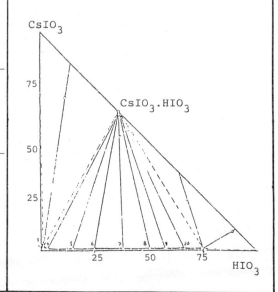

SOURCE AND PURITY OF MATERIALS:
Cesium iodate was prepared from iodic acid and cesium carbonate, and the product was recrystallized.

ESTIMATED ERROR:

Nothing specified.

COMPONENTS:	ORIGINAL MEASUREMENTS:
(1) Cesium iodate; $CsIO_3$; [13454-81-4]	Miyamoto, H.; Hasegawa, T.; Sano, H.
(2) N,N-Dimethylformamide; C_3H_7NO; [68-12-2]	J. *Solution Chem.* in press.
(3) Water; H_2O; [7732-18-5]	

VARIABLES:	PREPARED BY:
Solvent Composition	M. Salomon
Temperature	

EXPERIMENTAL VALUES:

$t/°C = 20$		$t/°C = 25$	
mass % dimethylformamide	$CsIO_3/mol\ dm^{-3}$	mass % dimethylformamide	$CsIO_3/mol\ dm^{-3}$
0	0.0747	0	0.0852
4.79	0.0570	5.12	0.0670
9.47	0.0458	9.48	0.0536
20.84	0.0231	20.09	0.0289
30.22	0.0132	29.71	0.0162
41.99	0.0054	40.02	0.0077

AUXILIARY INFORMATION

METHOD/APPARATUS/PROCEDURE:	SOURCE AND PURITY OF MATERIALS:
Same as in reference (1).	Extra pure grade Cs_2CO_3 and guaranteed grade HIO_3 used as received. $CsIO_3$ pptd by addn of excess HIO_3 sln to aq Cs_2CO_3 sln while heating. After stirring for 5 h, the sln was allowed to settle for 1 day, and the ppt washed with cold water until the dried salt produced a constant soly. The salt was stored in the dark.
	Guaranteed grade dimethylformamide (Wako) was stored over BaO for two days, and then distilled three times under reduced pressure.
	Doubly distilled water had an electrolytic conductance of 9.8×10^{-7} S cm^{-1}.

ESTIMATED ERROR:
Soly: standard deviation between 0.0002 and 0.001.
Temp: not stated.

REFERENCES:

1. Miyamoto, H.; Shimura, H.; Sasaki, K.
    J. *Solution Chem.* 1985, *14*, 485.

COMPONENTS:	ORIGINAL MEASUREMENTS:
(1) Cesium iodate; $CsIO_3$; [13454-81-4]	Miyamoto, H.; Hasegawa, T.; Sano, H.
(2) N,N-Dimethylformamide; C_3H_7NO; [68-12-2]	J. *Solution Chem.* in press.
(3) Water; H_2O; [7732-18-5]	

EXPERIMENTAL VALUES: (Continued)

$t/°C = 30$

mass % dimethylformamide	$CsIO_3$/mol dm^{-3}
0	0.0990
5.53	0.0750
11.49	0.0563
19.81	0.0341
29.79	0.0190
40.33	0.0086

For the binary $CsIO_3$-H_2O system, measured densities of saturated solutions permits conversions from mol dm^{-3} to mol kg^{-1} and mole fraction units.

$t/°C$	density/g cm^{-3}	c/mol dm^{-3}	m/mol kg^{-1}	mole fraction
20	1.019	0.0747	0.0750	0.00135
25	1.020	0.0852	0.0857	0.00154
30	1.022	0.0990	0.0998	0.00180

COMPONENTS:	ORIGINAL MEASUREMENTS:
(1) Cesium iodate; $CsIO_3$; [13454-81-4]	Miyamoto, H.; Hasegawa, T.; Sano, H.
(2) Dimethylsulfoxide ; C_2H_6OS; [67-88-5]	J. *Solution Chem.* in press.
(3) Water; H_2O; [7732-18-5]	

VARIABLES:	PREPARED BY:
Solvent composition	
Temperature	M. Salomon

EXPERIMENTAL VALUES:

$CsIO_3$ soly/mol dm^{-3}

mass % dimethylsulfoxide	t/°C =	20	25	30
0		0.0747	0.0852	0.0990
5.03		0.0580	0.0684	0.0806
10.02		0.0461	0.0550	0.0643
20.09		0.0270	0.0328	0.0378
30.01		0.0149	0.0182	0.0211
40.03		0.0076	0.0092	0.0106

AUXILIARY INFORMATION

METHOD/APPARATUS/PROCEDURE:	SOURCE AND PURITY OF MATERIALS:
Same as in reference (1).	Extra pure grade Cs_2CO_3 and guaranteed grade HIO_3 used as received. $CsIO_3$ pptd by addn of excess HIO_3 sln to aq Cs_2CO_3 sln while heating. After stirring for 5 h, the sln was allowed to settle for 1 day, and the ppt washed with cold water until the dried salt produced a constant soly. The salt was stored in the dark.
	Guaranteed grade dimethyl sulfoxide (Wako) was distilled three times under reduced pressure.
	Doubly distilled water had an electrolytic conductance of 9.8×10^{-7} S cm^{-1}.

ESTIMATED ERROR:
Soly: stnd deviation betweeen 0.0002 and 0.001
Temp: not stated.

REFERENCES:
1.   Miyamoto, H.; Shimura, H.; Sasaki, K. J. *Solution Chem.* 1985, *14*, 485.

COMPONENTS:	EVALUATOR:
(1) Ammonium iodate; NH_4IO_3; [13446-09-8] (2) Water; H_2O; [7732-18-5]	Hiroshi Miyamoto Department of Chemistry Niigata University Niigata, Japan June 1984

CRITICAL EVALUATION:

THE BINARY SYSTEM

Data for the solubility of NH_4IO_3 in water were reported in 6 publications (1-6). The study of Opalovskii and Kuznetsova (2) deals with the solubility of ammonium iodate in water at various temperatures. The remaining five studies (1, 3-6) deal with ternary systems, and the solubility in the binary system is given as 1 point on a phase diagram.

In the five publications dealing with ternary systems (1, 3-6), the stable solid phase in equilibrium with the saturated solutions was simply anhydrous ammonium iodate. The composition of the solid phase was determined by Schreinemakers' method of residues (1, 3, 5, 6), by X-ray diffraction, thermography, and infrared spectroscopy (5). Opalovskii and Kuzunetsova (2) reported the existence of $NH_4IO_3 \cdot 0.75H_2O$ crystals which was determined by thermogravimetry and X-ray analysis, but this result has not been confirmed by any other investigator.

In many cases, the iodate content was determined by iodometric titration, and the determination of the ammonium content was carried out by a distillation method (2, 4). Other studies employed the bromate method (3) and gravimetry using sodium tetraphenylborate (5, 6).

EVALUATION OF DATA

Some investigators reported the solubility in mass % units which the evaluator converted to units of mol kg^{-1} using 1977 IUPAC recommended atomic masses. Opalovskii and Kuznetsova (2) reported the solubility of NH_4IO_3 in terms of the I_2O_5 content, and the evaluator made the conversions to mol kg^{-1} units.

Solubility at 298.2 K. The solubility has been reported in 4 publications (2, 4-6). In 2 publications by Tarasova, Vinogradov and Lepeshkov (5, 6), identical solubility values of 0.200 mol kg^{-1} were reported. The data of Opalovskii and Kuznetsova (2) were rejected, and the arithmetic mean of 2 independent results from (4, 5), and for which the solid phase is NH_4IO_3, is 0.199 mol kg^{-1}. This mean is designated as a recommended value.

Solubility at 303.2 K. Only one value of 0.227 mol kg^{-1} was reported by Meerburg (1), and the composition of the stable solid was NH_4IO_3. The value of 0.227 mol kg^{-1} is designated as a tenative result.

Solubility at 323.2 K. The solubility has been reported in 2 publications (2, 3). The value of Opalovskii and Kuznetsov (2) is 0.389 mol kg^{-1}, and that of Tatrinov (3) is 0.428 mol kg^{-1}. The difference between two reported data is large. The solid phase reported in the former study was $NH_4IO_3 \cdot 0.75H_2O$, and that of the latter author was NH_4IO_3. Therefore, the evaluator is unable to average these two values. The result of Tatarinov is designated as a tenative value because their identification of an anhydrous solid phase is consistent with most other data at various temperatures. The results from (2) are rejected.

The recommended and tenative values of solubilities of ammonium iodate in water are given in Table 1.

Table 1. Recommended and tenative solubilities in the binary NH_4IO_3-H_2O system

T/K	m_1/mol kg^{-1}	Solid phase
298.2[a]	0.199	NH_4IO_3
303.2	0.227	"
323.2	0.428	"

[a]Recommended value

COMPONENTS:	EVALUATOR:
(1) Ammonium iodate; NH_4IO_3; [13446-09-8]	Hiroshi Miyamoto
(2) Water; H_2O; [7732-18-5]	Department of Chemistry Niigata University Niigata, Japan
	June, 1984

CRITICAL EVALUATION:

TERNARY SYSTEMS

The data for the solubility in ternary systems were reported in 4 publications (3-6). The phase diagrams of the ternary systems, $NH_4IO_3-NH_4F-H_2O$ (4) and $NH_4IO_3-Mg(IO_3)_2H_2O$ (6) are simple eutonic types, and no double salts are formed.

The dominant feature in the ternary systems $NH_4IO_3HIO_3-H_2O$ (3) and $NH_4IO_3-LiIO_3-H_2O$ (5) is the existence of double salts of the type $NH_4IO_3.2MIO_3(M = H, Li)$.

REFERENCES:

1. Meerburg, P.A. *Z. Anorg. Allg. Chem.* <u>1905</u>, *45*, 324.

2. Opalovskii, A.A.; Kuznetsova, Z.M. *Izv. Sib. Otd. Akad. Nauk SSR* <u>1962</u>, No. 3, 64.

3. Tatarinov, V.A. *Uch. Zap. Yarostov. Gos. Pedagog. Inst.* <u>1971</u>, No. 95, 113.

4. Kuznetsova, Z.M.; Samoilov, P.P.; Fedotova, T.D.; Fedorov, V.E. *Izv. Sib. Otd. Akad. Nauk SSR Ser. Khim. Nauk* <u>1972</u>, (1), 99.

5. Tarasova, G.N.; Vinogradov, E.E.; Lepeshkov, I.N. *Zh. Neorg. Khim.* <u>1976</u>, *21*, 3373; *Russ. J. Inorg. Chem. (Engl. Transl.)* <u>1976</u>, *21*, 1858.

6. Tarasova, G.N.; Vinogradov, E.E.; Lepeshkov, I.Nv *Zh. Neorg. Khim.* <u>1977</u>, *22*, 809; *Russ. J. Inorg. Chem. (Engl. Transl.)* <u>1977</u>, *22*, 488. Note that the compilation for this reference can be found in the first volume of this series (7).

7. Miyamoto, H.; Salomon, M.; Clever, H.L. *IUPAC SOLUBILITY DATA SERIES, VOLUME 14: ALKALINE EARTH METAL HALATES.* Pergamon Press, London, 1983.

COMPONENTS:	ORIGINAL MEASUREMENTS:
(1) Ammonium iodate; NH_4IO_3; [13446-09-8]	Opalovskii, A.A.; Kuznetsova, Z.M.
(2) Water; H_2O; [7732-18-5]	*Izv. Sib. Otd. Akad. Nauk SSSR* 1962, No. 3. 64-9.

VARIABLES:	PREPARED BY:
T/K = 273 to 358	Hiroshi Miyamoto

EXPERIMENTAL VALUES:

t/°C	I_2O_5 mass %	NH_4IO_3[a] mol kg^{-1}	Nature of the solid phase
0	2.03	2.35	$NH_4IO_3.0.75H_2O$
25	3.30	3.81	"
50	6.04	6.98	"
85	8.86	10.2	"

[a]Molalities calculated by the compiler using 1977 IUPAC recommended atomic masses.

AUXILIARY INFORMATION

METHOD/APPARATUS/PROCEDURE:	SOURCE AND PURITY OF MATERIALS:
Probably the isothermal method was used. The ammonia content was determined by a distillation method, and iodate was determined iodometrically. The composition of the solid phase was determined by thermography and X-ray analysis.	Ammonium iodate was prepared by treating ammonium fluoride with iodic acid.
	ESTIMATED ERROR: Nothing specified.
	REFERENCES:

COMPONENTS:	ORIGINAL MEASUREMENTS:
(1) Ammonium fluoride; NH_4F; [12125-01-8] (2) Ammonium iodate; NH_4IO_3; [13446-09-8] (3) Water; H_2O; [7732-18-5]	Kuznetsova, Z.M.; Samoilov, P.P.; Fedotova, T.D.; Fedorov, V.E. *Izv. Sib. Otd. Akad. Nauk SSR Ser. Khim. Nauk* 1972, (1), 99-104.
VARIABLES: T/K = 298 composition	PREPARED BY: Hiroshi Miyamoto

EXPERIMENTAL VALUES:

Composition of saturated solutions

NH_4F		NH_4IO_3		Nature of
mass %	mol % (compiler)	mass %	mol % (compiler)	the solid phase[a]
–	–	3.67[b]	0.355	A
9.48	4.88	0.72	0.071	"
16.14	8.603	0.51	0.052	"
31.86	18.59	0.28	0.031	"
43.08	29.98	0.23	0.028	"
46[c]	29.3	–	–	B

[a] A = NH_4IO_3;   B = NH_4F

[b] Value obtained from ref 1.

For the binary system the compiler computes the following:

soly of NH_4IO_3 = 0.198 mol kg^{-1}

[c] Value obtained from ref 2.

AUXILIARY INFORMATION

METHOD/APPARATUS/PROCEDURE:

The isothermal method was used. Mixtures of salts and water were stirred in sealed Teflon tubes placed in a thermostat. After equilibrium was established, aliquots of the liquid phases were withdrawn. The ammonia content was determined by distillation method (ref 3). Fluorine was determined with lanthanium nitrate by potentiometric titration using a fluoride ion selective electrode. The iodate concentration was determined iodometrically. The method used to determine composition of the solid phases was not specified.

SOURCE AND PURITY OF MATERIALS:

"Analytical" or chemically "pure" grade salts were used.

ESTIMATED ERROR:

Nothing specified.

REFERENCES:
1. Kirgintsev, A.N.; Trushiova, L.N.; Lavrenteva, V.G. *Rastvorinost Neorganicheskikh Veshchestv v Vode (Solubilities of Inorganic Substances in Water)*
2. Yatlov, V.S.; Polyakova, E.M. *Zh. Obshch. Khim.* 1945, 15, 724.
3. Kolthoff, I.M.; Sandell, E.B. *Textbook of Quantitative Inorganic Analysis.* Macmillan Co. N.Y. 1953.

COMMENTS AND/OR ADDITIONAL DATA:
The phase diagram is given below (based on mass %).

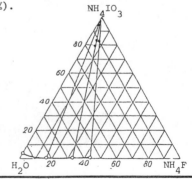

COMPONENTS:	ORIGINAL MEASUREMENTS:
(1) Ammonium iodate; NH_4IO_3; [13446-09-8]	Meerburg, F.A.
(2) Iodic acid; HIO_3; [7782-68-5]	Z. Anorg. Allg. Chem. 1905, 45, 324-44.
(3) Water; H_2O; [7732-18-5]	

VARIABLES:	PREPARED BY:
T/K = 303 Composition	Hiroshi Miyamoto

EXPERIMENTAL VALUES: Composition of saturated solutions at 30°C

Iodic Acid		Ammonium Iodate		Nature of
mass %	mol % (compiler)	mass %	mol % (compiler)	the solid phase[a]
0	0	4.20[b]	0.408	A
2.54	0.276	3.89	0.386	"
4.52	0.501	3.83	0.387	A+C
4.51	0.500	3.86	0.390	"
4.56	0.505	3.75	0.379	"
4.73	0.523	3.53	0.356	C
6.57	0.729	1.94	0.196	"
8.45	0.947	1.09	0.111	"
9.12	1.026	0.91	0.091	"
24.00	3.155	0.62	0.074	"
36.01	5.479	0.41	0.057	"
44.43	7.613	0.39	0.061	"
58.21	12.57	0.37	0.073	"
76.35	25.07	0.31	0.093	C+B
76.70[b]	25.21	0	0	B

[a] $A = NH_4IO_3$; $B = HIO_3$; $C = NH_4IO_3.2HIO_3$

[b] For binary systems the compiler computes the following:

$$\text{soly of } HIO_3 = 18.71 \text{ mol kg}^{-1}$$
$$\text{soly of } NH_4IO_3 = 0.227 \text{ mol kg}^{-1}.$$

METHOD/APPARATUS/PROCEDURE:	SOURCE AND PURITY OF MATERIALS:
A mixture of NH_4IO_3, HIO_3 and water was placed in a bottle, and the bottle agitated in a thermostat for a week or more at a desired temperature. Equilibrium was established from supersaturation.	Nothing specified.

SOURCE AND PURITY OF MATERIALS:

Nothing specified.

ESTIMATED ERROR:

Nothing specified.

METHOD/APPARATUS/PROCEDURE:

A mixture of NH_4IO_3, HIO_3 and water was placed in a bottle, and the bottle agitated in a thermostat for a week or more at a desired temperature. Equilibrium was established from supersaturation.

The iodic acid and ammonium iodate contents were determined by iodometric titration, and the details of the analytical method were probably similar to those of KIO_3-HIO_3-H_2O system. (See the compilation for this system.)

The composition of the solid phase was determined by the method of residues.

COMMENTS AND/OR ADDITIONAL DATA:

The phase diagram is given below (based on mass %).

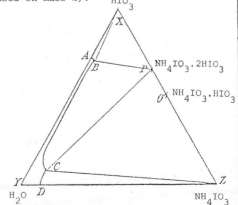

AMH—P*

COMPONENTS:	ORIGINAL MEASUREMENTS:
(1) Ammonium iodate; NH_4IO_3; [13446-09-8]	Tatarinov, V.A.
(2) Iodic acid; HIO_3; [7782-68-5]	*Uch. Zap. Yavostov. Gos. Pedagog. Inst.* 1971, No. 95, 113-5.
(3) Water; H_2O; [7732-18-5]	

VARIABLES:	PREPARED BY:
T/K = 322 composition	Hiroshi Miyamoto

EXPERIMENTAL VALUES:

Composition of saturated solutions

Ammonium Iodate mass %	mol % (compiler)	Iodic Acid mass %	mol % (compiler)	Nature of the solid phase
7.62[b]	0.764	–	–	A
6.51	0.671	3.95	0.447	"
6.32	0.670	6.93	0.806	A+C
6.31	0.669	6.96	0.810	"
5.07	0.547	9.89	1.16	C
0.43	0.105	68.41	18.34	"
0.42	0.124	75.82	24.60	B+C
–	–	76.53[b]	25.03	B

[a] A = NH_4IO_3; B = HIO_3; C = $NH_4IO_3.2H_2O$.

[b] For binary systems the compiler computes the following:

soly of NH_4IO_3 = 0.428 mol kg^{-1}

soly of HIO_3 = 18.54 mol kg^{-1}

AUXILIARY INFORMATION

METHOD/APPARATUS/PROCEDURE:

The isothermal method was used. Equilibrium was reached in 24 hours. Aliquots of the liquid and solid phases were used for analysis of NH_4^+ and IO_3^-. NH_4^+ was determined by the bromate method (ref 1), and IO_3^- determined iodometrically.

The composition of the solid phase was determined by Schreinemakers' method and chemical analyses.

SOURCE AND PURITY OF MATERIALS:

"Chemically pure" grade iodic acid was recrystallized from water. Ammonium iodate was made from iodic acid and ammonium carbonate. The product was washed with a large quantity of cold water and then recrystallized.

REFERENCES:

1. Levy, B. *Z. Anal. Chem.* 1931, *84*, 98.

ESTIMATED ERROR:

Nothing specified.

COMMENTS AND/OR ADDITIONAL DATA:

The phase diagram is given below.

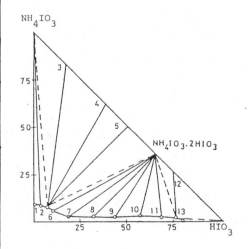

COMPONENTS:	EVALUATOR:
(1) Iodic acid; HIO_3; [7782-68-5] (2) Water; H_2O; [7732-18-5]	H. Miyamoto Niigata University Niigata, Japan and M. Salomon US Army ET & DL Fort Monmouth, NJ, USA September, 1985

CRITICAL EVALUATION:

THE BINARY SYSTEM

Data for the solubility of HIO_3 in pure water have been reported in 17 publications (1-17). Studies involving ternary systems over the temperature range of 273 - 333 K, confirmed the solid phase to be anhydrous HIO_3. Groschuff (2) reported that the eutectic point of ice and HIO_3 is about 259 K, that the $HIO_3 \rightarrow HIO_3I_2O_5$ transition occurs at 383 K, and that the conversion of $HIO_3 \cdot I_2O_5$ (or HI_3O_8) to iodic pentoxide, I_2O_5 occurs between 463 K and 473 K. It is quite surprising that the phase diagram for the HIO_3-H_2O system reported in detail by Groschuff in 1905 has never been restudied to confirm both the accuracy and precision of Groschuff's results.

A number of compilations containing solubility data for both binary and ternary systems can be found in other chapters in this volume, or in the earlier volume to this series (27). The location of these compilations are reviewed in Table 1 below.

Table 1. Location of compilations containing HIO_3 solubility data.

System	Reference	Compilation found in
$LiIO_3-HIO_3-H_2O$	5,10,12,13,18,24	$LiIO_3$ chapter
$NaIO_3-HIO_3-H_2O$	1,21	$NaIO_3$ chapter
$KIO_3-HIO_3-H_2O$	1,4	KIO_3 chapter
$RbIO_3-HIO_3-H_2O$	9	$RbIO_3$ chapter
$CsIO_3-HIO_3-H_2O$	11	$CsIO_3$ chapter
$NH_4IO_3-HIO_3-H_2O$	1,8	NH_4IO_3 chapter
HIO_3 + alkaline earth metal iodates + H_2O	6,15,22	SDS Volume 14 (27)

The Ice Polytherm

The only experimental solubility data along the ice polytherm are those of Groschuff, and the evaluators were unsuccessful in fitting all these data to the smoothing equation. This problem is due to the large standard error of estimate σ_x obtained using all reported data points. This error could be reduced slightly, but too many data points must be ignored (or rejected), and the resulting smoothing equation becomes trivial. The original data for the ice polytherm can be found in the compilation of reference (2).

The HIO_3 Polytherm

While the data of Groschuff (2) still dominate this part of the phase diagram, there are sufficient data from other studies (see Table 2) which permit fitting of all data to the smoothing equation. As seen from the summary in Table 2, a number of data points were rejected, notably from references (2, 3, 7-9, 12). The remaining data, treated as 23 independent solubility determinations as indicated in Table 2, were fitted to the following smoothing equation:

$$Y_x = 8079/(T/K) + 45.062 \, ln \, (T/K) - 269.85 - 0.05330(T/K)$$

$$\sigma_y = 0.023 \qquad\qquad \sigma_x = 0.0037$$

Smoothed solubilities calculated from this smoothing equation are given in Table 3, and all calculated solubilities are designated as *tentative* values.

COMPONENTS:	EVALUATOR:
(1) Iodic acid, HIO_3; [7782-68-5]	H. Miyamoto Niigata University Niigata, Japan
(2) Water; H_2O; [7732-18-5]	and Mark Salomon US Army ET & DL Fort Monmouth, NJ, USA September, 1985

CRITICAL EVALUATION:

Table 1.   Experimental solubilities in the HIO_3-H_2O system

T/K	mass %	mole fraction	Reference
259.2[b]	72.8	0.215	2
273.2	73.56	0.2217	4
273.2	74.1	0.227	2
273.2[a]	75.89	0.2438	7
286.7	74.10	0.227	2
289.2[a]	75.8	0.241	2
291.2	74.55	0.231	2
293.2[a]	68.72	-----	3
293.2[a]	75.8	0.243	2
298.2	75.10	0.2360	13
298.2	75.25	0.2374	11
298.2	75.32	0.2381	6
298.2	75.33	0.2382	6
298.2	75.40	0.2389	5
298.2	75.40	0.2389	14
298.2	75.40	0.2389	17
298.2	75.56	0.2405	4
303.2	76.70	0.2521	1
313.2[a]	73.70	0.2230	12
313.2	77.7	0.263	2
323.2[a]	76.53	0.2503	8,9
323.2	77.69	0.2629	16
323.2	78.62	0.2736	10,15
323.2	78.78	0.2755	4
333.2	80.0	0.291	2
353.2	82.5	0.326	2
358.2	83.0	0.333	2
374.2	85.2	0.371	2
383.2[c]	86.5	0.396	2

[a]Rejected data. Solid phase is HIO_3 except as noted below.

[b]Solid phase is ice + HIO_3.

[c]Solid phase is HIO_3 + HI_3O_8.

COMPONENTS:	EVALUATOR:
(1) Iodic acid, HIO_3; [7782-68-5] (2) Water; H_2O; [7732-18-5]	H. Miyamoto Niigata University Niigata, Japan and M. Salomon US Army ET & DL Fort Monmouth, NJ, USA September, 1985

CRITICAL EVALUATION:

Table 3.   Tentative solubilities in the HIO_3-H_2O system
calculated from the smoothing equation[a]

T/K	mass %	mole fraction
259.2[b]	73.04	0.217
273.2	73.45	0.221
283.2	74.10	0.227
293.2	74.98	0.235
298.2	75.48	0.240
303.2	76.03	0.245
313.2	77.20	0.257
323.2	78.46	0.272
333.2	79.78	0.288
343.2	81.13	0.306
353.2	82.48	0.325
363.2	83.82	0.347
373.2	85.14	0.370
383.2[c]	86.42	0.394

[a]Solid phase is HIO_3 except as noted.

[b]Solid phase is ice + HIO_3.

[c]Solid phase is HIO_3 + HI_3O_8.

COMPONENTS:	EVALUATOR:

(1) Iodic acid; HIO_3; [7782-68-5]
 	
(2) Water; H_2O; [7732-18-5]
	H. Miyamoto
	Niigata University
	Niigata, Japan
	                      and
	M. Salomon
	US Army ET & DL
	Fort Monmouth, NJ, USA
	                              September, 1985

CRITICAL EVALUATION:

The $HIO_3 \cdot I_2O_5$ Polytherm

Solubility data for HIO_3 for which the solid phase is $HIO_3 \cdot I_2O_5$ (or HI_3O_5) were reported only by Groschuff in 1905 (2). The four data points reported in (2) over the temperature range 383 - 433 K are given in the compilation of Groschuff's paper.

The phase diagram for the binary system over the entire experimental temperature range of 254 K to 433 K is given in Figure 1.

TERNARY SYSTEMS

Systems With One Saturating Component

The solubility of iodine pentoxide in sulfuric acid solutions containing 50 to 106 mass % acid at 279.92 K was reported by Lamb and Phillips: note that the mass % sulfuric acid in excess of 100 % represents the mass of H_2SO_4 equivalent to 100 g of the acid. Excess SO_3 accounts for mass % values greater than 100 %: e.g. the acid content of 106 % contained 29 mass % SO_3. All data (both the "initial" and "final" sets of data as given in the compilation) were used to plot the phase diagram for this system. The phase diagram is given in Figure 2. According to the authors (19), the "initial" set of data correspond to HIO_3 solubilities in which there is a slow transformation to a less soluble substance. The shape of the lower isotherm (based on the "final" set of solubility data) was attributed by Lamb and Phillips to the solubility of I_2O_5 and anhydro iodic acid (HI_3O_5). It would appear that this simple explanation to the complex phase diagram in Figure 2 is in fact too simple, and that new studies are required to correctly identify all solid phases present in this system.

The solubilities in HNO_3 and HF systems were reported in (2,3) and (7), respectively. In the latter work (7), several solutions of high HF content yielded a solid phase containing the compound $2HIO_3 \cdot 3HF$.

Ternary Systems Containing Two Saturating Components

Saturated solutions containing HIO_3 and an alkali metal iodate have been summarized in Table 1 above. There does not appear to be any major disagreement in any of these works. However, it should be noted that for the $NaIO_3-HIO_3-H_2O$ system, Meerburg (1) found the compounds $Na_2I_4O_{11}$ and $NaH_2I_3O_9$ but he did not report NaI_3O_8 which was found in the work of Shibuya and Watanabe (21).

The solubility of HIO_3 in solutions saturated with alkaline earth iodates and with transition and rare earth metal iodates are summarized in Tables 4 and 5, respectively. Note that all the compilations for the systems summarized in Table 4 were previously given in the earlier volume to this series (27).

QUATERNARY SYSTEMS

Two quaternary systems have been reported which are:

$$HIO_3 - LiIO_3 - KIO_3 - H_2O \qquad \text{at 323 K (24)}$$

and

$$HIO_3 - LiIO_3 - Al(IO_3)_3 - H_2O \qquad \text{at 298 K (25).}$$

The compilations for both (24 and 25) can be found in the $LiIO_3$ chapter in this volume.

COMPONENTS:	EVALUATOR:
(1) Iodic acid; HIO_3; [7782-68-5] (2) Water; H_2O; [7732-18-5]	H. Miyamoto Niigata University Niigata, Japan and M. Salomon US Army ET & DL Fort Monmouth, NJ, USA September, 1985

CRITICAL EVALUATION:

Table 4. Summary of ternary systems with alkaline earth iodates

Ternary system	T/K	Solid phase	Reference
$HIO_3 - Mg(IO_3)_2 - H_2O$	298	HIO_3; $Mg(IO_3)_2 \cdot 4H_2O$	(6)
$HIO_3 - Mg(IO_3)_2 - H_2O$	323	HIO_3; $Mg(IO_3)_2 \cdot 4H_2O$	(22)
$HIO_3 - Sr(IO_3)_2 - H_2O$	323	HIO_3; $Sr(IO_3)_2 \cdot H_2O$; $Sr(IO_3)_2 \cdot HIO_3 \cdot H_2O$	(15)
$HIO_3 - Ba(IO_3)_2 - H_2O$	298	$I_2O_5 \cdot H_2O(HIO_3)$; $Ba(IO_3)_2 \cdot H_2O$; $Ba(IO_3)_2 \cdot I_2O_5$	(6)

Table 5. Summary of ternary system with transition and rare earth metal iodates

Ternary system	T/K	Solid phase	Reference
$HIO_3 - Al(IO_3)_3 - H_2O$	298	HIO_3; $Al(IO_3)_3 \cdot 6H_2O$ $Al(IO_3)_3 \cdot 2HIO_3 \cdot 6H_2O$	(13)
$HIO_3 - Zn(IO_3)_2 - H_2O$	323	HIO_3; $Zn(IO_3)_2 \cdot 2H_2O$	(16)
$HIO_3 - Cd(IO_3)_2 - H_2O$	323	HIO_3; $Cd(IO_3)_2$; $2HIO_3 \cdot Cd(IO_3)_2$	(16)
$HIO_3 - La(IO_3)_3 - H_2O$	298	HIO_3; $La(IO_3)_2 \cdot 2.5H_2O$; $La(IO_3)_3$	(23)
$HIO_3 - Sc(IO_3)_3 - H_2O$	298	HIO_3; $Sc(IO_3)_3 \cdot 18H_2O$; $Sc(IO_3)_3 \cdot 4HIO_3 \cdot 18H_2O$	(14)
$HIO_3 - Nd(IO_3)_3 - H_2O$	298	HIO_3; $Nd(IO_3)_3$; $Nd(IO_3)_3 \cdot HIO_3 \cdot 2H_2O$; $Nd(IO_3)_3 \cdot 3HIO_3 \cdot 2H_2O$	(17)

COMPONENTS:	EVALUATOR:
(1) Iodic acid; HIO_3; [7782-68-5]	H. Miyamoto Niigata University Niigata, Japan
(2) Water; H_2O; [7732-18-5]	
	September, 1985

CRITICAL EVALUATION:

REFERENCES

1. Meerburg, P. A. *Z. Anorg. Alleg. Chem.* <u>1905</u>, *45*, 324.

2. Groschuff, E. *Z. Anorg. Alleg. Chem.* <u>1905</u>, *47*, 331.

3. Guichard, M. *C. R. Hebd. Seances. Acad. Sci.* <u>1909</u>, *148*, 923.

4. Smith, S. B. *J. Am. Chem. Soc.* <u>1947</u>, *69*, 2285.

5. Ricci, J. E.; Amron, I. *J. Am. Chem. Soc.* <u>1951</u>, *73*, 3613.

6. Ricci, J. E.; Freedman, A. J. *J. Am. Chem. Soc.* <u>1952</u>, *74*, 1769.

7. Nikolaev, N. S.; Buslaev, Y. A. *Zh. Neorg. Khim.* <u>1956</u>, *1*, 1672; *Russ. J. Inorg. Chem. (Engl. Transl.)* <u>1956</u>, *1*, 230.

8. Tatarinov, V. A. *Uch. Zap. Yarosl. Gas. Pedagog. Inst.* <u>1971</u>, No. 95, 113.

9. Tatarinov, V. A. *Uch. Zap. Yarosl. Gas. Pedagog. Inst.* <u>1972</u>, No. 103, 83.

10. Azarova, L. A.; Vinogradov, E. E.; Mikhailova, E. M.; Pakhomov, V. I. *Zh. Neorg. Khim.* <u>1973</u>, *18*, 239; *Russ. J. Inorg. Chem. (Engl. Transl.)* <u>1973</u>, *18*, 124.

11. Tatarinov, V. A. *Uch. Zap. Yarostav. Gas. Pedagog. Inst* <u>1973</u>, No. 120, 71.

12. Shklovskaya, R. M.; Arkhipov, S. M.; Kidyarov, B. I.; Mitnitskii, P. L.; *Izv. Sib. Otd. Akad. Nauk SSSR Ser. Khim. Nauk* <u>1976</u>, (6), 89.

13. Shklovskaya, R. M.; Arkhipov, S. M.; Kidyarov, B. I.; Kuzina, V. A.; Tsibulrvskaya, K. A. *Zh. Neorg. Khim.* <u>1977</u>, *22*, 1372; *Russ. J. Inorg. Chem. (Engl. Transl.)* <u>1977</u>, *22*, 747.

14. Vinogradov, E. E.; Lepeshkov, I. N.; Tarasova, G. N. *Zh. Neorg. Khim.* <u>1977</u>, *22*, 2858; *Russ. J. Inorg. Chem. (Engl. Transl.)* <u>1977</u>, *22*, 1552.

15. Vinogradov, E. E.; Azarova, L. A.; Pakhomov, V. I. *Zh. Neorg. Khim.* <u>1978</u>, *23*, 534; *Russ. J. Inorg. Chem. (Engl. Transl.)* <u>1978</u>, *23*, 297.

16. Lepeshkov, I. N.; Vinogradov, E. E.; Karataeva, I. M. *Zh. Neorg. Khim.* <u>1979</u>, *24*, 2540; *Russ. J. Inorg. Chem. (Engl. Transl.)* <u>1979</u>, *24*, 1412.

17. Tarasova, G. N.; Vinogradov, E. E.; Kudinov, I. B. *Zh. Neorg. Khim.* <u>1982</u>, *27*, 505; *Russ. J. Inorg. Chem. (Engl. Transl.)* <u>1982</u>, *27*, 287.

18. Lukasiewicz, T.; Pietaszewska, J.; Zmija, J. *Biul. Wojsk. Acad. Teck.* <u>1979</u>, *28*(12) 85.

19. Lamb, A. B.; Phillips, A. W. *J. Am. Chem. Soc.* <u>1923</u>, *45*, 108.

20. Moles, E.; Vitoria, A. P. *Ann. Soc. Esp. Fis. Quim.* <u>1932</u>, *30*, 200.

21. Shibuya, M.; Watanobe, T. *Denki Kagaku* <u>1967</u>, *35*, 550.

22. Vinogradov, E. E.; Azarova, L. A. *Zh. Neorg. Khim.* <u>1977</u>, *22*, 1666; *Russ. J. Inorg. Chem. (Engl. Transl.)* <u>1977</u>, *22*, 903.

23. Lyalina, R. B.; Soboleva, L. V. *Zh. Neorg. Khim.* <u>1975</u>, *20*, 2568; *Russ. J. Inorg. Chem. (Engl. Transl.)* <u>1975</u>, *20*, 1424.

24. Azarova, L. A.; Vinogradov, E. E.; Lepeshkov, I. M. *Zh. Neorg. Khim.* <u>1978</u>, *23*, 1952; *Russ. J. Inorg. Chem. (Engl. Transl.)* <u>1978</u>, *23*, 1072.

25. Shklovskaya, R. M.; Arkhipov, S. M.; Kidyarov, B. I.; Tsibulevskaya, K. A. *Zh. Neorg. Khim.* <u>1979</u>, *24*, 253; *Russ. J. Inorg. Chem. (Engl. Transl.)* <u>1979</u>, *24*, 141.

COMPONENTS:	EVALUATOR:
(1) Iodic acid; HIO₃; [7782-68-5] (2) Water; H₂O; [7732-18-5]	Hiroshi Miyamoto Department of Chemistry Niigata University Niigata, Japan September, 1985

CRITICAL EVALUATION:

REFERENCES (Continued)

26. Erkasov, R. Sh.; Bermzhanov, B. A.; Nurakhmetov, N. N. *Zh. Neorg. Khim.* <u>1981</u>, *26*, 1441-4; *Russ. J. Inorg. Chem. (Engl. Transl.)* <u>1981</u>, *26*, 776-8.

27. Miyamoto, H.; Salomon, M.; Clever, H. L. *IUPAC Solubility Data Series Volume 14: Alkaline Earth Metal Halates.* Pergamon Press, London, 1983.

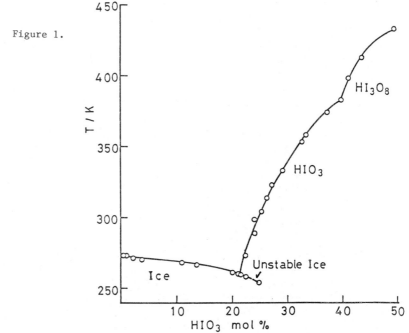

Figure 1.

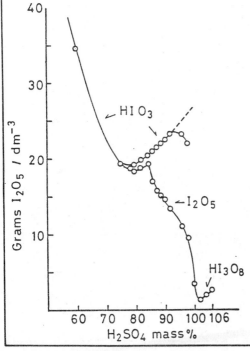

Figure 2.

COMPONENTS:	ORIGINAL MEASUREMENTS:
(1) Iodic acid; HIO$_3$; [7782-68-5] (2) Water; H$_2$O; [7732-18-5]	Groschuff, E. Z. Anorg. Alleg. Chem. 1905, 47, 331-52.

VARIABLES:	PREPARED BY:
Temperature: 254.2 - 433.2 K	Michelle C. Uchiyama

EXPERIMENTAL VALUES:

t/°C	T/K	mass %	mol %[a]	Nature of the solid phase
- 0.30	272.85	1.78	0.185	Ice
- 0.67	272.48	4.35	0.464	"
- 1.01	272.14	7.17	0.785	"
- 1.90	271.25	17.66	2.149	"
- 2.38	270.77	27.65	3.766	"
- 4.72	268.43	54.19	10.81	"
- 6.32	266.83	60.72	13.67	"
-12.25	260.90	71.04	20.08	"
-13.5	259.7	72.2	21.0	"
-14[b]	259.2	72.8	21.5	Ice + HIO$_3$
-15	258.2	73.8	22.4	Unstable ice
-19	254.2	76.2	24.7	"
0	273.2	74.1	22.7	HIO$_3$
13.5[c]	286.7	74.1	22.7	"
16	289.2	75.6	24.1	"
18[c]	291.2	74.55	23.08	"
40	313.2	77.7	26.3	"
60	333.2	80.0	29.1	"
80	353.2	82.5	32.6	"
85	358.2	83.0	33.3	"
101	374.2	85.2	37.1	"
110	383.2	86.5	39.6	HIO$_3$ + HI$_3$O$_8$
125	398.2	87.2	41.1	HI$_3$O$_8$
140	413.2	88.3	43.6	"
160	433.2	90.5	49.4	"

AUXILIARY INFORMATION

METHOD/APPARATUS/PROCEDURE:	SOURCE AND PURITY OF MATERIALS:
Below 0°C. Synthetic method. Solutions of known concn cooled to ppt ice, then warmed to determine the temperature of disappearance of ice. 0°C to 100°C. Isothermal method. Excess powdered HIO$_3$ and water sealed in glass tube and agitated for several hours (several days at 0°C). After settling, aliquots analyzed by thiosulfate titration. Above 100°C. Isothermal as for 0-100°C. Satd slns rapidly cooled to 0°C before aliquots taken for analyses. Author states no pptn occurs in this process of cooling before analyses.	Nothing specified.

Solid phases analyzed gravimetrically. Solid dried between filter paper, washed with alcohol, dried at the experimental temperature. Weight loss determined by heating to 190-195°C. Footnotes to data table: [a] compiler's calculation [b] extrapolated eutectic point [c] Synthetic method used for these two points	**ESTIMATED ERROR:** Author stated solubilities 1-2% higher by isothermal method. Nothing else specified. **REFERENCES:**

COMPONENTS:	ORIGINAL MEASUREMENTS:
(1) Iodic acid; HIO_3; [7782-68-5] (2) Nitric acid; HNO_3; [7697-37-2] (3) Water; H_2O; [7732-18-5]	Groschuff, E. Z. Anorg. Alleg. Chem. 1905, 47, 331-52.
VARIABLES: Concentration of HNO3 at 273 - 333 K	PREPARED BY: Michelle C. Uchiyama

EXPERIMENTAL VALUES:

Temperature (t/°C)	0°	20°	40°	60°
Water	74.1	75.8	77.7	80.0
27.73 per cent HNO_3	18	21	27	38
40.88 per cent HNO_3	9	10	14	18

AUXILIARY INFORMATION

METHOD/APPARATUS/PROCEDURE:	SOURCE AND PURITY OF MATERIALS:
Isothermal method. No other information given, but probably similar to method used for binary solutions (see compilation on page 474).	Nothing specified
	ESTIMATED ERROR: Nothing specified.
	REFERENCES:

COMPONENTS:	ORIGINAL MEASUREMENTS:
(1) Iodine oxide; I_2O_5; [2029-98-0] (2) Nitric acid; HNO_3; [7697-37-2] (3) Water; H_2O; [7732-18-5]	Guichard, M. *Hebd. Seances Acad. Sci.* 1909, *148*, 923-5.[1]

VARIABLES:	PREPARED BY:
Concentration of HNO_3 at 293 K	M. Salomon and K. Salomon

EXPERIMENTAL VALUES:

The solubility of I_2O_5 in pure water at 20°C was given as 187.4 g in 100 g water. This is equivalent to 65.205 mass % (compilers).

Solubilities at 20°C in nitric acid solutions are given below.

density of HNO_3 solution g/cm^3	solubility of I_2O_5	
	g_1 in 100 g acid sln	mass % [a]
1.27	9.1	8.34
1.33	5.5	5.21
1.4	0.67	0.666

[a]Calculated by the compilers.

AUXILIARY INFORMATION

METHOD/APPARATUS/PROCEDURE:	SOURCE AND PURITY OF MATERIALS:
Nothing specified, but the compilers assume that saturated solutions were evaporated and the residue dried and weighed. COMMENTS AND/OR ADDITIONAL DATA: The major objective of this work was to prep highly purified I_2O_5. Previous preps said to involve pptn of HIO_3 from solutions of $BA(IO_3)_2$ + H_2SO_4 followed by recryst of HIO_3. Author claims this method cannot eliminate impurities: $Ba(IO_3)_2$ when this salt is used in excess or $BaSO_4$ and H_2SO_4 when sulfuric acid is used in excess. Author determined that a solution of 96 g I_2O_5 in 100 g H_2O will dissolve 0.15 g $BaSO_4$ at 15°C. Author also states that $Ba(IO_3)_2$, $BaSO_4$ and H_2SO_4 impurities can be significantly reduced by recrystallizing the impure HIO_3 from concentrated nitric acide solution. Starting with an initial impurity level of 0.3 mass %, and recrystallizing five times from concentrated nitric acid, the impurity level was reduced to 0.008 mass %.	I_2O_5 prepd by oxidn of I_2 with N_2O_5. Dry or preferably moist I_2 treated with N_2O_5 prepd from pre-cooled mixt of fuming HNO_3 + P_2O_5 followed by slow heating to 90°C. The product was dissolved in water, and the water was then evaporated and the solid dried at 220°C. The yield of I_2O_5 is 20 g per each 100 g of fuming HNO_3. "High purity" I_2 used: source and purity of water not specified.

ESTIMATED ERROR:
Nothing specified.

REFERENCES:
1. Guichard, M. *Memoires Presentes a la Societe Chimique* 1909, 722-7.

COMPONENTS:	ORIGINAL MEASUREMENTS:
(1) Iodic acid; HIO_3; [7782-68-5] (2) Nitric acid; HNO_3; [7697-37-2] (3) Water; H_2O; [7732-18-5]	Moles, E.; Perez, V. A. *Am. Soc. Esp. Fis. Quim.* <u>1932</u>, *30*, 200-207.
VARIABLES: Concentration of HNO_3 at 298 K	PREPARED BY: R. Herrera, M. Salomon, H. Miyamoto

EXPERIMENTAL VALUES:

Table 1. Experimental results for the ternary system at 25°C.

Solubility of HIO_3[a]

HNO_3	after 24 h	after 48 h	after 48 h	density
mass %	mass %	mass %	mol kg^{-1}	g cm^{-3}
65.30	1.40$_6$	1.41	0.241	1.400
58.66	3.14	3.24	0.483	1.366
50.71	5.74	5.73	0.749	1.324
43.32	10.01	10.08	1.230	1.273
35.28	14.91	15.20	1.745	1.223
28.00	21.94	21.74	2.459	1.173
20.23	35.08	35.09	4.465	1.123

Table 2. Interpolated results based upon data from Table 1

H_2O	HNO_3	solubility of HIO_3[a]	
mass %	mass %	mass %	mol kg^{-1}
34.60	64.00	1.40	0.230
40.02	56.78	3.20	0.455
46.46	47.80	5.74	0.702
51.05	38.90	10.05	1.119
55.00	30.00	15.00	1.550
56.24	21.96	21.80	2.204
51.70	13.30	35.00	3.848

[a] Molalities calculated by the compilers.

AUXILIARY INFORMATION

METHOD/APPARATUS/PROCEDURE:	SOURCE AND PURITY OF MATERIALS:
Mixtures of varying composition were placed in an electric thermostat at 25°C and constantly agitated. Samples of the saturated solution were taken over 24 h intervals. The samples of saturated sln were rapidly filtered in a porous plaque-funnel inside the thermostat, and the HIO_3 content determined gravimetrically after evaporation of HNO_3 and water. In their original Table 1, the authors included solubility data of Groschuff (1) and Guichard (2). These data were omitted from the above Table 1, but have been compiled elsewhere in this volume COMMENTS AND/OR ADDITIONAL DATA The authors state that the data in Table 2 were calculated from the experimental results in Table 1. No other details were given, and the compilers assume that the data in Table 2 referred to as "interpolated" are averages or close to average values.	Nothing specified.
	ESTIMATED ERROR: Soly: nothing specified, but errors in accuracy may be as high as 3 %. Temp: nothing specified.
	REFERENCES: 1. Groschuff, E. *Z. Anorg. Chem.* <u>1905</u>, *47*, 343. 2. Guichard, M. *Bull. Chem. Soc. Fr.* <u>1909</u>, *5*, 722.

COMPONENTS:	ORIGINAL MEASUREMENTS:
(1) Iodic acid: HIO_3; [7782-68-5]	Niolaev, N. S.; Buslav, Yu. A.
(2) Hydrofluoric acid; HF; [7664-39-3]	*Zh. Neorg. Khim.* 1956, *1*, 1672-5; *Russ. J. Inorg. Chem. (Engl. Transl.)* 1956 *1*, 230-5.
(3) Water; H_2O; [7732-18-5]	

VARIABLES:	PREPARED BY:
T/K = 273 Concentration of HF	Hiroshi Miyamoto

EXPERIMENTAL VALUES:

Hydrofuoric Acid		Iodine oxide		Nature of the solid phase[a]
mass %	mol % (compiler)	mass %	mol % (compiler)	
0.00	0.00	72.00	12.19	A
2.43	5.69	64.81	9.098	A
4.91	9.66	56.83	6.704	A
5.14	9.84	55.45	6.363	A
8.15	14.2	50.38	5.277	A
11.34	18.16	45.07	4.325	A
15.28	23.32	41.74	3.819	A
17.76	27.45	42.25	3.914	A
18.28	31.90	49.23	5.148	A
18.67	35.18	53.22	6.010	A
19.16	39.49	57.51	7.106	A
21.37	43.98	57.21	7.057	B
22.18	45.19	56.65	6.917	B
25.53	51.10	55.46	6.653	B
27.96	56.98	56.06	6.848	B
28.57	58.67	56.35	6.936	B
30.02	62.12	56.55	7.014	B
30.36	63.33	56.88	7.111	B
30.61	70.32	61.05	8.405	B
27.78	75.07	67.56	10.94	B
21.40	70.74	74.66	14.79	C
23.00	84.15	77.27	16.94	C

[a]A = HIO_3 B = $2HIO_3 \cdot 3HF$;   C = I_2O_5.

AUXILIARY INFORMATION

METHOD/APPARATUS/PROCEDURE:	COMMENTS AND/OR ADDITIONAL DATA:
The soly vessels were of "florplast-4" and fitted with stirrers through a lid. Stirrers also made of florplast-4, and were lubricated with a polyfluoride oil. The vessels were equilibrated in an ice bath. Aliquots of satd sln and residue withdrawn with a Pt sampler, and weighed at low temp (about 0°C). Total acid dtd by alkali titrn using phenolphthalein idicator, and iodic acid detd by iodometric titrn. HF concn detd by difference. Composition of the solid phase detd by Schreinemakers method.	The phase diagram is given below (mass % units). 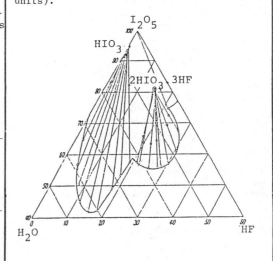

ESTIMATED ERROR:

Soly: the relative error in the determination of HF and I_2O_5 did not exceed 1%.
Temp: nothing specified.

SOURCE AND PURITY OF MATERIALS:

HIO_3 and I_2O_5 were recrystallized. HF was purified by distillation.

COMPONENTS:	ORIGINAL MEASUREMENTS:
(1) Iodine oxide; I_2O_5; [12029-98-0] (2) Sulfuric acid; H_2SO_4; [7664-93-9] (3) Water; H_2O; [7732-18-5]	Lamb, A. B.; Phillips, A. W. J. Am. Chem. Soc. 1923, 45, 108-12.

VARIABLES:	PREPARED BY:
One temperature: 279.92 K Concentration of sulfuric acid	Hiroshi Miyamoto

EXPERIMENTAL VALUES:

(1) With 50-78 mass % solutions of sulfuric acid, constant values of the solubilities were rapidly established as indicated by Table 1.

Table 1

Concentration of H_2SO_4 mass %

Time Days	50.0	60.0	75.0	78.0
	solubility of I_2O_5/g cm^{-3}			
1	48.86	34.84	19.46	----
2	----	----	19.46	----
3	----	----	19.54	----
5	54.82	34.68	19.44	----
9	54.82	34.58	----	18.73
12	54.74	34.50	----	----
19	----	----	----	18.63
22	----	34.77	----	18.63
26	----	----	----	18.63
Av[a]	54.79	34.68	19.48	18.66

[a]The average values are listed in "Initial" of Table 3 (see next page).

AUXILIARY INFORMATION

METHOD/APPARATUS/PROCEDURE:	SOURCE AND PURITY OF MATERIALS:
Mixtures of excess iodine pentoxide (5-8g) with 100-150 ml of the various solutions of sulfuric acid contained in 200 ml bottles having carefully ground stoppered and tightly fitting protective caps, were rotated in a water thermostat. The samples for analysis were withdrawn with a special filter-pipet. The filter consisted of a plug of asbestos wool packed in a bulb 1 cm in diameter on an extension tube which was attached to the tip of the pipet by a ground glass joint. The pipet was operated by an efficient water pump. The filtered 10-20ml samples were diluted to 250-500 ml. Aliquot portions were then treated with an excess of potassium iodide and titrated with 0.1 N sodium thiosulfate solution.	Very pure iodine pentoxide was prepared by the chloric acid method (ref. 1). The water content of the product was 0.55 % corresponding to 10.7 % conversion into iodic acid (HIO_3). The solutions of sulfuric acid were prepared by weight from a large stock sample of pure sulfuric acid. The concentration of this stock acid was ascertained by comparison of a diluted, weighed sample with a solution of 1 N hydrochloric acid.
	ESTIMATED ERROR: Soly: nothing specified. Temp: precision 0.005 K
	REFERENCES: 1. Lamb, A. B.; Bray, W. C.; Geldard, W. J. J. Am. Chem. Soc. 1920, 42, 1636.

COMPONENTS:	ORIGINAL MEASUREMENTS:
(1) Iodine oxide; I_2O_5; [12029-98-0]	Lamb, A. B.; Phillips, A. W.
(2) Sulfuric acid; H_2SO_4; [7664-93-9]	J. Am. Chem. Soc. 1923, 45, 108-12.
(3) Water; H_2O; [7732-18-5]	

EXPERIMENTAL VALUES: (continued)

(2) With acid of higher concentrations (82-96 mass %) definite "initial" values of the solubilities are rapidly established, as shown by the results collected in Table 2.

Table 2.

Time Hours	H_2SO_4 / mass %	82.0	86.0	90.3	95.96	95.96
		Solubility of iodine oxide (g dm^{-3})				
1		19.51[a]	20.98	22.63[a]	----	----
2		19.60	21.03	22.80	22.94	23.15
4		19.78	21.08	22.66	23.22	23.56
6		19.87	21.07	22.62	23.07	23.40
24		19.70	21.07	22.60	----	----
Av		19.74	21.4	22.67	23.08	23.37
		19.9[b]	21.0[b]	22.7[b]	23.2[b]	

[a]These determinations were made independently of the others on fresh samples of sulfuric acid.

[b]The values are listed in "Initial" of Table 3

(3) After the mixtures of iodine pentoxide and water were rotated for 40 days, the final solubilities were obtained. The values are listed in Table 3.

The authors reported that the initial values represent solubilities of iodic acid (HIO_3) and the final values represent solubilities of iodine pentoxide and of anhydro iodic acid (HI_3O_8).

Table 3

Concn of H_2SO_4 mass %	Initial g dm^{-3}	mol dm^{-3} (compiler)	Final g dm^{-3}	mol dm^{-3} (compiler)
50.0	54.79	0.1641	54.79	0.1641
60.0	34.68	0.1039	34.68	0.1039
75.0	19.48	0.05836	19.48	0.05836
78.0	18.66	0.05590	18.66	0.05590
79.6	19.0	0.0569	18.5	0.0554
82.0	19.9	0.0596	18.8	0.0568
84.6	20.5	0.0614	19.3	0.0578
86.0	21.0	0.0629	17.1	0.0512
87.4	21.5	0.0644	15.8	0.0473
89.0	22.1	0.0662	15.1	0.0452
90.3	22.7	0.0680	14.5	0.0434
92.0	23.4	0.0701	13.5	0.0404
96.0	(23.2)	0.0695	11.0	0.0330
98.0	(22.0)	0.0659	9.5	0.0285
99.9	----	----	3.48	0.0104
102.0[a]	----	----	1.28	0.00384
104.0	----	----	1.90	0.00569
106.0	----	----	2.67	0.00800

a: This percentage represents weights of 100 % H_2SO_4 equivalent to 100 g of the acid in question. The 106.0 % of acid, therefore, contained 29.0 % of free SO_2.

COMPONENTS:	ORIGINAL MEASUREMENTS:
(1) Iodic acid; HIO_3; [7782-68-5] (2) Cadmium iodate; $Cd(IO_3)_2$; [7790-81-0] (3) Water; H_2O; [7732-18-5]	Lepeshkov, I. N.; Vinogradov, E. E.; Karataeva, I. M. *Zh. Neorg. Khim.* 1979, *24*, 2540-4; *Russ. J. Inorg. Chem. (Engl. Transl.)* 1979, *24*, 1412-4.

VARIABLES:	PREPARED BY:
T/K = 323 Composition	Hiroshi Miyamoto

EXPERIMENTAL VALUES: Composition of saturated solutions

Iodic Acid		Cadmium Iodate		Nature of the solid phase[a]
mass %	mol % (compiler)	mass %	mol % (compiler)	
77.69[b]	26.29	–	–	A
77.83	26.46	0.022	0.0028	A + B
73.81	22.41	0.023	0.0027	B
64.83	15.88	0.012	0.0011	B
56.37	11.69	0.027	0.0021	B
44.33	7.544	0.032	0.0021	B
20.61	2.590	0.019	0.00091	B
22.02	2.813	0.082	0.0040	C
8.00	0.883	0.071	0.0030	C
--	--	0.069[b]	0.0027	C

[a]A = HIO_3;   B = $2HIO_3 \cdot Cd(IO_3)_2$;   C = $Cd(IO_3)_2$.

[b]For binary systems, the compiler computes the following:

Soly of HIO_3 = 19.80 mol kg^{-1}

Soly of $Cd(IO_3)_2$ = 1.5 x 10^{-3} mol kg^{-1}

AUXILIARY INFORMATION

METHOD/APPARATUS/PROCEDURE:	COMMENTS AND/OR ADDITIONAL DATA:
The compiler assumes that the system was studied by the isothermal method. Equilibrium was established in about a month. Specimens of liquid and solid phases were analyzed for iodic acid, and cadmium and iodate ions. The cadmium ion concentration was detd by titrn with EDTA. Solid phases were investigated by chemical, thermal, thermogravimetric, X-ray diffraction analyses, and infrared spectroscopy.	The phase diagram is given below (based on mass % units). 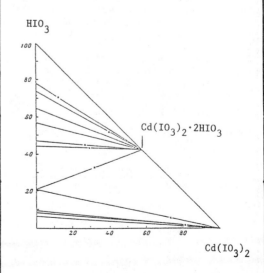
SOURCE AND PURITY OF MATERIALS: Chemically pure grade iodic acid was used. Cadmium iodate was made from cadmium chloride and iodic acid.	
ESTIMATED ERROR: Nothing specified.	

COMPONENTS:	ORIGINAL MEASUREMENTS:
(1) Iodic acid: HIO_3; [7782-68-5] (2) Zinc iodate; $Zn(IO_3)_2$; [7790-37-6] (3) Water; H_2O; [7732-18-5]	Lepeshkov, I. N.; Vinogradov, E. E.; Karataeva, I. M. *Zh. Neorg. Khim.* 1979, *24*, 2540-4; *Russ. J. Inorg. Chem. (Engl. Transl.)* 1979, *24*, 1412-4.

VARIABLES:	PREPARED BY:
T/K = 323 Composition	Hiroshi Miyamoto

EXPERIMENTAL VALUES:

Composition of saturated solutions

HIO_3		Zn(IO_3)_2		Nature of the solid phase
mass %	mol % (compiler)	mass %	mol % (compiler)	
77.69[b]	26.29	--	--	A
76.33	25.10	0.36	0.050	A + B
75.57	24.93	1.17	0.164	B
53.00	10.44	0.44	0.037	B
43.64	7.373	0.22	0.016	B
36.64	5.613	0.27	0.018	B
21.95	2.810	0.30	0.016	B
6.57	0.720	0.68	0.032	B
---	---	0.68[b]	0.030	B

[a] A = HIO_3;   B = $Zn(IO_3)_2.2H_2O$

[b] For binary systems the compiler computes the following

Soly of HIO_3 = 19.80 mol kg^{-1}

Soly of $Zn(IO_3)_2$ = 0.016 mol kg^{-1}

AUXILIARY INFORMATION

METHOD/APPARATUS/PROCEDURE:	COMMENTS AND/OR ADDITIONAL DATA:
The compiler assumes that the system was studied by the isothermal method. Equilibrium was established in about a month. Specimens of liquid and solid phases were analyzed for iodic acid, zinc and iodate ions. Zinc ion concn detd by titrn with EDTA. Solid phases were investigated by chemical, thermal, thermogravimetric, X-ray diffraction analyses, and infrared spectroscopy.	The phase diagram is given below (based on mass % units). 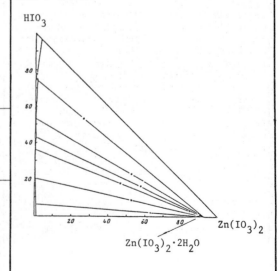

SOURCE AND PURITY OF MATERIALS:

Chemically pure grade iodic acid was used. Zinc iodate was prepd from zinc nitrate and iodic acid.

ESTIMATED ERROR:

Nothing specified.

COMPONENTS:	ORIGINAL MEASUREMENTS:
(1) Iodic acid; HIO_3; [7782-68-5] (2) Aluminum iodate; $Al(IO_3)_3$; [15123-75-8] (3) Water; H_2O; [7732-18-5]	Shklovskaya, R. M.; Arkhipov, S. M. Kidyarov, B. I.; Kuzina, V. A. Tsibulevskaya, K. A. *Zh. Neorg. Khim.* 1977, *22*, 1372-5; *Russ. J. Inorg. Chem. (Engl. Transl.)* 1977, *22*, 747-8.

VARIABLES:	PREPARED BY:
T/K = 298 Composition	Hiroshi Miyamoto

EXPERIMENTAL VALUES: Composition of saturated solutions

$Al(IO_3)_3$		HIO_3		Nature of the
mass %	mol % (compiler)	mass %	mol % (compiler)	solid phase[a]
5.70[b]	0.197	--	--	A
4.95	0.178	4.79	0.540	A
5.30	0.196	7.37	0.855	A
5.40	0.207	10.52	1.263	A + B
4.14	0.158	11.59	1.387	B
3.35	0.135	17.83	2.261	B
3.65	0.161	24.79	3.421	B
4.25	0.205	31.31	4.730	B
5.51	0.300	38.64	6.597	B
5.71	0.338	43.86	8.151	B
6.33	0.422	50.02	10.46	B
6.30	0.476	56.43	13.36	B
4.31	0.357	62.81	16.30	B
4.09	0.370	66.76	18.93	B
3.98	0.402	71.06	22.48	B + C
1.62	0.161	73.10	22.81	C
--	--	75.10[b]	23.60	C

[a] A = $Al(IO_3)_3 \cdot 6H_2O$;   B = $Al(IO_3)_3 \cdot 2HIO_3 \cdot 6H_2O$;   C = HIO_3

[b] For binary systems the compiler computes the following:

soly of $Al(IO_3)_3$ = 0.110 mol kg^{-1};   soly of HIO_3 = 17.15 mol kg^{-1}

AUXILIARY INFORMATION

METHOD/APPARATUS/PROCEDURE:	COMMENTS AND/OR ADDITIONAL DATA
Isothermal method used. Equilibrium reached in 20-30 days. Total iodate ion concn in the liquid phase detd by iodometric titrn, and aluminum detd by the complexometric method. Iodic acid found by difference. Compositions of the solid phases were detd by the method of residues and checked by X-ray diffraction.	The phase diagram below is in mass % units. 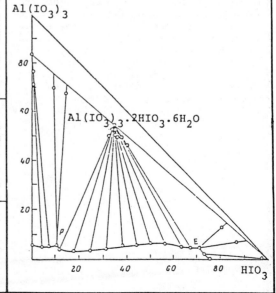

SOURCE AND PURITY OF MATERIALS:
Aluminum iodate prepd from iodic acid and freshly pptd aluminum hydroxide. Chemically pure grade iodic acid was recrystallized from aqueous solution before use.

ESTIMATED ERROR:
Nothing specified.

COMPONENTS:	ORIGINAL MEASUREMENTS:
(1) Iodic acid; HIO_3; [7782-68-5] (2) Scandium iodate; $Sc(IO_3)_3$; [42096-67-3] (3) Water; H_2O; [7732-18-5]	Vinogradov, E. E.; Lepeshkov, I. N.; Tarasova, G. N. Zh. Neorg. Khim. 1977, 22, 2858-61; Russ. J. Inorg. Chem. (Engl. Transl.) 1977, 22, 1552-4

VARIABLES:	PREPARED BY:
T/K = 298 Composition	Hiroshi Miyamoto

EXPERIMENTAL VALUES: Composition of saturated solutions

Scandium Iodate		Iodic Acid		Nature
mass %	mol % (compiler)	mass %	mol % (compiler)	of the solid phase[a]
-	-	75.40^b	23.99	A
0.001	8×10^{-5}	69.44	18.99	A + C
0.001	8×10^{-5}	69.42	18.86	A + C
0.001^c	8×10^{-5}	69.47	18.90	A + C
0.03	2×10^{-3}	65.60	16.35	C
0.02	1×10^{-3}	52.23	10.07	C
0.05	3×10^{-3}	45.26	7.813	C
0.02	9×10^{-4}	39.86	6.358	C
0.06	3×10^{-3}	32.19	4.640	C
0.05	2×10^{-3}	27.13	3.675	C
0.06	2×10^{-3}	22.86	2.948	C
0.08	3×10^{-3}	17.20	2.085	C
0.08	3×10^{-3}	10.12	1.141	C
0.09	3×10^{-3}	5.79	0.626	C
0.10	3.3×10^{-3}	3.25	0.343	C
0.15	4.8×10^{-3}	0.53	0.054	C
0.19	6.0×10^{-3}	0.16	0.016	C + B
0.19	6.0×10^{-3}	0.16	0.016	C + B
0.21	6.7×10^{-3}	-	-	B

[a]A = HIO_3;   B = $Sc(IO_3)_3.18H_2O$;   C = $Sc(IO_3)_3.4HIO_3.18H_2O$.

continued.....

AUXILIARY INFORMATION

METHOD/APPARATUS/PROCEDURE:	SOURCE AND PURITY OF MATERIALS:
The isothermal method was used. Equilibrium of the system $Sc(IO_3)_3-HIO_3-H_2O$ was reached in 7-10 days. Both the liquid and solid phases were analyzed: scandium was determined complexmetrically, and the iodate determined iodometrically. The iodic acid concentration was determined by titration with 0.01 mol dm^{-3} KOH solution using methyl red as an indicator. The composition and nature of the solid phases was determined by Schreinemakers' method of residues, X-ray diffraction, thermography, and thermogravimetry.	Chemically pure grade iodic acid was used. Scandium iodate was made by dissolving scandium oxide in hot concentrated nitric acid, and the $Sc(NO_3)_3.6H_2O$ obtained recrystallized several times. Scandium iodate was prepd by mixing the scandium nitrate and lithium iodate. The purity of the product was checked by chemical and X-ray diffraction analyses. The formula was found to be $Sc(IO_3)_3.18H_2O$.
	ESTIMATED ERROR: Nothing specified.
	REFERENCES:

COMPONENTS:	ORIGINAL MEASUREMENTS
(1) Iodic acid; HIO_3; [7782-68-5] (2) Scandium iodate; $Sc(IO_3)_3$; [42096-67-3] (3) Water; H_2O; [7732-18-5]	Vinogradov, E. E.; Lepeshkov, I. N.; Tarasova, G. N. *Zh. Neorg. Khim.* <u>1977</u>, *22*, 2858-61; *Russ. J. Inorg. Chem. (Engl. Transl.)* <u>1977</u> *22*, 1552-4.

EXPERIMENTAL VALUES: (Continued)

[b]For binary systems the compiler computes the following:

> soly of HIO_3 = 17.47 mol kg^{-1}

> soly of $Sc(IO_3)_3$ = 3.7 x 10^{-3} mol kg^{-1}

[c]The value is not given in the original paper, but the compiler presumes that the value is 0.001 mass % $Sc(IO_3)_3$.

COMMENTS AND/OR ADDITIONAL DATA:

The phase diagram is given below (based on mass % (?)).

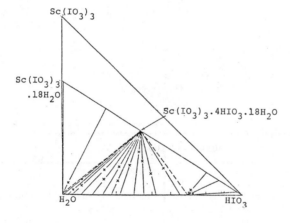

COMPONENTS:	ORIGINAL MEASUREMENTS:
(1) Iodic Acid; HIO_3; [7782-68-5] (2) Lanthanium iodate; $La(IO_3)_3$; [13870-19-4] (3) Water; H_2O; @7732-18-5]	Lyalina, R. B.; Soboleva, L. V. *Zh. Neorg. Khim.* 1975, 20, 2568-9; *Russ. J. Inorg. Chem. (Engl. Transl.)* 1975, 20, 1424-5.

VARIABLES:	PREPARED BY:
T/K = 298 Composition	Hiroshi Miyamoto

EXPERIMENTAL VALUES:

Composition of saturated solutions

Iodic Acid mass %	Lanthanium iodate		Nature of the solid phase[a]
	mass %	mol % (compiler)	
3.31	Very low solubility		A
9.66	"		A
11.36	"		A
18.84	"		A
25.50	"		A
31.93	"		A
32.51	"		A
35.91	"		A
39.55	Very low solubility		B
45.04	"		B
49.52	"		B
55.21	"		B
61.49	"		B
62.66	Very low solubility		B + C
69.27	"		B + C
72.00	"		B + C
77.54	"		B + C

[a]$A = La(IO_3)_3 \cdot 2.5H_2O$; $B = La(IO_3)_3$; $C = HIO_3$

AUXILIARY INFORMATION

METHOD/APPARATUS/PROCEDURE:	COMMENTS AND/OR ADDITIONAL DATA:
The compiler assumes that the isothermal method was used. The attainment of equilibrium was deduced from the constancy of the iodate ion concentration in the liquid phase. The time required for equilibrium was 40 hours. Iodic acid in the liquid phase was determined by iodometric titration. Lanthanium concentration in solution could not be determined owing to the very low solubility	The phase diagram is given below (based on mass % units). 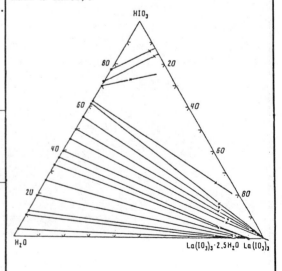

SOURCE AND PURITY OF MATERIALS:

Chemically pure grade iodic acid was used. Lanthanium iodate was synthesized from lanthanium nitrate and potassium iodate.

ESTIMATED ERROR:

Nothing specified.

COMPONENTS:	ORIGINAL MEASUREMENTS:
(1) Iodic acid; HIO_3; [7782-68-5] (2) Neodymium iodate; $Nd(IO_3)_3$; [14732-16-2] (2) Water; H_2O; [7732-18-5]	Tarasova, G. N.; Vinogradov, E. E. Kudinov, I. B. Zh. Neorg. Khim. 1982, 27, 505-12; Russ. J. Inorg. Chem. (Engl. Transl.) 1982, 27, 287-92.

VARIABLES:	PREPARED BY:
T/K = 298 Composition	Hiroshi Miyamoto

EXPERIMENTAL VALUES: Composition of saturated solutions

Neodymium Iodate		Iodic Acid		Nature
mass %	mol % (compiler	mass %	mol % (compiler)	of the solid phase[a]
<0.15[b]	4×10^{-3}	–	–	A
<0.01	3×10^{-4}	0.29	0.030	A
<0.01	3×10^{-4}	1.33	0.138	A
<0.01	3×10^{-4}	1.76	0.183	A
<0.01	3×10^{-4}	3.75	0.397	A
<0.01	3×10^{-4}	4.22	0.449	A
<0.01	3×10^{-4}	5.89	0.637	A + B
<0.01	3×10^{-4}	5.89	0.637	A + B
<0.01	3×10^{-4}	9.30	1.039	B
<0.01	3×10^{-4}	10.93	1.241	B
<0.01	3×10^{-4}	11.39	1.299	B
<0.01	3×10^{-4}	14.87	1.758	B
<0.01	3×10^{-4}	16.34	1.961	B
<0.01	3×10^{-4}	16.15	1.935	B + C
<0.01	3×10^{-4}	16.13	1.932	B + C
<0.01	3×10^{-4}	16.17	1.937	C
<0.01	3×10^{-4}	25.34	3.360	C
<0.01	4×10^{-4}	27.01	3.652	C
<0.01	4×10^{-4}	35.48	5.332	C
<0.01	4×10^{-4}	41.65	6.813	C
<0.01	5×10^{-4}	47.28	8.413	C
<0.01	5×10^{-4}	50.82	9.572	C
<0.01	5×10^{-4}	55.61	11.37	C
<0.01	6×10^{-4}	61.17	13.89	C

contd..

AUXILIARY INFORMATION

METHOD/APPARATUS/PROCEDURE:	SOURCE AND PURITY OF MATERIALS:
The experiments were carried out in a water thermostat with an electric heater. Equilibrium was established in 18 to 21 days. The liquid phases were analyzed for Nd^{3+} and IO_3^- ions. The iodate concentration was determined by titration with sodium thiosulfate in the presence of sulfuric acid and potassium iodide. The neodymium content was determined by complexometric titration with hexamethylenetetramine and methyl thymol blue indicator. The composition of the solid phase was determined by Schreinemakers' method of "residues", and identified by X-ray diffraction.	Chemically pure grade iodic acid was used. Neodymium iodate was made from neodymium oxide and iodic acid.
	ESTIMATED ERROR: Soly: nothing specified. Temp: $\pm$ 0.1 K (authors).
	REFERENCES:

COMPONENTS:	ORIGINAL MEASUREMENTS
(1) Iodic acid; HIO_3; [7782-68-5]	Tatasova, G. N.; Vinogradov, E. E.; Kudinov, I. B.
(2) Neodymium iodate; $Nd(IO_3)_3$; [14732-16-2]	
(3) Water; H_2O; [7732-18-5]	*Zh. Neorg. Khim.* 1982, 27, 505-12; *Russ. J. Inorg. Chem. (Engl. Transl.)* 1982, 27, 287-92.

EXPERIMENTAL VALUES (Continued)

Composition of saturated solutions

| Neodymium Iodate | | Iodic Acid | | Nature |
mass %	mol % (compiler)	mass %	mol % (compiler)	of the solid phase[a]
0.01	7×10^{-4}	66.24	16.74	C
0.01	8×10^{-4}	73.30	21.95	C
0.01	8×10^{-4}	74.15	22.71	C + D
0.01	8×10^{-4}	74.15	22.71	C + D
–	–	75.40[b]	23.89	D

[a] A = $Nd(IO_3)_3$;   B = $Nd(IO_3)_3 \cdot HIO_3 \cdot 2H_2O$;   C = $Nd(IO_3)_3 \cdot 3HIO_3 \cdot 2H_2O$;   D = HIO_3

[b] For binary systems, the compiler computes the following:

   Soly of HIO_3 = 17.42 mol kg^{-1}

   Soly of $Nd(IO_3)_3$ = 2.2×10^{-3} mol kg^{-1}

COMMENTS AND/OR ADDITIONAL DATA:

The phase diagram is given below (based on mass % units).

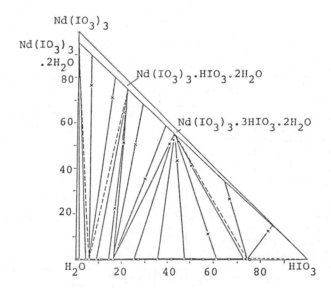

COMPONENTS:	ORIGINAL MEASUREMENTS:
(1) Iodic acid; HIO_3 [7782-68-5] (2) N-Phenylacetamide (acetanilide); C_8H_9NO; [103-84-4] (3) Water; H_2O; [7732-18-5]	Erkasov, R. Sh.; Beremzhanov, B. A.; Nurakhmhmetov, N. N. *Zh. Neorg. Khim.* <u>1981</u>, *26*, 1441-4; *Russ. J. Inorg. Chem. (Engl. Transl.)* <u>1981</u>, *26*, 776-8.
VARIABLES: T/K = 293 and 313	PREPARED BY: M. Salomon and H. Miyamoto

EXPERIMENTAL VALUES:

Numerical data given only for the two eutonic points at 20°C and 40°C. The phase diagram is given below (mole % units). below:

	C_8H_9NO			HIO_3	
t/°C	mass %	mol %[a]	mass %	mol %[a]	
20	0.88	0.149	22.84	2.971	
40	1.96	0.353	27.12	3.755	

[a]Calculated by compilers.

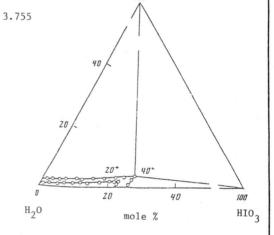

AUXILIARY INFORMATION

METHOD/APPARATUS/PROCEDURE:	SOURCE AND PURITY OF MATERIALS:
The solubility was studied by the isothermal method. Equilibrium in the system was reached after continuous stirring for 10-12 hours. Acetanilide was found from amount of nitrogen determined by the Kjeldahl method, or by titration with 0.05N potassium bromate solution (ref 1). Iodic acid was determined by titration with 0.1N sodium thiosulfate solution. The compositions of the solid phases were found by Schreinemakers' method of residues.	"Analytical reagent" grade acetanilide and "chemically pure" grade iodic acid were used.
	ESTIMATED ERROR: Nothing specified.
	REFERENCES: 1. Suslennikova, V. M.; Kiseleva, E. K. *Rukovodstvo po Prigotovleniya Titrovannykh Rastvorov (Handbook on the Preparation of Titrating Solutions)* Izd. Khimya, Leningrad <u>1973</u>.

SYSTEM INDEX

Page numbers preceded with E refer to evaluation texts, whereas those not preceded with E refer to compilation tables.

REGISTRY NUMBER INDEX

Page numbers preceded with E refer to evaluation texts, whereas those not preceded with E refer to compilation tables.

50-00-0	246
50-99-7	245
51-79-6	253
56-40-6	248, 413
56-81-5	153, 243
57-13-6	81, 253
60-29-7	244
60-35-5	253
64-17-5	151, 152, 243, 254, 328, 411
64-19-7	248
67-56-1	243, 373, 417, 418
67-64-1	154-156, 194, 247, 329
67-68-5	416, 419, 446
68-12-2	103, 160, 414, 415, 444, 445, 458, 459
69-65-8	245
71-23-8	243
75-12-7	253
75-85-4	243
76-64-1	E4, 23, 100
79-16-3	218, 256
79-20-9	250
98-01-1	255
103-84-4	469
107-15-3	101, 159
107-21-1	98, 157, 243
108-95-2	249
109-89-7	251
110-86-1	251
110-89-4	251
123-91-1	412
126-33-0	102
141-43-5	99, 158
144-55-8	202, 217
302-01-2	104, 219, 427
497-19-8	E32, 82, 85, 201, 216
554-13-2	293
645-09-2	252
1004-36-0	252
1310-58-3	147, 325-327, 405, 406, 409, 410
1310-65-2	313, 325-327
1310-73-2	E30, 67
1310-82-3	437
2029-98-0	476
3811-04-9	E29, 68-70, 90-92, E105-E112, 113-161, 411
4917-19-8	344, 345
7447-40-7	90-92, 125-140, 232, 391, 421, 422
7447-41-8	20-22, 254, 282
7550-35-8	283
7631-95-0	E30, 76, 77, 200, 343
7631-99-4	E30, 41, 203, 228, 346-353
7647-14-5	E31, E32, 82-84, 86-97, 121, 208, 209, 229, 357, 358, 423
7647-15-6	E28, 61, 210-213, 216, 217, 359-363
7647-17-8	95-97, 174, 175, 182, 183
7664-38-2	320
7664-39-3	478

AUTHOR INDEX

Page numbers preceded with E refer to evaluation texts, whereas those not preceded with E refer to compilation tables.